Kai Michels, Frank Klawonn, Rudolf Kruse, Andreas Nürnberger

Fuzzy Control

Studies in Fuzziness and Soft Computing, Volume 200

Further volumes of this series
can be found on our homepage:
springer.com

Vol. 184. Barry G. Silverman, Ashlesha Jain, Ajita Ichalkaranje, Lakhmi C. Jain (Eds.)
Intelligent Paradigms for Healthcare Enterprises, 2005
ISBN 3-540-22903-5

Vol. 185. Spiros Sirmakessis (Ed.)
Knowledge Mining, 2005
ISBN 3-540-25070-0

Vol. 186. Radim Bělohlávek, Vilém Vychodil
Fuzzy Equational Logic, 2005
ISBN 3-540-26254-7

Vol. 187. Zhong Li, Wolfgang A. Halang, Guanrong Chen (Eds.)
Integration of Fuzzy Logic and Chaos Theory, 2006
ISBN 3-540-26899-5

Vol. 188. James J. Buckley, Leonard J. Jowers
Simulating Continuous Fuzzy Systems, 2006
ISBN 3-540-28455-9

Vol. 189. Hans Bandemer
Mathematics of Uncertainty, 2006
ISBN 3-540-28457-5

Vol. 190. Ying-ping Chen
Extending the Scalability of Linkage Learning Genetic Algorithms, 2006
ISBN 3-540-28459-1

Vol. 191. Martin V. Butz
Rule-Based Evolutionary Online Learning Systems, 2006
ISBN 3-540-25379-3

Vol. 192. Jose A. Lozano, Pedro Larrañaga, Iñaki Inza, Endika Bengoetxea (Eds.)
Towards a New Evolutionary Computation, 2006
ISBN 3-540-29006-0

Vol. 193. Ingo Glöckner
Fuzzy Quantifiers: A Computational Theory, 2006
ISBN 3-540-29634-4

Vol. 194. Dawn E. Holmes, Lakhmi C. Jain (Eds.)
Innovations in Machine Learning, 2006
ISBN 3-540-30609-9

Vol. 195. Zongmin Ma
Fuzzy Database Modeling of Imprecise and Uncertain Engineering Information, 2006
ISBN 3-540-30675-7

Vol. 196. James J. Buckley
Fuzzy Probability and Statistics, 2006
ISBN 3-540-30841-5

Vol. 197. Enrique Herrera-Viedma, Gabriella Pasi, Fabio Crestani (Eds.)
Soft Computing in Web Information Retrieval, 2006
ISBN 3-540-31588-8

Vol. 198. Hung T. Nguyen, Berlin Wu
Fundamentals of Statistics with Fuzzy Data, 2006
ISBN 3-540-31695-7

Vol. 199. Zhong Li
Fuzzy Chaotic Systems, 2006
ISBN 3-540-33220-0

Vol. 200. Kai Michels, Frank Klawonn, Rudolf Kruse, Andreas Nürnberger
Fuzzy Control, 2006
ISBN 3-540-31765-1

Kai Michels
Frank Klawonn
Rudolf Kruse
Andreas Nürnberger

Fuzzy Control

Fundamentals, Stability and Design of Fuzzy Controllers

Dr. Kai Michels
Fichtner GmbH & Co.KG
Sarweystr. 3, 70191 Stuttgart
Germany
E-mail: kaimichels@hotmail.com

Professor Dr. Rudolf Kruse
Fak. Informatik, Abt. Wiss.-und
Sprachverarbeitung
Otto-von-Guericke-Universität
Universitätsplatz 2
39106 Magdeburg
Germany
E-mail: kruse@iik.cs.uni-magdeburg.de

Professor Dr. Frank Klawonn
FH Braunschweig/Wolfenbüttel
FB Informatik
Salzdahlumer Str. 46/48
38302 Wolfenbüttel
Germany
E-mail: klawonn@et-inf.fho-emden.de

Dr. Andreas Nürnberger
Computer Science Division
University California at Berkeley
Cory Hall
94720 Berkeley
USA
E-mail: anuernb@eecs.berkeley.edu

ISSN print edition: 1434-9922
ISSN electronic edition: 1860-0808
ISBN 978-3-642-06863-8 e-ISBN 978-3-540-31766-1

Springer is a part of Springer Science+Business Media
springer.com

Softcover reprint of the hardcover 1st edition 2006

Cover design: Erich Kirchner, Heidelberg

Preface

"Fuzzy Control - the revolutionary computer technology that is changing our world" - these and other headlines could be read when in the early 90's news from Japan came over telling us about the success of fuzzy controllers. The idea which was put into practice had been suggested by Lotfi A. Zadeh in Berkeley in 1965. It had been developed and tested in some practical applications, especially in Europe. In Japan fuzzy control was celebrated as a technology reflecting the Japanese way of thinking by its unsharpness and implicit overlapping of several statements. A new technology boom was predicted for Japan which would make Europe lose ground.

Consequently, this news created unrest. Research projects were initiated and development departments were engaged to translate fuzzy control into products. Adversaries and supporters hurried up to inform themselves and intensely discussed whether the "conventional" or the fuzzy control were the better alternative.

Finally, the excitement cooled down since in recent years fuzzy control was analyzed from the classical point of view. Thus, a more objective evaluation of its strong and weak points was possible. Furthermore, it was shown how fuzzy systems could be put to use in the steering level which is the level above the control loop, especially in interaction with other methods of soft computing and artificial intelligence. Based on these fundamentals, the aim of this book is to support the convenient use of fuzzy controllers and fuzzy systems in the branch of control engineering and automation systems. This book addresses both the control engineers, who should regard fuzzy controllers as additional option for solving control problems, and computer scientists, for introducing them into the world of control engineering and to show them some possibilities of applying methods of soft computing and artificial intelligence.

On the one hand, this book is a text book explaining the fundamentals which are required for people working with fuzzy controllers - it addresses students of engineering and computer science after their intermediate exam. On the other hand, this book serves as a comprehensive reference book to several aspects of fuzzy controllers and the current state of the art. The structure of this book was chosen according to these aims.

In the first chapter we give an introduction to the theory of fuzzy systems and describe i.e. the way of connecting fuzzy sets and their semantic background. Only the knowledge about this makes the user able to estimate really the possibilities of applying fuzzy systems.

In the second chapter we explain the fundamentals of control engineering which are important for people working with fuzzy controllers. Thus, this chapter is actually not for control engineers but also these could be interested in the subchapters about i.e. hyperstability theory or sliding mode controllers since, normally, these issues are not part of control engineering fundamentals.

In the third chapter we introduce the fuzzy controller; it can be quiet short because of the well-founded introduction of fuzzy systems in chapter one. The main issue of this chapter is to present the current types of fuzzy controllers. Furthermore, we give an interpretation of the Mamdani controller with similarity relations which help us to explain the ideas fuzzy controllers are based on. At the end of this chapter we discuss in how far fuzzy controllers are advantageous or disadvantageous with regard to classic controllers.

In the fourth chapter we discuss the stability analysis of fuzzy controllers. The stability is the most significant question for every controller and it is a field in which very interesting developments were made in the recent years. Therefore, it is the issue of a special chapter where we give an overview about the different approaches of stability analysis and discuss the pros and cons of the methods in order to provide some help for decision making in practical applications.

In the final chapter we describe approaches of evaluation and optimization of fuzzy controllers, that is, methods of supporting or even automating the design of fuzzy controllers, especially through the use of neural nets and evolutionary algorithms. Additionally to the fundamental descriptions of these issues we take recent results of research into account.

Finally, we would like to thank all the people who helped us writing this book, either by the work of their research projects, by student's works or by the discussions we benefited from. We are especially grateful to Prof. Werner Leonhard for initiating the research project about stability analysis and self adjustment of fuzzy conrollers, to Prof. Kai Müller for supporting us so excellently in questions of control engineering, to Prof. Lotfi A. Zadeh for many ideas and discussions, to Anne Kutzera for her support in translating parts of this book and to the series editor Janusz Kacprzyk for the good co-operation. We are also grateful to many others of our colleagues and friends who – directly or indirectly – supported us during the work on this book.

Germany
November 2005

Kai Michels
Frank Klawonn
Rudolf Kruse
Andreas Nürnberger

Contents

1

Fundamentals of Fuzzy Systems

Classical logic and mathematics assume that we can assign one of the two values, *true* or *false*, to each logical proposition or statement. If a suitable formal model for a certain problem or task can be specified, conventional mathematics provides powerful tools which help us to solve the problem. When we describe such a formal model, we use a terminology which has much more stringent rules than natural language. This specification often requires more work and effort, but by using it we can avoid misinterpretations. Furthermore, based on such models we can prove or reject hypotheses or derive unknown correlations.

However, in our everyday life formal models do not concern the inter-human communication. Human beings are able to assimilate easily linguistical information without thinking in any type of formalization of the specific situation. For example, a person will have no problems to accelerate slowly while starting a car, if he is asked to do so. If we want to automate this action, it will not be clear at all, how to translate this advice into a well-defined control action. It is necessary to determine a concrete statement based on an unambiguous value, i.e. step on the gas at the velocity of half an inch per second. On the other hand, this kind of information will not be adequate or very helpful for a person.

Therefore, automated control is usually not based on a linguistic description of heuristic knowledge or knowledge from one's own experience, but it is based on a formal model of the technical or physical system. This method is definitely a suitable approach, especially if there is a good model to be determined.

However, a completely different technique is to use knowledge formulated in natural language directly for the design of the control strategy. In this case, a main problem will be the translation of the verbal description into concrete values, i.e. assigning "step on the gas slowly" into "step on the gas at the velocity of a centimeter per second" as in the above mentioned example.

When describing an object or an action, usually use uncertain or vague concepts. In natural language we hardly ever find exactly defined concepts

K. Michels et al.: *Fuzzy-Control: Fundamentals, Stability and Design of Fuzzy Controllers*, StudFuzz **200**, 1–57 (2006)
www.springerlink.com

like supersonic speed for the velocity of a passing airplane. Supersonic speed characterizes an unambiguous set of velocities, because the speed of sound is a fixed entity and therefore it is unambiguously clear whether an airplane flies faster than sound or not. Frequently used vague concepts, like *fast, very big, small* and so on, make it impossible to decide unambiguously whether a given value satisfies such a vague concept or not. One of the reasons for this is that vague concepts are usually context dependent. Talking about airplanes *fast* has a different meaning than using this characteristic while referring to cars. But also if we agree that we are talking about cars it is not easy to distinguish clearly between fast and non-fast cars. The difficulty here is not to find a value telling us whether a car (or its top speed) is fast or not, but we had to presuppose that such a value does exist. It is more likely that we will be reluctant to fix such a value because there are velocities, we can classify as fast for a car and there are some we can classify as not fast, and in between there is a wide range of velocities which are considered as more or less fast.

1.1 Fuzzy Sets

The idea of fuzzy sets is to solve this problem by avoiding the sharp separation of conventional sets into two values - complete membership or complete non-membership. Instead, fuzzy sets can handle partial membership. So in fuzzy sets we have to determine to what degree or extend an element is a member of this fuzzy set. Therefore, we define:

Definition 1.1 *A fuzzy subset or simply a fuzzy set μ of a set X (the universe of discourse) is a mapping $\mu : X \to [0,1]$, which assigns to each element $x \in X$ a degree of membership $\mu(x)$ to the fuzzy set μ. The set of all fuzzy sets of X is denoted by $\mathcal{F}(X)$.*

A conventional set $M \subseteq X$ can be viewed as a special fuzzy set by identifying it with its *characteristic function* or *indicator function*.

$$I_M : X \to \{0,1\}, \qquad x \mapsto \begin{cases} 1 & \text{if } x \in M \\ 0 & \text{else} \end{cases}$$

Seen in this way, fuzzy sets can be considered as generalized characteristic functions.

Example 1.2 Figure 1.1 shows the characteristic function of the set of velocities which are higher than 170 km/h. This set does not represent an adequate model of all high velocities. The jump at the value of 170 causes that 169.9 km/h would not be a high velocity but 170.1 km/h would be. Therefore, a fuzzy set (figure 1.2) seems to be more adequate to model the concept *high velocity*. □

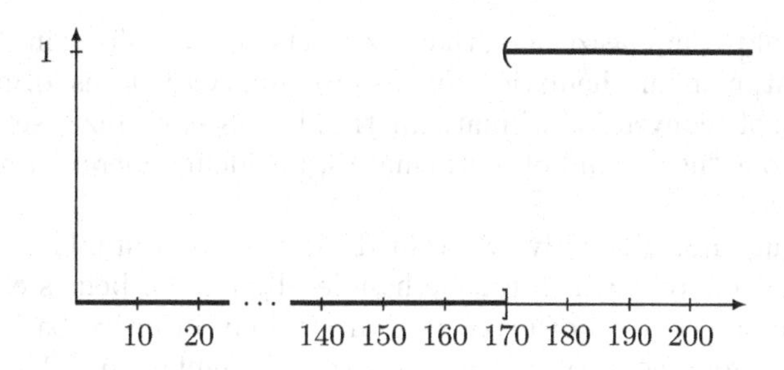

Fig. 1.1. The characteristic function of the set of velocities which are higher than 170 km/h

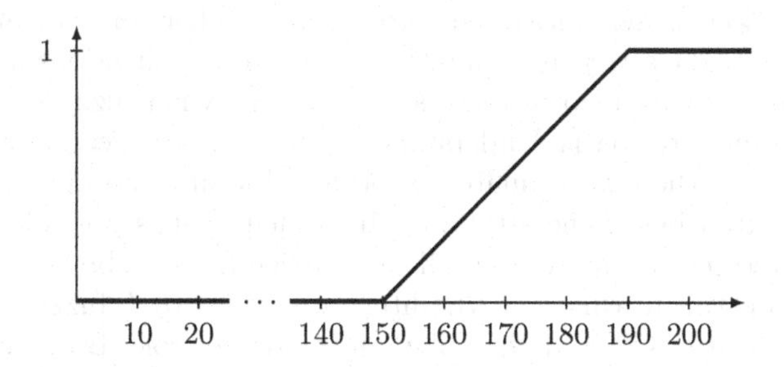

Fig. 1.2. The fuzzy set μ_{hv} of high velocities

Some authors use the term fuzzy set only for a vague concept $\mathcal{A}$ like *high velocity* and call the membership function $\mu_{\mathcal{A}}$, that models the vague concept, a characterizing or membership function of the fuzzy set or the vague concept $\mathcal{A}$. When operating with fuzzy sets, there is no advantage to make this distinction. Only from a philosophical point of view, one might be interested in distinguishing between an abstract vague concept and its concrete model in the form of a fuzzy sets. Since we do not want to initiate a philosophical discussion in this book, we stick to our restricted definition of a fuzzy set as a (membership) function, yielding values in the unit interval.

Besides the formal definition of a fuzzy set as a mapping to the unit interval, there are also other notations which are preferred by some authors, but we will not use them in this book. In some publications a fuzzy set is written as a set of pairs of the elements of the underlying set and the corresponding degrees of membership in the form of $\{(x, \mu(x)) \mid x \in X\}$ following the fact that in mathematics a function is usually formally defined as a set of pairs, each consisting of one argument of the function and the image of this argument. A little bid more misleading is the notation of a fuzzy set as a formal sum $\sum_{x \in X} x/\mu(x)$ with an at most countable reference set X or as an "integral" $\int_{x \in X} x/\mu(x)$ for an uncountable reference set X.

We want to emphasize here, that fuzzy sets are formalized in the framework of "conventional" mathematics, just as probability theory is formulated in the framework of "conventional" mathematics. In this sense fuzzy sets do not open the door to a "new" kind of mathematics, but define merely a new branch of mathematics.

Knowing that a strictly two-valued view is not suitable to model vague concepts adequately, which can be handled by human beings easily, we have introduced the concept of fuzzy sets on a purely intuitive basis. We did not specify precisely, how to interpret degrees of membership. The meanings of 1 as complete membership and 0 as complete non-membership are obvious, but we left open the question, how to interpret a degree of membership of 0.7 or why to prefer 0.7 over 0.8 as degree of membership for a certain element. These questions of semantics are often ignored. Therefore, a consistent interpretation of fuzzy sets is not maintained in some applications and this is one of the reasons, why inconsistencies may occur, when fuzzy sets are applied only on an intuitive basis. Understanding fuzzy sets as generalized characteristic functions, there is ultimate reason for choosing the unit interval as the canonical extension of the set $\{0, 1\}$. In principle, it is possible that any linearly ordered set or – more generally – a lattice L might be better suited than the unit interval. In this case, the literature refers to L fuzzy sets. However, in real applications they do not play an important role. But even if we agree on the unit interval as the set of possible degrees of membership, we should explain in which sense or as what kind structure we understand it.

The unit interval can be viewed as an ordinal scale. This means, only the linear ordering of the numbers is considered, i.e. for expressing preferences. In this case the interpretation of a number between 0 and 1 as a degree of membership makes sense only, when comparing it to another degree of membership. Thus we can express that the one element belongs more to a fuzzy set than another one. A problem resulting from this purely ordinal view of the unit interval is the incomparability of degrees of membership stated by different persons. The same difficulty appears comparing grades. Two examinees receiving the same grade from different examiners can have shown very different performances. Normally, the scale of grades is not used as a purely ordinal scale. Pointing out which performance or which amount of mistakes leads to which grade is an attempt to make it possible to compare grades given by different examiners.

With the canonical metric quantifying the distance between two numbers and operations like addition and multiplication the unit interval has a considerably richer structure than the linear ordering of the numbers. Therefore, in many cases it is better to understand the unit interval as a metric scale to obtain a more concrete interpretation of the degrees of membership. We will discuss the issue of semantics of degrees of membership and fuzzy sets in section 1.7. For the moment, we confine ourselves to a naive interpretation of degrees of memberships and say that the property of being an element of a fuzzy set can be satisfied gradually.

We want to emphasize here that gradual membership is a completely different idea than the concept of probability. It is clear that a fuzzy set μ must not be regarded as a probability distribution or density, because, in general, μ does not satisfy the condition

$$\sum_{x \in X} \mu(x) = 1 \quad \text{bzw.} \quad \int_X \mu(x)dx = 1$$

that is required in probability theory for density functions. Also the degree of membership $\mu(x)$ of an element x to the fuzzy set μ should not be interpreted as the probability that x belongs to μ.

To illustrate the difference between a gradual property and probability we take a look at the example below, following [16].

U denotes the "set" of non-toxic liquids. A person dying of thirst receives two bottles A and B and the information that bottle A belongs to U with a probability of 0.9 and bottle B has a degree of membership of 0.9 to U. From which of the bottles should the person drink? The probability of 0.9 for A could mean that the bottle was selected from among ten bottles in a room where nine were filled with water and one with cyanide. But the degree of membership of 0.9 means that the liquid is "reasonably" drinkable. For instance, B could contain a juice which has already past its best-before date. That is why the thirsty person should choose bottle B.

The liquid in bottle A has the property of being non-toxic either completely (with a probability of 0.9) or not at all (with a probability of 0.1). The liquid in bottle B satisfies the property of being non-toxic in a merely gradual way.

Probability theory and fuzzy sets serve us for modelling completely different phenomena – namely, on the one hand the quantification of the uncertainty whether an event may happen and on the other hand how much a property or statement is satisfied or to what degree a property is fulfilled.

1.2 Representation of Fuzzy Sets

After having introduced fuzzy sets formally as functions from a universe of discourse to the unit interval, we now want to discuss different methods for specifying concrete fuzzy sets and adequate techniques for representing fuzzy sets as well as store them in a computer.

1.2.1 Definition Using Functions

If the universe of discourse $X = \{x_1, \ldots, x_n\}$ is a finite, discrete set of objects x_i, a fuzzy set μ can, in general, only be specified by the degrees of membership $\mu(x)$ for each element $x \in X$ – i.e. in the form of $\mu \mathrel{\widehat{=}} \{(x_1, \mu(x_1)), \ldots, (x_n, \mu(x_n))\}$.

In most of the cases we will consider fuzzy sets here, the universe of discourse X will be the domain of a real-valued variable, i.e. a subset of the real line, usually an interval. Then a fuzzy set μ is a real function taking values in the unit interval and can be illustrated by drawing its graph. With a purely graphical definition of fuzzy sets membership degrees of the single elements can only be specified up to a certain, quite rough precision leading to difficulties and errors in further calculations. Thus the graphical representation is only suitable for illustration purposes.

Usually fuzzy sets are used for modelling expressions – sometimes also called *linguistic* expressions in order to emphasize the relation to natural language – like 'approximately 3', 'of medium height' or 'very tall' which describe a vague value or a vague interval. Fuzzy sets associated with such expressions should monotonically increase up to a certain value and monotonically decrease from this value. Such fuzzy sets are called *convex*.

Figure 1.3 shows three convex fuzzy sets which could model the expressions 'approximately 3', 'of medium height' and 'very tall'. In figure 1.4 we see a non-convex fuzzy set. Note that the convexity of a fuzzy set μ does not imply that μ is also convex as a real function.

For applications it is very often sufficient to consider only a few basic forms of convex fuzzy sets, so that a fuzzy set can be specified uniquely by a few parameters. Typical examples of such parametric fuzzy sets are *triangular functions* (cf. figure 1.5)

$$\Lambda_{a,b,c} : \mathbb{R} \to [0,1], \qquad x \mapsto \begin{cases} \frac{x-a}{b-a} & \text{if } a \leq x \leq b \\ \frac{c-x}{c-b} & \text{if } b \leq x \leq c \\ 0 & \text{else,} \end{cases}$$

where $a < b < c$ holds.

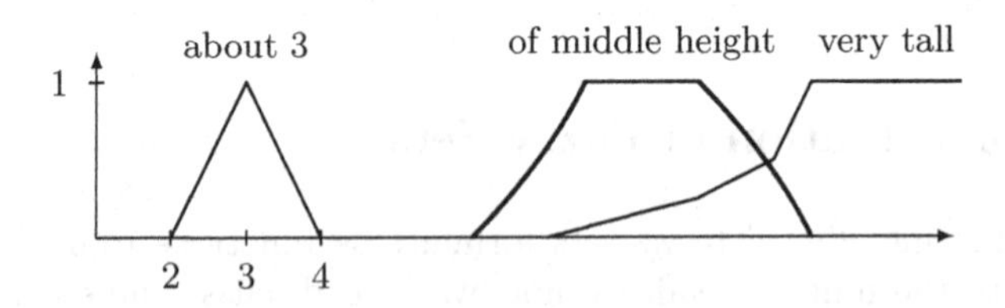

Fig. 1.3. Three convex fuzzy sets

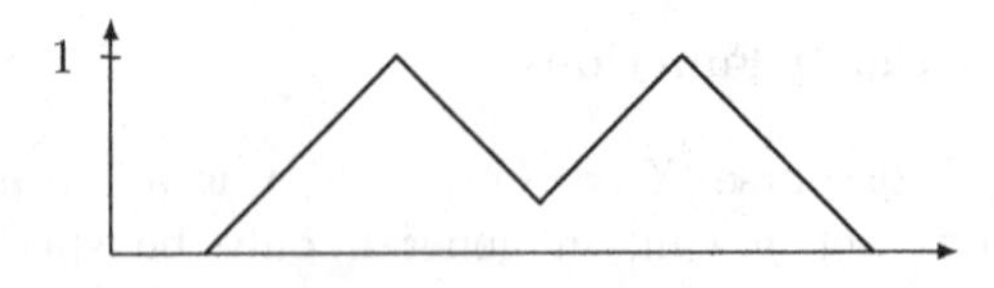

Fig. 1.4. A non-convex fuzzy set

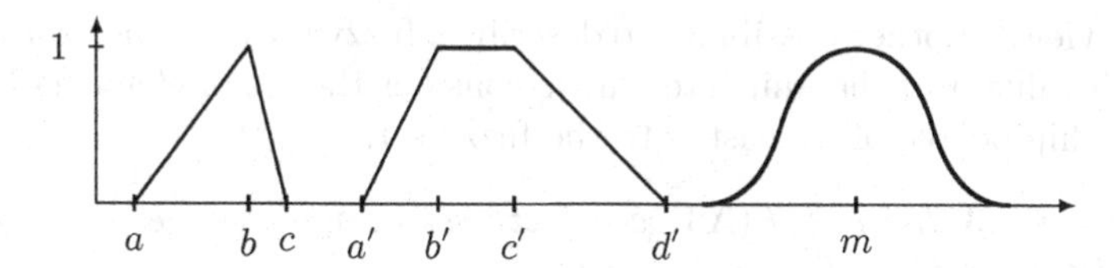

Fig. 1.5. The triangular functions $\Lambda_{a,b,c}$, the trapezoidal function $\Pi_{a',b',c',d'}$ and the bell curve $\Omega_{m,s}$

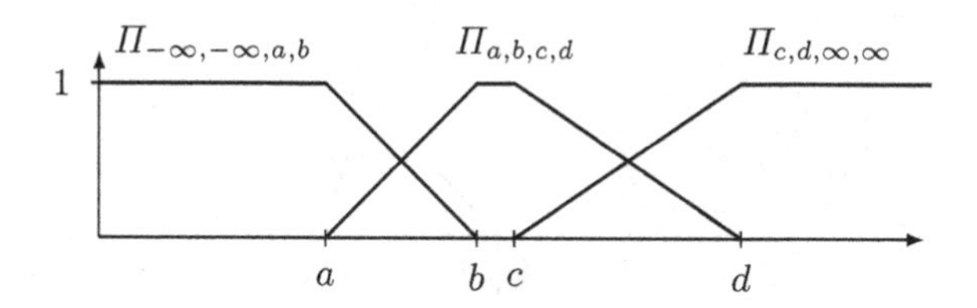

Fig. 1.6. The trapezoidal functions $\Pi_{-\infty,-\infty,a,b}, \Pi_{a,b,c,d}$ and $\Pi_{c,d,\infty,\infty}$

Triangular functions are special cases of *trapezoidal functions* (cf. figure 1.5)

$$\Pi_{a',b',c',d'} : \mathbb{R} \to [0,1], \qquad x \mapsto \begin{cases} \frac{x-a'}{b'-a'} & \text{if } a' \leq x \leq b' \\ 1 & \text{if } b' \leq x \leq c' \\ \frac{d'-x}{d'-c'} & \text{if } c' \leq x \leq d' \\ 0 & \text{else,} \end{cases}$$

where $a' < b' \leq c' < d'$ holds. We also permit the parameter combinations $a' = b' = -\infty$ or $c' = d' = \infty$. The resulting trapezoidal functions are shown in figure 1.6. For $b' = c'$ we have $\Pi_{a',b',c',d'} = \Lambda_{a',b',d'}$.

If we want to use smooth functions instead of piecewise linear functions like triangular or trapezoidal ones, *bell curves* in the form of

$$\Omega_{m,s} : \mathbb{R} \to [0,1], \qquad x \mapsto \exp\left(\frac{-(x-m)^2}{s^2}\right)$$

might be a possible choice. We have $\Omega_{m,s}(m) = 1$. The parameter s determines the width of the bell curve.

1.2.2 Level Sets

The representation of a fuzzy set as a function from the universe of discourse to the unit interval, assigning a membership degree to each element is called *vertical view*. Another possibility to describe a fuzzy set is the *horizontal view*. For each value α of the unit interval we consider the set of elements having a membership degree of at least α to the fuzzy set.

Definition 1.3 *Let $\mu \in \mathcal{F}(X)$ be a fuzzy set over the universe of discourse X and let $0 \leq \alpha \leq 1$. The (usual) set*

$$[\mu]_\alpha = \{x \in X \mid \mu(x) \geq \alpha\}$$

is called α-level set or α-cut of the fuzzy set μ.

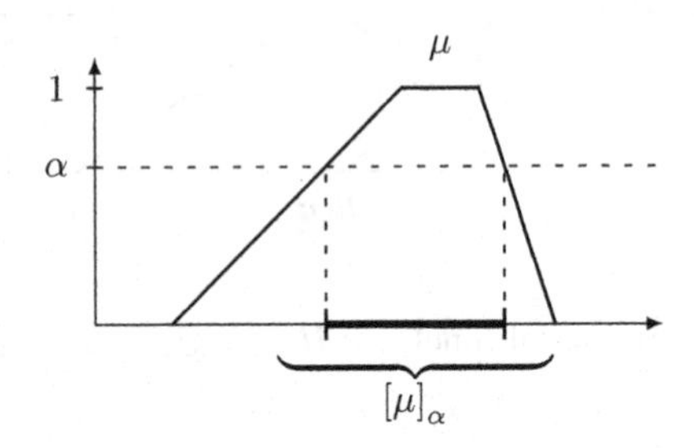

Fig. 1.7. The α-level set or α-cut $[\mu]_\alpha$ of the fuzzy set μ

Figure 1.7 shows the α-cut $[\mu]_\alpha$ of the fuzzy set μ for the case that μ is a trapezoidal function. In this case, the α-cut is a closed interval. For an arbitrary fuzzy set μ over the real numbers we have that μ is convex as a fuzzy set if all its level sets are intervals. Figure 1.8 shows an α-cut of a non-convex fuzzy set consisting of two disjoint intervals.

The level sets of a fuzzy set have the important property of characterizing the fuzzy set uniquely. When we know the level sets $[\mu]_\alpha$ of a fuzzy set μ for all $\alpha \in [0, 1]$, we can determine the degree of membership $\mu(x)$ of any element x to μ by the equation

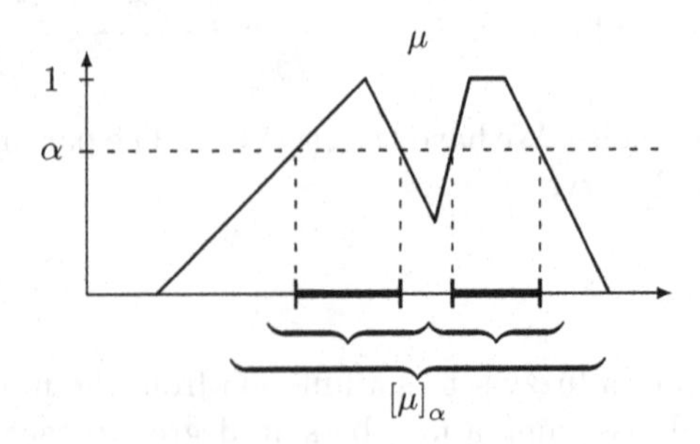

Fig. 1.8. An α-cut $[\mu]_\alpha$ of the fuzzy set μ consisting of two disjoint intervals

$$\mu(x) = \sup\{\alpha \in [0,1] \mid x \in [\mu]_\alpha\}. \tag{1.1}$$

Geometrically speaking, a fuzzy set is the upper envelope of its level sets.

Characterizing a fuzzy set through its level sets will allow us in section 1.4 and 1.5 to work levelwise with operations on fuzzy sets on the basis of usual sets.

The connection between a fuzzy set and its level sets is frequently used for the internal representation of fuzzy sets in computers. But only the α-cuts for a finite amount of selected values α, e.g. $\alpha = 0.25, 0.5, 0.75, 1$, are used and the corresponding level sets of the fuzzy set are saved. In order to determine the degree of membership of an element x to the fuzzy set μ the formula (1.1) can be used, where the supremum is only taken over a finite number of values for α. Thus we discretise the degrees of membership and obtain an approximation of the original fuzzy set. Figure 1.10 shows the level sets $[\mu]_{0.25}$, $[\mu]_{0.5}$, $[\mu]_{0.75}$ and $[\mu]_1$ of the fuzzy set μ defined in figure 1.9. If we only use these four level sets in order to represent μ, we obtain the fuzzy set

$$\tilde{\mu}(x) = \max\{\alpha \in \{0.25, 0.5, 0.75, 1\} \mid x \in [\mu]_\alpha\}$$

in figure 1.11 as an approximation for μ.

Confining us to a finite number of level sets in order to represent or save a fuzzy set corresponds to a discretisation of the membership degrees. Besides this vertical discretisation we can also use discretise the domain (horizontal discretisation). Depending on the considered problem, we have to choose how fine the discretisation should be chosen in both directions. Therefore, no general rules for discretisation can be specified. In general, a refined discretisation of the membership degrees seldom leads to significant improvements of a fuzzy system. One reason for this is that the fuzzy sets are usually determined heuristically or can only be specified roughly. Another reason is that human expert tend to use a limited amount of differentiation levels or degrees of acceptance or membership in order to judge a situation.

1.3 Fuzzy Logic

The notion *fuzzy logic* has three different meanings. In most cases the term fuzzy logic refers to fuzzy logic in the broader sense, including all applications and theories where fuzzy sets or concepts are involved. This includes also fuzzy controllers which are the subject of this book.

On the contrary, the second (and narrower) meaning of the term fuzzy logic focuses on the field of approximative reasoning where fuzzy sets are used and propagated within an inference mechanism as it is for instance common in expert systems.

Finally, fuzzy logic in the narrow sense, which is the topic of this section, considers fuzzy systems from the point of view of multi-valued logics and is

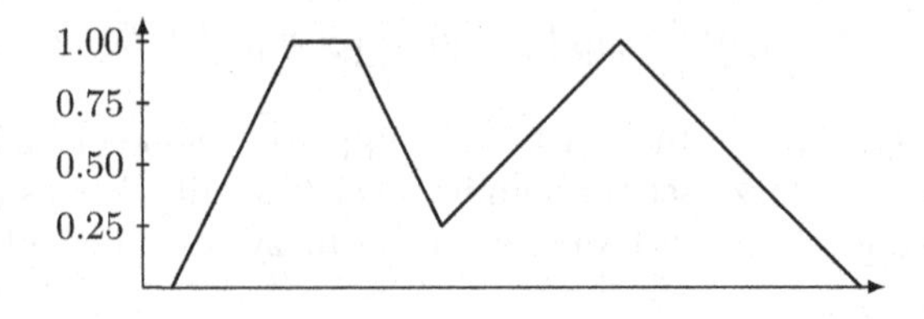

Fig. 1.9. The fuzzy set μ

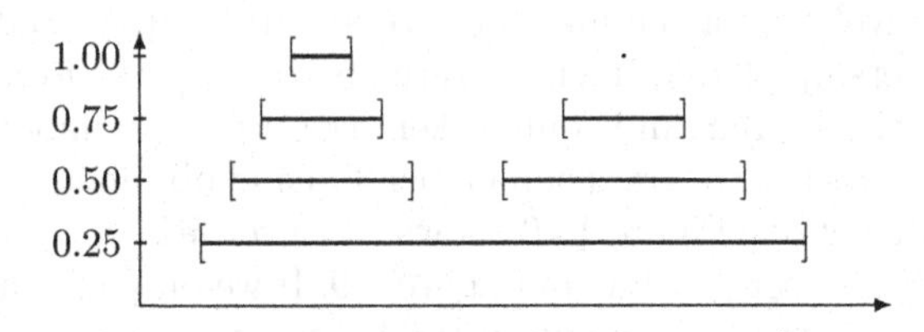

Fig. 1.10. The α-level sets of the fuzzy set μ for $\alpha = 0.25, 0.5, 0.75, 1$

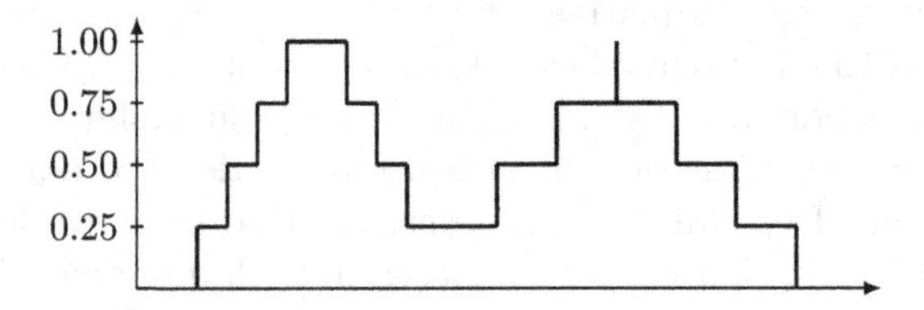

Fig. 1.11. The approximation of the fuzzy set μ resulting from the α-level sets

devoted to issues connected to logical calculi and the associated deduction mechanisms.

We cannot provide a complete introduction to fuzzy logic as a multi-valued logic. A detailed study of this aspect is found in [55]. In this section we will introduce those notions of fuzzy logic which are necessary or useful to understand fuzzy controllers. In section 3.3 about logic-based fuzzy controllers, we discuss some further aspects of fuzzy logic in the narrow sense. We mainly need the concepts of (fuzzy) logic to introduce the set theoretical operations for fuzzy sets. The basis for operations like union, intersection and complement are the logical connectives disjunction, conjunction and negation, respectively. Therefore we briefly repeat some fundamental concepts from classical logic in order to generalise them to the field of fuzzy logic.

1.3.1 Propositions and Truth Values

Classical propositional logic deals with the formal handling of statements (propositions) to which one of the two truth values 1 (for true) or 0 (for false) can be assigned. We represent these propositions by greek letters φ, ψ etc. Typical propositions, for which the formal symbols φ_1 and φ_2 may stand are

$$\varphi_1 : \text{Four is an even number.}$$
$$\varphi_2 : 2 + 5 = 9.$$

The truth value which is assigned to a proposition φ is denoted by $[\![\varphi]\!]$. For the above propositions we obtain $[\![\varphi_1]\!] = 1$ and $[\![\varphi_2]\!] = 0$. If the truth values of single propositions are known, we can determine the truth values of combined propositions using truth tables that define the interpretation of the corresponding logical connectives. The most important logical connectives are the logical AND $\wedge$ (conjunction), the logical OR $\vee$ (disjunction), the negation NOT $\neg$ and the IMPLICATION $\rightarrow$.

The conjunction $\varphi \wedge \psi$ of two propositions φ and ψ is true, if and only if both φ and ψ are true. The disjunction $\varphi \vee \psi$ of φ and ψ obtains the truth value 1 (true), if and only if at least one of the two propositions is true. The implication $\varphi \rightarrow \psi$ is only false, if the premise φ is true and the conclusion ψ is false. The negation $\neg\varphi$ of the proposition φ is false, if and only if φ is true. These definitions are shown in the truth tables for the conjunction, disjunction, implication and negation in Table 1.1.

$[\![\varphi]\!]$	$[\![\psi]\!]$	$[\![\varphi \wedge \psi]\!]$
1	1	1
1	0	0
0	1	0
0	0	0

$[\![\varphi]\!]$	$[\![\psi]\!]$	$[\![\varphi \vee \psi]\!]$
1	1	1
1	0	1
0	1	1
0	0	0

$[\![\varphi]\!]$	$[\![\psi]\!]$	$[\![\varphi \rightarrow \psi]\!]$
1	1	1
1	0	0
0	1	1
0	0	1

$[\![\varphi]\!]$	$[\![\neg\varphi]\!]$
1	0
0	1

Table 1.1. The truth tables for conjunction, disjunction, implication and negation

This definition implies that the propositions

Four is an even number AND $2 + 5 = 9$.

and

Four is an even number IMPLIES $2 + 5 = 9$.

are false, whereas the propositions

$$\text{Four is an even number } \texttt{OR } 2 + 5 = 9.$$

and

$$\texttt{NOT } 2 + 5 = 9.$$

are true. This means, formally expressed, $[\![\varphi_1 \wedge \varphi_2]\!] = 0$, $[\![\varphi_1 \rightarrow \varphi_2]\!] = 0$, $[\![\varphi_1 \vee \varphi_2]\!] = 1$ and $[\![\neg\varphi_2]\!] = 1$.

The assumption that a statement is either true or false is suitable for mathematical issues. But for many expressions formulated in natural language such a strict separation between true and false statements would be unrealistic and would lead to counterintuitive consequences. If somebody promises to come to an appointment at 5 o'clock, his statement would have been false, if he came one minute later. Nobody would call him a liar, although, strictly speaking, his statement was not true. Even more complicated is the statement of being at a party at about 5. The greater the difference between the arrival and 5 o'clock the "less true" the statement is. A sharp definition of an interval of time corresponding to "about 5" is impossible.

Humans are able to formulate such "fuzzy" statements, understand them, draw conclusions from them and work with them. If someone starts an approximately four-hour-drive at around 11 o'clock and is going to have lunch for about half an hour, we can use these imprecise pieces of information and conclude at what time more or less the person will arrive. A formalisation of this simple issue in a logical calculus, where statements can be either true or false only, is not adequate.

In natural language using fuzzy statements or information is not an exception but normal. In a recipe nobody would replace the statement "Take a pinch of salt" by "Take 80 corns of salt". A driver will not calculate the distance he will need for stopping his car abruptly on a wet road by using another friction constant in some mathematical formula to calculate this distance. He will consider the rule: the wetter the road, the longer the distance needed for breaking.

In order to model this human information processing in a more appropriate way, we use gradual truth values for statements. This means a statement can not only be true (truth value 1) or false (truth value 0) but also more or less true expressed by a value between 0 and 1.

The connection between fuzzy sets and imprecise or fuzzy statements can be described in the following way. A fuzzy set models a property that elements of the universe of discourse can have more or less. For example, let us consider the fuzzy set μ_{hG} of high velocities from figure 1.2 from page 3. The fuzzy set represents the property or the predicate *high velocity*. That means the degree of membership of a concrete velocity v to the fuzzy set of high velocities corresponds to the "truth value" which is assigned to the statement "v is a high velocity". In this sense, a fuzzy set determines the corresponding truth values for a set of statements – in our example for all statements we obtain, when we consider in a concrete velocity value for v. In order to understand how to operate with fuzzy sets, it is useful to consider first of all classical crisp propositions.

Dealing with combined propositions like "160 km/h is a high velocity AND the stopping distance is about 110 m" requires the extension of the truth tables for logical connectives like conjunction, disjunction, implication or negation. The truth tables shown in table 1.1 determine a truth function for each logical connective. For conjunction, disjunction and implication this truth function assigns to each combination of two truth values (the truth value assigned to φ and ψ) one truth value (the truth value of the conjunction, disjunction of φ and ψ or the implication $\varphi \rightarrow \psi$). The truth function assigned to the negation has only one truth value as argument. If we denote the truth function by w_* associated with the logical connective $* \in \{\wedge, \vee, \rightarrow, \neg\}$, w_* is a binary or unary function. That means

$$w_\wedge, w_\vee, w_\rightarrow : \{0,1\}^2 \rightarrow \{0,1\}, \;\; w_\neg : \{0,1\} \rightarrow \{0,1\}.$$

For fuzzy propositions, where the unit interval $[0,1]$ replaces the binary set $\{0,1\}$ as the set of possible truth values, we have to assign truth functions to the logical connectives accordingly. These truth functions have to be defined on the unit quadrat square or the unit interval.

$$w_\wedge, w_\vee, w_\rightarrow : [0,1]^2 \rightarrow [0,1], \;\; w_\neg : [0,1] \rightarrow [0,1]$$

A minimum requirement we demand of these function is that, limited to the values 0 and 1, they should provide the same values as the corresponding truth function associated with the classic logical connectives. This requirement says that a combination of fuzzy propositions which are actually crisp (non-fuzzy), because there truth values are 0 or 1, coincide with the usual combination of classical crisp propositions.

The most frequently used truth functions for conjunction and disjunction in fuzzy logic are the minimum or maximum. That means $w_\wedge(\alpha,\beta) = \min\{\alpha,\beta\}$, $w_\vee(\alpha,\beta) = \max\{\alpha,\beta\}$. Normally the negation is defined by $w_\neg(\alpha) = 1-\alpha$. In his seminal work [210] published in 1965, L. Zadeh introduced the concept of fuzzy sets and used these functions for operating with fuzzy sets.

The implication is often understood in the sense of the *Łukasiewicz implication*

$$w_\rightarrow(\alpha,\beta) = \min\{1-\alpha+\beta, 1\}$$

or the *Gödel implication*

$$w_\rightarrow(\alpha,\beta) = \begin{cases} 1 & \text{if } \alpha \leq \beta \\ \beta & \text{else.} \end{cases}$$

1.3.2 t-Norms and t-Conorms

Until now we have interpreted the truth values from the unit interval in a purely intuitive way as gradual truth. So choosing the truth functions for the logical connectives in the above mentioned way seems to be plausible but it is not unique at all. Instead of trying to find more or less arbitrary functions, we might better use an axiomatic approach where we define some reasonable properties a truth function should satisfy and thus confining the possible truth functions. We discuss this axiomatic approach in detail for the conjunction.

Let us consider the function $t : [0,1]^2 \to [0,1]$ as a potential candidate for the truth function of a conjunction for fuzzy propositions. The truth value of a conjunction of several propositions should not depend on the order in which the propositions are considered. In order to guarantee this property t has to be commutative and associative, that means

(T1) $t(\alpha, \beta) = t(\beta, \alpha)$
(T2) $t(t(\alpha, \beta), \gamma) = t(\alpha, t(\beta, \gamma))$.

should hold.

The truth value of the conjunction $\varphi \wedge \psi$ should not be less than the truth value of the conjunction $\varphi \wedge \chi$, if χ has a lower truth value than ψ. Therefore, we require some monotonicity condition of t:

(T3) $\beta \leq \gamma$ implies $t(\alpha, \beta) = t(\alpha, \gamma)$.

Because of the commutativity (T1), (T3) implies that t is non-decreasing in both arguments.

Furthermore, we require that the truth value of a proposition φ will be the same as the truth value of the conjunction of φ with any true proposition ψ. For the truth function t this leads to

(T4) $t(\alpha, 1) = \alpha$.

Definition 1.4 *A function* $t : [0,1]^2 \to [0,1]$ *is called a t-norm (triangular norm) , if the axioms* (T1) – (T4) *are satisfied.*

In the framework of fuzzy logic we should always choose a t-norm as the truth function for conjunction. From the property (T4) follows that for every t-norm t we have $t(1,1) = 1$ and $t(0,1) = 0$. From $t(0,1) = 0$ we obtain $t(1,0) = 0$ by the commutativity property (T1). Furthermore, because of the monotonicity property (T3) and $t(0,1) = 0$ we must have $t(0,0) = 0$. In this way every t-norm restricted to the values 0 and 1 coincides with the truth function given by the truth value table of the usual conjunction.

We can verify easily that the discussed truth function $t(\alpha, \beta) = \min\{\alpha, \beta\}$ for the conjunction is a t-norm. Other examples of t-norms are

Łukasiewicz t-norm: $t(\alpha, \beta) = \max\{\alpha + \beta - 1, 0\}$

algebraic product: $t(\alpha, \beta) = \alpha \cdot \beta$

drastic product: $t(\alpha, \beta) = \begin{cases} 0 & \text{if } 1 \notin \{\alpha, \beta\} \\ \min\{\alpha, \beta\} & \text{otherwise} \end{cases}$

These few examples show that the spectrum of t-norms is very broad. The limits are given by the drastic product, which is the smallest t-norm and which is discontinuous, and the minimum, which is the greatest t-norm. Besides this the minimum can be considered as special t-norm, since it is the only idempotent t-norm which means that only the minimum satisfies the property $t(\alpha, \alpha) = \alpha$ for all $\alpha \in [0, 1]$.

Only the idempotency of a t-norm can guarantee that the truth values of the proposition φ and the conjunction $\varphi \wedge \varphi$ coincide, which at first sight seems to be a canonical requirement, letting the minimum seem to be the only reasonable choice for the truth functions for the conjunction in the context of fuzzy logic. However, the following example shows that the imdempotency property is not always desirable.

A buyer has to decide between the houses A and B. The houses are very similar in most aspects, so he makes his decision considering the criteria good price and good location. After careful consideration he assigns the following "truth values" to the decisive aspects:

	proposition	truth value $[\![\varphi_i]\!]$
φ_1	The price of house A is good.	0.9
φ_2	The location of house A is good.	0.6
φ_3	The price of house B is good.	0.6
φ_4	The location of house B is good.	0.6

He chooses house $x \in \{A, B\}$ for which the proposition "The price of house x is good `AND` The location of house x is good" yields the greater truth value. That means, the buyer will choose house A, if $[\![\varphi_1 \wedge \varphi_2]\!] > [\![\varphi_3 \wedge \varphi_4]\!]$ holds, and house B otherwise. When we determine the truth value of the conjunction by the minimum, we would obtain the value 0.6 for both of the houses and thus the houses would be regarded as equally good. But this is counterintuitive, because house A has definitely a better price than house B and the locations are equally good. However, when choose a non-idempotent t-norm – like e.g. the algebraic product or the Łukasiewicz t-norm – as truth function for the conjunction, we will always favour house A.

Besides the discussed examples for the t-norms there are many others. In particular, there are whole families of t-norms which can be defined in a parametric way. For example the Weber family

$$t_\lambda(\alpha, \beta) = \max\left\{\frac{\alpha + \beta - 1 + \lambda\alpha\beta}{1 + \lambda}, 0\right\}$$

which determines a t-norm for each $\lambda \in (-1, \infty)$. For $\lambda = 0$ it results in the Łukasiewicz t-norm.

In most practical applications only the minimum, the algebraic product and the Łukasiewicz t-norm are chosen. Therefore, we will not consider the enormous variety of other t-norms here. For further readings on t-norms, we refer the reader to [25, 97, 102].

In the same way as we have have defined t-norms as possible truth functions for the conjunction, we can define candidates for truth functions for the disjunction. Just like the t-norms they should satisfy the properties (T1) – (T3). Instead of (T4) we ask for

(T4') $t(\alpha, 0) = \alpha$,

which means that the truth value of a proposition φ will be the same as the truth value of the disjunction of φ with any false proposition ψ.

Definition 1.5 *A function* $s : [0,1]^2 \to [0,1]$ *is called t-conorm (triangular conorm), if the axioms* (T1) – (T3) *and* (T4') *are satisfied.*

t-norms and t-conorms are dual concepts in the following sense. Each t-norm induces a t-conorm s by

$$s(\alpha, \beta) \;=\; 1 - t(1 - \alpha, 1 - \beta), \tag{1.2}$$

and vice versa, from a t-conorm s we obtain the corresponding t-norm by

$$t(\alpha, \beta) \;=\; 1 - s(1 - \alpha, 1 - \beta). \tag{1.3}$$

These equations (1.2, 1.3) correspond to DeMorgan's Law

$$[\![\varphi \vee \psi]\!] \;=\; [\![\neg(\neg\varphi \wedge \neg\psi)]\!] \quad \text{and} \quad [\![\varphi \wedge \psi]\!] \;=\; [\![\neg(\neg\varphi \vee \neg\psi)]\!],$$

if we compute the negation using the truth function $[\![\neg\varphi]\!] \;=\; 1 - [\![\varphi]\!]$.

The t-conorms we obtain from the t-norms minimum, Łukasiewicz t-norm, algebraic and drastic product by applying formula (1.2) are

maximum: $s(\alpha, \beta) = \max\{\alpha, \beta\}$

Łukasiewicz t-conorm: $s(\alpha, \beta) = \min\{\alpha + \beta, 1\}$

algebraic sum: $s(\alpha, \beta) = \alpha + \beta - \alpha\beta$

drastic sum: $s(\alpha, \beta) = \begin{cases} 1 & \text{if } 0 \notin \{\alpha, \beta\} \\ \max\{\alpha, \beta\} & \text{otherwise.} \end{cases}$

The duality between t-norms and t-conorms implies immediately that the drastic sum is the greatest, the maximum is the smallest t-conorm, and the maximum is the only idempotent t-conorm. Analogously to t-norms we can define parametric families of t-conorms. Such as

$$s_\lambda(\alpha, \beta) \;=\; \min\left\{\alpha + \beta - \frac{\lambda\alpha\beta}{1 + \lambda}, 1\right\}$$

which is the Weber family of t-conorms.

Operating with t-norms and t-conorms we should be aware that not all laws we know for classical conjunction and disjunction hold also for t-norms and t-conorms. For instance, minimum and maximum are not merely the only idempotent t-norms or t-conorms, but also the only pair defined by duality (1.2) which satisfies the distributive laws.

In the example of the man who wanted to buy a house we could see that the idempotency of a t-norm is not always desirable. The same holds for t-conorms. Let us consider the propositions $\varphi_1, \ldots, \varphi_n$ which shall be connected in a conjunctive or disjunctive manner. The significant disadvantage of the idempotency is the following. Applying the conjunction in terms of the minimum, the resulting truth value of the connection of propositions depends only on the truth value of the proposition to which the least truth value is assigned. Applying the disjunction in the sense of the maximum only the proposition with the greatest truth value determines the truth value of the disjunction of the propositions. We can avoid this disadvantage, if we give up idempotency. Another approach is the use of *compensatory operators* which are a compromise between conjunction and disjunction. An example for a compensatory operator is the *Gamma operator* [220]

$$\Gamma_\gamma(\alpha_1, \ldots, \alpha_n) = \left(\prod_{i=1}^{n} \alpha_i\right) \cdot \left(1 - \prod_{i=1}^{n}(1 - \alpha_i)\right)^\gamma$$

with parameter $\gamma \in [0, 1]$. For $\gamma = 0$ this results in the algebraic product, for $\gamma = 1$ we obtain the algebraic sum. Another compensatory operator is the arithmetical mean. Other suggestions for such operators can be found in [121]. A big disadvantage of these operator is the fact that they do not satisfy associativity. So we do not use them here.

In addition to the connection between t-norms and t-conorm, we can also find connections between t-norms and implications. A continuous t-norm t induces the *residuated implication* $\vec{t}$ by the formula

$$\vec{t}(\alpha, \beta) = \sup\{\gamma \in [0, 1] \mid t(\alpha, \gamma) \leq \beta\}.$$

In this way, by residuation we obtain the Łukasiewicz implication from the Łukasiewicz t-norm and the Gödel implication from the minimum.

Later we will need the corresponding *biimplication* $\overleftrightarrow{t}$ which is defined by the formula

$$\begin{aligned} \overleftrightarrow{t}(\alpha, \beta) &= \vec{t}\left(\max\{\alpha, \beta\}, \min\{\alpha, \beta\}\right) \\ &= t(\vec{t}(\alpha, \beta), \vec{t}(\beta, \alpha)) \\ &= \min\{\vec{t}(\alpha, \beta), \vec{t}(\beta, \alpha)\}. \end{aligned} \tag{1.4}$$

This formula is motivated by the definition of the biimplication or equivalence in classical logic in terms of

$$[\![\varphi \leftrightarrow \psi]\!] = [\![(\varphi \rightarrow \psi) \wedge (\psi \rightarrow \varphi)]\!].$$

Besides the logical connectives like conjunction, disjunction, implication or negation in (fuzzy) logic we have also the quantifiers $\forall$ (for all) and $\exists$ (exists).

The universal quantifier $\forall$ and the existential quantifier $\exists$ are closely related to the conjunction and the disjunction, respectively. Let us consider the universe of discourse X and the predicate $P(x)$. X could for instance be the set $\{2, 4, 6, 8, 10\}$ and $P(x)$ the predicate "x is an even number". If the set X is finite, e.g. $X = \{x_1, \ldots, x_n\}$, the statement $(\forall x \in X)(P(x))$ is equivalent to the statement $P(x_1) \wedge \ldots \wedge P(x_n)$. Therefore, in this case it is possible to define the truth value of the statement $(\forall x \in X)(P(x))$ on the basis of the conjunction which means

$$[\![(\forall x \in X)(P(x))]\!] \;=\; [\![P(x_1) \wedge \ldots \wedge P(x_n)]\!].$$

If we assign the minimum to the conjunction as truth function we obtain

$$[\![(\forall x \in X)(P(x))]\!] \;=\; \min\{[\![P(x)]\!] \mid x \in X\}$$

which can be extended to an infinite universe of discourse X by

$$[\![(\forall x \in X)(P(x))]\!] \;=\; \inf\{[\![P(x)]\!] \mid x \in X\}.$$

Other t-norms than the minimum are normally not used for the universal quantifier, since the non-idempotency property leads to easily to the truth value zero in the case of an inifinite universe of discourse.

The same consideration for the existential quantifier lead to the definition

$$[\![(\exists x \in X)(P(x))]\!] \;=\; \sup\{[\![P(x)]\!] \mid x \in X\}.$$

If the universe of discourse for the existential quantifier is finite, the propositions $(\exists x \in X)(P(x))$ and $P(x_1) \vee \ldots \vee P(x_n)$ are equivalent.

As an example we consider the predicate $P(x)$ with the interpretation "x is a high velocity". Let the truth value $[\![P(x)]\!]$ be given by the fuzzy set of the high velocities from figure 1.2 on page 3 which means $[\![P(x)]\!] = \mu_{hG}(x)$. So we have for instance $[\![P(150)]\!] = 0$, $[\![P(170)]\!] = 0.5$ and $[\![P(190)]\!] = 1$. Thus the statement $(\forall x \in [170, 200])(P(x))$ ("All velocities between 170 km/h and 200 km/h are high velocities") has the truth value

$$\begin{aligned} [\![(\forall x \in [170, 200])(P(x))]\!] &= \inf\{[\![P(x)]\!] \mid x \in [170, 200]\} \\ &= \inf\{\mu_{hG}(x) \mid x \in [170, 200]\} \\ &= 0.5. \end{aligned}$$

Analogously we obtain $[\![(\exists x \in [100, 180])(P(x))]\!] \;=\; 0.75$.

1.3.3 Basic Assumptions and Problems

In this section about fuzzy logic we have discussed various ways of combining fuzzy propositions. An essential assumption we have used in the section is that of *truth functionality*. This means that the truth value of the combination of several propositions depends only on the truth values of the propositions, but not on the individual propositions. This assumption holds in classical logic but not, for example, in the context of probability theory or probabilistic logic. In probability theory it is not enough to know the probability of two events in order to determine the probability that both events will occur simultaneously or at least one of them will occur. For this we also need information about the dependency of these events. In the case of independence, the probability that both events occur is the product of the single probabilities, and the probability that at least one of the events will occur is the sum of the probabilities. We cannot determine these probabilities without knowing the independence of the events.

We should be aware of the assumption of truth functionality in the framework of fuzzy logic. It is not always satisfied. Coming back to the example of the man buying a house, we gave reasons for using non-idempotent t-norms. If we use these t-norms, like for instance the algebraic product, for propositions like "The price of the house A is good `AND` ... `AND` the price of house A is good", this combined proposition can obtain a very small truth value. Depending on how we interpret the conjunction, this effect might be desirable or might lead to inconsistency. If we understand the conjunction in its classical sense, a conjunctive combination of a proposition with itself should be equivalent to itself which is not satisfied for non-idempotent t-norms. Another possibility is to understand the conjunction as a list of pro and con arguments for a thesis or as a proof. The repeated use of the same (fuzzy) argument within a proof might result in a loss of credibility and thus idempotency is not desirable, even for a conjunction of a proposition with itself.

Fortunately, for fuzzy control these consideration are of minor importance, because in this application area fuzzy logic is used in a more restricted context, where we do not have to worry about combining the same proposition with itself. More difficulties will show up, when we apply fuzzy logic in the framework of complex expert systems.

1.4 Operations on Fuzzy Sets

Sections 1.1 and 1.2 described how vague concepts can be modelled using fuzzy sets and how fuzzy sets can be represented. In order to operate with vague concepts or apply some kind of deduction mechanism to them, we need suitable operations for fuzzy sets. Therefore, in this section operations like union, intersection or complement well known from classical set theory will be extended to fuzzy sets.

1.4.1 Intersection

The underlying concept of generalising fundamental set-theoretic operations to fuzzy sets is explained in detail for the intersection of (fuzzy) sets. For the other operations, a generalisation can be carried out in a straight forward manner analogously to intersection. For two ordinary sets M_1 and M_2 we have that an element x belongs to the intersection of the two sets, if and only if it belongs to both M_1 and M_2. Whether x belongs to the intersection depends only on the membership of x to M_1 and M_2 but not on the membership of any other element $y \neq x$ to M_1 and M_2. Formally speaking, this means

$$x \in M_1 \cap M_2 \quad \Longleftrightarrow \quad x \in M_1 \ \wedge \ x \in M_2. \tag{1.5}$$

For two fuzzy sets μ_1 and μ_2 we also assume that the degree of membership of an element x to the intersection of the two fuzzy sets depends only on the membership degrees of x to μ_1 and μ_2. We interpret the degree of membership $\mu(x)$ of an element x to the fuzzy set μ as the truth value $[\![x \in \mu]\!]$ of the fuzzy proposition "$x \in \mu$", that x is an element of μ. In order to determine the membership degree of an element x to the intersection of the fuzzy sets μ_1 and μ_2, we have to calculate the truth value of the conjunction "x is an element of μ_1 AND x is an element of μ_2" following the equivalence in (1.5). The previously discussed concepts of fuzzy logic have told us, how we can define the truth value of the conjunction of two fuzzy propositions. Therefore, it is necessary to choose a suitable t-norm t as the truth function for the conjunction. Thus we define the intersection of two fuzzy sets μ_1 and μ_2 (w.r.t. the t-norm t) as the fuzzy set $\mu_1 \cap_t \mu_2$ with

$$(\mu_1 \cap_t \mu_2)(x) \;=\; t\left(\mu_1(x), \mu_2(x)\right).$$

If we interpret the degree of membership $\mu(x)$ of an element x to the fuzzy set μ as the truth value $[\![x \in \mu]\!]$ of the fuzzy proposition "$x \in \mu$", that x is an element of μ, the definition of the intersection of two fuzzy sets can be written in the following way

$$[\![x \in (\mu_1 \cap_t \mu_2)]\!] \;=\; [\![x \in \mu_1 \wedge x \in \mu_2]\!]$$

where we assign the t-norm t as the truth function for the conjunction.

By defining the intersection of fuzzy sets on the basis of a t-norm, the properties of the t-norm are inherited to the intersection operator: the axioms (T1) and (T2) make the intersecting of fuzzy sets commutative and associative. The monotonicity property (T3) guarantees that replacing a fuzzy set μ_1 by a larger fuzzy set μ_2, which means $\mu_1(x) \leq \mu_2(x)$ for all x, can only lead to a larger intersection:

$$\mu_1 \leq \mu_2 \text{ implies } \mu \cap_t \mu_1 \;\leq\; \mu \cap_t \mu_2.$$

Axiom (T4) guarantees that the intersection of a fuzzy set with an ordinary set, respectively its characteristic function, results in the original fuzzy set limited to the ordinary set with which we intersect it. If $M \subseteq X$ is an ordinary subset of X and $\mu \in \mathcal{F}(X)$ a fuzzy set of X, we have

$$(\mu \cap_t I_M)(x) \;=\; \begin{cases} \mu(x) & \text{if } x \in M \\ 0 & \text{otherwise.} \end{cases}$$

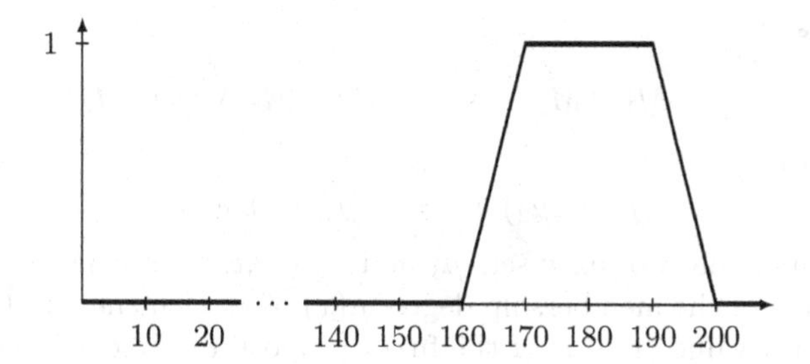

Fig. 1.12. The fuzzy set $\mu_{170-190}$ of the velocities not much less than 170 km/h and not much greater than 190 km/h

If not otherwise stated, the intersection of two fuzzy sets will be computed w.r.t. the minimum t-norm. In this case, or when it is clear to which t-norm we refer, we will write $\mu_1 \cap \mu_2$ instead of $\mu_1 \cap_t \mu_2$ for the case of $t = \min$.

As an example, we consider the intersection of the fuzzy set μ_{hv} of high velocities from figure 1.2 on page 3 and the fuzzy set $\mu_{170-190}$ of the velocities not much less than 170 km/h and not much greater than 190 km/h from figure 1.12. Both of them are trapezoidal functions:

$$\mu_{hv} = \Pi_{150,180,\infty,\infty}, \quad \mu_{170-190} = \Pi_{160,170,190,200}.$$

Figure 1.13 shows the intersection of the two fuzzy sets on the basis of the minimum (continuous line) and the Łukasiewicz t-norm (dashed line).

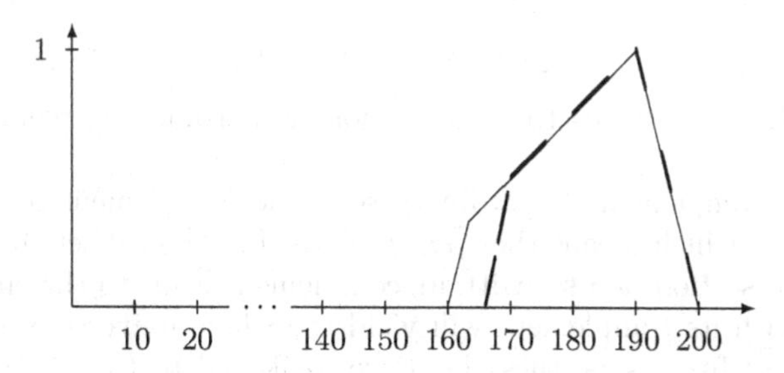

Fig. 1.13. The intersection $\mu_{hv} \cap_t \mu_{170-190}$ of the fuzzy sets μ_{hv} and $\mu_{170-190}$ computed using the minimum (continuous line) and the Łukasiewicz t-norm (dashed line)

1.4.2 Union

From the representation (1.5) we have derived the definition for the intersection of two fuzzy sets. Analogously we can define the union of two fuzzy sets on the basis of

$$x \in M_1 \cup M_2 \quad \Longleftrightarrow \quad x \in M_1 \ \vee \ x \in M_2.$$

This leads to

$$(\mu_1 \cup_s \mu_2)(x) \;=\; s\,(\mu_1(x), \mu_2(x))\,,$$

as the union of the two fuzzy sets μ_1 and μ_2 w.r.t. the t-conorm s. Using the interpretation of the membership degree $\mu(x)$ of an element x to the fuzzy set μ as the truth value $[\![x \in \mu]\!]$ of the fuzzy proposition "$x \in \mu$", we can define the union in the form of

$$[\![x \in (\mu_1 \cup_t \mu_2)]\!] \;=\; [\![x \in \mu_1 \vee x \in \mu_2]\!].$$

where we assign the t-conorm s as the truth function for the disjunction. As in the case of the intersection, we will write $\mu_1 \cup \mu_2$ instead of $\mu_1 \cup_s \mu_2$, when we use the maximum t-conorm or when it is clear to which t-conorm we refer.

1.4.3 Complement

The complement of a fuzzy set is derived from the formula

$$x \in \overline{M} \quad \Longleftrightarrow \quad \neg(x \in M)$$

for ordinary sets where $\overline{M}$ stands for the complement of the (ordinary) set M. If we assign the truth function $w_\neg(\alpha) = 1 - \alpha$ to the negation, we obtain the fuzzy set

$$\overline{\mu_1}(x) \;=\; 1 - \mu(x),$$

as the complement $\overline{\mu}$ of the fuzzy set μ. This is also in according with

$$[\![x \in \overline{\mu}]\!] \;=\; [\![\neg(x \in \mu)]\!].$$

Figure 1.14 illustrates the intersection, union and complement of fuzzy sets.

Like the complement for ordinary sets, the complement of fuzzy sets is an involution, which means that $\overline{\overline{\mu}} = \mu$ holds. In classical set theory we have that the intersection of a set with its complement leads to the empty set and the union with its complement will yield the whole universe of discourse. In the context of fuzzy sets, these laws are weakened to $(\mu \cap \overline{\mu})(x) \leq 0.5$ and $(\mu \cup \overline{\mu})(x) \geq 0.5$ for all x. Figure 1.15 illustrates this phenomenon.

If the intersection and the union are defined on the basis of minimum or maximum, we can use the representation of fuzzy sets by the level sets introduced in section 1.2, in order to compute the resulting fuzzy set. We have

$$[\mu_1 \cap \mu_2]_\alpha \;=\; [\mu_1]_\alpha \cap [\mu_2]_\alpha \qquad \text{and} \qquad [\mu_1 \cup \mu_2]_\alpha \;=\; [\mu_1]_\alpha \cup [\mu_2]_\alpha$$

for all $\alpha \in [0, 1]$. According to these equations the level sets of the intersection and the union of two fuzzy sets are the intersection or the union of the level sets of the single fuzzy sets.

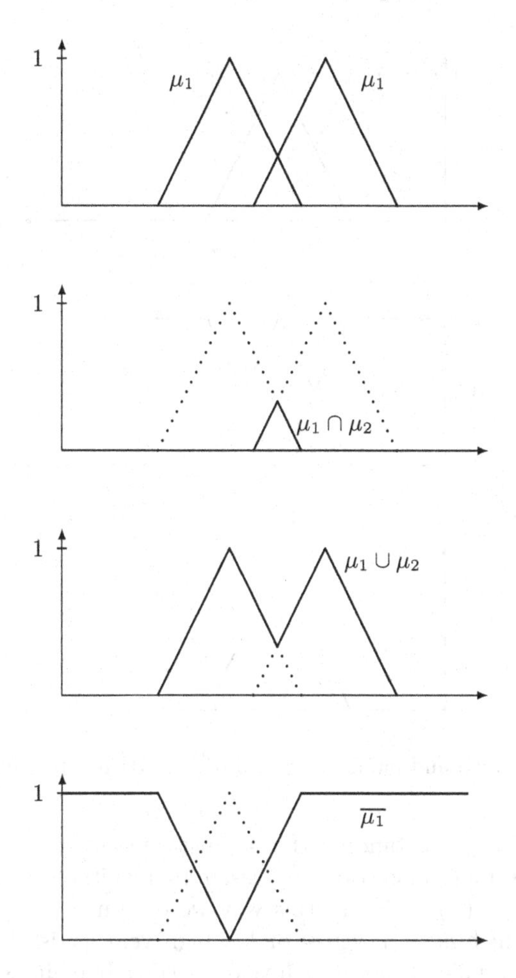

Fig. 1.14. Intersection, union and complement for fuzzy sets

1.4.4 Linguistic Modifiers

Besides the complement as a unary operation on fuzzy sets, derived from the corresponding operation for ordinary sets, there are more fuzzy set specific unary operations, that have no counterpart in classical set theory. Normally, a fuzzy set represents a vague concept like "high velocity", "young" or "tall". From such concepts we can derive other vague concepts applying *linguistic modifiers* or ("linguistic hedges") like "very" or "more or less".

As an example we consider the fuzzy set μ_{hv} of high velocities from figure 1.2 on page 3. How should the fuzzy set μ_{vhv} representing the concept of "*very* high velocities" look like? A very high velocity is also a high velocity, but not vice versa. Thus the membership degree of a specific velocity v to the fuzzy set μ_{vhv} should not exceed its membership degree to the fuzzy set μ_{hv}. We

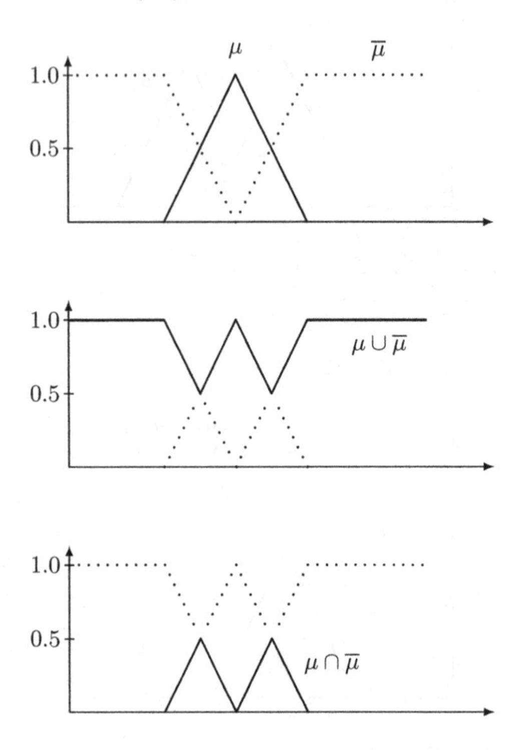

Fig. 1.15. Union and intersection of a fuzzy set with its complement

can achieve this by understanding the linguistic modifier "very", similar to the negation, as a unary operation and assigning a suitable truth function to it, for instance $w_{\text{very}}(\alpha) = \alpha^2$. In this way we obtain $\mu_{vhv}(x) = (\mu_{hv}(x))^2$. Now a velocity which is to a degree of 1 a high velocity is also a very high velocity. A velocity which is not a high velocity (membership degree 0) is also not a high velocity. If the membership degree of a velocity to μ_{hv} is between 0 and 1, it is also a very high velocity, but with a lower membership degree.

In the same way we can assign a truth function to the modifier "more or less". This truth function should increase the degree of membership degree, for instance $w_{\text{more or less}}(\alpha) = \sqrt{\alpha}$.

Figure 1.16 shows the fuzzy set μ_{hv} of high velocities and the resulting fuzzy sets μ_{vhv} of very high velocities and μ_{mhv} of more or less high velocities.

1.5 The Extension Principle

In the previous section we have discussed how set theoretic operations like intersection, union and complement can be generalised to fuzzy sets. This section is devoted to the issue of extending the concept of mappings or functions to fuzzy sets. These ideas allow us to define operations like addition,

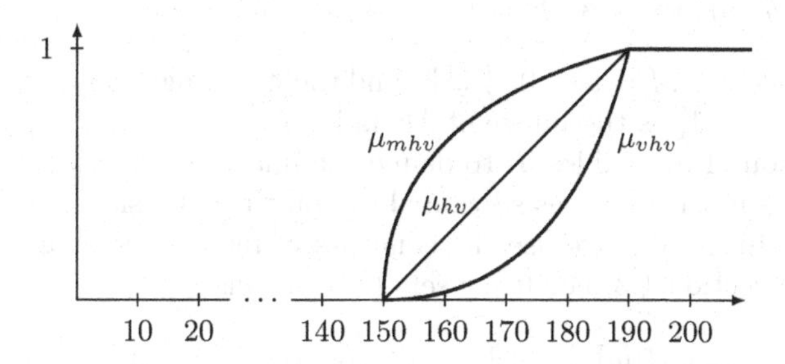

Fig. 1.16. The fuzzy sets μ_{hv}, μ_{vhv} and μ_{mhv} of high, very high and more or less high velocities

subtraction, multiplication, division or taking squares as well as set theoretic concepts like the composition of relations for fuzzy sets.

1.5.1 Mappings for Fuzzy Sets

As an example we consider the mapping $f : \mathbb{R} \to \mathbb{R}$, $x \mapsto |x|$. The fuzzy set $\mu = \Lambda_{-1.5,-0.5,2.5}$ shown in figure 1.17 models the vague concept "about -0.5".

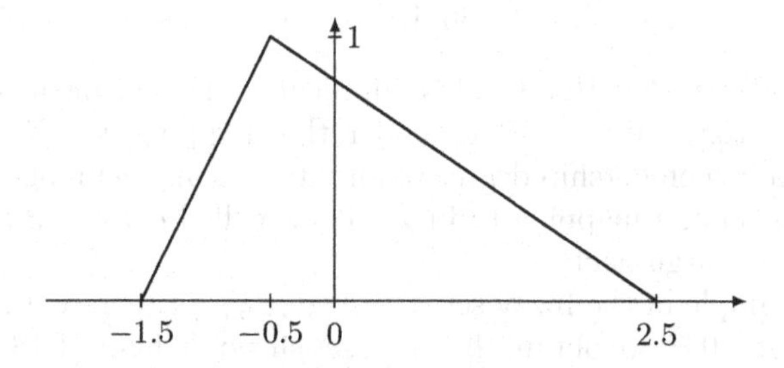

Fig. 1.17. The fuzzy set $\mu = \Lambda_{-1.5,-0.5,1.5}$ representing "about -0.5"

Which fuzzy set should represent "the absolute value of about -0.5" or in other words, which is the image $f[\mu]$ of the fuzzy set μ? For a usual subset M of the universe of discourse X the image $f[M]$ under the mapping $f : X \to Y$ is defined as the subset of Y that contains all images of elements of M. Technically speaking, this means

$$f[M] \;=\; \{y \in Y \mid (\exists x \in X)(x \in M \wedge f(x) = y)\},$$

or, in other words,

$$y \in f[M] \quad \iff \quad (\exists x \in X)(x \in M \wedge f(x) = y). \tag{1.6}$$

For instance, for $M = [-1, 0.5] \subseteq \mathbb{R}$ and the mapping $f(x) = |x|$ we obtain the set $f[M] = [0, 1]$ as the image of M under f.

The relation (1.6) enables us to define the image of a fuzzy set μ under a mapping f. As in the previous section above on the extension of set theoretic operations to fuzzy sets, we use the concepts of fuzzy logic again, that were introduced in section 1.3. For fuzzy sets (1.6) means

$$[\![y \in f[\mu]]\!] \;=\; [\![(\exists x \in X)(x \in \mu \wedge f(x) = y)]\!].$$

As explained in section 1.3 we evaluate the existential quantifier by the supremum and associate a t-norm t with the conjunction. Therefore, the fuzzy set

$$f[\mu](y) \;=\; \sup\left\{t\left(\mu(x), [\![f(x) = y)]\!]\right) \mid x \in X\right\} \tag{1.7}$$

represents the image of μ under f. The choice of the t-norm t does not play any role in this case, because the statement $f(x) = y$ is either true or false which means $[\![f(x) = y]\!] \in \{0, 1\}$, and therefore

$$t\left(\mu(x), [\![f(x) = y)]\!]\right) \;=\; \begin{cases} \mu(x) & \text{if } f(x) = y \\ 0 & \text{otherwise.} \end{cases}$$

Thus (1.7) simplifies to

$$f[\mu](y) \;=\; \sup\left\{\mu(x) \mid f(x) = y\right\}. \tag{1.8}$$

This definition says that the membership degree of an element $y \in Y$ to the image of the fuzzy set $\mu \in \mathcal{F}(X)$ under the mapping $f : X \to Y$ is the greatest possible membership degree of all x to μ that are mapped to y under f. This extension of a mapping to fuzzy sets is called *extension principle* (for a function in one argument).

For the example of the fuzzy set $\mu = \Lambda_{-1.5,-0.5,2.5}$ representing the vague concept "about -0.5" we obtain the fuzzy set shown in figure 1.18 as the image under the mapping $f(x) = |x|$. To further illustrate the underlying principle, we determine the membership degree $f[\mu](y)$ for $y \in \{-0.5, 0, 0.5, 1\}$. Because of $f(x) = |x| \geq 0$, no value is mapped to $y = -0.5$ under f, so we obtain $f[\mu](-0.5) = 0$. There is only one value that is mapped to $y = 0$, that is $x = 0$, hence we have $f[\mu](0) = \mu(0) = 5/6$. For $y = 0.5$ there are two values, $x = -0.5$ and $x = 0.5$, mapped to y, leading to

$$f[\mu](0.5) \;=\; \max\{\mu(-0.5), \mu(0.5)\} \;=\; \max\{1, 2/3\} \;=\; 1.$$

The values mapped to $y = 1$ are $x = -1$ and $x = 1$. Therefore we obtain

$$f[\mu](1) \;=\; \max\{\mu(-1), \mu(1)\} \;=\; \max\{0.5, 0.5\} \;=\; 0.5.$$

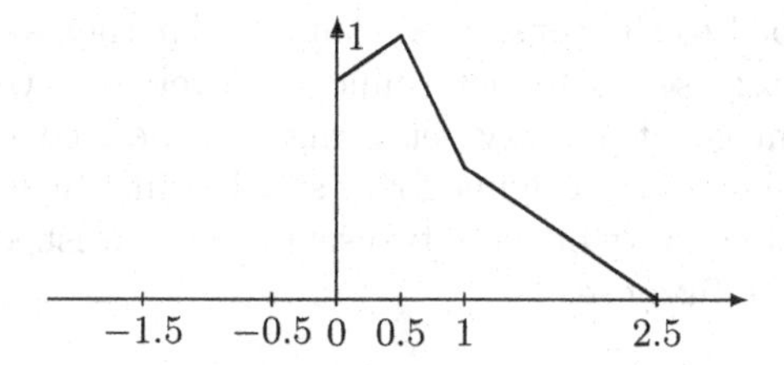

Fig. 1.18. The fuzzy set representing the vague concept "the absolute value of about −0.5"

Example 1.6 Let $X = X_1 \times \ldots \times X_n$, $i \in \{1, \ldots, n\}$.

$$\pi_i : X_1 \times \ldots \times X_n \to X_i, \quad (x_1, \ldots, x_n) \mapsto x_i$$

denotes the projection of the Cartesian product $X_1 \times \ldots \times X_n$ to the ith coordinate space X_i. The projection of a fuzzy set $\mu \in \mathcal{F}(X)$ to the space X_i is according to the extension principle (1.8)

$$\pi_i[\mu](x) = \sup\{\ \mu(x_1, \ldots, x_{i-1}, x, x_{i+1}, \ldots, x_n) \mid \\ x_1 \in X_1, \ldots, x_{i-1} \in X_{i-1}, x_{i+1} \in X_{i+1}, \ldots, x_n \in X_n\}.$$

Figure 1.19 shows the projection of a fuzzy set which has non-zero membership degrees in two different regions. □

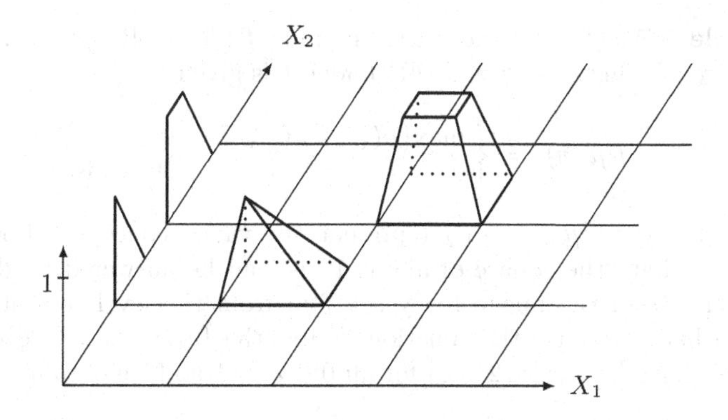

Fig. 1.19. Projection of a fuzzy set to the space X_2

1.5.2 Mapping of Level Sets

The membership degree of an element to the image of a fuzzy set can be computed on the basis of the membership degrees of all elements to the original one that are mapped to the considered element. Another way to characterize the image of a fuzzy set is to determine its level sets. Unfortunately, the level set of the image of a fuzzy set cannot be derived directly form the corresponding level set of the original fuzzy set. The inclusion $[f[\mu]]_\alpha \supseteq f[[\mu]_\alpha]$ is always valid, but equality of these two sets is only satisfied in special cases. For instance, for the fuzzy set

$$\mu(x) = \begin{cases} x & \text{if } 0 \leq x \leq 1 \\ 0 & \text{otherwise} \end{cases}$$

we obtain as an image under the mapping

$$f(x) = I_{\{1\}}(x) = \begin{cases} 1 & \text{if } x = 1 \\ 0 & \text{otherwise} \end{cases}$$

the fuzzy set

$$f[\mu](y) = \begin{cases} 1 & \text{if } y \in \{0,1\} \\ 0 & \text{otherwise.} \end{cases}$$

Hence, we have $[f[\mu]]_1 = \{0,1\}$ and $f[[\mu]_1] = \{1\}$ because of $[\mu]_1 = \{1\}$.

Provided that the universe of discourse $X = \mathbb{R}$ is the set of real numbers, the effect that the image of a level set is smaller than the corresponding level set of the image fuzzy set cannot happen, when the mapping f is continuous and for all $\alpha > 0$ the α-level sets of the original fuzzy set are compact. Therefore, in this case it is possible to characterize the image fuzzy set on the basis of level sets.

Example 1.7 Let us consider the mapping $f : \mathbb{R} \to \mathbb{R}, x \mapsto x^2$. Obviously, the image of a fuzzy set $\mu \in \mathcal{F}(\mathbb{R})$ under f is given by

$$f[\mu](y) = \begin{cases} \max\{\mu(\sqrt{y}), \mu(-\sqrt{y})\} & \text{if } y \geq 0 \\ 0 & \text{otherwise.} \end{cases}$$

Let the fuzzy set $\mu = \Lambda_{0,1,2}$ represent the vague concept "about 1". The question, what "the square of about 1" is, can be answered by determining the level sets of the image fuzzy set $f[\mu]$ from the level sets of μ. This is possible here, because the function f and the fuzzy set μ are continuous. Thus, we have $[\mu]_\alpha = [\alpha, 2-\alpha]$ for all $0 < \alpha \leq 1$ and we obtain

$$[f[\mu]]_\alpha = f[[\mu]_\alpha] = [\alpha^2, (2-\alpha)^2].$$

The fuzzy sets μ and $f[\mu]$ are shown in figure 1.20. Here we can observe that the vague concept "the square of about 1" is not match identical to the vague concept "about 1". The concept "the square of about 1" is "fuzzier" than "about 1". This effect is very similar to the increase of round-pff errors, when more and more computation steps are applied. □

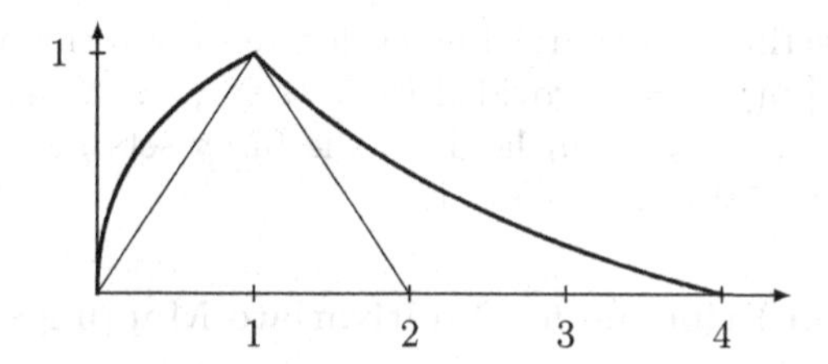

Fig. 1.20. The fuzzy sets μ and $f[\mu]$ representing the vague concepts "about 1" and "The square of about 1", respectively

1.5.3 Cartesian Product and Cylindric Extension

So far we have only extended mappings with one argument to fuzzy sets. In order to define operations like addition for fuzzy sets over real numbers, we need a concept how to apply a mapping $f : X_1 \times \ldots \times X_n \to Y$ to a tuple $(\mu_1, \ldots, \mu_n) \in \mathcal{F}(X_1) \times \ldots \times \mathcal{F}(X_n)$ of fuzzy sets. Since we can interpret addition as a function in two arguments $f : \mathbb{R} \times \mathbb{R} \to \mathbb{R}$, $(x_1, x_2) \mapsto x_1 + x_2$, this concept will enable us to extend addition and other algebraic operations to fuzzy sets over real numbers.

In order to generalize the extension principle described in equation (1.8) to mappings with several arguments, we introduce the concept of the Cartesian product of fuzzy sets. Consider the fuzzy sets $\mu_i \in \mathcal{F}(X_i)$, $i = 1, \ldots, n$. The *Cartesian product* of the fuzzy sets $\mu_1, \ldots, \mu_n$ is the fuzzy set

$$\mu_1 \times \ldots \times \mu_n \in \mathcal{F}(X_1 \times \ldots \times X_n)$$

with

$$(\mu_1 \times \ldots \times \mu_n)(x_1, \ldots, x_n) = \min\{\mu_1(x_1), \ldots, \mu_n(x_n)\}.$$

This definition is motivated by the property

$$(x_1, \ldots, x_n) \in M_1 \times \ldots \times M_n \quad \Longleftrightarrow \quad x_1 \in M_1 \wedge \ldots \wedge x_n \in M_n$$

of the Cartesian product of usual sets and corresponds to the formula

$$[\![(x_1, \ldots, x_n) \in \mu_1 \times \ldots \times \mu_n]\!] = [\![x_1 \in \mu_1 \wedge \ldots \wedge x_n \in \mu_n]\!],$$

where the minimum is chosen as the truth function for the conjunction.

A special case of a Cartesian product is the *cylindrical extension* of a fuzzy set $\mu \in \mathcal{F}(X_i)$ to a product space $X_1 \times \ldots \times X_n$. The cylindrical extension is the Cartesian product of μ with the remaining universes of discourse X_j, $j \neq i$, or their characteristic functions:

$$\hat{\pi}_i(\mu) = I_{X_1} \times \ldots \times I_{X_{i-1}} \times \mu \times I_{X_{i+1}} \times \ldots \times I_{X_n},$$

$$\hat{\pi}_i(\mu)(x_1, \ldots, x_n) = \mu(x_i).$$

Obviously, projecting a cylindrical extension results in the original fuzzy set which means $\pi_i[\hat{\pi}_i(\mu)] = \mu$ provided the sets $X_1, \ldots, X_n$ are non-empty. In general, $\pi_i[\mu_1 \times \ldots \times \mu_n] = \mu_i$ holds, if the fuzzy sets μ_j, $j \neq i$ are *normal* which means $(\exists x_j \in X_j)\big(\mu_j(x_j)\big) = 1$.

1.5.4 Extension Principle for Multivariate Mappings

Using the Cartesian product, the extension principle for mappings with several arguments can be simplified to the extension principle for functions with one argument. Consider mapping

$$f : X_1 \times \ldots \times X_n \to Y.$$

Then the image of the tuple

$$(\mu_1, \ldots, \mu_n) \in \mathcal{F}(X_1) \times \ldots \times \mathcal{F}(X_n)$$

of fuzzy sets under f is the fuzzy set

$$f[\mu_1, \ldots, \mu_n] \;=\; f[\mu_1 \times \ldots \times \mu_n]$$

in the universe of discourse Y. That means

$$\begin{aligned} & f[\mu_1, \ldots, \mu_n](y) && (1.9)\\ =& \sup_{(x_1,\ldots,x_n)\in X_1\times\ldots\times X_n} \big\{(\mu_1 \times \ldots \times \mu_n)(x_1, \ldots, x_n) f(x_1, \ldots, x_n) = y\big\} \\ =& \sup_{(x_1,\ldots,x_n)\in X_1\times\ldots\times X_n} \big\{\min\{\mu_1(x_1), \ldots, \mu_n(x_n)\} f(x_1, \ldots, x_n) = y\big\}. \end{aligned}$$

This formula represents the *extension principle* of Zadeh [214, 215, 216].

Example 1.8 Consider the mapping $f : \mathbb{R} \times \mathbb{R} \to \mathbb{R}$, $(x_1, x_2) \mapsto x_1 + x_2$ representing the addition. The fuzzy sets $\mu_1 = \Lambda_{0,1,2}$ and $\mu_2 = \Lambda_{1,2,3}$ model the vague concepts "about 1" and "about 2". Applying the extension principle, we obtain the fuzzy set $f[\mu_1, \mu_2] = \Lambda_{1,3,5}$ for the vague concept "about 1 + about 2" (cf. Figure 1.21). We can again observe the effect we know already from computing the square of "about 1" (see example 1.7 and Figure 1.20). The "Fuzziness" of the resulting fuzzy set is greater than these of the original fuzzy sets to be added. □

Analogously to the addition of fuzzy sets we can define subtraction, multiplication and division using the extension principle. These operations are continuous, therefore we can, like in example 1.7, calculate the level sets of the resulting fuzzy sets directly from the level sets of the given fuzzy sets, provided that these are continuous. When we have convex fuzzy sets, we are carry out interval arithmetics on the corresponding levels. Interval arithmetics [130, 131] allows us to operate with intervals instead of real numbers.

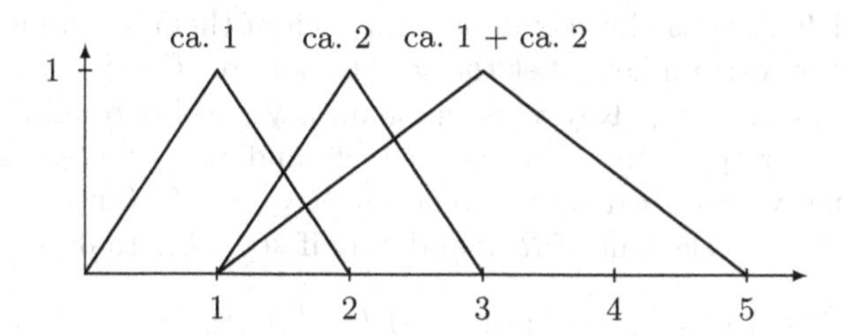

Fig. 1.21. The result based on the extension principle for "about 1 + about 2"

Applying the extension principle we should be aware that we carry out two generalization steps at the same time: the extension of single elements to sets and the change from crisp to fuzzy sets. The extension principle cannot preserve all properties of the original mapping. This effect is not necessarily caused by the extension from crisp to fuzzy sets. Most of the problems are caused by the step from extending a pointwise mapping to (crisp) sets. For example, in contrast to the standard addition, there is no inverse for the addition of fuzzy sets. There is no fuzzy set which added to the fuzzy set for "about 1 + about 2" from figure 1.21 will yield back the fuzzy set "about 1". This phenomenon already occurs in interval arithmetics, so not the "fuzzification" of the addition is the problem, but the extension of the addition from numbers (points) to sets.

1.6 Fuzzy Relations

Relations can be used to model dependencies, correlations or connections between variables, quantities or attributes. Technically speaking, a (binary) relation over the universes of discourse X and Y is a subset R of the Cartesian product $X \times Y$ of X and Y. The pairs $(x, y) \in X \times Y$ belonging to the relation R are linked by a connection described by the relation R. Therefore, a common notation for $(x, y) \in R$ is also xRy.

We generalize the concept of relations to fuzzy relations. Fuzzy relations are useful for representing and understanding fuzzy controllers that describe a vague connection between input and output values. Furthermore, we can establish an interpretation of fuzzy sets and membership degrees on the basis of special fuzzy relations called similarity relations. This interpretation plays a crucial role in the context of fuzzy controllers. Similarity relations will be discussed in section 1.7.

1.6.1 Crisp Relations

Before we introduce the definition of a fuzzy relation, we briefly review the fundamental concepts and mechanisms of crisp relations that are needed for understanding fuzzy relations.

Example 1.9 A house has six doors and each of them has a lock which can be unlocked by certain keys. Let the set of doors be $T = \{t_1, \ldots, t_6\}$, the set of keys $S = \{s_1, \ldots, s_5\}$. Key s_5 is the main key and fits to all doors. Key s_1 fits only to door t_1, s_2 to t_1 and t_2, s_3 to t_3 and t_4, s_4 to t_5. This situation can be formally described by the relation $R \subseteq S \times T$ ("fits to"). The pair $(s,t) \in S \times T$ is an element of R, if and only if key s fits to door t that means

$$\begin{aligned} R = \{ & (s_1,t_1),(s_2,t_1),(s_2,t_2),(s_3,t_3),(s_3,t_4),(s_4,t_5), \\ & (s_5,t_1),(s_5,t_2),(s_5,t_3),(s_5,t_4),(s_5,t_5),(s_5,t_6) \}. \end{aligned}$$

Another way of describing the relation R is shown in table 1.2. The entry 1 at position (s_i, t_j) indicates that $(s_i, t_j) \in R$ holds, 0 stands for $(s_i, t_j) \notin R$. □

R	t_1	t_2	t_3	t_4	t_5	t_6
s_1	1	0	0	0	0	0
s_2	1	1	0	0	0	0
s_3	0	0	1	1	0	0
s_4	0	0	0	0	1	0
s_5	1	1	1	1	1	1

Table 1.2. The relation R: "key fits to door"

Example 1.10 Let us consider a measuring instrument which can measure a quantity $y \in \mathbb{R}$ with a precision of ± 0.1. If x_0 is the measured value, we know that the true value y_0 lies within the interval $[x_0 - 0.1, x_0 + 0.1]$. This can be described by the relation

$$R = \{(x,y) \in \mathbb{R} \times \mathbb{R} \mid |x - y| \leq 0.1\}$$

A graphical representation of this relation is given in figure 1.22. □

Mappings or their graphs can be considered as special cases of relations. If $f : X \to Y$ is a mapping from X to Y, the graph of f is the relation

$$\text{graph}(f) = \{(x, f(x)) \mid x \in X\}.$$

In order to be able to interpret a relation $R \subseteq X \times Y$ as the graph of a function, we need that for each $x \in X$ there exists exactly one $y \in Y$ so that the pair (x, y) is contained in R.

1.6.2 Application of Relations and Deduction

So far we have used relations in a merely descriptive way. But similar to functions relations can also be applied to elements or sets. If $R \subseteq X \times Y$ is a

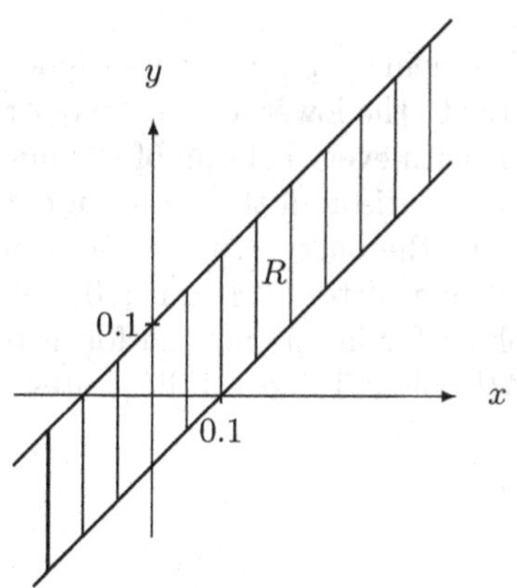

Fig. 1.22. The relation $y\hat{=}x \pm 0.1$

relation between the sets X and Y and $M \subseteq X$ is a subset of X, the image of M under R is the set

$$R[M] \;=\; \{y \in Y \mid (\exists x \in X)\big((x,y) \in R \wedge x \in M\big)\}. \tag{1.10}$$

$R[M]$ contains the elements from Y which are related to at least one element of the set M.

If $f : X \to Y$ is a mapping, applying the relation graph(f) to a one-element set $\{x\} \subseteq X$, we obtain the one-element set which contains the image of x under the function f:

$$\text{graph}(f)[\{x\}] \;=\; \{f(x)\}.$$

More generally, we have

$$\text{graph}(f)[M] \;=\; f[M] \;=\; \{y \in Y \mid (\exists x \in X)(x \in M \wedge f(x) = y)\}$$

for arbitrary subsets $M \subseteq X$.

Example 1.11 Now we use the relation R from example 1.9 in order to determine which doors can be unlocked, if we have keys $s_1, \ldots, s_4$. All we have to do, is to calculate all elements (doors) which are related (relation "fits to") to at least one of the keys $s_1, \ldots, s_4$. That means

$$R[\{s_1, \ldots, s_4\}] \;=\; \{t_1, \ldots, t_5\}$$

is the set of doors we want to know.

The set $R[\{s_1, \ldots, s_4\}]$ can be determined easily using the matrix in table 1.2 in the following way. We encode the set $M = \{s_1, \ldots, s_4\}$ as a row vector with five components which contains the entry 1 at the ith place, if $s_i \in M$ holds, and 0 in the case of $s_i \notin M$. Thus we obtain the vector $(1, 1, 1, 1, 0)$.

Analogously to the Falk scheme for matrix multiplication of a vector by a matrix, we write the vector to the lower left of the matrix. Then we transpose the vector and compare it with every column of the matrix. If we find at least one position during this comparison of the vector and a matrix column where vector and the matrix have the entry 1, we write a one under this column, otherwise a zero. The resulting vector $(1,1,1,1,1,0)$ below the matrix specifies the set $R[M]$ we are looking for in an encoded form: It contains a 1 at place i, if and only if $t_i \in R[M]$ holds. Table 1.3 illustrates this "Falk scheme" for relations.

□

					1	0	0	0	0	0
					1	1	0	0	0	0
					0	0	1	1	0	0
					0	0	0	0	1	0
					1	1	1	1	1	1
1	1	1	1	0	1	1	1	1	1	0

Table 1.3. Falk scheme for the calculation of $R[M]$

Example 1.12 We follow up example 1.10 and assume that we have the information that the measuring instrument indicated a value between 0.2 and 0.4. From this we can conclude that the true value is contained in the set $R\big[[0.2, 0.4]\big] = [0.1, 0.5]$ which is illustrated in figure 1.23.

In this figure we can see that we obtain the set $R[M]$ as the projection of the intersection of the relation with the cylindric extension of the set M which means

$$R[M] \;=\; \pi_y \left[R \cap \hat{\pi}_x(M)\right]. \tag{1.11}$$

□

Example 1.13 Logical deduction based on an implication of the form $x \in A \rightarrow y \in B$ can be modelled and computed by relations as well. All we have to do is to encode the rule $x \in A \rightarrow y \in B$ by the relation

$$R \;=\; \{(x, y) \in X \times Y \mid x \in A \rightarrow y \in B\} \;=\; (A \times B) \cup \bar{A} \times Y. \tag{1.12}$$

X and Y are the sets of possible values that x and y can assume. For the rule "If the velocity is between 90 km/h and 110 km/h the fuel consumption is between 6 and 8 litres" (as a logical formula: $v \in [90, 110] \rightarrow b \in [6, 8]$) we obtain the relation in figure 1.24.

If we know that the velocity has the value v, in the case $90 \leq v \leq 110$, we can conclude that for the consumption b we must have $6 \leq b \leq 8$. Otherwise and without further knowledge and information than just the rule and the

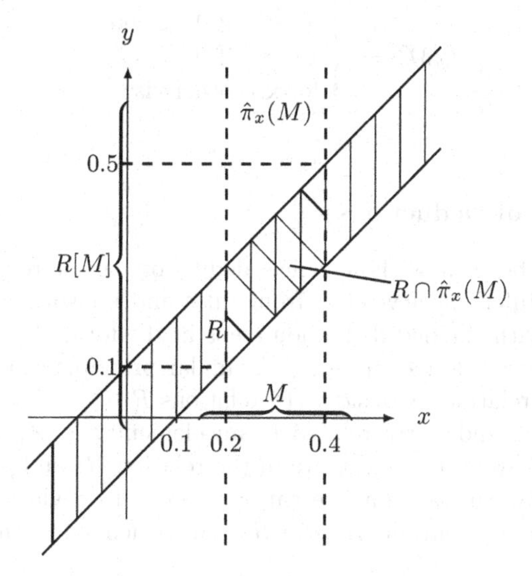

Fig. 1.23. How to determine the set $R[M]$ graphically

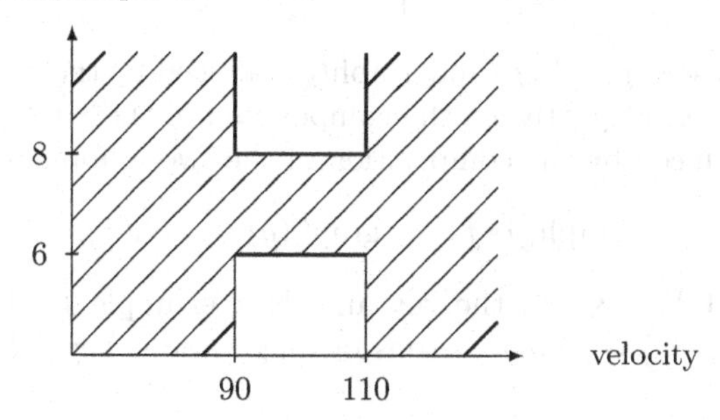

Fig. 1.24. The relation for the rule $v \in [90, 110] \rightarrow b \in [6, 8]$

value for v, we cannot say anything about the value for the consumption, which means we obtain $b \in [0, \infty)$. We obtain the same result by applying the relation R to the one-element set $\{v\}$:

$$R[\{v\}] = \begin{cases} [6, 8] & \text{if } v \in [90, 110] \\ [0, \infty) & \text{otherwise.} \end{cases}$$

More generally, we have: When we know that the velocity assumes a value in the set M, we can conclude in the case $M \subseteq [90, 110]$ that the consumption is between 6 and 8 litres, otherwise we only know $b \in [0, \infty)$. This coincides with the result we obtain by applying the relation R to the set M:

$$R[M] = \begin{cases} [6,8] & \text{if } M \subseteq [90,110] \\ \emptyset & \text{if } M = \emptyset \\ [0,\infty) & \text{otherwise.} \end{cases}$$

□

1.6.3 Chains of Deductions

The example above shows how logical deduction can be represented in terms of a relation. Inferring new facts from rules and known facts usually means that we deal with chained deduction steps in the form of $\varphi_1 \rightarrow \varphi_2$, $\varphi_2 \rightarrow \varphi_3$ from which we can derive $\varphi_1 \rightarrow \varphi_3$. A similar principle can be formulated in the context of relations. Consider the relations $R_1 \subseteq X \times Y$ and $R_2 \subseteq Y \times Z$. An element x is indirectly related to an element $z \in Z$, if there exists an element $y \in Y$, so that x and y are in the relation R_1 and y and z are in the relation R_2. We can say that we can go from x to z via y. In this way the composition of the relations R_1 and R_2 can be defined as the relation

$$R_2 \circ R_1 = \{(x,z) \in X \times Z \mid (\exists y \in Y)\big((x,y) \in R_1 \wedge (y,z) \in R_2\big)\} \quad (1.13)$$

between X and Z. Then we have for all $M \subseteq X$

$$R_2\Big[R_1[M]\Big] = (R_2 \circ R_1)[M].$$

For the relations graph(f) and graph(g) induced by the mappings $f : X \rightarrow Y$ and $g : Y \rightarrow Z$, respectively, the composition of these relations is equal to the relation induced by the composition of the two mappings:

$$\text{graph}(g \circ f) = \text{graph}(g) \circ \text{graph}(f).$$

Example 1.14 We extend the key and door example 1.9 by considering a set $P = \{p_1, p_2, p_3\}$ of three persons owning various keys. This is expressed by the relation

$$R' = \{(p_1,s_1),(p_1,s_2),(p_2,s_3),(p_2,s_4),(p_3,s_5)\} \subseteq P \times T.$$

$(p_i, s_j) \in R'$ means that person p_i owns the key s_j. The composition

$$\begin{aligned} R \circ R' = \{\ & (p_1,t_1),(p_1,t_2),(p_2,t_3),(p_2,t_4),(p_2,t_5), \\ & (p_3,t_1),(p_3,t_2),(p_3,t_3),(p_3,t_4),(p_3,t_5),(p_3,t_6)\ \} \end{aligned}$$

of the relations R' and R contains the pair $(p,t) \in P \times T$ if and only if person p can unlock door t. For example, using the relation $R \circ R'$ we can determine which doors can be unlocked, if the persons p_1 and p_2 are present. The corresponding set of doors is

$$(R \circ R')[\{p_1,p_2\}] = \{t_1,\ldots,t_5\} = R\Big[R'[\{p_1,p_2\}]\Big].$$

□

Example 1.15 In example 1.10 we used the relation $R = \{(x,y) \in \mathbb{R} \times \mathbb{R} \mid |x-y| \leq 0.1\}$ to model the fact that the measured value x represents the true value y with a precision of 0.1. When we can determine the quantity z from the quantity y with a precision of 0.2, we obtain the relation $R' = \{(y,z) \in \mathbb{R} \times \mathbb{R} \mid |x-y| \leq 0.2\}$. The composition of R' and R results in the relation $R' \circ R = \{(x,z) \in \mathbb{R} \times \mathbb{R} \mid |x-z| \leq 0.3\}$. If the measuring instrument indicates the value x_0, we can conclude that the value of the quantity z is in the set

$$(R' \circ R)[\{x_0\}] = [x_0 - 0.3, x_0 + 0.3].$$

□

Example 1.16 Example 1.13 demonstrated how an implication of the form $x \in A \to y \in B$ can be represented by a relation. When another rule $y \in C \to z \in D$ is known, in the case of $B \subseteq C$ we can derive the rule $x \in A \to z \in D$. Otherwise, knowing x does not provide any information about z in the context of these two rules. That means we obtain the rule $x \in X \to z \in Z$ in this case. Correspondingly, the composition of the relations R' and R representing the implications $x \in A \to y \in B$ and $y \in C \to z \in D$, respectively, results in the relation associated with the implication $x \in A \to z \in D$ or $x \in A \to z \in Z$, respectively:

$$R' \circ R = \begin{cases} (A \times D) \cup (\bar{A} \times Z) & \text{if } B \subseteq C \\ (A \times Z) \cup (\bar{A} \times Z) = X \times Z & \text{otherwise.} \end{cases}$$

□

1.6.4 Simple Fuzzy Relations

After giving a general idea about the fundamental terminology and concepts of crisp relations, we can now introduce fuzzy relations.

Definition 1.17 *A fuzzy set $\varrho \in \mathcal{F}(X \times Y)$ is called (binary) fuzzy relation between the universes of discourse X and Y.*

In this sense, a fuzzy relation is a generalized crisp relation where two elements can be gradually related to each other. The greater the membership degree $\varrho(x,y)$ the stronger is the the relation between x and y.

Example 1.18 Let $X = \{s, f, e\}$ denote a set of financial fonds, devoted to shares(s), fixed-interest stocks (f) and real estates (e). The set $Y = \{l, m, h\}$ contains the elements low (l), medium (m) and high (h) risk. The fuzzy relation $\varrho \in \mathcal{F}(X \times Y)$ in table 1.4 shows for each pair $(x,y) \in X \times Y$ how much the fond x is considered having the risk factor y.

For example, the entry in column m and row e means that the fond dedicated to real estates is considered to have a medium risk with a degree of 0.5. Therefore, $\varrho(e,m) = 0.5$. □

ϱ	l	m	h
s	0.0	0.3	1.0
f	0.6	0.9	0.1
e	0.8	0.5	0.2

Table 1.4. The fuzzy relation ϱ: "x is a financial fond with risk factor y"

Example 1.19 The measuring instrument from example 1.10 had a precision of ± 0.1. However, it is not very realistic to assume that, given the instrument shows the value x_0, all values from the interval $[x_0 - 0.1, x_0 + 0.1]$ are equally likely to represent the true value of the measured quantity. Instead of the crisp relation R from example 1.10 for representing this fact, we can use a fuzzy relation, for instance

$$\varrho : \mathbb{R} \times \mathbb{R} \to [0,1], \qquad (x,y) \mapsto 1 - \min\{10|x-y|, 1\},$$

yielding the membership degree one for $x = y$. The membership degree to the relation decreases linearly with increasing distance $|x - y|$ until the difference between x and y exceeds the value 0.1. □

In order to operate with fuzzy relations in a similar way as with usual relations, we have define what the image of a fuzzy set under a fuzzy relation is. This means we have to extend equation (1.10) to the framework of fuzzy sets and fuzzy relations.

Definition 1.20 *For a fuzzy relation $\varrho \in \mathcal{F}(X \times Y)$ and a fuzzy set $\mu \in \mathcal{F}(X)$ the image of μ under ϱ is the fuzzy set*

$$\varrho[\mu](y) \;=\; \sup\big\{ \min\{\varrho(x,y), \mu(x)\} \mid x \in X \big\} \tag{1.14}$$

over the universe of discourse Y.

This definition can be justified in several ways. If ϱ and μ are the characteristic functions of the crisp relation R and the crisp set M, respectively, then $\varrho[\mu]$ is the characteristic function of the image $R[M]$ of M under R. In this sense, the definition is a generalization of equation (1.10) for sets to fuzzy sets.

Equation (1.10) is equivalent to

$$y \in R[M] \iff (\exists x \in X)\big((x,y) \in R \wedge x \in M\big).$$

We obtain equation (1.14) for fuzzy relations from this equivalence by assigning the minimum as the truth function to the conjunction and evaluate the existential quantifier by the supremum. This means

$$\begin{aligned} \varrho[\mu](y) &= [\![y \in \varrho[\mu]]\!] \\ &= [\![(\exists x \in X)\big((x,y) \in \varrho \wedge x \in \mu\big)]\!] \\ &= \sup\big\{ \min\{\varrho(x,y), \mu(x)\} \mid x \in X \big\}. \end{aligned}$$

Definition 1.20 can also be derived by applying the extension principle. In order to do this, we consider the partial mapping

$$f : X \times (X \times Y) \to Y, \quad (x,(x',y)) \mapsto \begin{cases} y & \text{if } x = x' \\ \text{undefiniert} & \text{otherwise.} \end{cases} \qquad (1.15)$$

It is obvious that for a set $M \subseteq X$ and an relation $R \subseteq X \times Y$ we have

$$f[M,R] \;=\; f[M \times R] \;=\; R[M].$$

When we introduced the extension principle, we did not require that the mapping f, which has to be extended to fuzzy sets, must be defined everywhere. Therefore the extension principle can also be applied to partial mappings. The extension principle for the mapping (1.15), which can be used to compute the image of a set under a relation, leads to the formula specified in definition 1.20, the formula for the image of a fuzzy set under a fuzzy relation.

Another justification of definition 1.20 is based on the idea, that was exploited in example 1.12 and figure 1.23. There we computed the image of a set under a relation as the projection of the intersection of the cylindric extension of the set with the relation (cf. equation (1.11)). Having this equation in mind, we replace the set M by a fuzzy set μ and the relation R by a fuzzy relation ϱ and again obtain formula (1.14), if the intersection of fuzzy sets is computed using the minimum and the projection and the cylindric extension for fuzzy sets are calculated as in section 1.5.

Example 1.21 On the basis of the fuzzy relation from example 1.18 we want to estimate the risk of a mixed fond which concentrates on shares, but also invests a smaller part of its money into real estates. We represent this mixed fond over the universe of discourse $\{s, f, e\}$ as a fuzzy set μ with

$$\mu(s) = 0.8, \qquad \mu(f) = 0, \qquad \mu(e) = 0.2.$$

In order to determine the risk of this mixed fond we compute the image of the fuzzy set μ under the fuzzy relation ϱ from table 1.4. We obtain

$$\varrho[\mu](l) = 0.2, \qquad \varrho[\mu](m) = 0.3, \qquad \varrho[\mu](h) = 0.8.$$

Analogously to example 1.11 the fuzzy set $\varrho[\mu]$ can be determined using a modified Falk scheme. The zeros and ones in table 1.3 have to be replaced by the corresponding membership degrees. Below each column of the fuzzy relation, we obtain the membership degree of the corresponding value to the image fuzzy set $\varrho[\mu]$ in the following way. We first take the componentwise minimum of the vector representing the fuzzy set μ and the corresponding column of the matrix representing the fuzzy relation ϱ and then we compute the maximum of these minima.

In this sense, the calculation of the image of a fuzzy set μ under a fuzzy relation ϱ is similar to matrix multiplication of a matrix with a vector where the multiplication of the components is replaced by the minimum and the addition by the maximum. □

Example 1.22 We have the information that the measuring instrument from example 1.19 indicated a value of "about 0.3" which we represent by the fuzzy set $\mu = \Lambda_{0.2,0.3,0.4}$. For the true value y we obtain the fuzzy set

$$\varrho[\mu](y) = 1 - \min\{5|y - 0.3|, 1\}$$

as the image of the fuzzy set μ under the relation ϱ from example 1.19. □

Example 1.23 Example 1.13 illustrated how logical deduction on the basis of an implication of the form $x \in A \to y \in B$ can be represented using a relation. We generalize this method for the case that the sets A and B are replaced by fuzzy sets μ and ν. Following equation (1.12) and using the formula $[\![(x, y) \in \varrho]\!] = [\![x \in \mu \to y \in \nu]\!]$ where we choose the Gödel implication as the truth function for the implication, we define the fuzzy relation

$$\varrho(x, y) = \begin{cases} 1 & \text{if } \mu(x) \leq \nu(y) \\ \nu(y) & \text{otherwise.} \end{cases}$$

The rule "If x is about 2 then y is about 3" leads to the fuzzy relation

$$\varrho(x, y) = \begin{cases} 1 & \text{if } \min\{|3 - y|, 1\} \leq |2 - x| \\ 1 - \min\{|3 - y|, 1\} & \text{otherwise,} \end{cases}$$

if we model "about 2" by the fuzzy set $\mu = \Lambda_{1,2,3}$ and "about 3" by the fuzzy set $\nu = \Lambda_{2,3,4}$. Knowing "x is about 2.5" represented by the fuzzy set $\mu' = \Lambda_{1.5,2.5,3.5}$, we obtain for y the fuzzy set

$$\varrho[\mu'](y) = \begin{cases} y - 1.5 & \text{if } 2.0 \leq y \leq 2.5 \\ 1 & \text{if } 2.5 \leq y \leq 3.5 \\ 4.5 - y & \text{if } 3.5 \leq y \leq 4.0 \\ 0.5 & \text{otherwise,} \end{cases}$$

shown in figure 1.25.

The membership degree of an element y_0 to this fuzzy set should be interpreted as how much one can believe that it is possible that the variable y takes the value y_0. This interpretation is a generalization of what we have obtained

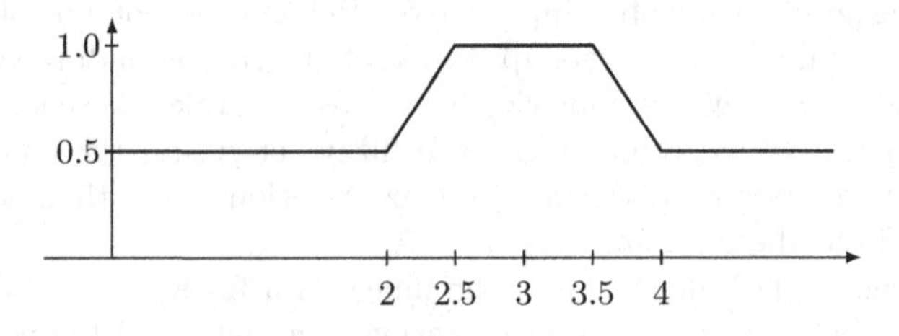

Fig. 1.25. The fuzzy set $\varrho[\mu']$

results for the implication based on crisp sets in example 1.13. In that case, only two sets were possible results of applying the deduction scheme: either the entire universe of discourse, if we could not guarantee that the premise of the implication is satisfied, or the set determined specified in the conclusion part of the implication for the case that the premise is satisfied. In the first case tells us that – only knowing the single rule – all values for y are still possible, since the rule is not applicable in this case. In the second case we know that the rule is applicable and only those values specified in the conclusion part of the rule are considered possible for Y. Extending this framework from crisp to fuzzy sets, the premise and the conclusion part of the implication can be partially satisfied. The consequence is that not only the whole universe of discourse and the fuzzy set in the conclusion part are possible results, but also fuzzy sets in between. In our example, all values y have a membership degree of at least 0.5 to the fuzzy set $\varrho[\mu']$. The reason for this is that there exists a value, namely $x_0 = 2.0$, which has a membership degree of 0.5 to the fuzzy set μ' and a membership degree of 0 to μ. This means that the variable x can assume a value with a degree of 0.5, for which we cannot apply the rule, i.e. y can assume any value. The membership degree 1 of the value $x_0 = 2.5$ to the fuzzy set μ' leads to the fact that all values of the interval $[2.5, 3.5]$ have a membership degree of 1 to $\varrho[\mu']$. For $x_0 = 2.5$ we obtain $\mu(2.5) = 0.75$ which means the premise of the implication is satisfied with a degree of 0.75. This implies that in order to satisfy the implication, a membership degree of at least 0.75 is required for the conclusion part. And the values in the interval $[2.5, 3.5]$ are those with a membership degree of at least 0.75 to the fuzzy set ν.

In a similar way, we can treat membership degrees between 0 and 1 to justify or compute the fuzzy set $\varrho[\mu']$ or its level sets. □

1.6.5 Composition of Fuzzy Relations

Now we are able to discuss the composition of fuzzy relations. The definition of an image of a fuzzy set under a fuzzy relation was motivated by equation (1.10) for crisp sets. Analogously, we define the composition of fuzzy relations based on equation (1.13) that describes composition in the case of crisp relations.

Definition 1.24 *Let $\varrho_1 \in \mathcal{F}(X \times Y)$ and $\varrho_2 \in \mathcal{F}(Y \times Z)$ be fuzzy relations. The composition of the two fuzzy relations is the fuzzy relation*

$$(\varrho_2 \circ \varrho_1)(x, z) = \sup\big\{ \min\{\varrho_1(x, y), \varrho_2(y, z)\} \mid y \in Y\big\} \tag{1.16}$$

between the universes of discourse X and Z.

These definition can be obtained from the equivalence

$$(x, z) \in R_2 \circ R_1 \iff (\exists y \in Y)\big((x, y) \in R_1 \land (y, z) \in R_2\big),$$

assigning the minimum as the truth function to the conjunction and evaluating the existential quantifier by the supremum so that we obtain

$$\begin{aligned}(\varrho_2 \circ \varrho_1)(x,z) &= [\![(x,y) \in (\varrho_2 \circ \varrho_1)]\!] \\ &= [\![(\exists y \in Y)\big((x,y) \in R_1 \wedge (y,z) \in R_2\big)]\!] \\ &= \sup\big\{ \min\{\varrho_1(x,y), \varrho_2(y,z)\} \mid y \in Y \big\}.\end{aligned}$$

Equation (1.16) can also be derived by applying the extension principle to the partial mapping

$$f : (X \times Y) \times (Y \times Z) \to (X \times Y),$$

$$\big((x,y),(y',z)\big) \mapsto \begin{cases} (x,z) & \text{if } y = y' \\ \text{undefined} & \text{otherwise} \end{cases}$$

on which the composition of crisp relations is based, since we have

$$f[R_1, R_2] \;=\; f[R_1 \times R_2] \;=\; R_2 \circ R_1.$$

If ϱ_1 and ϱ_2 are the characteristic functions of the crisp relations R_1 and R_2, respectively, then $\varrho_2 \circ \varrho_1$ is the characteristic function of the relation $R_2 \circ R_1$. In this sense the definition 1.24 generalizes the composition of crisp relations to fuzzy relations.

For every fuzzy set $\mu \in \mathcal{F}(X)$ we have

$$(\varrho_2 \circ \varrho_1)[\mu] \;=\; \varrho_2\big[\varrho_1[\mu]\big].$$

Example 1.25 Let us come back to example 1.21 analysing the risk of financial fonds. Now we extend the risk estimation of fonds by the set $Z = \{hl, ll, hp, lp\}$. The elements stand for "high loss", "low loss", "high profit", "low profit". The fuzzy relation $\varrho' \in \mathcal{F}(Y \times Z)$ in table 1.5 determines for each tuple $(y,z) \in Y \times Z$ the possibility to have a profit or loss of z under the risk y. The fuzzy relation resulting from the composition of the fuzzy relations ϱ and ϱ' is shown in table 1.6.

In this case, where the universes of discourse are finite and the fuzzy relations can be represented as tables or matrices, the computation scheme for the composition of fuzzy relations is similar to matrix multiplication where

ϱ'	hl	ll	lp	hp
l	0.0	0.4	1.0	0.0
m	0.3	1.0	1.0	0.4
h	1.0	1.0	1.0	1.0

Table 1.5. The fuzzy relation ϱ': "Given the risk y the profit/loss z is possible"

ϱ'	hl	ll	lp	hp
s	1.0	1.0	1.0	1.0
f	0.3	0.9	0.9	0.4
e	0.3	0.5	0.8	0.4

Table 1.6. The fuzzy relation $\varrho' \circ \varrho$: "With the yield object x the profit/loss z is possible"

we have to replace the componentwise multiplication by the minimum and the addition by the maximum. For the mixed fond from example 1.21, which was represented by the fuzzy set μ

$$\mu(s) = 0.8, \qquad \mu(f) = 0, \qquad \mu(e) = 0.2,$$

we obtain

$$(\varrho' \circ \varrho)[\mu](hl) = (\varrho' \circ \varrho)[\mu](ll) = (\varrho' \circ \varrho)[\mu](lp) = (\varrho' \circ \varrho)[\mu](hp) = 0.8$$

as the fuzzy set describing the possible profit or loss. □

Example 1.26 The precision of the measuring instrument from example 1.19 was described by the fuzzy relation $\varrho(x, y) = 1 - \min\{10|x - y|, 1\}$ which determines in how far the value y is the true value, if x is the value indicated by the measuring instrument. We assume that we cannot read off the value from the (analog) instrument exactly and therefore use the fuzzy relation $\varrho'(a, x) = 1 - \min\{5|a - x|, 1\}$. $\varrho'(a, x)$ tells us in how far the value x is value indicated by the measuring instrument, when we read off the value a. In order to estimate which value could be the true value y of the measured quantity, given we have read off the value a, we have to compute the composition of the fuzzy relations ϱ' and ϱ.

$$(\varrho \circ \varrho')(a, y) \;=\; 1 - \min\left\{\frac{10}{3}|a - y|, 1\right\}.$$

Assuming we have read off $a = 0$, we obtain the fuzzy set

$$(\varrho \circ \varrho')[I_{\{0\}}] \;=\; \Lambda_{-0.3,0,0.3}$$

for the true value y. □

1.7 Similarity Relations

In this section we discuss a special type of fuzzy relations, called similarity relations. They play an important role in interpreting fuzzy controllers and, more generally, can be used to characterize the inherent indistinguishability or vagueness of a fuzzy systems.

Similarity relations are fuzzy relations that specify for pairs of elements or objects how indistinguishable or similar they are. From a similarity relation we should expect that it is reflexive and symmetric which means that every element is similar to itself (with degree 1) and that x is as similar to y as y to x. In addition to these two minimum requirements, we also ask for the following weakened transitivity condition: If x is similar to y to a certain degree and y is similar to z to a certain degree, then x should also be similar to z to a certain (maybe lower) degree. Technically speaking, we define a similarity relation as follows:

Definition 1.27 *A similarity relation $E : X \times X \to [0,1]$ with respect to the t-norm t on the set X is a fuzzy relation over $X \times X$ which satisfies the conditions*

$$\begin{aligned} &(\text{E1})\ E(x,x) = 1, && \text{(reflexivity)} \\ &(\text{E2})\ E(x,y) = E(y,x), && \text{(symmetry)} \\ &(\text{E3})\ t\big(E(x,y), E(y,z)\big) \ \le\ E(x,z) && \text{(transitivity)} \end{aligned}$$

for all $x, y, z \in X$.

The transitivity condition for similarity relations can be understood in terms of fuzzy logic, as it was presented in section 1.3, as follows: The truth value of the proposition

$$x \text{ and } y \text{ are similar } \texttt{AND } y \text{ and } z \text{ are similar}$$

should be at most as great as the truth value of the proposition

$$x \text{ and } z \text{ are similar}$$

where the t-norm t is used as the truth function for the conjunction AND.

In example 1.19 we have already seen an example for a similarity relation, the fuzzy relation

$$\varrho : \mathbb{R} \times \mathbb{R} \to [0,1], \qquad (x,y) \mapsto 1 - \min\{10|x-y|, 1\}$$

that indicates how indistinguishable two values are using a measuring instrument. We can prove easily that this fuzzy relation is a similarity relation with respect to the Łukasiewicz t-norm $t(\alpha,\beta) = \max\{\alpha+\beta-1, 0\}$. More generally, an arbitrary pseudo-metric, i.e. a distance measure $\delta : X \times X \to [0,\infty)$ satisfying the symmetry condition $\delta(x,y) = \delta(y,x)$ and the triangular inequality $\delta(x,y) + \delta(y,z) \ge \delta(x,z)$, induces a similarity relation with respect to the Łukasiewicz t-norm by

$$E^{(\delta)}(x,y) \;=\; 1 - \min\{\delta(x,y), 1\}$$

and vice versa, any similarity relation E with respect to the Łukasiewicz t-norm defines a pseudo-metric by

$$\delta^{(E)}(x,y) = 1 - E(x,y).$$

Furthermore, we have $E = E^{(\delta^{(E)})}$ and $\delta(x,y) = \delta^{(E^{(\delta)})}(x,y)$, if δ is bounded by one, i.e. $\delta(x,y) \leq 1$ holds for all $x, y \in X$, so that similarity relations and pseudo-metrics (bounded by one) represent dual concepts.

Later on, we other examples will provide a motivation to consider similarity relations with respect to other t-norms than the Łukasiewicz t-norm in order to characterize the vagueness or indistinguishability that is inherent in a fuzzy systems.

1.7.1 Fuzzy Sets and Extensional Hulls

If we assume that a similarity relation characterises a certain indistinguishability, we should expect that elements that are (almost) indistinguishable should behave similar or have similar properties. For fuzzy systems the (fuzzy) property of being an element of a (fuzzy) set is essential. Therefore, those fuzzy sets play an important role that are coherent with respect to a given similarity relation in the sense that similar elements have similar membership degrees. This property of a fuzzy set is called extensionality and is formally defined as follows:

Definition 1.28 *Let $E : X \times X \to [0,1]$ be a similarity relation with respect to the t-norm t on the set X. A fuzzy set $\mu \in \mathcal{F}(X)$ is called extensional with respect to E, if*

$$t\big(\mu(x), E(x,y)\big) \leq \mu(y)$$

holds for all $x, y \in X$.

In the view of fuzzy logic the extensionality condition can be interpreted in the sense that the truth value of the proposition

x is an element of the fuzzy set μ AND
x and y are similar (indistinguishable)

should not be smaller than the truth value of the proposition

y is an element of the fuzzy set μ

where the t-norm t is used as the truth function for the conjunction AND.

A fuzzy set can always be made extensional by adding all elements which are similar to at least one of its elements. If we formalize this idea, we obtain the following definition.

Definition 1.29 *Let $E : X \times X \to [0,1]$ be a similarity relation with respect to the t-norm t on set X. The extensional hull $\hat{\mu}$ of the fuzzy set $\mu \in \mathcal{F}(X)$ (with respect to the similarity relation E) is given by*

$$\hat{\mu}(y) = \sup\big\{t\big(E(x,y), \mu(x)\big) \mid x \in X\big\}.$$

If t is a continuous t-norm, then the extensional hull $\hat{\mu}$ of μ is the smallest extensional fuzzy set containing μ in the sense of $\mu \leq \hat{\mu}$.

In principle, we obtain the extensional hull of a fuzzy set μ under the similarity relation E as the image of μ under the fuzzy relation E as in definition 1.20. For the extensional hull the minimum in equation (1.14) in definition 1.20 is replaced by the t-norm t.

Example 1.30 We consider the similarity relation $E : \mathbb{R} \times \mathbb{R} \to [0,1]$, $E(x,y) = 1 - \min\{|x-y|, 1\}$ with respect to the Łukasiewicz t-norm, which is induced by the usual metric $\delta(x,y) = |x-y|$ on the real numbers. A (crisp) set $M \subseteq \mathbb{R}$ can be viewed as a special type of fuzzy set, when we consider its characteristic function I_M. In this way, we can also compute extensional hulls of crisp sets.

The extensional hull of a point x_0, which means a one-element set $\{x_0\}$, with respect to the similarity relation E mentioned above is a fuzzy set with a triangular membership function $\Lambda_{x_0-1,x_0,x_0+1}$. The extensional hull of the interval $[a,b]$ is the trapezoidal function $\Pi_{a-1,a,b,b+1}$ (cf. Figure 1.26).

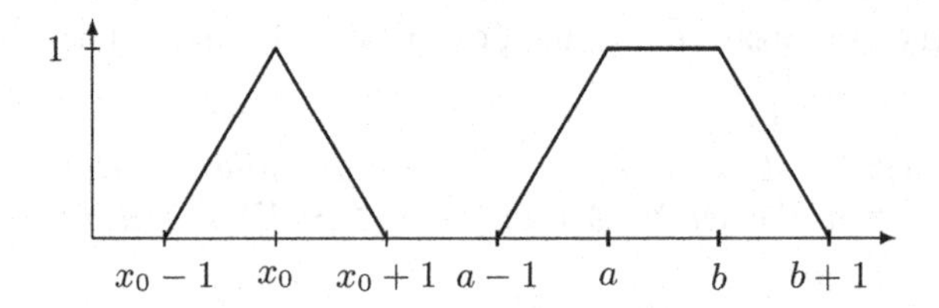

Fig. 1.26. The extensional hulls of the point x_0 and the interval $[a,b]$

□

This example establishes an interesting connection between fuzzy sets and similarity relations: Triangular and trapezoidal membership functions, that are very popular in most applications of fuzzy systems, can be interpreted as extensional hulls of points or intervals, that means as fuzzy points or fuzzy intervals in a vague environment which is characterized by a similarity relation induced by the standard metric on the real numbers.

1.7.2 Scaling Concepts

The standard metric on the real numbers allows only very limited forms of triangular and trapezoidal functions as extensional hulls of points and intervals: the slopes where the membership degrees increase and decrease are always 1 and -1, respectively. However, it is reasonable to allow a scaling of the standard metric so that other forms of fuzzy sets as extensional hulls can result. This scaling can have two different meanings.

The degree of similarity of two measured values depends on the measuring unit. Two values measured in kilo-units can have a small distance and might be considered as almost indistinguishable or very similar, while the very same quantities measured in milli-units have a much greater distance and are distinguishable. Of course, the degree of similarity should not depend on the measurement unit. The height of two persons should have the same degree of similarity, no matter whether we measure the height in centimetres, metres or inches. In order to adapt a similarity relation to a measuring unit, the distance or the real axis has to be scaled by a constant factor $c > 0$, as in example 1.19. In this way we obtain scaled metric $|c \cdot x - c \cdot y|$ which induces the similarity relation $E(x, y) = 1 - \min\{|c \cdot x - c \cdot y|, 1\}$.

An extension of the scaling concept is the use of varying scaling factors allowing a local scaling fitting to the problem.

Example 1.31 We want to describe the behaviour of an air conditioner using fuzzy rules. It is neither necessary nor desirable to measure the room temperature, to which the air conditioner will react, as exactly as possible. However, the temperature values are of different importance for the air conditioner. Temperatures of 10°C or 15°C, for example, are much too cold, so the air conditioner should heat at full power, and the values 27°C or 32°C are much too hot and the air conditioner should cool at full power. Therefore, there is no need to distinguish between the values 10°C and 15°C or 27°C and 32°C for the control of the room temperature. Since we do not have to distinguish between 10°C and 15°C, we can choose a very small positive scaling factor – in the extreme case even the factor 0, then these temperatures are not distinguished at all. However, it would not be correct, if we chose a small scaling factor for the entire range of the temperature, because the air conditioner has to distinguish very well between the cold temperature 18.5°C and the warm one 23.5°C.

Instead of a global scaling factor we should choose individual scaling factors for different ranges of the temperature, so that we can make a fine distinction between temperatures which are near the optimum room temperature and a more rough one between temperatures which are much too cold or much too warm. Table 1.7 shows an example of a partition into five temperature ranges, each one having its individual scaling factor.

These scaling factors define a transformation of the range and the distances for the temperature, that can be used to define a similarity relation. Table 1.8 shows the transformed distances and the resulting similarity degrees for some pairs of values of the temperature. For each pair the two values lie in a range where the scaling factor does not change. In order to understand how to determine the transformed distance and the resulting similarity degree for two temperatures that are not in a range with a constant scaling factor, we first analyse the effect of a single scaling factor.

Let us consider an interval $[a, b]$ where we measure the distance between two points based on the scaling factor c. Computing the scaled distance means

temperature (in °C)	scaling factor	interpretation
< 15	0.00	exact value insignificant (much too cold temperature)
15-19	0.25	too cold, but not too far away from the desired temperature, not very sensitive
19-23	1.50	very sensitive, near the desired temperature
23-27	0.25	too warm, but not too far away from the desired temperature, not very sensitive
> 27	0.00	exact value insignificant (much too hot temperature)

Table 1.7. Different sensitivities and scaling factors for controlling the room temperature

pair of values (x, y)	scaling factor c	transf. distance $\delta(x, y) = \lvert c \cdot x - c \cdot y \rvert$	similarity degree $E(x, y) = 1 - \min\{\delta(x, y), 1\}$
(13,14)	0.00	0.000	1.000
(14,14.5)	0.00	0.000	1.000
(17,17.5)	0.25	0.125	0.875
(20,20.5)	1.50	0.750	0.250
(21,22)	1.50	1.500	0.000
(24,24.5)	0.25	0.125	0.875
(28,28.5)	0.00	0.000	1.000

Table 1.8. Transformed distances based on the scaling factors and the induced similarity degrees

that we apply stretching ($c > 1$) or shrinking ($0 \leq c < 1$) to the interval according to the factor c and calculate the distances between the points in the transformed (stretched or shrunk) interval. In order to take individual scaling factors for different ranges into account, we have to stretch or shrink each range, where the scaling factor is constant, correspondingly. Gluing these transformed ranges together, we can now measure the distance between points that do not lie in a region with a constant scaling factor. The induced transformation is piecewise linear as shown in figure 1.27.

On the basis of the following three examples, we explain and illustrate the calculation of the transformed distance and the resulting similarity degrees. We determine the similarity degree between the values 18 and 19.2. The value 18 is in the interval $[15, 19]$ with constant scaling factor 0.25. This interval of length 4 is now shrunk to an interval of length 1. Therefore, the distance of the value 18 to the right boundary of the interval, 19, is also shrunk by the factor

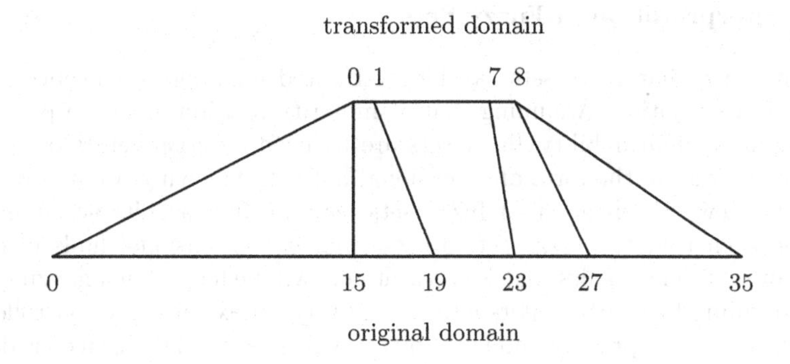

Fig. 1.27. Transformation of a domain using scaling factors

0.25, so that the transformed distance between 18 and 19 is exactly 0.25. In order to calculate the transformed distance between 18 and 19.2, we have to add the transformed distance between 19 and 19.2. In this range the scaling factor is constantly 1.5, so the distance between 19 and 19.2 is stretched by the factor 1.5 and the resulting transformed distance is 0.3. Thus the transformed distance between the values 18 and 19.2 is $0.25 + 0.3 = 0.55$ which leads to a similarity degree of $1 - \min\{0.55, 1\} = 0.45$.

In the second example we consider the values 13 and 18. Because of the scaling factor 0, the transformed distance between 13 and 15 is also 0. In the range between 15 and 18 the scaling factor is 0.25 and therefore the transformed distance between 15 and 18 is 0.75. Since the transformed distance between 13 and 15 is zero, the overall transformed distance between 13 and 18 is 0.75 and the similarity degree between 13 and 18 is 0.25.

Finally, we determine the transformed distance and the similarity degree between the values 22.8 and 27.5. Here we have to take three ranges with different scaling factors into account: the scaling factor between 22.8 and 23 is 1.5, between 23 and 27 it is 0.25 and between 27 and 27.5 it is constantly 0. Therefore, the transformed distances for the pairs (22.8,23), (23,27) and (27,27.5) are 0.3, 1 and 0, respectively. The sum of these distances and therefore the transformed distance between 22.8 and 27.5 is 1.3. The degree of similarity is $1 - \min\{1.3, 1\} = 0$. □

The idea of using different scaling factors for different ranges can be extended by assigning a scaling factor to each single value which determines how to distinguish in the direct environment of this value. Instead of a piecewise constant scaling function as in example 1.31, any (integrable) scaling function $c : \mathbb{R} \to [0, \infty)$ can be used. The transformed distance between the values x and y under such a scaling function can be computed using the following equation [92]:

$$\left| \int_x^y c(s)\, ds \right| . \tag{1.17}$$

1.7.3 Interpretation of Fuzzy Sets

We have seen that fuzzy sets can be interpreted as induced concepts based on similarity relation. Assuming that a similarity relation models a problem-specific indistinguishability, fuzzy sets are induced by crisp values or sets in a canonical way in the form of extensional hulls. In following, we analyse the opposite view. Given a set of fuzzy sets, can we find a suitable similarity relation such that the fuzzy sets are nothing but extensional hulls of crisp values or sets? The results that we present here will be helpful in analyzing and understanding fuzzy controllers later on. In the context of fuzzy controllers a number of vague expression modelled by fuzzy sets are used for each considered variable. So for each domain X a set $\mathcal{A} \subseteq \mathcal{F}(X)$ of fuzzy sets is given. As we will see later the inherent indistinguishability of these fuzzy sets can be characterized by similarity relations. The coarsest (greatest) similarity relation for which all fuzzy sets in $\mathcal{A}$ are extensional plays an important role. The following theorem [94] describes how to compute this similarity relation.

Theorem 1.32 *Let t be a continuous t-norm and $\mathcal{A} \subseteq \mathcal{F}(X)$ a set of fuzzy set. Then*

$$E_{\mathcal{A}}(x, y) = \inf\{ \overleftrightarrow{t}(\mu(x), \mu(y)) \mid \mu \in \mathcal{A}\} \tag{1.18}$$

is the coarsest (greatest) similarity relation with respect to the t-norm t for which all fuzzy sets in $\mathcal{A}$ are extensional. $\overleftrightarrow{t}$ is the biimplication from equation (1.4) associated with the t-norm t.

Coarsest similarity relation means that for every similarity relation E for which all fuzzy sets in $\mathcal{A}$ are extensional we have $E_{\mathcal{A}}(x, y) \geq E(x, y)$ for all $x, y \in X$.

Equation (1.18) for the similarity relation $E_{\mathcal{A}}$ can be understood and interpreted within the framework of fuzzy logic. We interpret the fuzzy sets in $\mathcal{A}$ as representations of vague properties. Two elements x and y are similar with respect to a single property, if x has the property if and only if y has the property. Modelling this idea within fuzzy logic, this means that we have to interpret "x has the property associated with the fuzzy set μ" as the membership degree of x to μ. Then the similarity degree of x and y, taking only the property associated with the fuzzy set μ into account, is defined by $\overleftrightarrow{t}(\mu(x), \mu(y))$. When we can use all properties associated with the fuzzy sets in $\mathcal{A}$ to distinguish x and y, this means: x and y are similar, if they are similar with respect to all properties in $\mathcal{A}$. Evaluating "for all" by the infimum, we obtain equation (1.18) for the similarity degree of two elements.

We have seen in example 1.30 that typical fuzzy sets with triangular membership functions can be interpreted as extensional hulls of single points or values. This interpretation of fuzzy sets as vague values will be very useful in the context of fuzzy control. This is why we also study the issue, when a set $\mathcal{A} \subseteq \mathcal{F}(X)$ of fuzzy sets fuzzy sets can be interpreted as extensional hulls of points.

Theorem 1.33 *Let t be a continuous t-norm and let $\mathcal{A} \subseteq \mathcal{F}(X)$ a set of fuzzy sets. Let each fuzzy set $\mu \in \mathcal{A}$ have the property such that there exists an $x_\mu \in X$ with $\mu(x_\mu) = 1$. There exists a similarity relation E, such that for all $\mu \in \mathcal{A}$ the extensional hull of the point x_μ is equal to the fuzzy set μ, if and only if*

$$\sup_{x \in X}\{t(\mu(x), \nu(x))\} \leq \inf_{y \in X}\{\overleftrightarrow{t}(\mu(y), \nu(y))\} \tag{1.19}$$

holds for all $\mu, \nu \in \mathcal{A}$. In this case $E = E_{\mathcal{A}}$ is the coarsest similarity relation for which all fuzzy sets in $\mathcal{A}$ are extensional hulls of points.

Condition (1.19) says that the degree of non-disjointness of any two fuzzy sets $\mu, \nu \in \mathcal{A}$ must not be exceed their degree of equality. The corresponding formulae are obtained by interpreting the following conditions in terms of fuzzy logic:

- Two sets μ and ν are non-disjoint, if and only if
$$(\exists x)(x \in \mu \wedge x \in \nu)$$
holds.
- Two sets μ and ν are equal, if and only if
$$(\forall y)(y \in \mu \leftrightarrow y \in \nu)$$
holds.

Condition (1.19) in theorem 1.33 is definitely satisfied, when the fuzzy sets μ and ν are disjoint with respect to the t-norm t, that is, we have $t(\mu(x), \nu(x)) = 0$ for all $x \in X$. A proof for this theorem can be found in [102].

Most of the variables in the context of fuzzy control are numerical or continuous variables. Similarity relations on the real line can be defined in a simple and understandable way using scaling functions based on the concept of distance, as described in equation (1.17). When we require that the similarity relation in theorem 1.33 should be definable in terms of a scaling function, the following result proved in [92] is important.

Theorem 1.34 *Let $\mathcal{A} \subseteq \mathcal{F}(\mathbb{R})$ be a non-empty, at most countable set of fuzzy sets such that for each $\mu \in \mathcal{A}$ the following conditions are satisfied:*

- *There exists an $x_\mu \in \mathbb{R}$ such that $\mu(x_\mu) = 1$.*
- *μ (as real-valued function) is non-decreasing in the interval $(-\infty, x_\mu]$.*
- *μ is non-increasing in the interval $[x_\mu, \infty)$.*
- *μ is continuous.*
- *μ is differentiable almost everywhere.*

There exists a scaling function $c : \mathbb{R} \to [0, \infty)$ such that for all $\mu \in \mathcal{A}$ the extensional hull of the point x_μ with respect to the similarity relation

$$E(x,y) = 1 - \min\left\{\left|\int_x^y c(s)\,ds\right|, 1\right\}$$

is equal to the fuzzy set μ, if and only if the condition

$$\min\{\mu(x),\nu(x)\} > 0 \Rightarrow \left|\frac{d\mu(x)}{dx}\right| = \left|\frac{d\nu(x)}{dx}\right| \tag{1.20}$$

is satisfied for all $\mu,\nu \in \mathcal{A}$ almost everywhere. In this case

$$c : \mathbb{R} \to [0,\infty), \qquad x \mapsto \begin{cases} \left|\frac{d\mu(x)}{dx}\right| & \textit{if } \mu \in \mathcal{A} \textit{ and } \mu(x) > 0 \\ 0 & \textit{otherwise} \end{cases}$$

can be chosen as the (almost everywhere well-defined) scaling function.

Example 1.35 In order to illustrate, how extensional hulls of points with respect to a similarity relation induced by a piecewise constant scaling function might look like, we rcall the scaling function

$$c : [0,35) \to [0,\infty), \qquad s \mapsto \begin{cases} 0 & \text{if} \quad 0 \le s < 15 \\ 0.25 & \text{if} \quad 15 \le s < 19 \\ 1.5 & \text{if} \quad 19 \le s < 23 \\ 0.25 & \text{if} \quad 23 \le s < 27 \\ 0 & \text{if} \quad 27 \le s < 35. \end{cases}$$

from example 1.31. Figure 1.28 shows the extensional hulls of the points 15, 19, 21, 23 and 27 with respect to the similarity relation induced by the scaling function c.

These extensional hulls have triangular or trapezoidal membership functions. The reason for this is that the points are chosen in such a way that – moving away from the corresponding point – the membership degree drops to zero before the scaling function changes its value. When we choose a point close to a point where the scaling function jumps from one value to another we will not obtain a triangular or trapezoidal membership function for the extensional hull. In the case of a piecewise scaling function we can only guarantee that we obtain a piecewise linear membership functions as extensional hulls as they are shown in figure 1.29.

It is quite common to use sets of fuzzy sets for fuzzy controllers as they are illustrated in figure 1.30. There, we choose values $x_1 < x_2 < \ldots < x_n$ and use triangular functions of the form $\Lambda_{x_{i-1},x_i,x_{i+1}}$ or the trapezoidal functions $\Pi_{-\infty,-\infty,x_1,x_2}$ and $\Pi_{x_{n-1},x_n,\infty,\infty}$ at the limits x_1 and x_n of the corresponding range, i.e.

$$\mathcal{A} = \{\Lambda_{x_{i-1},x_i,x_{i+1}} \mid 1 < i < n\} \cup \{\Pi_{-\infty,-\infty,x_1,x_2}, \Pi_{x_{n-1},x_n,\infty,\infty}\}.$$

In this case we can always define a scaling function c such that the fuzzy sets can be interpreted as the extensional hulls of the points $x_1,\ldots,x_n$, namely

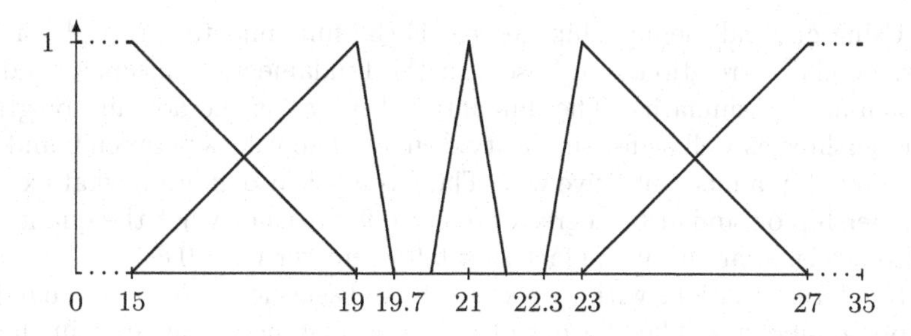

Fig. 1.28. The extensional hulls of the points 15, 19, 21, 23 and 27, respectively

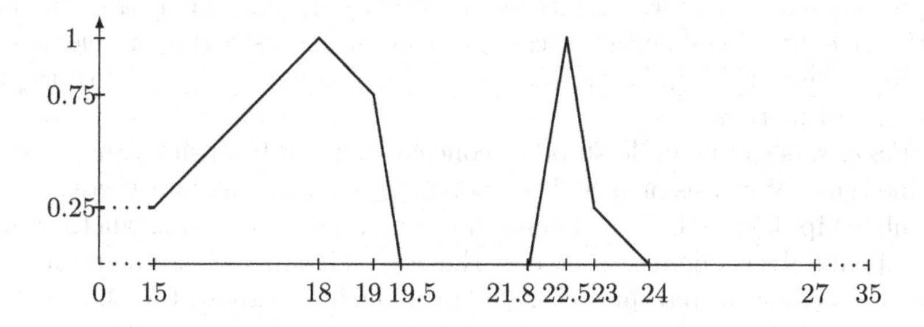

Fig. 1.29. The extensional hulls of the points 18.5 and 22.5, respectively

$$c(x) \quad = \quad \frac{1}{x_{i+1} - x_i} \qquad \text{if } x_i < x < x_{i+1},$$

□

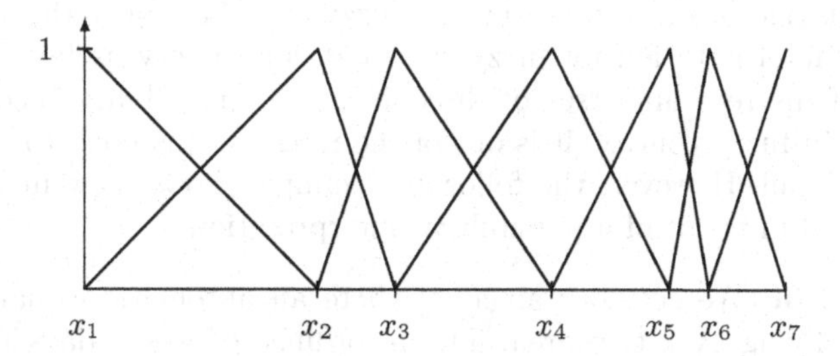

Fig. 1.30. Fuzzy sets for which a scaling function can be defined

Now that we have introduced the basic concepts of similarity relations and their relation to fuzzy sets, we are able to discuss in more detail the conceptual background of fuzzy sets.

Using gradual membership degrees is the fundamental principle of fuzzy sets. Similarity relations are based on the fundamental concept of indistinguishability or similarity. The unit interval serves as the domain for gradual memberships as well as for similarity degrees. The values between 0 and 1 are interpreted in a mere intuitive way. There is no clear definition what exactly a membership or similarity degree of 0.8 or 0.9 means or what the quantitative difference is, apart from the fact that 0.9 is greater than 0.8.

Similarity relations with respect to the Łukasiewicz t-norm can be induced by pseudo-metrics. The concept of metric or distance is, at least in the case of the real line, elementary and does not need any further explanation or motivation.

In this sense, similarity relations, induced by the canonic metric – perhaps with some additional suitable scaling – can be understood as an elementary concept where the similarity degrees are interpreted dual to the concept of distance of metrics.

Fuzzy sets can be understood as concepts derived from similarity relations in the sense of extensional hulls of points or sets. In this interpretation, the membership degrees have a concrete meaning. The question is whether fuzzy sets should always be interpreted in this way. The answer is yes and no. Yes, because lacking an interpretation for membership degrees, the choice of the fuzzy sets and operations like t-norms becomes more or less arbitrary and a pure problem of parameter optimisation. Yes, because at least in the context of fuzzy control, in most cases we have to deal with real numbers and the fuzzy sets usually fit quite well to the notion of an imprecise value, i.e. an extensional hull of a point with respect to some suitable similarity relation. Also the results on how fuzzy sets can be induced by similarity relations and how we can compute suitable similarity relations for given fuzzy sets speak in favour of an interpretation in terms of similarity relations.

Nevertheless, similarity relations are not the only and unique way to provide a more rigorous interpretation for fuzzy sets. Possibilty theory offers an alternative interpretation for fuzzy sets that is also very useful especially in the case of discrete universes of discourse. Since possibility theory plays a minor role in fuzzy control, it is out of the range of this book to explain this theory in detail. However, the following example shows, how fuzzy sets can be motivated in terms of a possibilistic interpretation.

Example 1.36 We consider an area where an automatic camara observes aeroplanes flying over this area. The recordings of a few days provide the following information: 20 planes of the type A, 30 of type B and 50 of type C were observed. So next time, when we hear an aeroplane (before we can actually see it), we would think that with probability 20%, 30% and 50% it is of type A, B and C, respectively.

Now we modify this purely probabilistic example slightly in order to explain the meaning of possibility distributions. In addition to the camera we now have a radar unit and a microphone. The microphone has recorded 100 aeroplanes, but we are not able to say anything about the type of aeroplane just by the sound. Because of poor visibility (fog or clouds) the camara could only identify 70 aeroplanes, 15 of type A, 20 of type B and 35 of type C. The radar unit was able to register 20 of the missing 30 aeroplanes, but failed in the remaining 10 cases. From the radar unit we know that 10 of these 20 aeroplanes were of type C, because the radar unit can distinguish type C from type A and B, since type C is much smaller than type A and B. The other 10 of the 20 aeroplanes registered by the radar unit are of type A or B, but we cannot tell which of the two types, since the radar image is to rough to distinguish between them.

This sample of 100 observed aeroplanes is very different from the 100 observations above where we were able to identify the type of aeroplane for each observation. Now a single observations can only be presented as a set of possible aeroplanes. How often each set was observed is shown in table 1.9.

set	$\{A\}$	$\{B\}$	$\{C\}$	$\{A, B\}$	$\{A, B, C\}$
observed amount	15	20	45	10	10

Table 1.9. Set-valued observations of aeroplane types

Without further knowledge or assumptions, we cannot estimate a probability for the single aeroplanes based on these observations without additional asumptions about the distribution of the aeroplane types of the observations in $\{A, B\}$ and $\{A, B, C\}$. In this case (non-normalised) possibility distributions provide an alternative approach. Instead of a probability in the sense of a relative frequency we determine a *possibility degree*, . The possibility degree for a certain type of an aeroplane is the quotient of the cases, where the occurence of the corresponding aeroplane is possible according to the observed set, and the total number of all observations. Therefore, the corresponding possibility degrees are $35/100$ for A, $40/100$ for B and $55/100$ for C, respectively. Then this "fuzzy set" over the universe of discourse {A,B,C} is called a *possibility distribution*. □

This example illustrates the difference between a possibilistic interpretation of fuzzy sets and one based on similarity relations. The possibilistic view is based on a form of uncertainty where the probabilistic concept of relative frequency is replaced by degrees of possibility. The underlying principle of similarity relations is not based on an uncertainty concept, but on an idea of indistinguishability or similarity, especially as a dual concept to the notion of distance. In fuzzy control the main issue is to model imprecision or vagueness in terms of "small distances" or similar values. Therefore, we do not further elaborate the details of possibility theory.

2

Fundamentals of Control Theory

This chapter is dedicated to the reader with no prior knowledge in control theory. On the one hand, we will explain the fundamentals of control theory, which are necessary to understand a fuzzy controller or any advanced problems related to them. On the other hand, we will present an overview of the possibilities that classical control theory offers in order to enable the reader, when confronted with real applications, to judge whether to solve the problem using a fuzzy controller or a classical controller. A complete introduction to the fundamentals of control theory, however, will not be given, as the chapter would then exceed the proportions of the book. More thorough introductions can be found in the basic literature of classical control theory.

2.1 Basic Concepts

Control theory is concerned with influencing systems in order to achieve that certain output quantities take a desired course. These can be technical systems, like the heating of a room with the output quantity *temperature*, a ship with the output quantities *course* and *speed*, or a power plant with the output quantity *electrical power*. The systems could as well be social, chemical or biological, as, for instance, the system of a *national economy* with the output quantity *rate of inflation*. The nature of the system does not matter. Only its dynamic behavior is of importance to the control engineer. We can describe this behavior by differential equations, difference equations or other functional equations. In classical control theory, which is mostly concerned with technical systems, the system that will be influenced is called the *(controlled) plant*.

In what kinds of ways can we influence the system? Every plant consists not only of output quantities, but as well of input quantities. For the heating of a room, these will, for example, be the position of the valve, for the ship the power of the engine and the angle of the rudder. These input variables have to be adjusted in a way that the output variables take the desired course, and they are called *actuating variables*. In addition to the actuating variables, the

K. Michels et al.: *Fuzzy-Control: Fundamentals, Stability and Design of Fuzzy Controllers*, StudFuzz **200**, 57–235 (2006)
www.springerlink.com

disturbance variables affects the plant, too. For example, a heating system, where temperature will be influenced by the number of people in the room or an open window, or a ship, whose course will be affected by water currents.

The desired course of the output variables is defined by the *reference variables*. They can be defined by humans, but they can also be defined by another system. For example, the autopilot of an aeroplane computes the reference values for the altitude, course, and speed of the plane. But we do not want to discuss the generation of the reference variables here. In the following, we take them for granted. We just have to take into account, that reference variables do not necessarily have to be constant, they can also be time-varying.

What information do we need to calculate the actuating variables to make the output variables of the system follow the reference variables? Obviously the reference values for the output quantities, the behavior of the plant and the time-dependent behavior of the disturbance variables have to be known. With this information, we can theoretically compute the values for the actuating variables, which then will influence the system in a way that the output quantities will take the desired course. This is the principle of a *steering mechanism* (Fig. 2.1). The input variable of the steering mechanism is the reference variable w, its output quantity the actuating variable u, which again - together with the disturbance variable d - forms the input value of the plant. y stands for the output value of the system.

The disadvantage of this method is obvious. If the behavior of the plant is not in accordance with the assumptions that we made about it, or if unforeseen disturbances occur, then the output quantities will not continue to take the desired course. A steering mechanism cannot react to this deviation, as it does not know about the output quantities of the plant.

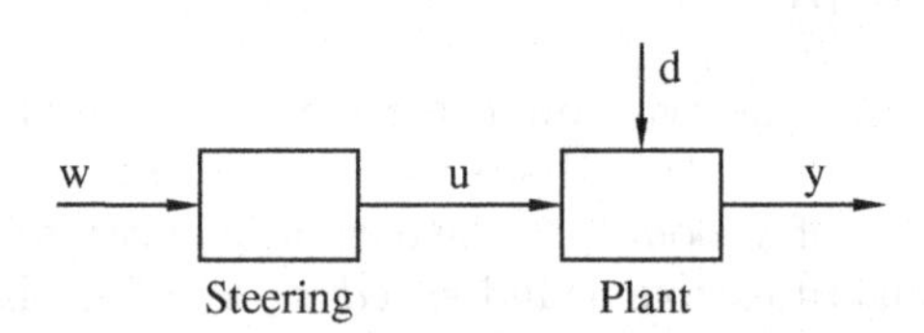

Fig. 2.1. Principle of a steering mechanism

One improvement that can immediately be made is the principle of an *(automatic) control* (Fig. 2.2). Inside the automatic control, the reference variable w is compared to the measured output variable of the plant y (*control variable*), and a suitable output quantity u of the controller (actuating variable) is computed inside the control unit from the difference Δy (*control error*). In former times the control unit itself was called controller, but modern controllers, including, among others, the fuzzy controllers, show a structure where the calculation of the difference between actual and desired output value and the computations of the control algorithm cannot be distinguished in the way

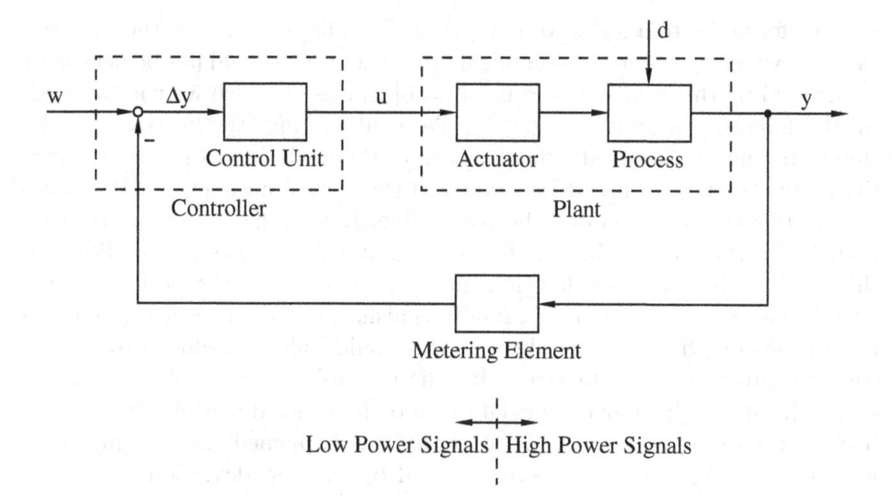

Fig. 2.2. Elements of a control loop

just described. Because of this, the tendency today is towards giving the name *controller* to the section in which the actuating variable is obtained from the reference variable and the measured control variable.

The quantity u is usually given as signal of low power, e.g. as a digital signal. But with low power, it is not possible to take action against a physical process. How for example could a ship be made to change its course by a digitally computed rudder angle, that means a sequence of zeros and ones at a voltage of 5 V? As this is not possible directly, a static converter and an electrical rudder drive are necessary, that can influence the rudder angle and the course of the ship. If the position of the rudder is seen as actuating variable of the system, the static converter, the electrical rudder drive and the rudder itself form the *actuator* of the system. This actuator transforms the controller output, a signal of low power, into the actuating variable, a signal of high power, that can directly influence the plant.

Alternatively, the output of the static converter, that means the armature voltage of the rudder drive, could be seen as actuating variable. In that case, the actuator would consist only of the static converter, while the the rudder drive and the rudder would have to be added to the plant. These different views already show that a strict separation between actuator and process is not possible. But it is not necessary either, as for the design of the controller, we will have to take every transfer characteristic from the controller output to the control variable into consideration anyway. Thus, we will treat the actuator as part of the plant, and from now on we will use the term actuating variable to refer to the output quantity of the controller.

For the feedback of the control variable to the controller the same problem holds, this time only in the opposite direction: a signal of high power has to be transformed into a signal of low power. This happens inside the metering element, which again shows dynamical properties that should not be neglected.

Caused by this feedback, a crucial problem arises, which we will illustrate by the following example (Fig. 2.3). We could formulate the strategy of a ship's automatic control like this: the bigger the deviation from the course is, the more the rudder should be steered in the opposite direction. At a quick glance, this strategy seems to be reasonable. If for some reason a deviation occurs, the rudder is adjusted. By steering into the opposite direction, the ship receives a rotary acceleration into the direction of the desired course. The deviation is reduced until it finally vanishes, but the rotating speed does not vanish together with the deviation, it could only be reduced to zero by steering into the other direction. In this example, because of the rotating speed the ship will receive a deviation into the other direction after getting back to the desired course. Only after this has happened the rotating speed will be reduced by counter-steering caused by the new deviation. But as we already have a new deviation, the whole procedure starts again, only the other way round. The new deviation might be even bigger than the first one. The ship will start zigzagging its way, if worst comes to worst, with ever increasing deviations. This latter case is called *instability*. If the amplitude of the vibration remains the same, it is called *borderline case of stability*. Only if the amplitudes decrease the system is *stable*. In order to receive an acceptable control algorithm for the example given, one should have taken the dynamics of the plant into account when designing the control strategy. A suitable controller would produce a counter-steering with the rudder right in time, in order to reduce the rotating speed to zero at the same time the ship gets back on course.

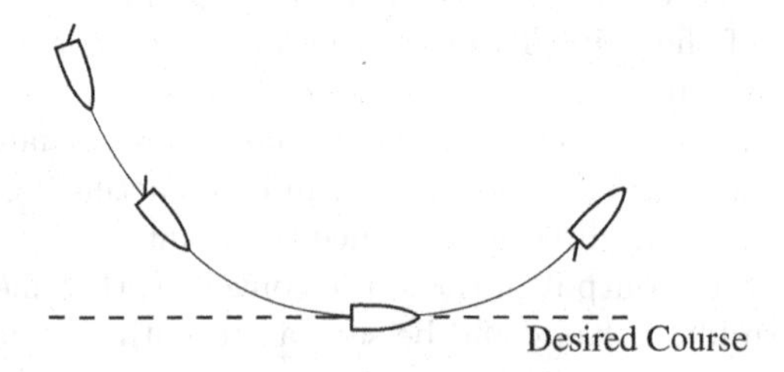

Fig. 2.3. Automatic cruise control of a ship

This example illustrates the demands on controlling devices. One requirement is accuracy, i.e. the control error should be as small as possible once all initial transients are finished and a stationary state is reached. Another requirement is speed, i.e. in the case of a changing reference value or a disturbance, the control error should be eliminated as soon as possible. This is called the *response behavior* or *disturbance response*. The third and most important requirement is the one for stability of the entire system. We will see that these requirements contradict each other, thus forcing every kind of controller (and therefore fuzzy controllers, too) to be a compromise between the three.

2.2 The Model of the Plant

2.2.1 General Task

In classical control theory the design of an automatic control consists of two steps. In the first step the plant gets analyzed and a mathematical model, describing the plant, is constructed. During step two the automatic controller is developed using the model as a basis.

For the description of a dynamic system, the use of differential equations or difference equations seems the obvious choice. The aim is to describe the dynamic behavior by a few equations which should be as simple as possible. These equations can then be illustrated by graphical symbols, which again can be summarized in a diagram showing the overall structure, the so-called *block diagram*. The graphical representation has the advantage of showing the way in which the single variables effect each other in a relatively clear manner, which will of course make the design of an automatic control easier.

Let us illustrate this process by a simple example (fig. 2.4): We are given a body of mass m, connected to a wall via a spring and freely movable in one direction on an even surface. This body is affected by the moving force f_m as well as by the force f_f, caused by the friction on the surface, and the restoring force of the spring f_r. Friction and restoring force work against the moving force. The quantity l describes the length of displacement of the body with respect to its position of rest. The goal is to find a model that describes the connection between the moving force and the length of displacement.

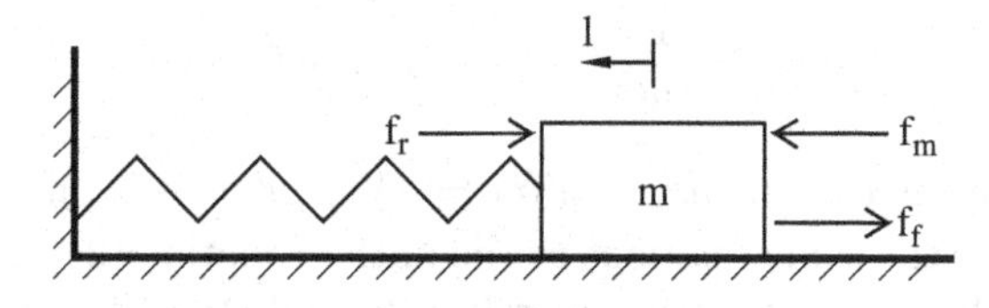

Fig. 2.4. Spring-mass system

Now we need to find the equations describing the system. First we consider the Newtonian Equation of Motion, which describes the relation between the forces affecting the body and the resulting acceleration a:

$$a(t) = \frac{1}{m}(f_m(t) - f_r(t) - f_f(t)) \tag{2.1}$$

Furthermore, the relations defined by differentiation or integration between acceleration and velocity v, or between acceleration and path length l are respectively:

$$a(t) = \frac{dv(t)}{dt} \qquad \text{or} \qquad v(t) = \int\limits_{\tau=0}^{t} a(\tau)d\tau + v(0) \tag{2.2}$$

$$v(t) = \frac{dl(t)}{dt} \qquad \text{or} \qquad l(t) = \int\limits_{\tau=0}^{t} v(\tau)d\tau + l(0) \tag{2.3}$$

Let the restoring force of the spring be proportional to the displacement; also let the force caused by the friction be proportional to the velocity of the body:

$$f_r(t) = c_r\, l(t) \tag{2.4}$$

$$f_f(t) = c_f\, v(t) \tag{2.5}$$

2.2.2 Scaling

As further considerations will require all quantities used in these equations to be dimensionless, we first of all need to scale them. To do this, for example in equation (2.1) we have to fix a constant acceleration a_0, a constant mass m_0 and a constant force $f_0 = a_0 m_0$. These quantities should, in order to avoid unnecessary calculations, all have the value 1, i.e. $a_0 = 1\frac{m}{s^2}$, $m_0 = 1kg$ and $f_0 = 1\frac{kgm}{s^2}$. Afterwards, we divide the equation by a_0:

$$\frac{a(t)}{a_0} = \frac{1}{\frac{m}{m_0}} \frac{1}{f_0}(f_m(t) - f_r(t) - f_f(t)) \tag{2.6}$$

We obtain a new equation

$$a'(t) = \frac{1}{m'}(f'_m(t) - f'_r(t) - f'_f(t)) \tag{2.7}$$

with the dimensionless quantities $a'(t) = \frac{a(t)}{a_0}$, $m' = \frac{m}{m_0}$ and $f'(t) = \frac{f(t)}{f_0}$, which still shows the same relations as the initial equation (2.1). This example shows that scaling in principle is merely a formal step. If working with the units of the MKS system, scaling is achieved by simply removing the units.

There may be cases where it might seem reasonable to scale some quantities with a value other than one. For instance, if the input quantities of an equation have different orders of magnitude, which might cause numerical problems, or if the range of values should be limited to the unit interval, which is often the case for fuzzy controllers. The time t, however, will be an exception to this rule. It should always be scaled with $t_0 = 1s$ in order to still allow an estimation of the course of events after scaling.

In the following sections we will assume that all the occurring physical quantities are suitably scaled, and we will therefore refrain from making special remarks concerning this topic. As the basis for further considerations, we can still use the equations (2.1) to (2.5), assuming that all quantities involved are dimensionless.

2.2.3 Fundamental Linear Transfer Elements

The single equations now have to be represented by suitable symbols. We will therefore need graphical representations for the additive superposition of different signals, multiplication with a constant factor and integration over a certain variable. These so-called *transfer elements* are shown in fig. 2.5. The output value of the *summing point* is the sum of its two input values, the output value of the *proportional element* is equal to the input value multiplied with k, and the output value of the *integrator* is the integral of the input variable from $t = 0$ to the current time t.

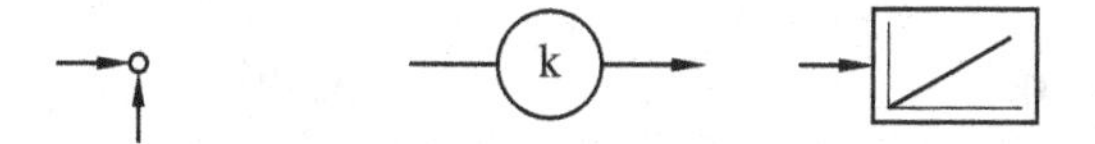

Summing Point Proportional Element Integrator

Fig. 2.5. Elements of block diagrams

Using these elements, the block diagram 2.6 for the plant can be given. The summing point and the proportional element $\frac{1}{m}$ represent the Newtonian Equation of Motion (2.1), in which the negative signs of f_r and f_f are represented by the minus sign at the upper summing point. The first integrator represents equation (2.2), the second integrator equation (2.3). The proportional elements, with coefficients c_r and c_f, describe equations (2.4) and (2.5).

This block diagram demonstrates why the use of integrators makes more sense than the use of differentiators. For example, the input quantity for the first *integrator* is the acceleration and the output quantity the velocity. This

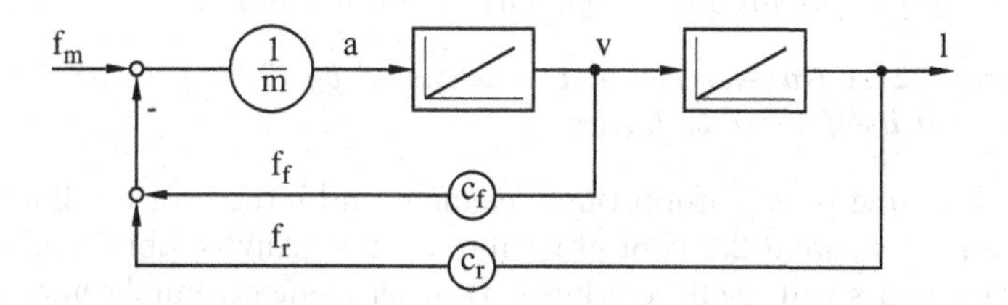

Fig. 2.6. Block diagram of a spring-mass system

corresponds to the physical facts, as the velocity is a result of the acceleration or the moving force, but not vice versa. The usage of a differentiator would require swapping input and output quantities. Doing so, the flow of signals would no longer correspond to the idea of acceleration causing velocity.

We represent the integrator by a block in which the graph of a function is shown. This is characteristic of the integrator; it is the so-called *step response*. If the input signal of an integrator switches from zero to one at the instant $t = 0$ (fig. 2.7), then the output signal of the integrator will, because of the integration, take the ramp-like course shown, assuming that it was initially zero. We can think of an integrator as a storehouse of energy, its contents being adjusted by an input quantity. Provided this input quantity does not grow to infinity, the contents of the storehouse can change only continuously.

It is common in control theory literature to represent linear transfer elements, i.e. transfer elements which can be described by a linear differential equation, by a block showing the corresponding step response. We will follow this convention throughout the rest of this book with the only exceptions *summing point* and *proportional element*.

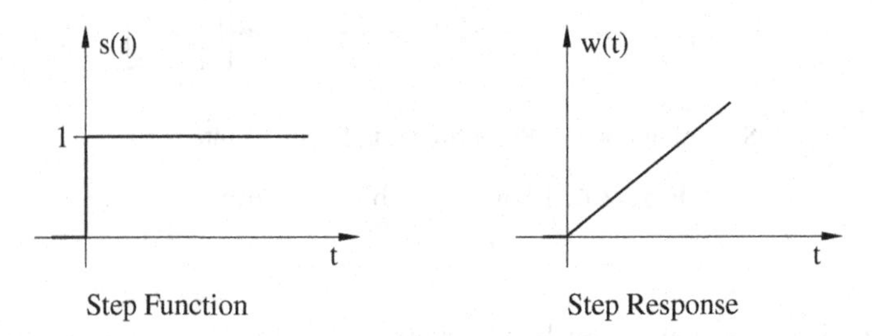

Fig. 2.7. Step response of an integrator

Linear transfer elements have some remarkable features which we can characterize by the following two theorems:

Theorem 2.1 *(Superposition Theorem) A transfer element produces the output signal $y_1(t)$ given the input signal $x_1(t)$, and the output signal $y_2(t)$ given the input signal $x_2(t)$. It is linear if and only if it produces the output signal $a_1y_1(t) + a_2y_2(t)$ for the input signal $a_1x_1(t) + a_2x_2(t)$*

Theorem 2.2 *A transfer element constructed by linking up linear transfer elements will itself again be linear.*

The summing point, proportional element and integrator are linear transfer elements. Theorem 2.2 is of great importance: Any combination of linear transfer elements will itself be a linear transfer element. Furthermore, we are now able to explain why we picked the step response to characterize linear elements. According to theorem 2.1, if steps of different amplitudes occur in the input signal of a linear transfer element, the output signal will differ only for the same amplitude value, but not for the principle course. In this respect, the step response is a relatively universally applicable feature of a system, and to use it to characterize the system seems justified.

Besides the three transfer elements introduced in figure 2.5, four other fundamental elements for block diagrams exist. The first is the *delay element*. It is the graphical representation of the equation $y(t) = x(t - T_D)$, i.e. the output signal in the course of time corresponds to the input signal delayed by the delay time T_D (fig. 2.8). The delay element is also a linear transfer element and is characterized by its step response. Delay times occur, for example, in long-distance communication. Someone steering a robot in space should not get impatient if the movements of the robot seem hesitant to him. The slow reaction is not the result of a badly constructed machine, but of long delays of the signals between the human and the robot. An automatic controller must take the delays of a plant into consideration in the same way.

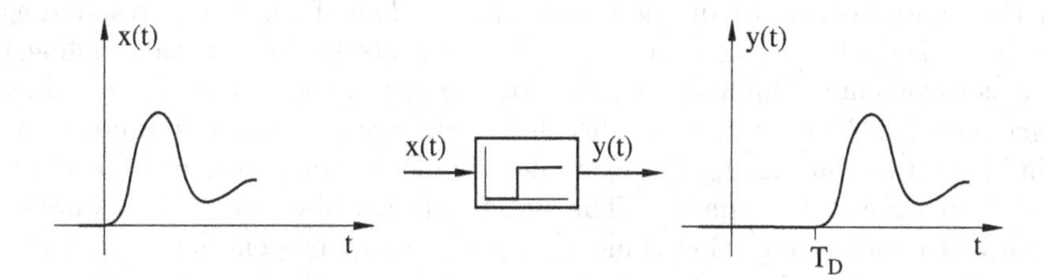

Fig. 2.8. Delay element

2.2.4 Fundamental Nonlinear Transfer Elements

Further elements are the *multiplication* and the *division* (fig. 2.9). In contrast to the proportional element, a multiplication multiplies two time-variant signals instead of multiplying one signal with a constant factor. An example here is the formation of the torque T_a inside a dc machine. The torque is proportional to the product of the armature current i_a and the excitation flux Φ_e, both of which can be regulated independently:

$$T_a(t) = c\, i_a(t)\Phi_e(t) \tag{2.8}$$

Analogous to the multiplication, inside a division two time-variant signals are divided. A division by zero, of course, has to be avoided. An example is the reckoning of the angular velocity ω around the main axis of a robot. Of similar structure as the Newtonian Equation of Motion (2.1), the correlation

$$\frac{d\omega}{dt} = \frac{1}{J(t)} T_m(t) \tag{2.9}$$

holds for rotational motions. Here, $T_m(t)$ is the moving torque at the main axis, and $J(t)$ is the inertia of the rotating body. During one rotation of the robot around its main axis the position of the other swivel joints—i.e. the arm position and the inertia of the rotating body—may change. We must therefore treat $J(t)$ in this equation as a time-dependent quantity. In any case, though, it will be different from zero.

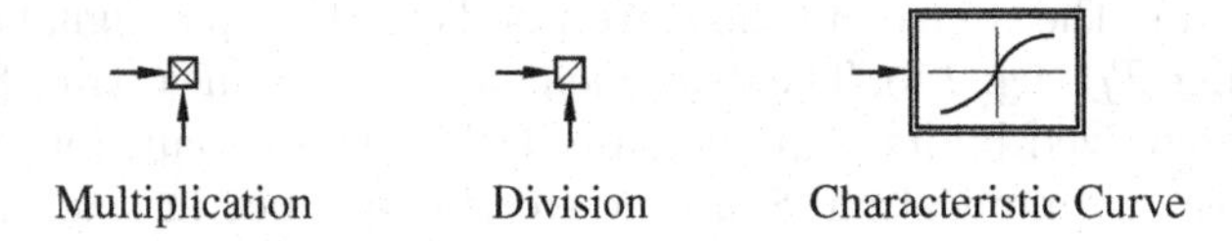

Fig. 2.9. Further fundamental transfer elements

The equations (2.8) and (2.9) are nonlinear differential equations because of the multiplication or division, respectively. Therefore, the corresponding transfer elements are, in contrast to the elements treated so far, nonlinear transfer elements. This holds as well for the last element, the *characteristic curve element*. With the help of this element we are now able to represent any kind of static relationship. For example, the block shown in figure 2.9 stands for the relation $y(t) = \sin x(t)$. The sine function will occur, if, for example, we need to find the equation of motion for a free-moving pendulum (fig. 2.10). The equation describing the equilibrium of forces in the tangential direction of motion is given by:

$$g\, m \sin\alpha(t) + ml\frac{d^2\alpha(t)}{dt^2} = 0 \tag{2.10}$$

or

$$\frac{d^2\alpha(t)}{dt^2} = \frac{-g}{l}\sin\alpha(t) \tag{2.11}$$

where g is the gravitational acceleration, l the length of the rope of the pendulum and m its mass, considered to be punctiform. The corresponding block diagram is given by figure 2.11. The negative sign of $\frac{g}{l}$ designates that the body will slow down for a positive angle and accelerate into the opposite direction. Furthermore, we will neglect effects caused by friction, so that our model will be one of an ideal pendulum, which, once started, will continue to swing *ad infinitum* if not being influenced from outside.

In contrast to the linear transfer elements, we mark the characteristic curve element by a block with doubled borders, with the characteristic curve drawn in its middle. The doubled edges should make it immediately clear that the function drawn in the block is not a step response but a characteristic curve.

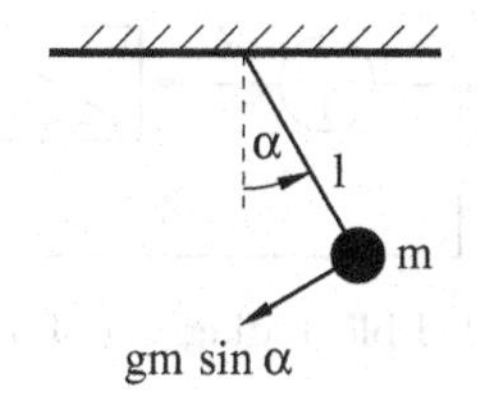

Fig. 2.10. Pendulum

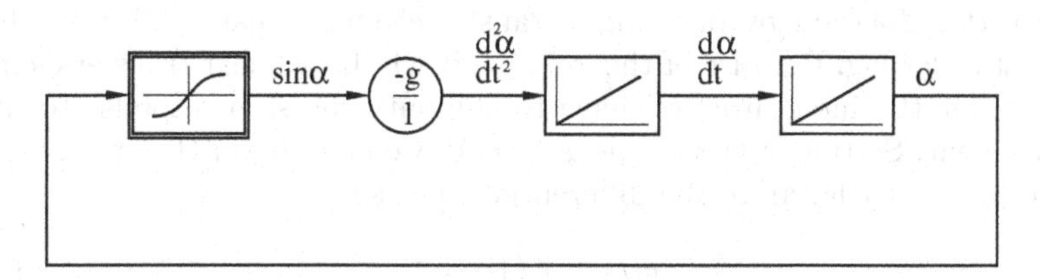

Fig. 2.11. Block diagram of a pendulum

2.2.5 First and Second Order Lags

It often happens that structures consisting of several fundamental transfer elements are combined into and replaced by a single unit in order to increase readability. Two of these composite structures occur relatively frequently, the *first order lag* and the *second order lag*, which we describe now.

If we hypothetically remove the spring in the spring-mass system discussed before, we get a system with a body being accelerated by a moving force on a surface with friction. Let us examine the relation between the moving force f_m and the velocity v. If we leave out the spring's restoring force in equation (2.1) and use the equations (2.2) and (2.5) for substitution, we obtain the desired differential equation

$$\frac{dv(t)}{dt} + \frac{c_f}{m}v(t) = \frac{1}{m}f_m(t) \tag{2.12}$$

Using $T = \frac{m}{c_f}$, $V = \frac{1}{c_f}$, $y(t) = v(t)$ and $x(t) = f_m(t)$ leads to

$$T\frac{dy(t)}{dt} + y(t) = Vx(t) \tag{2.13}$$

or, alternatively,

$$\frac{dy(t)}{dt} = \frac{1}{T}(Vx(t) - y(t)) \tag{2.14}$$

In the same way as $y(t) = kx(t)$ describes the transfer behavior of a proportional element, these equations describe the transfer behavior of a first order lag in normal form.

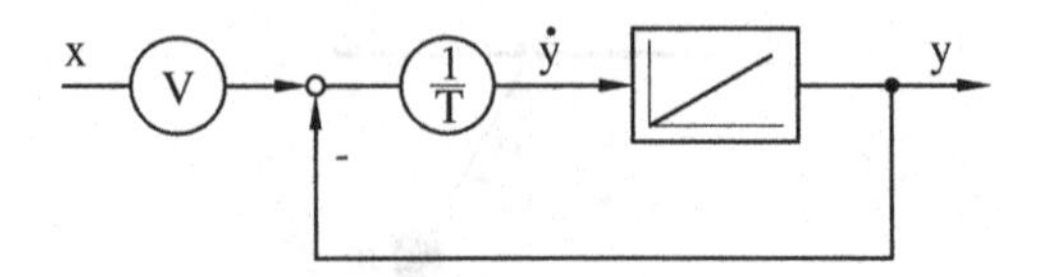

Fig. 2.12. Scaled block diagram of a first order lag

Figure 2.12 shows the corresponding block diagram. Now, we want to replace this diagram by one single transfer element, that shall characterize the first order lag. Because of theorem 2.2 it will be a linear transfer element.

Like for the integrator, we have to compute the step response to define this element. Setting $x(t) = 1$ and $y(0) = 0$, we can obtain the step response of the first order lag from the differential equation:

$$y(t) = V(1 - e^{\frac{-t}{T}}) \tag{2.15}$$

This is given in figure 2.13, as well as the drawing of the first order lag, that replaces the block diagram in fig. 2.12.

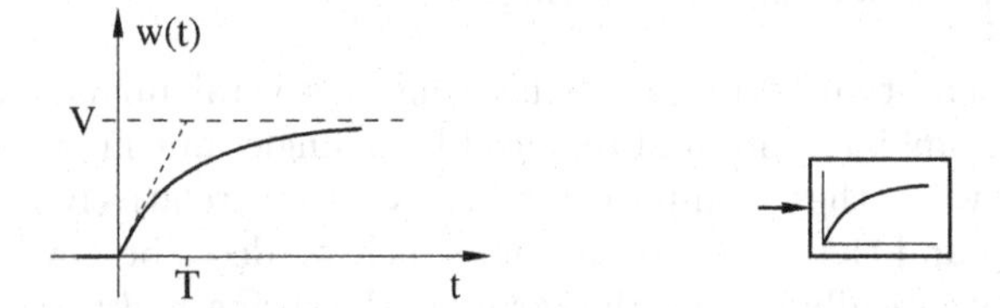

Fig. 2.13. Step response of a first order lag

Analyzing the step response we see, that the less the delay time T, the faster the curve approaches the limit, without ever reaching it in finite time. The limit is, by the factor V, proportional to the input quantity, which is a constant 1 in the case of the step response. We can easily explain the course of the step response for the accelerated body: If the moving force is set to the constant value of 1, the velocity will rise. Because of this, the force caused by the friction f_f working in the opposite direction of the moving force will rise as well. This will make the sum of the forces, i.e. the acceleration, smaller and smaller, which will also diminish the increase in velocity. After a certain period of time (longer than T), the limit will be approximately reached.

In order to explain the *second order lag*, we have to examine the entire spring-mass system. Substituting equation (2.2) to (2.5) into equation (2.1) allows elimination of the quantities v and a, giving a differential equation of second order describing the length of the path l:

$$\frac{m}{c_r}\frac{d^2l(t)}{dt^2} + \frac{c_f}{c_r}\frac{dl(t)}{dt} + l(t) = \frac{1}{c_r}f_m(t) \tag{2.16}$$

With $\omega_0 = \sqrt{\frac{c_r}{m}}$, $D = \frac{\omega_0 c_f}{2c_r}$, $V = \frac{1}{c_r}$, $y(t) = l(t)$ and $x(t) = f_m(t)$, we obtain the scaled block diagram 2.14 and the normal form of the differential equation of a second order lag:

$$\frac{1}{{\omega_0}^2}\frac{d^2y(t)}{dt^2} + \frac{2D}{\omega_0}\frac{dy(t)}{dt} + y(t) = Vx(t) \tag{2.17}$$

or, alternatively,

$$\frac{d^2y(t)}{dt^2} = \omega_0^2\left[Vx(t) - \frac{2D}{\omega_0}\frac{dy(t)}{dt} - y(t)\right] \tag{2.18}$$

Compared to the initial differential equation for the spring-mass system we only had to rename some parts. The block diagram should therefore show a structure similar to the one in figure 2.6.

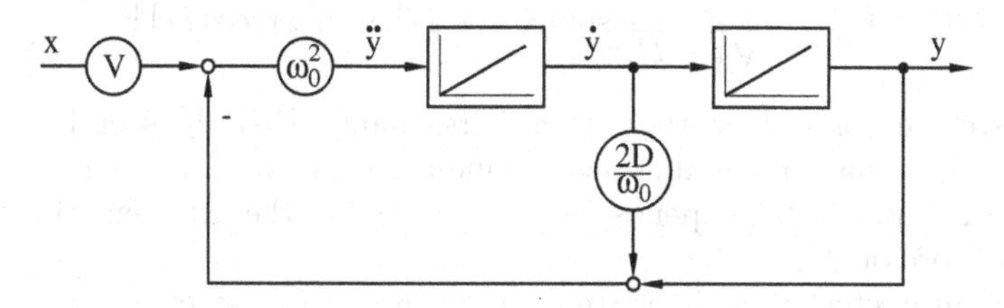

Fig. 2.14. Scaled block diagram of a second order lag

Analogously to the first order lag, this block diagram shall be replaced by one single linear transfer element, a representation of the second order lag. In order to achieve this, we have to compute the step response again.

The characteristic polynomial for the homogeneous differential equation (2.17) is

$$\frac{1}{{\omega_0}^2}s^2 + \frac{2D}{\omega_0}s + 1 \tag{2.19}$$

With its zeros $s_{1,2} = \omega_0\left[-D \pm \sqrt{D^2-1}\right]$, initial conditions $y(0) = \dot{y}(0) = 0$ and $x(t) = 1$, we obtain the step response of the second order lag (2.17) as:

$$y(t) = \begin{cases} V(1 + \frac{s_2}{s_1 - s_2}e^{s_1 t} + \frac{s_1}{s_2 - s_1}e^{s_2 t}) & : \quad D \neq 1 \\ V\left[1 - (1 - s_1 t)e^{s_1 t}\right] & : \quad D = 1 \end{cases} \tag{2.20}$$

With this step response we can draw the second order lag (fig. 2.15).

The so-called *damping ratio*, D, determines the form of the initial transient. For $D > 0$ both s_1 and s_2 show a negative real part. Because of this, the exponential functions of the step response decline and converge to the limit V. For $D \geq 1$ both s_1 and s_2 are purely real numbers, and for this case the step response will look similar to the first order lag. Only the initial slope is zero. This is called the *aperiodic case*.

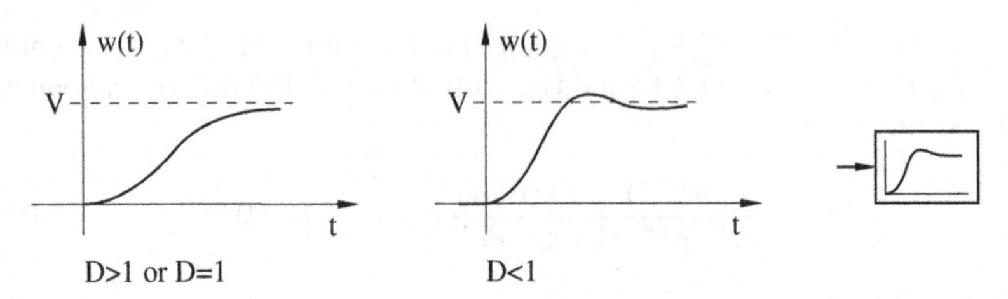

Fig. 2.15. Step response of a second order lag

In the case of $0 \leq D < 1$, we can write $s_{1,2} = \omega_0 \left[-D \pm j\sqrt{1-D^2}\right]$. The exponents in (2.20) are now complex values. In this case, some transformations seem reasonable, finally leaving us with

$$y(t) = V \left[1 - \frac{e^{-D\omega_0 t}}{\sqrt{1-D^2}} \sin\left(\sqrt{1-D^2}\omega_0 t + \arccos D\right)\right] \tag{2.21}$$

as the step response. Now the system is resonant, as can be seen by the sine function. How fast the oscillations decline after an excitation, for example a step of the input value, depends on the exponent of the exponential function, and therefore on D.

The parameter ω_0 is the system's *angular self frequency*. From equation (2.21) we can see that for $D = 0$ the step response of the system is a sinusoidal oscillation with a constant amplitude and precisely this frequency ω_0. For $0 < D < 1$ we get a declining oscillation with a smaller *angular natural frequency* $\omega_n = \omega_0\sqrt{1-D^2}$.

Interestingly enough, there exists a simple geometrical relation between the quantity D, which determines the system's transient behavior, and the positions of the zeros $s_{1,2}$ (the *eigenvalues* of the system) in the complex plane (fig. 2.16):

$$\cos\alpha_0 = \frac{|\mathrm{Re}(s_1)|}{\sqrt{(\mathrm{Re}(s_1))^2 + (\mathrm{Im}(s_1))^2}} = \frac{\omega_0 D}{\omega_0\sqrt{D^2 + 1 - D^2}} = D \tag{2.22}$$

The wider the angle α_0 gets, the smaller the damping will be. For $\alpha_0 = 0$ the system will show an aperiodic transient response, while for $\alpha_0 = \pi/2$ the oscillations will never decline.

Having introduced the second order lag, we can now replace the entire block diagram 2.6 by a single second order lag with the input variable f_m and the output variable l. The inner quantities *acceleration* and *velocity* do not occur explicitly any longer.

First order lag and second order lag are, according to theorem 2.2, obviously linear transfer elements. They are frequently used to give an approximate description of more complicated structures. If, for example, we have to design a control for the spring-mass system described at the beginning of

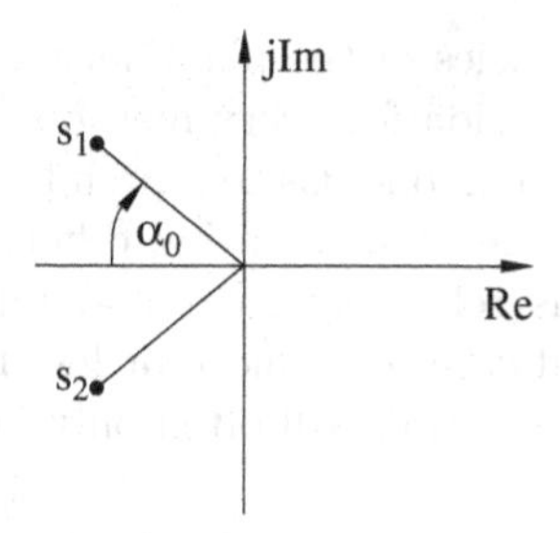

Fig. 2.16. Eigenvalues of a second order lag

this chapter, we should also take the generation of the moving force f_m in a machine into account. Here, using a first or second order lag to give an approximate description of the transfer characteristic—from the actuating variable of the controller (i.e. the machine's input quantity) to the mechanical force—will be easier and might make more sense than explicitly modeling the series of events taking place inside the machine. Because of the speed with which the transient reactions inside the machine are happening—compared to the dynamic process of the spring-mass system—the fault built into the model will be relatively small. Later on we will describe how to actually do such a simplification.

2.2.6 Fields of Application

Although the examples we gave in this chapter were of purely mechanical nature, we may use the now well-known transfer elements as well to describe dynamic processes occurring in, for example, electrical engineering or hydro mechanics. In principle, any system can be described which shows the following properties:

- The system is *time-invariant*, i.e. the parameters of the plant remain constant during the flow of time. An example of a time-variant system is an airplane, where the contents of the tank are used up during the flight. This will change the weight, which is a parameter of the plant, and therefore the dynamic behavior of the airplane.
- The system is *continuous*, therefore the values of the signals are given for every instant of time. *Time-discrete* systems, where the value of a signal is known only for certain instants, form a contrast to this. If, for example, microprocessors are used to control a plant, the required analog-to-digital (A/D) conversion will produce a time-discrete signal out of a continuous one. As information is lost in this process, this fact has to be taken into account when designing the controller. This topic is well documented in control literature, but will not be treated in this book, since time discretization can be neglected if the A/D-converter's scanning rate is high

compared to the frequencies of the plant's signals. But in principle there exists a time-discrete version for every method presented in the following subsections. To apply them, one has to switch from using differential equations for the model of the plant to using difference equations. For linear systems, this change is relatively easy, and might be done by control-engineering specific software tools more or less by pressing a button. For nonlinear systems, the solution will often only be approximative.

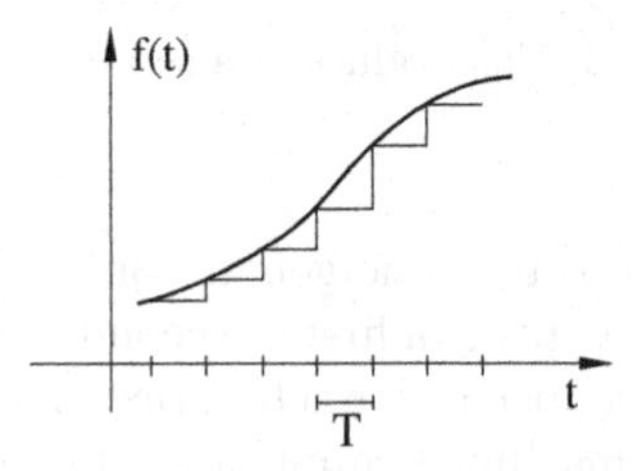

Fig. 2.17. Time discretization of a signal

- The parameters of the system are *concentrated*. The temperature inside a room, for example, depends on the time as well as on the place. In order to describe this process, partial differential equations are required. Every element of volume represents a small storehouse of energy, and interacts with adjacent elements. In trying to design a block diagram for this system using only the transfer elements introduced so far, we would need infinitely many elements, as every single element of volume would have to be modeled. The parameters of this system are not concentrated. In cases like this, an approximate solution, where the system is modeled by only a few storehouses of energy and therefore a finite number of transfer elements, will have to be accepted instead of an exact model.

2.2.7 Linearization

In contrast, linearity of the transfer characteristic is not a requirement for the use of block diagrams. Linear as well as nonlinear differential equations can be represented by transfer elements or block diagrams. But, as further steps in the work process—especially the design of the controller—are much easier for linear systems, the control engineer will always try to describe the given system by linear differential equations.

On the other hand, in most of the real-world plants non-linearities occur which can not be neglected. One way to cope with this is to *linearize* nonlinear plants: An *operating point* is defined, and in and around this point the nonlinear function is approximated by a linear one. This is done by developing the corresponding Taylor's Series and truncating after the first term of the sequence. Let, for example, a nonlinear element be given with the transfer function $y = f(x)$ and the operating point $(x_0, f(x_0))$. We can describe the displacement to the working point by $\Delta x = x - x_0$ and $\Delta y = \Delta f = f(x) - f(x_0)$. Transformation of $f(x)$ into a Taylor's Series yields:

$$f(x) = f(x_0) + \frac{\partial f}{\partial x}(x_0)\, \Delta x + r(x) \tag{2.23}$$

Here, $r(x)$ represents the residual term of the sequence, containing the higher derivations of the function $f(x)$, which should be neglected. Focusing on the differences Δf and Δx between the quantities and the operating point instead of examining the quantities f and x itself, we can find a relation between input and output variable of the transfer element, which is approximately linear:

$$\Delta f = f(x) - f(x_0) \approx \frac{\partial f}{\partial x}(x_0)\, \Delta x = k\, \Delta x \tag{2.24}$$

Let us use the example of the pendulum (fig. 2.10) once more to illustrate this process. We linearize the behavior of the system at the operating point $\alpha_0 = 0$. To do this, we have to replace the sine function in equation (2.11) by a linear transfer element. Setting $f(\alpha) = \sin \alpha$, the following equation holds:

$$\Delta f \approx \frac{\partial f}{\partial \alpha}(\alpha_0)\, \Delta \alpha = \cos 0\, (\alpha - 0) = 1\, \alpha \tag{2.25}$$

According to this, we can replace the sine function in the vicinity of the working point by a proportional element with the factor 1. This again implies that the characteristic curve element in block diagram 2.11 can be dropped, leaving a linear system.

We will discuss linearization of nonlinear transfer elements with multiple input and output quantities in greater detail in chapter 2.8.2. Even in that case, the method can be schematized, and is not very difficult. But one thing that should always be kept in mind is that a model of a plant won by linearization is valid only in a limited range around the operating point.

2.2.8 Closing Comments

It can easily be seen that several different block diagrams may be found for one single plant. Depending on the choice of intermediate quantities and transfer blocks, totally different structures may emerge, although they should all be equivalent to one another. The decision of how to split up the block diagram will in most cases depend on which quantities can actually be measured, or which of them have a special physical meaning. In these cases, the block diagram will reflect the real structure of a system relatively well, and the relations and mutual influences between the single physical variables will be more comprehensible than a set of equations. This is of help to the control engineer, as, even with all the possibilities which the use of computers in the field of control theory offers, there is still a certain amount of intuition and an understanding of the overall interdependencies needed when designing an automatic control.

Statistical methods are available in control literature which can be used to obtain information about the structure of a plant from measured values, which then again can be used to determine the block diagrams needed. The use of these methods is mandatory if the dependencies among the physical quantities of the plant are not as easy to understand as in the examples given in this chapter, or if only the structure, but not the single parameters of the differential equations are known. Once the block diagram has been found, the possibility of simulating the plant with the aid of numerical integration methods exists. Hereby, further insight into the behavior of the plant can be won, and last, but not least, the controller can be tested.

In principle, all the ideas we presented in this chapter can be extended to transfer elements with multiple input and output variables, although we will avoid this extension for the next chapters, in order to make the topic more comprehensible. Only from chapter 2.7 on we will focus on multi-input-multi-output (MIMO) systems.

2.3 Transfer Function

For this and the following chapters, the class of transfer elements or systems should be subject to even further restrictions: we will consider only purely linear systems with only one input and output variable. Therefore, before applying the methods being given, we may have to perform a linearization of nonlinear transfer elements.

2.3.1 Laplace Transformation

First of all, let us introduce the *Laplace transformation*, as it provides an easy and elegant way for solving problems of linear control theory. The Laplace transformation can be applied to a complex valued function $f(t)$ of a real-valued variable t, if the function meets the following criteria:

- $f(t)$ is defined for $t \geq 0$.
- $f(t)$ can be integrated in $(0, \infty)$.
- $f(t)$ is limited by an exponential growth:

$$|f(t)| \leq Ke^{ct} \tag{2.26}$$

This makes the Laplace transformation applicable to most of the signal functions occurring in a control system. Using the complex variable s, the Laplace transform $f(s)$ for the function $f(t)$ is defined by

$$f(s) = \mathcal{L}\{f(t)\} = \int_0^\infty e^{-st} f(t) dt \tag{2.27}$$

The integral converges absolutely for $\mathrm{Re}(s) > c$, c from (2.26). Inside this convergence half-plane, $F(s)$ is an analytic function in s. When applying the transformation to signal functions, the dimension of t is *time*. Accordingly, the dimension of s has to be $time^{-1}$, because of the exponential function in (2.27), and s is a complex frequency. Therefore, in control theory the domain of the Laplace transform is called the *frequency domain* and the domain of the original function the *time domain*. In order to keep things simple, we will treat t and s as dimensionless variables.

Under certain circumstances a retransformation is possible:

$$\begin{aligned} f(t) &= \mathcal{L}^{-1}\{f(s)\} = \frac{1}{2\pi j}\int_{c-j\infty}^{c+j\infty} e^{st} f(s)ds \qquad \text{for } t \geq 0 \\ f(t) &= 0 \qquad \text{for } t < 0 \end{aligned} \tag{2.28}$$

The parameter c has to be chosen in a way that the path of integration lies inside the convergence half-plane and that c is greater than the real parts of all singular points of $f(s)$.

The following theorems hold:

1. **Addition Theorem** (Superposition Theorem)

$$\mathcal{L}\{a_1 f_1(t) + a_2 f_2(t)\} = a_1 \mathcal{L}\{f_1(t)\} + a_2 \mathcal{L}\{f_2(t)\} \tag{2.29}$$

2. **Integration Theorem**

$$\mathcal{L}\left\{\int_0^t f(\tau)d\tau\right\} = \frac{1}{s}\mathcal{L}\{f(t)\} \tag{2.30}$$

3. **Differentiation Theorem**

$$\mathcal{L}\left\{\frac{d^n f(t)}{dt^n}\right\} = s^n \mathcal{L}\{f(t)\} - \sum_{i=1}^{n} s^{n-i} \lim_{\substack{t\to 0\\ t>0}} \frac{d^{i-1} f(t)}{dt^{i-1}} \tag{2.31}$$

4. **Delay Theorem**

$$\mathcal{L}\{f(t - T_L)\} = e^{-T_L s}\mathcal{L}\{f(t)\} \tag{2.32}$$

5. **Convolution Theorem**

$$\mathcal{L}\left\{\int_0^t f_1(t-\tau) f_2(\tau)d\tau\right\} = \mathcal{L}\{f_1(t)\}\,\mathcal{L}\{f_2(t)\} \tag{2.33}$$

6. **Limit Theorems**

$$\lim_{t\to\infty} f(t) = \lim_{s\to 0} sf(s) \qquad \text{if } \lim_{t\to\infty} f(t) \text{ exists} \tag{2.34}$$

$$\lim_{\substack{t\to 0\\ t>0}} f(t) = \lim_{s\to\infty} sf(s) \qquad \text{if } \lim_{\substack{t\to 0\\ t>0}} f(t) \text{ exists} \tag{2.35}$$

2.3.2 Computation of Transfer Functions

We will now discuss the problem of how to compute the corresponding output signal if the function of a plant's input signal is given. In principle, we can find a solution using the differential equation of the plant, but this would require an extremely high amount of computation. To apply Laplace transformation seems a reasonable idea. First, we transform the input signal according to (2.27) or using the correspondence table in A. Next, using the theorems given above, we calculate the output signals in the frequency domain for the single transfer elements in the same sequence as they are passed by the signal. The output signal of the last transfer element is the output of the plant, that finally has to be retransformed into the time domain. Again, the correspondence table in A can be used, leaving the computation of the output signal in the frequency domain as the only problem.

For linear transfer elements, this calculation is fairly easy. The integration of a signal in the frequency domain is reduced to a multiplication by $\frac{1}{s}$, using the Integration Theorem. Corresponding to this, with fading initial values the differentiation is replaced by a multiplication by s, according to the Differentiation Theorem. Addition of two signals and multiplication with a constant factor stay the same because of the Addition Theorem, and a delay element is taken into consideration by using the factor $e^{-T_L s}$, according to the Delay Theorem. In any case, a transformed input signal $x(s)$ is multiplied by a function $G(s)$ depending on s, giving the output signal $y(s)$. $G(s)$ is called the *transfer function* (fig. 2.18):

$$y(s) = G(s)x(s) \tag{2.36}$$

Instead of having to solve a complicated set of differential equations in the time domain, our problem is reduced to finding a transfer function and multiplying it by the input signal in the frequency domain.

The transfer function for the integrator is

$$G(s) = \frac{1}{s} \tag{2.37}$$

for the delay element

$$G(s) = e^{-T_L s} \tag{2.38}$$

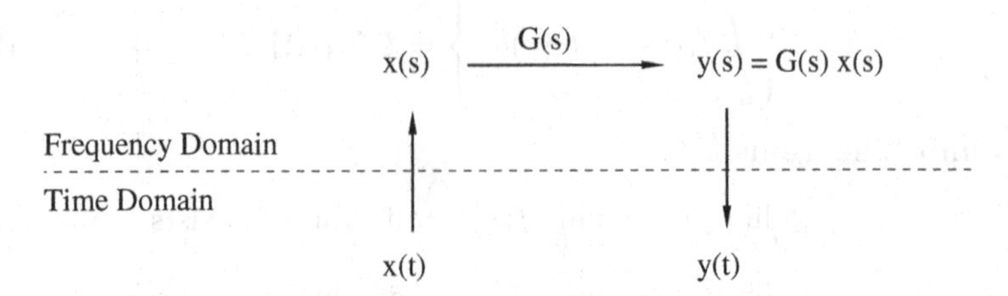

Fig. 2.18. Application of the Laplace transformation

and for the proportional element

$$G(s) = k \tag{2.39}$$

For the first order and second order lag, transfer functions can also be given: applying the Differentiation Theorem to equation (2.13) (assuming fading initial values of $y(t)$) yields

$$Tsy(s) + y(s) = Vx(s) \tag{2.40}$$

From this, we can obtain the transfer function for the first order lag:

$$G(s) = \frac{y(s)}{x(s)} = \frac{V}{Ts+1} \tag{2.41}$$

Analogously, we can transform the equation for the second order lag (2.17) into

$$\frac{1}{{\omega_0}^2}s^2y(s) + \frac{2D}{\omega_0}sy(s) + y(s) = Vx(s) \tag{2.42}$$

and obtain the transfer function

$$G(s) = \frac{y(s)}{x(s)} = \frac{V}{\frac{1}{\omega_0^2}s^2 + \frac{2D}{\omega_0}s + 1} \tag{2.43}$$

We can see, that we can take the coefficients of the differential equation directly as coefficients of the transfer function. The denominator of the transfer function corresponds to the characteristic polynomial of the homogeneous differential equation.

If a block diagram consists only of integrators, summing points and proportional elements, we obtain a purely rational transfer function by summing up the single terms,

$$G(s) = \frac{y(s)}{x(s)} = \frac{b_m s^m + b_{m-1}s^{m-1} + ... + b_1 s + b_0}{a_n s^n + a_{n-1}s^{n-1} + ... + a_1 s + a_0} \qquad m \leq n \tag{2.44}$$

where the degree of the denominator will always be higher than or the same as the degree of the numerator. Once again it has to be emphasized that such a transfer function can only be obtained if the initial values of the single signals and, if necessary, their derivatives, are equal to zero. Otherwise the use of the Differentiation Theorem would produce additional terms. In the following, we assume, that the initial signal values and their derivatives are zero.

Furthermore, one should keep in mind that a transfer function can only be given for linear transfer elements—not for nonlinear transfer elements. They even have to be handled with care, as a multiplication or division in the time domain does not correspond to a multiplication or division in the frequency domain. It is also not possible to transform a characteristic curve directly from the time domain into the frequency domain. For this reason, computations concerning nonlinear transfer elements still have to be done in the time domain.

2.3.3 Interpretation of the Transfer Function

In this section, we will present a few ideas which should make the interpretation of the term *transfer function* easier. In order to do this, we first have to introduce the *impulse (function)* $\delta(t)$ (fig. 2.19). It is defined in an approximating way by

$$\delta(t) = \lim_{\varepsilon \to 0} \frac{1}{\varepsilon}(s(t) - s(t - \varepsilon)) \tag{2.45}$$

where $s(t)$ is the step function. We can think of the impulse function as being the derivation of the step function. The area between the impulse function and the abscissa is equal to 1.

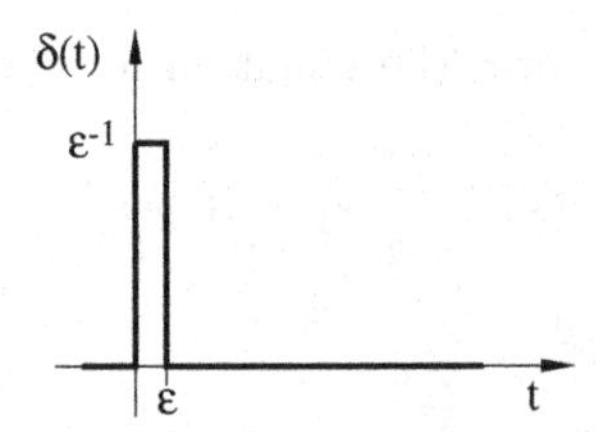

Fig. 2.19. Impulse

The Laplace transform of the impulse function is

$$\begin{aligned}
\mathcal{L}\{\delta(t)\} &= \int_0^\infty e^{-st} \lim_{\varepsilon \to 0} \frac{1}{\varepsilon}(s(t) - s(t - \varepsilon))dt \\
&= \lim_{\varepsilon \to 0} \frac{1}{\varepsilon} \int_0^\infty e^{-st}(s(t) - s(t - \varepsilon))dt \\
&= \lim_{\varepsilon \to 0} \frac{1}{\varepsilon} \int_0^\varepsilon e^{-st} dt = \lim_{\varepsilon \to 0} \frac{1}{\varepsilon} \frac{1}{s}(1 - e^{-s\varepsilon}) = 1
\end{aligned} \tag{2.46}$$

thus allowing to write:

$$G(s) = G(s)\, 1 = G(s)\mathcal{L}\{\delta(t)\} \tag{2.47}$$

A comparison with equation (2.36) shows that we might as well interpret the transfer function $G(s)$ as being the Laplace transform of the output signal of the plant, activated by an impulse. Therefore, $G(s)$ is the Laplace transform of a hypothetical *impulse response* $g(t)$.

This interpretation leads to a very clear explanation, together with the help of the Convolution Theorem. In order to calculate the output signal in the frequency domain, we have to multiply the Laplace transform of an input signal by the transfer function, according to (2.36). Using the Convolution Theorem, this corresponds to a convolution in the time domain:

$$y(s) = G(s)x(s) \longleftrightarrow y(t) = \int_0^t g(t-\tau)x(\tau)d\tau \tag{2.48}$$

The function $g(t)$ represents the impulse response. Figure 2.20 illustrates this formula. We can think of the input signal as being approximately a sequence of pulses with height $x(\tau)$ and width $d\tau$, each of these pulses causing an impulse response at the plant output. The impulse responses do not interfere with each other because of the linearity of the plant, so that the instantaneous value at the plant output is a superposition of all impulse responses. The pulse at the plant input at the instant τ, with the area $x(\tau)d\tau$, causes the impulse response $g(t-\tau)x(\tau)d\tau$. This is just the contribution of this impulse's response to the instantaneous value $y(t)$.

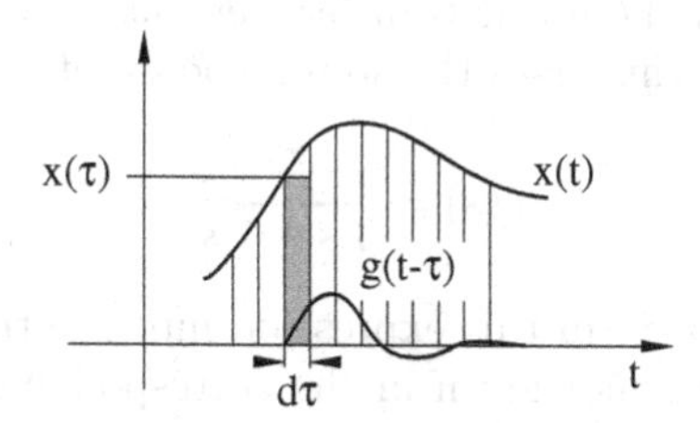

Fig. 2.20. An illustration of the convolution integral

2.3.4 Computation of the Step Response

First, with the aid of the transfer function, we can now set up rules about how to restructure block diagrams of linear plants without changing the model's transfer characteristic. In figure 2.21, each two block diagrams equivalent to one another are given. The equivalence can immediately be shown by computing the corresponding transfer functions, with using all computation rules like commutativity and associativity. The first example shows that linear transfer elements can be put in any order. But we should keep in mind that such changes will normally lead to losing the relation to internal, real physical quantities.

Let us now demonstrate how to determine the functions of the output signals for rational transfer functions by a simple example. With the aid of the correspondence table in appendix A, this will be no problem. For example, the Laplace transform for the step function is $\frac{1}{s}$. Using the general formula $y(s) = G(s)x(s)$, we obtain

$$y(s) = G(s)\frac{1}{s} \tag{2.49}$$

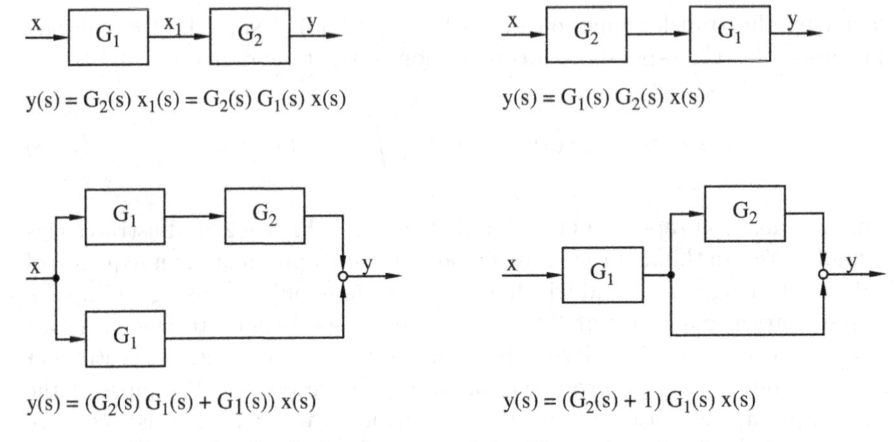

Fig. 2.21. Equivalent restructuring of linear transfer elements

as the step response of the linear transfer element. Inserting the equation for the first order lag into this gives the step response of

$$y(s) = \frac{V}{Ts+1}\frac{1}{s} \tag{2.50}$$

We only need to retransform this expression into the time domain. Unfortunately, this expression is not given in the correspondence table A. A partial fraction expansion, though, leads to

$$y(s) = V\left[\frac{1}{s} - \frac{1}{s+\frac{1}{T}}\right] \tag{2.51}$$

According to the Addition Theorem, we can retransform the two summands inside the brackets separately into the time domain. $\frac{1}{s}$ is the Laplace transform of the step function $s(t) = 1$, and $\frac{1}{s+\frac{1}{T}}$ the transform of the function $s(t)e^{-t/T} = e^{-t/T}$, as can be found in the correspondence table. This gives the already known function for the step response of the first order lag

$$y(t) = V\left[1 - e^{-\frac{t}{T}}\right] \qquad \text{for } t \geq 0 \tag{2.52}$$

Regarding the step response of a transfer element for a rational transfer function, a general term can be given. First we need to split up both the polynomial for the numerator and for the denominator from (2.44) into linear factors:

$$G(s) = \frac{b_m}{a_n}\frac{\prod\limits_{\mu=1}^{m}(s-n_\mu)}{\prod\limits_{\nu=1}^{n}(s-p_\nu)} \qquad \text{with } m \leq n \tag{2.53}$$

Since all the coefficients of $G(s)$ are real-valued, all the poles p_ν and zeros n_μ will be either real-valued or complex conjugated pairs.

The following equation holds for the step response $y(s)$ (because of $y(s) = \frac{1}{s}G(s)$):

$$y(s) = \frac{b_m}{a_n} \frac{\prod_{\mu=1}^{m} (s - n_\mu)}{\prod_{\nu=1}^{n+1} (s - p_\nu)} \qquad \text{with } p_{n+1} = 0 \tag{2.54}$$

For further discussions we assume that the polynomials of the numerator and denominator of $y(s)$ are prime to each other. Otherwise we would have to reduce them to their lowest terms. In $y(s)$,the numerator's degree will in any case be lower than the denominator's degree, because of the additional pole $p_{n+1} = 0$. If we take into account that multiple poles may also occur in $y(s)$, then—after suitably renaming the poles—our partial fraction expansion will be

$$y(s) = \sum_{\lambda=1}^{i} \sum_{\nu=1}^{n_\lambda} \frac{r_{\lambda\nu}}{(s - s_\lambda)^\nu} \qquad \text{with } n + 1 = \sum_{\lambda=1}^{i} n_\lambda \tag{2.55}$$

i is the number of different poles, and n_λ their corresponding multiplicity. With the aid of the correspondence table A, we can perform the retransformation into the time domain. For each single summand using the Addition and Delay Theorems, we obtain the following term:

$$\frac{r_{\lambda\nu}}{(s - s_\lambda)^\nu} \longleftrightarrow \frac{r_{\lambda\nu}}{(\nu - 1)!} t^{\nu-1} e^{s_\lambda t} \tag{2.56}$$

This way, every single pole s_λ contributes a term

$$\sum_{\nu=1}^{n_\lambda} \frac{r_{\lambda\nu}}{(\nu - 1)!} t^{\nu-1} e^{s_\lambda t} = h_\lambda(t) e^{s_\lambda t} \tag{2.57}$$

to the step response in the time domain, according to its multiplicity n_λ. This term is a product of a polynomial $h_\lambda(t)$ of degree $n_\lambda - 1$, and an exponential function. Hence, we get the step response for the overall plant

$$y(t) = \sum_{\lambda=1}^{i} h_\lambda(t) e^{s_\lambda t} \tag{2.58}$$

For purely real-valued poles with negative real part, the term $h_\lambda(t)e^{s_\lambda t}$ will vanish for increasing values of t, as the exponential function converges faster towards zero than any finite power of t. In contrast to this, if $\mathrm{Re}(s_\lambda) > 0$, then the expression will exceed all limits. For any complex conjugated pole pair we can sum up the corresponding terms in a way similar to the one for the second order lag (equation (2.21)). This way, every pair of poles of this type characterizes an oscillating part of the system. In a similar way as for the purely real-valued poles, those oscillations will be increasing or decaying, depending on the real parts of the pole pairs.

For poles $s_\lambda = 0$, the exponential function will deliver the value 1, and can therefore be neglected. Only the polynomial remains. If zero is not one of the plant's poles, then, because of the step function $\frac{1}{s}$, $y(s)$ will have only one single pole for this value. The corresponding polynomial $h_\lambda(t)$ will therefore be of degree zero, i.e. a constant value. Accordingly, the corresponding expression $h_\lambda(t)e^{s_\lambda t}$ will be constant, too. In the case that all other poles have a negative real part—where, as already stated, the amount contributed to the step response vanishes for increasing values of t— this constant value will be the final value of the step response. On the other hand, if the plant contains at least one pole at $s_\lambda = 0$, the degree of $h_\lambda(t)$ is greater than zero and the expression will exceed all limits.

A pair of poles with $\mathrm{Re}(s_\lambda) = 0$ and $\mathrm{Im}(s_\lambda) \neq 0$ will cause an oscillation with a constant amplitude, according to equation (2.21). If this pair of poles occurs with a multiplicity greater than 1, then the degree $h_\lambda(t)$ for the polynomial will be greater than zero, and the product $h_\lambda(t)e^{s_\lambda t}$ will exceed all limits for this case, too.

The system's transient behavior is obviously completely characterized by the poles of the transfer function. Summing up the points mentioned above, we can say that the step response will always converge towards a finite value if all the poles of the transfer function have a negative real part.

Interestingly, we can as well compute the initial and final value of the step response using the Limit Theorems of the Laplace transformation. Using the Limit Theorem of the Laplace transformation (2.34), the generalized transfer function (2.44) and the formula for the step response of a linear transfer element (2.49) the final value of a step response can be given by:

$$\lim_{t\to\infty} y(t) = \lim_{s\to 0} sy(s) = \lim_{s\to 0} s\frac{1}{s}G(s) = \lim_{s\to 0} G(s) = \frac{b_0}{a_0} \tag{2.59}$$

and analogously for the initial value:

$$\lim_{t\to 0} y(t) = \lim_{s\to\infty} G(s) = 0 \qquad \text{for } m < n \tag{2.60}$$

We can therefore obtain the final value of the step response as well by reading off the coefficients of the transfer function.

In later chapters we will show that the computation of signal functions is neither a requirement for designing controllers nor for analyzing control loops, as already the analysis of the transfer function will deliver all the necessary information concerning the plant's stability and transient behavior. But anyway, having mentioned the form of the step responses is necessary to understand the later discussions.

2.3.5 Simplification of a Transfer Function

We close this chapter with a discussion of the possibility of simplifying a transfer function. Especially if the transfer function will be used for the design of a controller, approximating it by a function that is as simple as possible might give advantages for the controller design. Let a transfer function consisting of a rational part and a delay element be given. Let the rational part have poles with negative real parts only, i.e. the step response has a finite final value:

$$G(s) = \frac{b_m s^m + b_{m-1}s^{m-1} + ... + b_1 s + b_0}{a_n s^n + a_{n-1}s^{n-1} + ... + a_1 s + a_0} e^{-T_L s} \qquad \text{with } m < n \tag{2.61}$$

Resolving this into linear factors gives

$$G(s) = \frac{b_0}{a_0} \frac{\prod\limits_{\mu=1}^{m} (T_{z\mu} s + 1)}{\prod\limits_{\nu=1}^{n} (T_{n\nu} s + 1)} e^{-T_L s} \tag{2.62}$$

We can now expand the exponential function and the factors of the numerator into power series:

$$G(s) = \frac{b_0}{a_0} \frac{1}{\left[\prod\limits_{\mu=1}^{m} (1 + \sum\limits_{i=1}^{\infty} (-T_{z\mu} s)^i)\right] \left[\prod\limits_{\nu=1}^{n} (T_{n\nu} s + 1)\right]} \frac{1}{1 + \sum\limits_{\lambda=1}^{\infty} \frac{(T_L s)^\lambda}{\lambda!}} \tag{2.63}$$

Multiplying out produces a new series for the denominator

$$G(s) = \frac{b_0}{a_0} \frac{1}{1 + (-\sum\limits_{\mu=1}^{m} T_{z\mu} + \sum\limits_{\nu=1}^{n} T_{n\nu} + T_L)s + ...} \tag{2.64}$$

Truncating after the first term of the sequence gives an approximating first order lag with the *equivalent time constant* T_e:

$$G(s) \approx \frac{b_0}{a_0} \frac{1}{1 + T_e s} \tag{2.65}$$

where

$$T_e = \sum_{\nu=1}^{n} T_{n\nu} - \sum_{\mu=1}^{m} T_{z\mu} + T_L = a_1 - b_1 + T_L \tag{2.66}$$

With this formula, it is easy to compute the equivalent time constant using the coefficients of the transfer function without resolving the function into linear factors.

If we look at the step response of this approximating first order lag, we can achieve a good comparison with the original plant (fig. 2.22). It can be shown that the so-called *integral control error*

$$\int_0^\infty \left[\frac{b_0}{a_0} - y(t)\right] dt \qquad \text{with } y(t) = \text{step response} \tag{2.67}$$

i.e. the area between the step response and its final value, will be the same for both cases. Because of this fact it is possible to construct an approximating first order lag if only the measured step response but not the transfer function of the original plant is available.

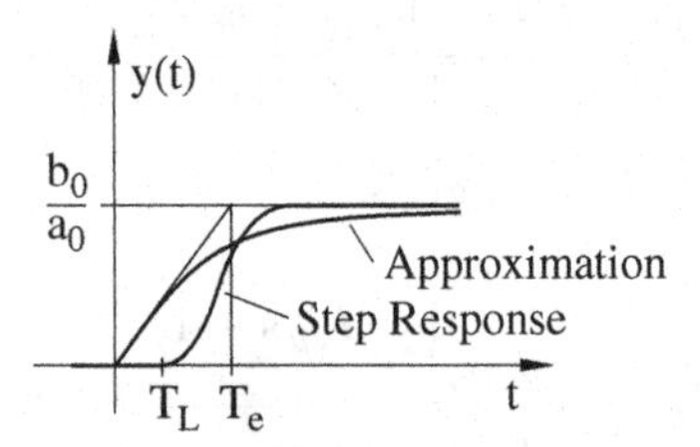

Fig. 2.22. Illustration to the equivalent time constant

Depending on how accurate the approximation has to be, we might as well truncate the sequence in equation (2.64) after a higher term. However, we should keep in mind that it will still be merely an approximation. Dynamic processes occurring in a plant of degree n cannot be fully characterized by a transfer function of a degree less than n. It might have severe effects, especially regarding the stability of the system. While a first order lag will cause no problems, the plant behind this approximation might be close to instability, and a controller, which will work fine for the first order lag might form an instable system with the real plant. We should therefore handle the given approximation with great care. Usually, a plant of lower degree will be a good approximation for a plant of higher degree only for the range of low signal frequencies. The higher the frequencies, the more inexact the approximation will be. But if for a controlled system the range of frequencies used lies within the range of lower frequencies, then using this approximation will help to come up with a more comprehensible transfer function and to a much more simple controller design.

2.4 Frequency Response

2.4.1 Introduction to the Frequency Response

If we know a plant's transfer function, it is easy to construct a suitable controller using this information. If we cannot develop the transfer function by theoretical considerations, we might as well use statistical methods on the basis of a sufficient amount of measured values to determine it. This method requires the use of a computer, a means which was not available in former times. Therefore, in those days a different method was frequently used in order to describe a plant's dynamic behavior, the *frequency response*. As we will explain in the following, the frequency response can easily be measured. Its good graphical representation leads to a clear method in the design process for simple PID controllers. Not to mention that several criteria for the stability, which are as well used in connection with fuzzy controllers, root in the frequency response based characterization of a plant's behavior.

The easiest way might be to define the frequency response to be the transfer function of a linear transfer element with purely imaginary values for s. Therefore we only have to replace the complex variable s of the transfer function by a purely imaginary variable $j\omega$: $G(j\omega) = G(s)|_{s=j\omega}$. The frequency response is therefore a complex-valued function of the parameter ω. Due to the restriction of s to purely imaginary values, the frequency response represents only a part of the transfer function, but a part with special properties, as the following theorem shows:

Theorem 2.3 *If a linear transfer element has the frequency response $G(j\omega)$, then its response to the input signal $x(t) = a \sin \omega t$ will be—after all initial transients have settled down—the output signal*

$$y(t) = a\,|G(j\omega)|\,\sin\left(\omega t + \varphi(G(j\omega))\right) \tag{2.68}$$

if the following equation holds:

$$\int_0^\infty |g(t)|dt < \infty \tag{2.69}$$

$|G(j\omega)|$ is obviously the ratio of the output sine amplitude to the input sine amplitude ((transmission) gain or amplification). $\varphi(G(j\omega))$ is the phase of the complex quantity $G(j\omega)$ and shows the delay of the output sine in relation to the input sine (phase lag). $g(t)$ is the impulse response of the plant. In case the integral given in (2.69) does not converge, we have to add the term $r(t)$ to the right hand side of (2.68), which will, even for $t \to \infty$, not vanish.

The proof can be found in [43]. Examination of this theorem makes clear what kind of information about the plant the frequency response gives: The frequency response characterizes the system's behavior for any frequence of the input signal. Because of the linearity of the transfer element, the effects caused by the single frequencies of the input signal do not interfere with each other. This way, we are now able to predict the resulting effects at the system output for every single signal component separately, and we can finally superimpose these effects to predict the overall system output.

In contrast to the coefficients of a transfer function, we can measure amplitude and phase lag of the frequency response directly: The plant gets excited by a sinusoidal input signal of a certain frequency and amplitude. After all initial transients have settled down we get a sinusoidal signal at the plant output, whose phase position and amplitude differ from the input signal. Both quantities can be measured, and according to (2.68), this will also instantly provide the amplitude and phase lag of the frequency response $G(j\omega)$. In this manner we can construct a table for different input frequencies which will give the principle curve of the frequency response. Taking measurements for negative values of ω, i.e. for negative frequencies, is of course not possible; but it is not necessary either, as for rational transfer functions with real coefficients as well as for delay elements, $G(j\omega)$ will be conjugate complex to $G(-j\omega)$. Now, knowing that the function $G(j\omega)$ for $\omega \geq 0$ already contains all the information needed, we can omit an examination of negative values of ω.

2.4.2 Nyquist Plot

An extraordinary good representation of the frequency response is the one of a curve in the complex plain. This is the so-called *Nyquist plot*. With regard to the facts just mentioned, we can restrict the representation of $G(j\omega)$ to positive values of ω. We now demonstrate appearance and interpretation of some Nyquist plots by a few examples (Fig. 2.23).

The frequency response of a first order lag is, according to (2.41):

$$G(j\omega) = \frac{V}{Tj\omega + 1} = \frac{V}{\sqrt{\omega^2 T^2 + 1}} \, e^{-j \arctan \omega T} \tag{2.70}$$

The corresponding Nyquist plot is the semi circle shown in figure 2.23. Every point of the plot represents the complex value $G(j\omega_1)$ for a certain frequency ω_1. But it can also be interpreted as the ending point of a vector, that starts in the origin. Any vector has got a certain length, that is equal to the transmission gain of the transfer element for the frequency ω_1, and a phase, that is equal to the phase lag for that frequency.

How can we interpret this curve? For example, it can be seen that the gain for dc signals of frequency $\omega = 0$ will have the value V. Because of equation (2.59), this is also the final value for the step response. This is not a mere coincidence, as for a step in the input signal a pure dc signal transfer will occur after all the initial transients have settled down. Furthermore, the Nyquist plot shows that for higher frequencies the gain will diminish more and more, until finally approximating zero, which is equal to the initial value of the step response according to (2.60). All in all, the first order lag represents a low-pass filter because of the decrease in amplification for higher frequencies. Besides this, the higher the frequencies, the greater the phase lag caused by the first order lag, what can be easily seen from the plot with taking into account, that all curve points are ending points of vectors as mentioned before. For $\omega = 0$, $G(j\omega)$ will be purely real and the phase lag will be zero, whereas for increasing frequencies the phase lag will converge towards $-\pi/2$. Therefore,

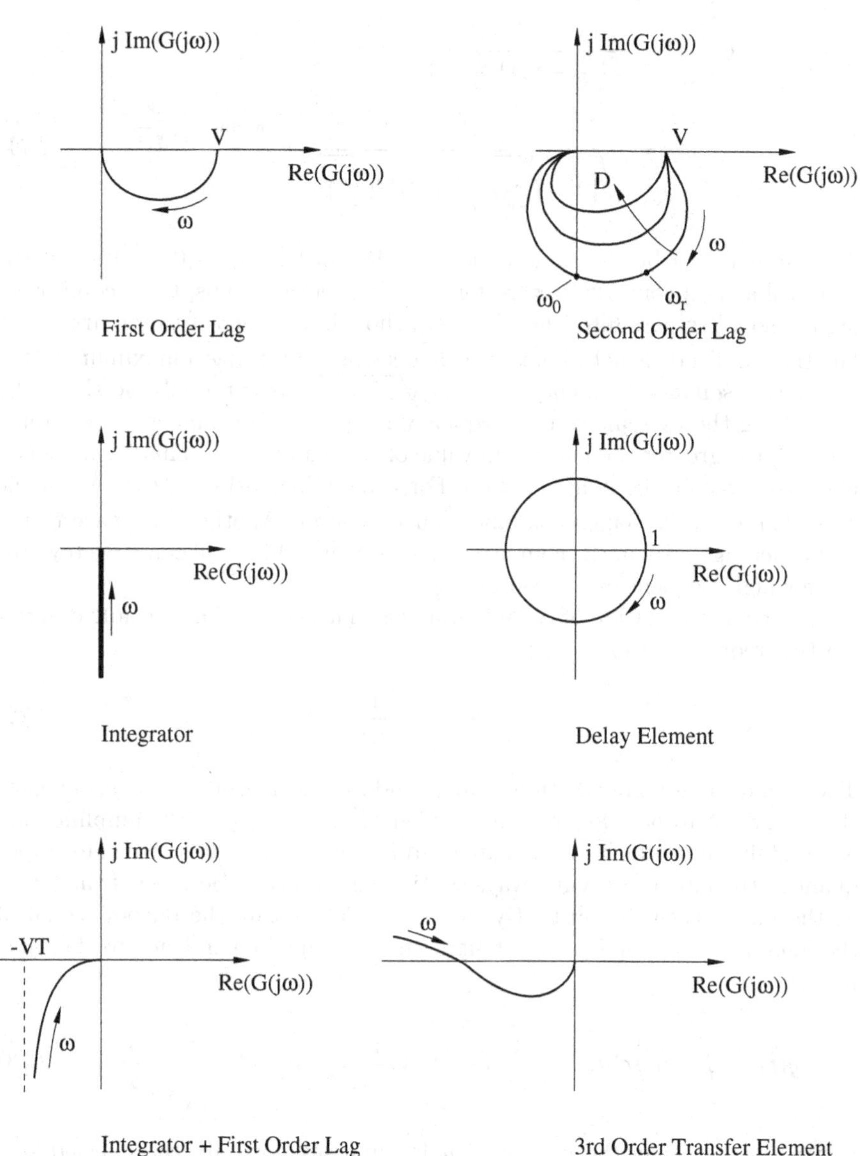

Fig. 2.23. Nyquist plots of linear transfer elements

a sinusoidal oscillation of high frequency will become delayed by almost a one-fourth period.

From equation (2.43) it follows that the frequency response of the second order lag is given by:

$$
\begin{aligned}
G(j\omega) &= \frac{V}{-\frac{\omega^2}{\omega_0^2} + j2D\frac{\omega}{\omega_0} + 1} \\
&= \frac{V}{\sqrt{\left(1 - (\frac{\omega}{\omega_0})^2\right)^2 + 4D^2(\frac{\omega}{\omega_0})^2}} e^{-j \arctan \frac{2D\frac{\omega}{\omega_0}}{1-(\frac{\omega}{\omega_0})^2}}
\end{aligned} \tag{2.71}
$$

The dc gain will be the same as for the first order lag, $G(0) = V$, and the gain will as well converge to zero for high frequencies. Thus, the second order lag is also a low-pass filter. For $D < \frac{1}{\sqrt{2}}$, though, in contrast to the first order lag, the amplitude is not monotonously decreasing, but has a maximum at the so-called *resonance frequency* $\omega_r = \omega_0\sqrt{1 - 2D^2}$, as can easily be shown by calculating the extreme of the frequency response. The smaller the damping ratio D, the greater the maximum value of the gain will be. This phenomenon is called a *resonance magnification.* For a damping ratio of $D \geq \frac{1}{\sqrt{2}}$ on the other hand, no resonance magnification will occur. Another difference to the first order lag exists in the course of the phase lag, which will converge towards $-\pi$ for high frequencies instead of $-\frac{\pi}{2}$.

The characterization of the integrator is much easier. The formula describing the frequency response is:

$$G(j\omega) = \frac{1}{j\omega} \tag{2.72}$$

The gain decreases corresponding to ω, and for this reason we can also regard the integrator to be a low-pass filter. Concerning dc signals the amplification is infinitely high, which can be explained by the fact that for a constant input quantity the integrator will always continue integrating. Because of the factor $\frac{1}{j}$, the phase lag will constantly be $-\pi/2$. Calculating the output signal of the integrator for a sine as input signal in the time domain leads to the same result:

$$y(t) = \int_0^t \sin(\omega\tau)d\tau = -\frac{1}{\omega}\cos(\omega t) + \frac{1}{\omega} = \frac{1}{\omega}\sin(\omega t - \frac{\pi}{2}) + \frac{1}{\omega} \tag{2.73}$$

It can be seen that the output signal comprises not only the delayed sine with reduced amplitude, but also a constant term $\frac{1}{\omega}$, even though according to theorem 2.3 we should expect only a pure sine at the plant output. The reason for this is that the integrator is the only case of the given examples which does not meet the prerequisites of the theorem, as the integral of its impulse response does not converge. $\frac{1}{\omega}$ therefore represents the remainder $r(t)$, which we introduced in theorem 2.3. It can be neglected, though, as it represents a constant value for the given case, and will have no further effects.

The analysis of the Nyquist plot of the delay element will be interesting. $G(s) = e^{-T_D s}$ causes the frequency response to be

$$G(j\omega) = e^{-jT_D\omega} \tag{2.74}$$

In this way, the gain will always be 1, and the phase lag will depend on the frequency and the time delay. We can easily explain this behavior: if we impress a stationary sine on a delay element then the amplitude of the signal will appear unchanged at the system output, but delayed by the delay time T_D. If we express this delay in terms of the angular distance, it will of course be the greater the higher the frequency of the signal is.

The next example is the Nyquist plot of a series connection of an integrator and a first order lag:

$$\begin{aligned} G(s) &= \frac{1}{s}\frac{V}{Ts+1} \\ G(j\omega) &= \frac{1}{j\omega}\frac{V}{Tj\omega+1} = \frac{-VT}{\omega^2T^2+1} + j\frac{-V}{\omega(\omega^2T^2+1)} \end{aligned} \tag{2.75}$$

For $\omega \to 0$, the real part converges towards $-VT$, and the imaginary part towards $-\infty$. For $\omega \to \infty$, both parts converge towards zero. This is just another low-pass filter.

Finally, we examine a slightly more complicated, rational transfer function:

$$G(s) = \frac{(s+s_1)^2}{(s+s_2)s^2} \qquad \text{with } s_1 \gg s_2 > 0 \tag{2.76}$$

or, alternatively,

$$G(j\omega) = \frac{s_1^2+\omega^2}{\omega^2\sqrt{\omega^2+s_2^2}}\, e^{2\arctan\frac{\omega}{s_1} - \arctan\frac{\omega}{s_2} - \pi} \tag{2.77}$$

The evaluation of the second formula is easy, if we consider every factor of the transfer function separately. For example, the factor $(j\omega + s_1)$ contributes $\sqrt{\omega^2+s_1^2}$ to the gain and $\arctan\frac{\omega}{s_1}$ to the phase lag of the frequency response. The single contributions to the gain have to be multiplicated, and the contributions to the phase lag have to be summarized.

The course of values which the gain takes obviously starts at infinity and ends at zero. The phase of the function will be smaller than $-\pi$ for small values of ω, since the function $\arctan\frac{\omega}{s_2}$ grows faster than $2\arctan\frac{\omega}{s_1}$ because of $s_1 \gg s_2$. For low frequencies, the curve runs through the second quadrant. For $\omega \to \infty$, though, the arctan functions converge towards $\frac{\pi}{2}$, letting the entire phase converge to $2\frac{\pi}{2} - \frac{\pi}{2} - \pi = -\frac{\pi}{2}$.

Having discussed these examples, we can state a few general results concerning the transfer function according to equation (2.61) (i.e. rational transfer functions with a delay time): If the degree of the numerator polynomial is less than the degree of the denominator polynomial, then the curves will always end at the origin of the complex plane, as the gain converges obviously to zero. If no pole exists at zero, the starting point will always lie at a finite real value, as it is the case for the first order or second order lag. If there exists at least one pole at zero, then the Nyquist plot will start at infinity, embracing an angle of $-k\frac{\pi}{2}$, where k is the pole's order. The courses of values which the phase lag and gain take do not always have to be monotonously decreasing, though. An example for this was already given in our examination of the second order lag. The course of the Nyquist plot depends on the distribution of the poles and zeros of the transfer function.

2.4.3 Bode Diagram

The disadvantage of Nyquist plots is the missing possibility of cross referencing single points of the curve and certain frequencies. Here, an alternative for representing the frequency response exists, the *Bode diagram* ([19]). Gain and phase lag are given in two separate diagrams, one on top of the other one, both as functions of the frequency. Gain and frequency are given in logarithmic scales, the phase lag in linear scale.

In order to give an example, figure 2.24 shows the Bode diagrams of the first order and second order lag. Initial and final values of the phase lag courses can be seen clearly. We can as well obtain these courses immediately from equation (2.70) and (2.71), or the corresponding Nyquist plots. The same holds for the initial values of the gain courses. Zero, the final value of these courses, lies at negative infinity, because of the logarithmic representation. Furthermore, for the second order lag the resonance magnification for smaller dampings can be see at the resonance frequency ω_0.

Examining a dynamic system by analyzing the Nyquist plot and the Bode diagram will always be of help, if it is only possible for us to measure the frequency response of an unknown plant, but no model of the plant is available.

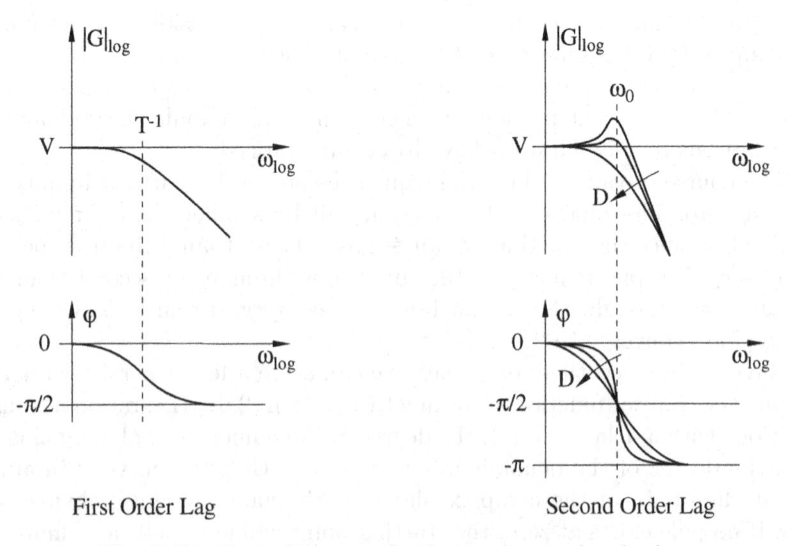

Fig. 2.24. Bode diagram of a first and second order lag

Even with a transfer function being given, a graphical analysis using these two diagrams might be clearer, and of course it can be tested more easily than, for example, a numerical analysis done by a computer. It will almost always be easier to estimate the effects of changes in the values of the parameters of the system, if we use a graphical approach instead of a numerical one. For this reason, today every control design software tool provides the possibility of computing the Nyquist plot or the Bode diagram for a given transfer function by merely clicking on a button.

Another important tool for the analysis of linear controlled plants should not be unmentioned, the *root locus*. In this method, the positions of the poles in the complex plane with respect to certain parameters are represented. The behavior of the plant is therefore characterized by a direct examination of the poles, not indirectly as in the case of Nyquist plot and Bode diagram, where the frequency response is used for this purpose. In contrast to those methods, root loci can not be used for the analysis of nonlinear control systems. And since these systems are the main field of application for fuzzy controllers, we will refrain from discussing root loci in this book.

2.5 Stability of Linear Systems

2.5.1 A Definition of Stability

In this chapter, we discuss the term *stability* with regard to linear systems. Let the linear system be given by the transfer function

$$\begin{aligned} G(s) &= \frac{b_m s^m + b_{m-1}s^{m-1} + ... + b_1 s + b_0}{a_n s^n + a_{n-1}s^{n-1} + ... + a_1 s + a_0} e^{-T_L s} \qquad \text{where } m \le n \\ &= \frac{b_m}{a_n} \frac{\prod\limits_{\mu=1}^{m} (s - n_\mu)}{\prod\limits_{\nu=1}^{n} (s - p_\nu)} e^{-T_L s} = V \frac{\prod\limits_{\mu=1}^{m} (\frac{-s}{n_\mu} + 1)}{\prod\limits_{\nu=1}^{n} (\frac{-s}{p_\nu} + 1)} e^{-T_L s} \end{aligned} \tag{2.78}$$

with the gain

$$V = \frac{b_0}{a_0} \tag{2.79}$$

First of all we have to explain what we mean by stability of a system. Several possibilities to define the term exist, two of which we are going to discuss now. A third definition by the Russian mathematician Lyapunov will be presented later on. The first definition is based on the system's step response:

Definition 2.4 *A system is called stable if, for $t \to \infty$, its step response converges towards a finite value. Otherwise it is said to be instable.*

That the unit step function was chosen to stimulate the system causes no restrictions, since if the height of the step is changed by the factor k, then the values at the system output will change by the same factor k, too, according to the linearity of the system. Convergence towards a finite value is therefore preserved.

A motivation for this definition may be the following illustrating idea: If a system converges towards a finite value after the strong stimulation that a step in the input signal represents, it can be assumed that it will not get stuck in permanent oscillations for other kinds of stimulations.

It is plain to see that according to this definition the first order and the second order lag will be stable, and that the integrator will be instable.

Another definition pays attention to the possibility that the input quantity may be subject to permanent variations:

Definition 2.5 *A linear system is called stable if for an input signal with limited amplitude, its output signal will also show a limited amplitude. This is the BIBO-Stability (bounded input - bounded output).*

Immediately the question about the connection between the two definitions arises, which we now examine briefly. Starting point for the discussion is the convolution integral (see eq. (2.48)), which gives the relation between the system's input and output quantity ($g(t)$ is the impulse response):

$$y(t) = \int_{\tau=0}^{t} g(t-\tau)x(\tau)d\tau = \int_{\tau=0}^{t} g(\tau)x(t-\tau)d\tau \tag{2.80}$$

$x(t)$ is bounded if and only if $|x(t)| \leq k$ holds (with $k > 0$) for all t. This implies:

$$|y(t)| \leq \int_{\tau=0}^{t} |g(\tau)||x(t-\tau)|d\tau \leq k \int_{\tau=0}^{t} |g(\tau)|d\tau \tag{2.81}$$

Now, with absolute convergence of the integral of the impulse response,

$$\int_{\tau=0}^{\infty} |g(\tau)|d\tau = c < \infty \tag{2.82}$$

$y(t)$ will be limited by kc, too, and therefore the entire system will be BIBO-stable. In the same way it can be shown that the integral (2.82) converges absolutely for every BIBO-stable system. BIBO stability and the absolute convergence of the impulse response integral are equivalent system properties.

Now we have to find the conditions on which the system will be stable in the sense of a finite step response (Definition 2.4): Concerning the step response of a system in the frequency domain,

$$y(s) = G(s)\frac{1}{s}$$

holds (cf. 2.49)). If we interpret the factor $1/s$ as an integration (instead of being the Laplace transform of the step signal), we obtain

$$y(t) = \int_{\tau=0}^{t} g(\tau)d\tau \tag{2.83}$$

in the time domain for $y(0) = 0$. $y(t)$ converges towards a finite value only if the integral is convergent:

$$\int_{\tau=0}^{\infty} g(\tau)d\tau = c < \infty \tag{2.84}$$

Convergence is obviously a weaker criterion than absolute convergence. Hence, every BIBO-stable system will have a finite step response. To treat stability always in the sense of BIBO-stability is tempting, as this stronger definition makes further differentiations unnecessary. On the other hand, we can simplify the following considerations a lot if we use the finite-step-response-based definition of stability. In addition to this, both definitions are equivalent regarding purely rational transfer functions anyway. Therefore, from now on we will think of stability as characterized in definition 2.4.

Sometimes stability is also defined by demanding that the impulse response $g(t)$ has to converge towards zero for $t \to \infty$. One look at the integral in (2.84) shows that this criterion is necessary but not sufficient for stability according to definition 2.4, so that 2.4 is the stronger definition. If one can prove a finite step response, then the impulse response will definitely converge towards zero.

2.5.2 Stability of a Transfer Function

If we want to avoid having to explicitly compute the step response of a system in order to prove its stability, then a direct examination of the system's transfer function, trying to derive criteria for the stability, seems to suggest itself. This is relatively easy regarding all the ideas we developed so far concerning the step response of a rational transfer function. The following theorem holds:

Theorem 2.6 *A transfer element with a rational transfer function is stable in the sense of definition 2.4 if and only if all poles of the transfer function have a negative real part.*

According to equation (2.58), the step response of a rational transfer element is given by:

$$y(t) = \sum_{\lambda=1}^{i} h_\lambda(t)e^{s_\lambda t} \tag{2.85}$$

For every pole s_λ of multiplicity n_λ we obtain a corresponding summand $h_\lambda(t)e^{s_\lambda t}$, where $h_\lambda(t)$ is a polynomial of degree $n_\lambda - 1$. For a pole with a negative real part, this summand vanishes for increasing t, as the exponential function converges faster towards zero than the polynomial $h\lambda(t)$ can increase. If all poles of the transfer function have a negative real part, then all corresponding terms disappear. Only the summand $h_i(t)e^{s_i t}$ for the simple pole $s_i = 0$ remains, due to the step function. The polynomial $h_i(t)$ is of degree $n_i - 1 = 0$, i.e. a constant, and the exponential function is also reduced to a constant. This way, this summand forms the finite final value of the step function, and the system is stable.

We omit the proof in opposite direction, i.e. that a system is instable if at least one pole has a nonnegative real part, as this would not lead to further insights. It is interesting that theorem 2.6 holds as well for systems with delay time according to (2.78). The proof for this last statement will also be omitted.

In general, the form of the initial transients as a reaction to excitations from outside will also be of interest besides the fact of stability. If a plant has, among others, a complex conjugate pair of poles $s_\lambda, \bar{s_\lambda}$, then according to equation (2.22) the ratio $|\mathrm{Re}(s_\lambda)|/\sqrt{\mathrm{Re}(s_\lambda)^2 + \mathrm{Im}(s_\lambda)^2}$ is equal to the damping ratio D and therefore responsible for the form of the initial transient corresponding to this pair of poles. In practical applications one will therefore pay attention not only to that the system's poles have a negative real part, but also to the damping ratio D having a sufficiently high value, i.e. that a complex conjugate pair of poles lies at a reasonable distance to the axis of imaginaries.

2.5.3 Stability of a Control Loop

The system whose stability should be determined will in most cases be a closed control loop, as shown in figure 2.2. A simplified structure is given in figure 2.25. Let the transfer function of the control unit be $K(s)$, the plant be given by $G(s)$ and the metering element by $M(s)$. In order to keep further derivations simple, we set $M(s)$ to 1, i.e. we neglect the dynamic behavior of the metering element; for single cases, though, it should usually be no problem to take the metering element also into consideration.

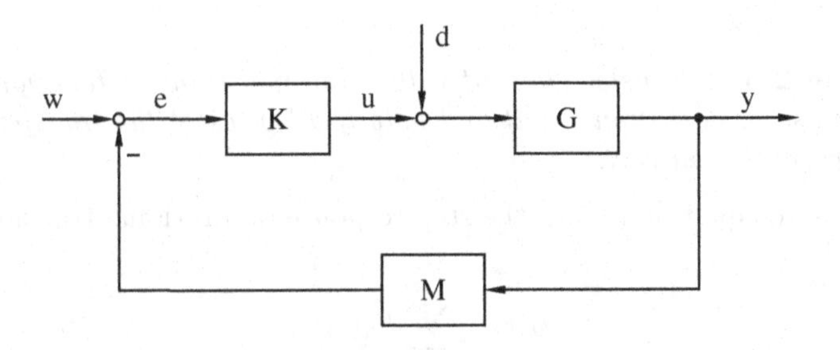

Fig. 2.25. Closed-loop system

We sum up disturbances, which might affect the closed-loop system at virtually any point, into a single disturbance load which we impress at the plant input. This step simplifies the theory without making the situation for the controller easier than it would be in practical applications. Choosing the plant input as the point where the disturbance affects the plant is most disadvantageous: the disturbance can affect the plant and no countermeasures can be applied, as the controller can only counteract after changes at the system output.

In order to be able to apply the criteria of stability to this system we first have to compute the transfer function that describes the transfer characteristic of the entire system between the input quantity w and the output quantity y. This is the transfer function of the closed loop, which is sometimes referred to as the *reference (signal) transfer function.* To calculate it, we initially set d to zero. In the frequency domain we get

$$\begin{aligned} y(s) &= G(s)u(s) = G(s)K(s)(w(s) - y(s)) \\ T(s) &= \frac{y(s)}{w(s)} = \frac{G(s)K(s)}{G(s)K(s)+1} \end{aligned} \tag{2.86}$$

Analogously, we can calculate a *disturbance transfer function*, that describes the transfer characteristic between the disturbance d and the output quantity y:

$$S(s) = \frac{y(s)}{d(s)} = \frac{G(s)}{G(s)K(s)+1} \tag{2.87}$$

The term $G(s)K(s)$ has a special meaning: if we remove the feedback loop, then this term represents the transfer function of the resulting open circuit. Therefore, $G(s)K(s)$ is sometimes called the *open-loop transfer function.* The gain V of this function (see eq. (2.78)) is called *open-loop gain.*

It can be seen that the reference transfer function as well as the disturbance transfer function have the same denominator $G(s)K(s)+1$. On the other hand, according to theorem 2.6, it is the denominator of the transfer function that determines stability. From this it follows that only the open-loop transfer function affects the stability of a system, but not the point of application of an input quantity. We can therefore restrict an analysis of stability to an examination of the term $G(s)K(s) + 1$.

Now, since both the numerator and the denominator of the two transfer functions $T(s)$ and $S(s)$ are obviously relatively prime to each other, the zeros of $G(s)K(s) + 1$ are the poles of these functions, and as a direct consequence of theorem 2.6 we can state:

Theorem 2.7 *A closed-loop system with the open-loop transfer function $G(s)K(s)$ is stable if and only if all solutions of the characteristic equation*

$$G(s)K(s) + 1 = 0 \tag{2.88}$$

have a negative real part.

Computing these zeros in an analytic way will no longer be possible if the degree of the plant is greater than two, or if an exponential function forms a part of the open-loop transfer function. The exact positions of the zeros, though, are not needed in the analysis of stability. Only the fact whether the solutions have a positive or negative real part is of importance. For this reason, in the history of control theory criteria of stability were developed which could be used to determine precisely this without having to do complicated calculations.

2.5.4 Criterion of Cremer, Leonhard and Michailow

Let us first discuss a criterion which was developed independently by Cremer [33], Leonhard [109] and Michailow [123] during the years 1938 to 1947. It is usually named after those scientists. The focus of interest is the phase shift of the Nyquist plot of a polynomial with respect to the polynomial's zeros. Let a polynomial of the form

$$P(s) = s^n + a_{n-1}s^{n-1} + ... + a_1 s + a_0 = \prod_{\nu=1}^{n}(s - s_\nu) \tag{2.89}$$

be given. Setting $s = j\omega$ and substituting we obtain

$$\begin{aligned} P(j\omega) &= \prod_{\nu=1}^{n}(j\omega - s_\nu) = \prod_{\nu=1}^{n}(|j\omega - s_\nu|e^{j\varphi_\nu(\omega)}) \\ &= \prod_{\nu=1}^{n}|j\omega - s_\nu| \; e^{j\sum_{\nu=1}^{n}\varphi_\nu(\omega)} = |P(j\omega)|e^{j\varphi(\omega)} \end{aligned} \tag{2.90}$$

We can see, that the frequency response $P(j\omega)$ is the product of the vectors $(j\omega - s_\nu)$, where the phase $\varphi(\omega)$ is given by the sum of the angles $\varphi_\nu(\omega)$ of those vectors. Figure 2.26 shows the situation corresponding to a pair of complex conjugated zeros with negative real part and only one zero with a positive real part.

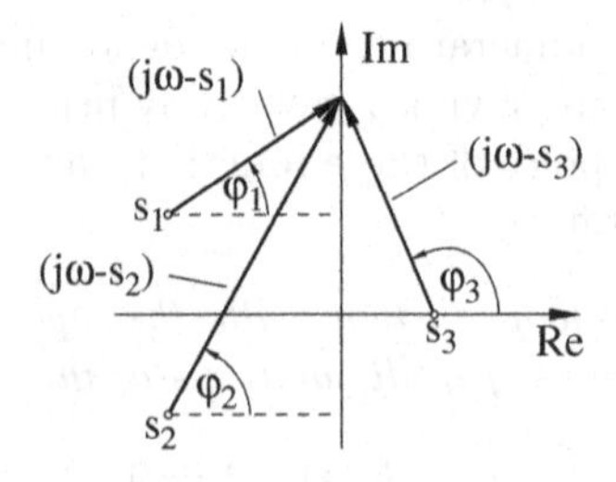

Fig. 2.26. Illustration to the Cremer-Leonhard-Michailow criterion

If the parameter ω traverses the interval $(-\infty, \infty)$, it causes the end point of the vectors $(j\omega - s_\nu)$ to move along the axis of imaginaries in positive direction. For zeros with negative real part, the corresponding angle φ_ν traverses the interval from $-\frac{\pi}{2}$ to $+\frac{\pi}{2}$, for zeros with positive real part the interval from $+\frac{3\pi}{2}$ to $+\frac{\pi}{2}$. For zeros lying on the axis of imaginaries the corresponding angle φ_ν initially has the value $-\frac{\pi}{2}$ and switches to the value $+\frac{\pi}{2}$ at $j\omega = s_\nu$.

We now analyze the phase shift of the frequency response, i.e. the entire course which the angle $\varphi(\omega)$ takes. This angle is just the sum of the angles $\varphi_\nu(\omega)$. Therefore every zero with a negative real part contributes an angle of $+\pi$ to the phase shift of the frequency response, and every zero with a positive real part the angle $-\pi$. Nothing can be said about zeros lying on the axis of imaginaries because of the discontinuous course which the values of the phase take. But we can immediately decide whether such zeros exist or not by looking at the Nyquist plot of the polynomial $P(s)$. If it has got a purely imaginary zero $s = s_\nu$, the corresponding Nyquist plot has to pass through the origin at the frequency $\omega = |s_\nu|$. This leads to the following theorem:

Theorem 2.8 *A polynomial $P(s)$ of degree n with real coefficients will have only zeros with negative real part if and only if the corresponding Nyquist plot does not pass through the origin of the complex plane and the phase shift $\Delta\varphi$ of the frequency response is equal to $n\pi$ for $-\infty < \omega < +\infty$. If ω traverses the interval $0 \leq \omega < +\infty$ only, then the phase shift needed will be equal to $\frac{n}{2}\pi$.*

We can easily prove the fact that for $0 \leq \omega < +\infty$ the phase shift needed is only $\frac{n}{2}\pi$—only half the value:

For zeros lying on the axis of reals, it is obvious that their contribution to the phase shift will be only half as much if ω traverses only half of the axis of imaginaries (from 0 to ∞). The zeros with an imaginary part different from zero are more interesting. Because of the polynomial's real-valued coefficients, they can only appear as a pair of complex conjugated zeros. Figure 2.27 shows such a pair with $s_1 = \bar{s_2}$ and $\alpha_1 = -\alpha_2$. For $-\infty < \omega < +\infty$ the contribution to the phase shift by this pair is 2π. For $0 \leq \omega < +\infty$, the contribution of s_1 is $\frac{\pi}{2} + |\alpha_1|$ and the one for s_2 is $\frac{\pi}{2} - |\alpha_1|$. Therefore, the overall contribution of this pair of poles is π, so also for this case the phase shift is reduced by one half if only the half axis of imaginaries is taken into consideration.

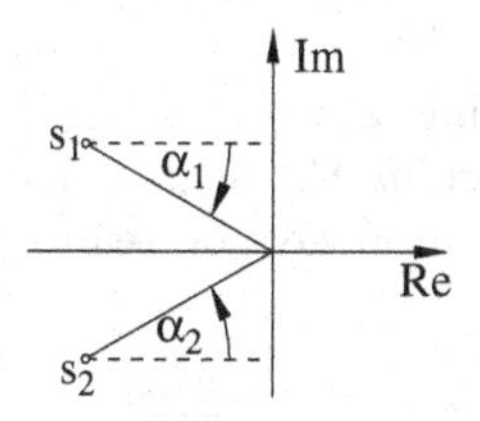

Fig. 2.27. Illustration to the phase shift for a complex conjugated pair of poles

Figure 2.28 gives an example of two Nyquist plots of fifth order polynomials. Since the Nyquist plot is the graphical representation of the frequency response, we can easily obtain the phase shift directly from this curve. To do so, we have to draw a vector connecting the origin with the Nyquist plot, as shown for curve 1. Then we have to find out how many times this vector revolves around the origin when following the course of the entire Nyquist plot. The angle received this way is the desired phase shift of the frequency response.

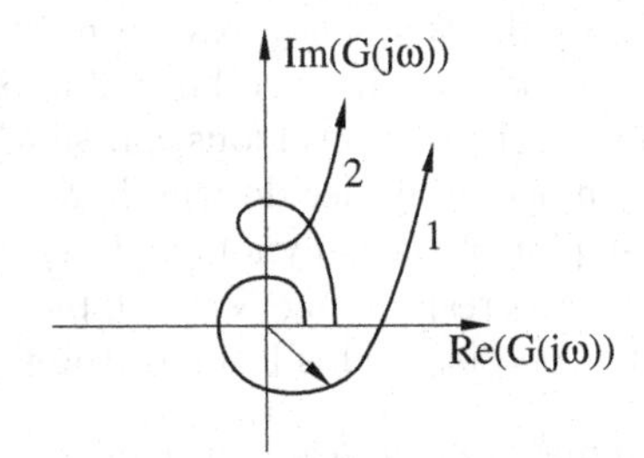

Fig. 2.28. Nyquist plot of fifth order polynomials

The total phase shift $\Delta\varphi$ of curve 1 is $5\frac{\pi}{2}$. The corresponding polynomial will therefore have zeros with a negative real part only. The phase shift of curve 2 is only $\frac{\pi}{2}$, although its initial and final angle are equal to those of curve 1. But as no revolutions around the origin occur in between those two angles, the total amount can be only the difference between the initial and the final value. The corresponding polynomial of this curve must have zeros with positive real part.

2.5.5 Nyquist Criterion

From theorems 2.7 and 2.8, we can now derive a very famous and elegant stability criterion, the *Nyquist criterion.* In the Nyquist criterion [151], the Nyquist plot of the open-loop transfer function $G(s)K(s)$ — which we can also measure easily, if the function is not available in an analytic form—is directly examined. Let

$$\frac{Z_k(s)}{N_k(s)} = G(s)K(s) \tag{2.91}$$

be given, with two polynomials $Z_k(s)$ and $N_k(s)$, which are relatively prime to each other. Let furthermore m, the degree of $Z_k(s)$, be at most equal to n, the degree of $N_k(s)$, which can always be realized for any kind of real-world physical systems. Because of

$$T(s) = \frac{G(s)K(s)}{1 + G(s)K(s)} = \frac{\frac{Z_k(s)}{N_k(s)}}{1 + \frac{Z_k(s)}{N_k(s)}} = \frac{Z_k(s)}{Z_k(s) + N_k(s)} \tag{2.92}$$

$N_g(s) = Z_k(s) + N_k(s)$ will be the denominator of the transfer function for the closed-loop system $T(s)$, and it will be of degree n, too. From this it follows that

$$1 + G(s)K(s) = 1 + \frac{Z_k(s)}{N_k(s)} = \frac{N_g(s)}{N_k(s)} \tag{2.93}$$

The phase of the frequency response $1+G(j\omega)K(j\omega)$ is the difference between the phase responses of the numerator and denominator polynomials:

$$\varphi_{1+GK}(\omega) = \varphi_{N_g}(\omega) - \varphi_{N_k}(\omega) \tag{2.94}$$

We therefore obtain

$$\Delta\varphi_{1+GK} = \Delta\varphi_{N_g} - \Delta\varphi_{N_k} \tag{2.95}$$

for the entire phase shift.

For the calculation of the phase shifts $\Delta\varphi_{N_g}$ and $\Delta\varphi_{N_k}$ according to theorem 2.8, we need to know the distribution of the zeros of the polynomials $N_g(s)$ and $N_k(s)$ in the complex plane. The zeros of $N_g(s)$ are the poles of the closed-loop transfer function. Let r_g out of those n poles lie in the right half of the complex s-plane, i_g lie on the axis of imaginaries and $n - r_g - i_g$ lie in the left half. Correspondingly, the zeros of $N_k(s)$ are just the poles of the open-loop transfer function. Out of these—again n—poles, let r_k lie to the right of the axis of imaginaries, i_g on and $n - r_k - i_k$ to the left of it.

Since i_g as well as i_k can be different from zero, the phases $\varphi_{N_g}(\omega)$ and $\varphi_{N_k}(\omega)$ may show a discontinuous curve for the values which the phase takes. This was already shown in the derivation of theorem 2.8. In order to avoid some problems, we examine only the continuous part of the phase shift. According to theorem 2.8, every zero with negative real part contributes $\frac{\pi}{2}$ to the entire phase shift of the Nyquist plot with $0 < \omega < \infty$, and every zero with a positive real part contributes $-\frac{\pi}{2}$:

$$\begin{aligned} \Delta\varphi_{N_g,\text{cont}} &= [(n - r_g - i_g) - r_g]\frac{\pi}{2} \\ \Delta\varphi_{N_k,\text{cont}} &= [(n - r_k - i_k) - r_k]\frac{\pi}{2} \end{aligned} \tag{2.96}$$

This gives

$$\begin{aligned} \Delta\varphi_{1+GK,\text{cont}} &= [(n - r_g - i_g) - r_g]\frac{\pi}{2} - [(n - r_k - i_k) - r_k]\frac{\pi}{2} \\ &= [2(r_k - r_g) + i_k - i_g]\frac{\pi}{2} \end{aligned} \tag{2.97}$$

for the continuous part of the phase shift of $1 + G(j\omega)K(j\omega)$. If we ask for stability of the closed-loop system, only poles with negative real part may occur. This implies $r_g = i_g = 0$, and accordingly

$$\Delta\varphi_{1+GK,\text{cont}} = r_k\pi + i_k\frac{\pi}{2} \tag{2.98}$$

Whether the open-loop transfer function is stable or instable is unimportant. We only need to know the number of poles to the right and lying on the axis of imaginaries.

We have to keep in mind that only the continuous part of the phase shift is covered by this formula. Zeros of $N_g(s)$ lying on the axis of imaginaries cause discontinuous changes in the phase shift. We can therefore exclude that $N_g(s)$ has zeros with positive real parts by analyzing the continuous phase shift according to (2.98), but not that purely imaginary zeros occur.

Because of equation (2.93), the zeros of $N_g(s)$ correspond to the zeros of $G(s)K(s)+1$. Thus, a purely imaginary zero of $N_g(s)$ also causes the frequency response $G(j\omega)K(j\omega)+1$—whose argument $j\omega$ is purely imaginary—to have a zero at the corresponding frequency. This again implies that the Nyquist plot $G(j\omega)K(j\omega)+1$ passes through the origin. Therefore, for a closed-loop system to be stable not only equation (2.98) has to be fulfilled, but also the Nyquist plot $G(j\omega)K(j\omega)+1$ may not pass through the origin.

We can also focus on the Nyquist plot of the open-loop transfer function $G(j\omega)K(j\omega)$, which can be measured directly, instead of focusing on the Nyquist plot of $1+G(j\omega)K(j\omega)$. If we do so, in all ideas developed so far we have to take the point -1 as the point of reference instead of the origin of the complex plane, which is illustrated in figure 2.29. This leads to the following theorem:

Theorem 2.9 *(Nyquist criterion) A closed-loop system is stable if and only if the continuous phase shift of the Nyquist plot of its open-loop transfer function $G(s)K(s)$ around the point -1 is given by*

$$\Delta\varphi_{GK,cont} = r_k\pi + i_k\frac{\pi}{2} \tag{2.99}$$

and the curve does not pass through the point -1. Here, i_k is the number of poles of the open-loop transfer function lying on the axis of imaginaries of the s-plane, and r_k is the number of poles lying to the right of it.

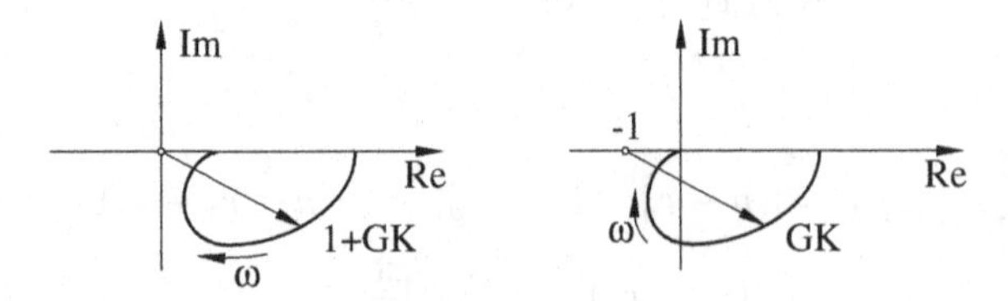

Fig. 2.29. Transition from the (1+GK)-curve to the (GK)-curve

For the application of the Nyquist criterion, it is of importance that r_k and i_k are known. Furthermore, it should be emphasized that the Nyquist criterion holds as well if the open-loop transfer function contains delay elements, but a proof of this will here be omitted.

Figure 2.30 shows three examples for the application of the Nyquist criterion. The left Nyquist plot results from a series connection of an integrator and a first order lag. r_k therefore is zero and i_k is one. The phase shift around the point -1, required for stability, will therefore be $\frac{\pi}{2}$. It can be seen that the pointer from -1 to the Nyquist plot points down initially and then turns to the right. This counterclockwise quarter revolution corresponds to the required angle $\frac{\pi}{2}$. According to this, a closed-loop system consisting of an integrator and a first order lag would be stable. Changing the open-loop gain V would lead to stretching or compressing the Nyquist plot, but the basic course would remain the same, and so would the phase shift, i.e. even for a change in V the system is stable. This is of course not true for all systems, as our next two examples show.

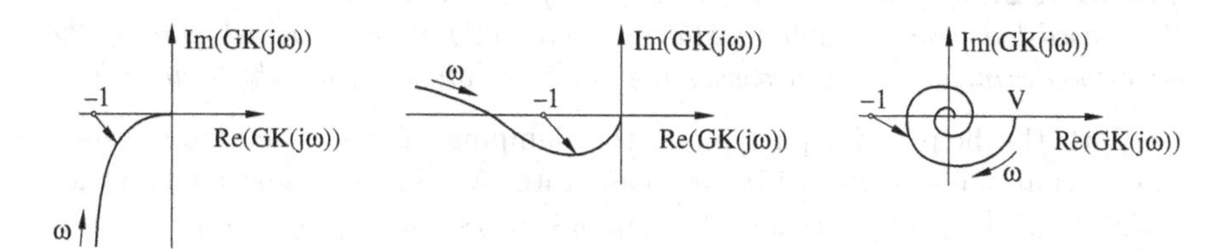

Fig. 2.30. Examples for the Nyquist criterion

The middle of the figure shows the already known Nyquist plot of the open-loop transfer function

$$G(s)K(s) = V\frac{(s+s_1)^2}{(s+s_2)s^2} \qquad 0 < s_2 < s_1 \tag{2.100}$$

Because of $i_k = 2$ and $r_k = 0$, the phase shift with respect to the point -1 required for a stable system is π. The Nyquist plot starts in negative infinity. The pointer from -1 to the Nyquist plot therefore initially points to the left and then turns counterclockwise to the right. This corresponds to a phase shift of π. For this case, a closed-loop system with this open-loop transfer function would be stable, too. In contrast to the first example, reducing V would here lead to a compression, and the curve would finally pass above the point -1. The phase shift would then be $-\pi$ instead of π, and the resulting closed-loop system would be instable.

The third example shows the series connection of a first order lag and a delay element:

$$G(s)K(s) = \frac{V}{Ts+1}e^{-T_D s} \tag{2.101}$$

Just like for the first order lag on its own, the curve starts at the axis of reals, but then continues to spiral into the origin, as the phase of the frequency response gets continuously diminished by the delay element, and the gain by the first order lag. Depending on the choice of values for the parameters V

and T_D, one, some or no revolutions at all around the point -1 are made. Not revolving it as shown in the figure, the phase shift is zero, as the pointer from -1 to the Nyquist plot steadily oscillates from positive to negative angles and back, but the total balance is not affected by this, as the initial value as well as the final value lies to the right-hand side of -1 on the real axis and no revolutions around the point -1 occur. As we have $r_k = i_k = 0$, a closed-loop system would be stable. Increasing the value of V, so that the curve is stretched and revolves around the point -1, would lead to an instable closed-loop system.

Regarding these examples, it should be comprehensible that we can also formulate the Nyquist criterion for stable open-loop transfer functions in more simple terms:

Theorem 2.10 *If the open-loop transfer function $G(s)K(s)$ is stable, then the closed-loop system will be stable if and only if the Nyquist plot of the open-loop transfer function passes the point -1 on its own right hand side.*

With the help of Nyquist plots, the damping of the closed-loop system can be characterized, too. First we can state: A system's transient behavior is determined by the position of the poles of its transfer function—the further the distance between a pair of complex conjugated poles and the axis of imaginaries, the greater the damping ratio of the corresponding oscillation will be (cf. (2.22)). Furthermore, the poles of a closed-loop system are the zeros of the equation $G(s)K(s) + 1 = 0$. Accordingly, the function $G(s)K(s)$ maps all poles onto the point -1. In contrast to this, any point $s = j\omega$ with vanishing real part is mapped onto $G(j\omega)K(j\omega)$. From this it follows that the axis of imaginaries of the complex plane is mapped onto the Nyquist plot $G(j\omega)K(j\omega)$. But if -1 is the counterpart of all poles and the Nyquist plot is the counterpart of the axis of imaginaries, then for a continuous mapping function the distance between the Nyquist plot and the point -1 will also be a measure for the distance between the poles and the axis of imaginaries, and therefore for the damping of the closed loop.

Two other criteria for stability should be mentioned briefly, the criteria by Hurwitz [69] and Routh [170]. Both refer to coefficients of the denominator of the transfer function and are some kind of numerical criteria. But as today computers can be used to calculate the zeros of polynomials, criteria like those are actually without relevance.

2.6 PID Controller

2.6.1 Demands on a Controller

Now that we have presented the fundamental principles needed for the analysis of dynamic systems, in this chapter we discuss the design process for controllers. As a recapitulation, we once more present the standard configuration of a closed-loop system (fig. 2.31), but in contrast to fig. 2.25, we now neglect the metering element. The reference and disturbance transfer functions are

$$T(s) = \frac{y(s)}{w(s)} = \frac{G(s)K(s)}{G(s)K(s)+1} \tag{2.102}$$

$$S(s) = \frac{y(s)}{d(s)} = \frac{G(s)}{G(s)K(s)+1} \tag{2.103}$$

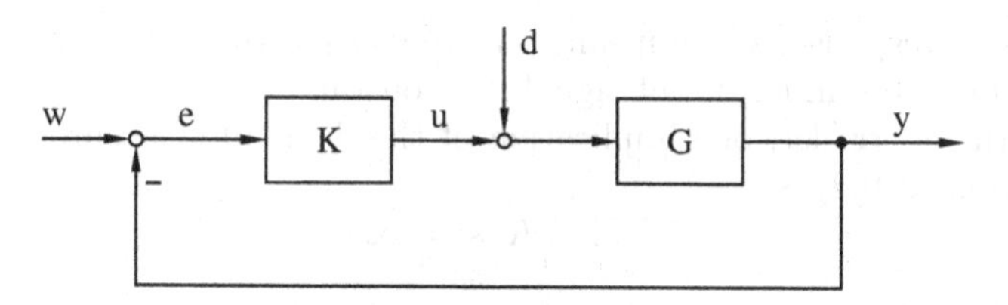

Fig. 2.31. Closed-loop system

Our task is to find a suitable controller $K(s)$ for a given plant $G(s)$. What demands have to be fulfilled? It would obviously be best to have $T(j\omega) = 1$ and $S(j\omega) = 0$. The meaning of the first demand is that the control variable y should always be equal to the (desired) reference variable w, independently from the frequency ω. This would of course imply that the system is BIBO-stable. The other demand corresponds to the total suppression of the influence which the disturbance d has on the control variable. All in all, the two criteria represent the demand for *accuracy* of the automatic control. Unfortunately, a control engineer will not be able to realize this heavenly state since, according to equation (2.102), for a given frequency response of the plant $G(j\omega)$, the function $T(j\omega)$ can only be a constant 1 if the controller's frequency response $K(j\omega)$ takes infinitely great values for all frequencies. The demand of total suppression of the disturbance leads to the same result. A controller like this cannot be realized, and its output quantities would exceed the possibilities of any kind of actuator.

For real applications, on the other hand, it is not necessary that those demands are fulfilled for all frequencies. It is sufficient to obtain accuracy for the actual frequency range used, i.e. for those frequencies which are contained in the input signal. Those frequencies are usually the low frequencies, including zero, which corresponds to a dc signal. Especially for these dc signals accuracy is required most, as for a constant input signal we expect the output variable of a controlled system to hold the same value as the input or reference variable after all initial transients have settled down. Therefore, we reduce our demands on an automatic control to that extend that the optimal criteria given above only have to be fulfilled for dc signals (frequency $s = 0$):

$$\lim_{s\to 0} T(s) \stackrel{!}{=} 1 \qquad \text{and} \qquad \lim_{s\to 0} S(s) \stackrel{!}{=} 0 \tag{2.104}$$

And because of the continuity of the transfer functions $T(s)$ and $S(s)$, the demands are fulfilled at least in an approximative way for small values of s or ω, and therefore for the frequency range used. If the equations (2.104) hold, this is sometimes called *steady-state accuracy*.

A system which is steady-state accurate is in any case stable in the sense of definition 2.4, i.e. it has a finite step response, as we obtain

$$\lim_{t\to\infty} y(t) = \lim_{s\to 0} s\frac{1}{s}T(s) = \lim_{s\to 0} T(s) = 1 \tag{2.105}$$

for the step response, which means that the output quantity of the controlled system for a step in the input signal is a constant 1.

For the controller, one implication of the demand of accuracy according to equation (2.102) is

$$\lim_{s\to 0} K(s) = \infty \tag{2.106}$$

If we assume $K(s)$ to be a rational function, then $K(s)$ necessarily has to have one pole at $s = 0$. If $G(s)$ does not have a zero at $s = 0$, then the product $G(s)K(s)$ becomes infinitely great for $s = 0$, and $T(s)$ converges towards 1. On the other hand, if $G(s)$ does have such a zero, then $\lim_{s\to 0} G(s)K(s)$ takes a finite value, and $\lim_{s\to 0} T(s)$ does not converge towards 1. The degree of the pole of $K(s)$ at $s = 0$ obviously has to exceed the degree of the zero of $G(s)$ by at least 1.

We should mention a special case: If the plant has an integrating effect, then we can write the transfer function alternatively in the form $G(s) = \frac{1}{s}\tilde{G}(s)$, where $\tilde{G}(0) \neq 0$, which is equivalent to a series connection of an integrator and the $\tilde{G}(s)$-part of the plant. If furthermore the affecting point of the disturbance variable d lies behind the integrator (fig. 2.32), then we get

$$\begin{aligned} T(s) &= \frac{\tilde{G}(s)K(s)}{\tilde{G}(s)K(s) + s} \\ S(s) &= \frac{s\tilde{G}(s)}{\tilde{G}(s)K(s) + s} \end{aligned} \tag{2.107}$$

for $T(s)$ and $S(s)$. In this case, we reach the limits which we demanded for achieving steady-state accuracy (see (2.104)) already if $K(0) \neq 0$ holds. We can set $K(s) = 1$, which means that in principle we do not need a controller at all. To put it a different way: We can treat the integrator as a part of the controller, so that $\lim_{s\to 0} K(s) = \infty$ holds.

This convenient situation can happen, if the actuator, which can be seen as a part of the plant, has some integrating effect. One example is a motor-driven valve, that shall be used to steer the media flow through a pipe. The input variable of the motor, whose internal transient behaviour shall be neglected, is the actuating variable of the controller. For that constellation, the opening

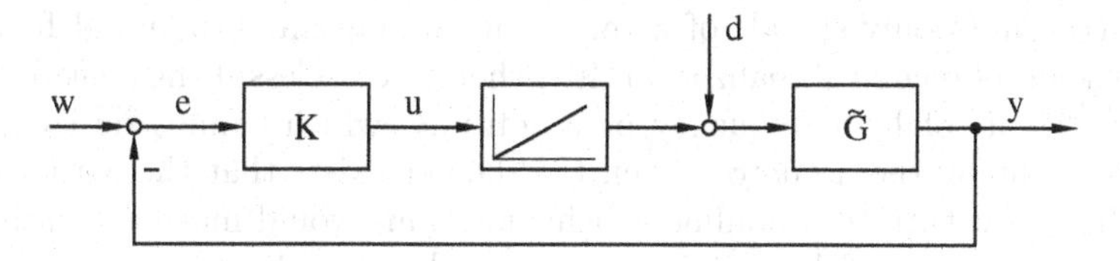

Fig. 2.32. Splitting-off an integrator of a plant

diameter of the valve changes continuously for a step function input, just like the output variable of an integrator.

The fulfillment of equation (2.104), i.e. steady-state accuracy, is almost always a fundamental prerequisite of an automatic control. In addition to this, other criteria exist which we should not neglect. On the one hand, there is the demand for a sufficient *control rate.* As an example, people can not be expected to wait for minutes for an elevator to reach the next floor. As later examples will show, this demand is often contrary to the demand of steady-state accuracy—and therefore contrary to stability, forcing an automatic control to become a compromise between the two. Another criterion is a sufficiently high damping ratio of the system. Again we can use an elevator as an example: here, an overshot is unacceptable. If a certain floor has to be reached, then nobody would like the elevator to go up a little bit too far and then start oscillating until it gets the right position. In this case, we would like an *aperiodic transient behavior*, i.e. the damping ratio D has to be greater than 1. For systems where a little overshooting can be tolerated, in most cases a damping ratio of $D = \frac{1}{\sqrt{2}}$ is desired, which is is the lowest damping possible for which still no resonant magnification occurs (see fig. 2.24).

For concrete cases further criteria can be defined. Depending on the application, for example the amplitude of the overshot of a step response might be relevant, or the time needed for a signal to lie within a certain tolerance of the step response. Accordingly, a variety of quality indices can be defined for an automatic control. For example, one quality measure which is frequently used is

$$Q = \int_0^\infty \left[(e(t))^2 + k(u(t))^2\right] dt \qquad \text{with } k > 0 \tag{2.108}$$

This measure has to be minimized. In order for Q to be minimal, on the one hand the mean squared control error e, and therefore the deviation between reference variable and control variable, should be as small as possible. On the other hand the second term ensures that this aim will be achieved with small values of the actuating variable, in order to protect the actuator of the control system.

We have to consider the behavior of the actuator twice anyway when designing the controller. On the one hand as a part of the controlled plant, when we handle stability, damping ratio and control rate, and on the other hand as it transmits only signals of a certain maximum amplitude and frequency because of its technical realization. It is therefore useless if the controller generates signals of high frequency or amplitude, which cannot be transferred to the plant by the actuator. Even the danger exists that the actuator may overdrive and turn to a nonlinear behavior. This would make the entire controller design doubtful, as it is based on a linear behavior of each transfer element. A simple example here is the rudder of a ship. It cannot be adjusted further than a maximum angle, which is determined by technical boundary conditions. While in normal use the angle of the rudder is proportional to the input quantity of the rudder unit, it cannot react to a further increase of the input quantity once the maximum angle has been reached. The linear transfer element with proportional behavior has converted into a nonlinear element, whose characteristic curve can be seen in figure 2.33. Such a curve is typical for many actuators.

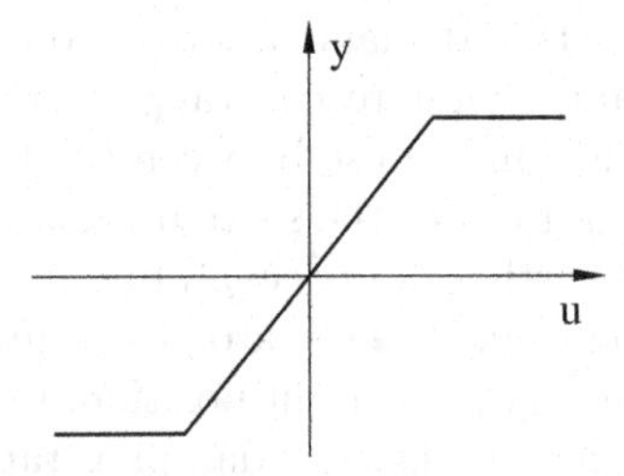

Fig. 2.33. Characteristic curve of an actuator

Now that we have given a picture of the demands put on a controller, we will describe some of the standard controllers in the following. They are easy to understand, to realize and, most of all, to parameterize. It is for this reason that most of the automatic controls occurring in real applications are realized using these controller types.

2.6.2 Types of Controllers

Proportional Controller

The *proportional controller* (*P controller*) surely represents the easiest approach. Since the input quantity of the controller is the control error and its output quantity the actuating variable, this controller produces an actuating variable proportional to the control error using the transfer function

$$K(s) = P \tag{2.109}$$

The bigger the difference between reference and control value is, the bigger the value of the actuating variable will be. The only adjustable parameter for this controller is the *control gainP*. Steady-state accuracy and the total suppression of disturbances cannot be achieved with a proportional controller, as then $K(0) = P$ would have to take an infinite value, according to the considerations we made earlier. In order to get the controller to work at least approximatively accurate, the value for P should be as large as possible. This increases the accuracy as well as the control rate, as for a given control error a bigger value for P obviously leads to an increase of the actuating variable. And as a result of this stronger excitation of the plant we achieve a faster approach of the control variable towards the reference value, even though this value will never be reached exactly.

But there are limits to the values of P, for reason of stability. This will be illustrated by two examples. Let two low-pass plants of degree two and three be given, i.e. a series connection of two or three first order lags respectively. The corresponding Nyquist plots are given in fig. 2.34. If we use a P controller to control the plant, then the open-loop transfer function of the entire system is given by $G(s)K(s) = G(s)P$. The corresponding Nyquist plots can be obtained by multiplying the given Nyquist plots by P, which is, for $P > 1$, equal to stretching the given curves. This also illustrates why an increase of P might be dangerous: According to the Nyquist criterion, the permissible phase shift with respect to -1 is zero for both cases. In other words, the Nyquist plots must not revolve around -1 to the left and should also not come too close to -1 in order to guarantee sufficient damping. The closed-loop system for the third order plant might therefore get instable if P increases too much. The closed-loop system for the second order plant remains stable for any increase of P, but the Nyquist plot gets closer and closer to the point -1, which corresponds to an unreasonable small damping ratio of the closed-loop system.

In real applications, the upper bounds given by the demands for stability or damping will normally be so small that steady-state accuracy cannot be achieved for the permissible values of P, not even approximately. Still, there are enough cases where steady-state accuracy is unimportant, and instead a decision according to the argument of low costs will be made in favor of the proportional controller. Finally, the P controller can always be used if the actuator or the plant respectively includes an integrator, as mentioned before.

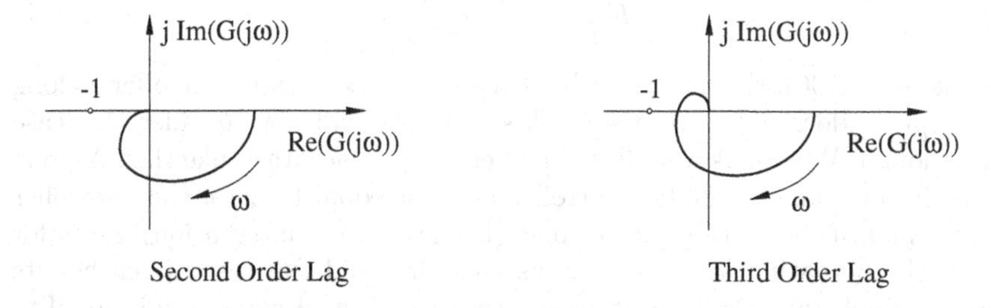

Fig. 2.34. Nyquist plots of low-pass plants of second and third order

In order to help us get a clearer picture of the behavior of the proportional controller (and other controllers, which we will introduce in the following), figure 2.35 shows the step responses of a closed-loop system with different controllers and a third order low-pass as the plant:

$$G(s) = \frac{1}{(T_1 s + 1)(T_2 s + 1)(T_3 s + 1)} \tag{2.110}$$

The curve denoted by P characterizes the step response of a system with a proportional controller. It can be seen clearly that, after all initial transients have settled down, a constant control error remains. If we increase the value of the control gain we would diminish the control error, but at the same time we would have to accept bigger oscillations at the beginning and finally even instability.

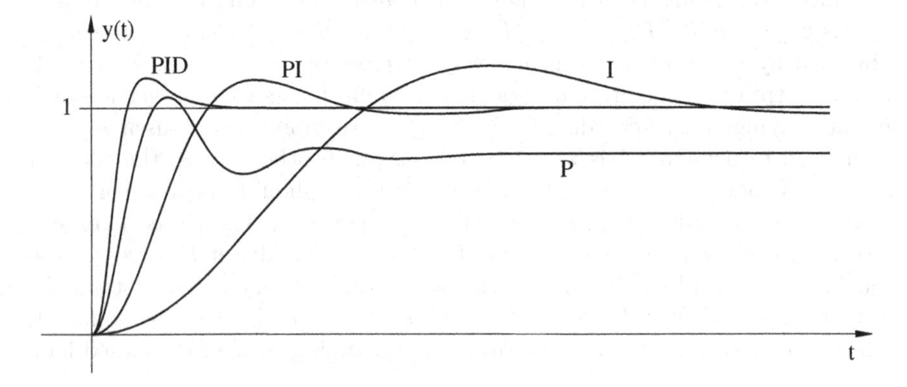

Fig. 2.35. A comparison of different controller types

Integral Controller

Much better results in controlling can be achieved using an *integral controller*(*I controller*):

$$K(s) = I\frac{1}{s} = \frac{1}{T_i s} \tag{2.111}$$

Because of (2.106), this is obviously a steady-state accurate controller as long as the plant does not have a zero at $s = 0$. We will not consider this case here, though. We can as well illustrate steady-state accuracy like this: As long as the input quantity of the controller e is not equal to zero, the controller output quantity u continues to change (because of the integration). Only for $e = 0$ the actuating variable u stops changing and the system reaches its constant final state. And $e = 0$ implies that the control variable y is equal to the reference variable w.

The parameter T_i is called integration time. The shorter this integration time, the faster the actuating variable changes for a given control error. With regard to a high control rate, T_i should be set to the smallest value possible. However, as for the P controller, the demand for stability limits the possible values here, too. To give an example, we try to control a second order lag by an integral controller. In this case, the open-loop transfer function is:

$$G(s)K(s) = \frac{V}{\frac{s^2}{\omega_0^2} + \frac{2D}{\omega_0}s + 1} \frac{1}{T_i s} = \frac{V}{T_i} \frac{1}{\frac{s^3}{\omega_0^2} + \frac{2D}{\omega_0}s^2 + s} \tag{2.112}$$

with the open-loop gain $\frac{V}{T_i}$. The corresponding Nyquist plot is shown in figure 2.36 (left curve). The permissible phase shift of the Nyquist plot with regard to the point -1 is $\frac{\pi}{2}$ because of the integrator. Therefore, if the Nyquist plot of the open-loop transfer function follows the curve in the figure, then the closed-loop system will be stable. If we decrease the integration time T_i in order to increase the control rate, then we also automatically increase the open-loop gain and stretch the Nyquist plot until it finally revolves counterclockwise around the point -1. This would lead to an instable closed loop.

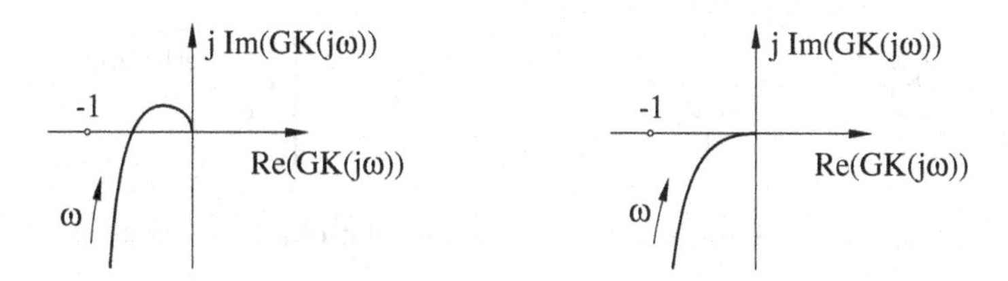

Fig. 2.36. Nyquist plot of a second order lag with I- and PI controller

Regarding stability, the I controller is not really favourable. In order to obtain the open-loop transfer function, we have to multiply the plant transfer function by the factor $\frac{1}{T_i s}$. This means that the phase of the open-loop transfer function is the sum of the phase of the plant $G(s)$ and the constant angle $-\frac{\pi}{2}$, and the Nyquist plot is bend clockwise towards the point -1. This obviously increases the danger of instability. Furthermore, because of its configuration, the integral controller is a slow type of controller, independent of its integral time. If, for example, a sudden control error occurs, then the controller first has to integrate this quantity before the actuating variable can reach an significant value. This effect can as well be seen in figure 2.35. The step response of the third order lag controlled by an integral controller (the curve marked with an I) reaches the correct final value, but increases only very slowly in the beginning.

PI Controller

We can fix these disadvantages by combining a proportional controller with an integral controller. This way we obtain the most frequently used type of controller, the *proportional integral controller* or *PI controller* Its transfer function is given by:

$$K(s) = P + I\frac{1}{s} = V_R\frac{T_{pi}s + 1}{T_{pi}s} \tag{2.113}$$

Accordingly, we can regard a PI controller to be a parallel connection of a P- and an I controller, or as a series connection of the lead $(T_{pi}s + 1)$ and the integrator $\frac{1}{T_{pi}s}$. This second representation is more convienient to develop the open-loop transfer function $G(s)K(s)$, as phase lag and gain can be computed more easily, while the first one helps us better in finding the step response and the Nyquist plot of the controller (fig. 2.37).

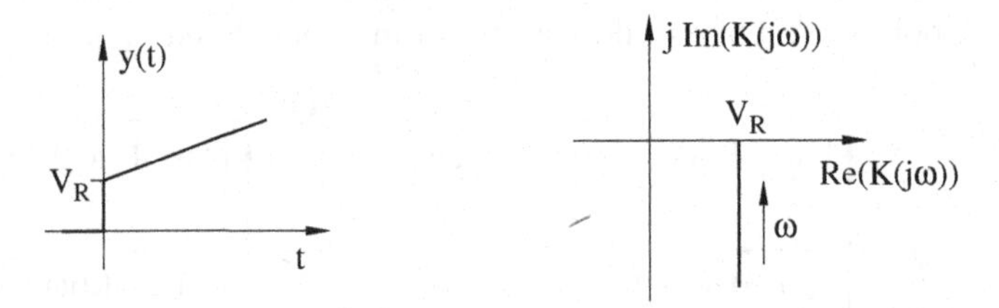

Fig. 2.37. Step response and Nyquist plot of a PI controller

Looking at the step response we see why this type of controller reacts faster than the integral controller: The first reaction of this controller to a step-like control error is a value different from zero, which only has to be adjusted by the integrator. In contrast to the integral controller, the actuating variable here does not have to be slowly integrated. The improvement in the overall behavior of the controller can as well be seen in figure 2.35 (curve PI).

We can see the better performance with respect to the criterion of stability—compared to the integral controller—directly when looking at the Nyquist plot. For low frequencies, the phase is almost $-\frac{\pi}{2}$, too, but for higher frequencies it converges towards zero. While the integral controller rotates the Nyquist plot of the plant by $-\frac{\pi}{2}$ for every frequency range, the rotation produced by the PI controller will be the smaller, the higher the frequency gets. Now, especially in ranges of high frequencies the Nyquist plots of many real plants get close to the point -1. If, in addition to this, also the controller itself shows a considerable phase lag for this frequency range, it can easily happen that the Nyquist plot for the open-loop transfer function of the plant and the controller revolves around the point -1, which makes the closed-loop system instable. For this reason, it will be an advantage for the stability of the closed-loop system that a PI controller causes only a small phase lag for this frequency range. We can see this improvement clearly in figure 2.36. The Nyquist plot to the right shows little phase shifts for high frequencies and is therefore no thread to stability.

PID Controller

The actuating variable of a PI controller consists of two parts, one part contributed by the integral, in charge of the controller's accuracy, and a proportional part in order to increase the control rate. We can expect a further improvement of the control performance if counteractions against a control error are taken not when the error occurred, as a proportional controller does, but already when it is starting to occur. For this purpose, a PI controller can be extended by adding a differential part. We obtain the proportional integral derivative controller (*PID controller*):

$$K(s) = P + I\frac{1}{s} + Ds \tag{2.114}$$

An ideal differentiator with transfer function s can neither be realized nor is it desired, as the factor s occurring in a summand means that this summand takes the bigger values, the higher the frequency is. Because of this high-pass characteristic, in real applications the ideal differentiator would amplify the always occurring high-frequency noise. This should, of course, be avoided. For this reason, the differential part of a proportional integral derivative controller is delayed by the time constant T_v:

$$K(s) = P + I\frac{1}{s} + D\frac{s}{T_v s + 1} = V_R \frac{T_1 s + 1}{T_1 s}\,\frac{T_2 s + 1}{T_v s + 1} \tag{2.115}$$

As we can see, we can as well think of a PID controller as a series connection of a PI controller and a rational first order transfer function. In principle, the two zeros T_1 and T_2 might be complex conjugated, but we will not discuss this case here. The advantage of a PID controller compared to a PI controller can be explained using the step response and the Nyquist plot, which are shown in figure 2.38. The curves for the ideal PID controller (of eq. (2.114)) are given by the dashed lines.

The step response shows that the controller behaves exactly in the desired way: During its initial phase, a control error—here of a step-like form—is fighted vehemently in its first phase by the differential component, while for the following phases the controller acts like a PI controller. In addition to this, we can observe a further improvement of the stability compared to the PI controller regarding the Nyquist plot: The phase of the PI controller converges towards zero for high frequencies, assuring that the plant's Nyquist plot will not be twisted towards the point -1 for high frequencies. The frequency response of the PID controller, on the other hand, shows a positive phase for higher frequencies. For this frequency range, a given Nyquist plot of a plant can therefore even be bend away counterclockwise from the point -1.

We should pay attention to the fact that, because of the higher number of adjustable parameters, the PID controller is more difficult to design than a PI controller. While PI controllers may often be adjusted by hand, without any calculations, this is almost impossible for a PID controller, especially not if an optimal adjustment is desired which makes full use of all the possibilities the controller offers.

Figure 2.35 shows clearly that the PID controller is the best of the controllers introduced so far, regarding the results which the controllers achieve. In principle we can say that the results which are achieved when using a certain type of controller will be the better, the higher the complexity of the controller is. This should not strike us, as more degrees of freedom are available, which can be used to fulfill the contradicting demands of stability and sufficient damping on the one hand, and high control rate on the other hand.

Looking at equation (2.114), we can see that the PID controller comprises the other controllers, which we already introduced, as special cases. Depending on whether I, P or D are set to zero, a proportional, integral or PI controller results. For this reason, the term *PID controller* is often used as a collective term for all the types of controllers being introduced so far.

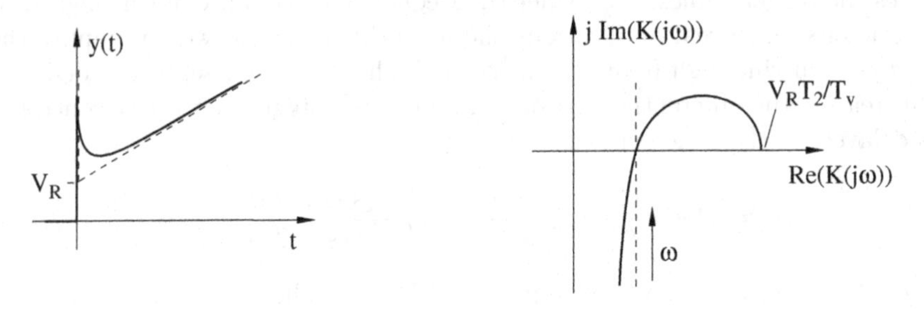

Fig. 2.38. Step response and Nyquist plot of a PID controller

PD Controller

In this discussion, we should not leave the *proportional derivative controller* or *PD controller* unmentioned. It consists, as already expressed in the name, of a proportional and a differential component. As the differential part cannot be realized in the ideal way, just like for the PID controller, the transfer function of this controller is given by:

$$K(s) = P + D\frac{s}{T_v s + 1} = V_R \frac{T_1 s + 1}{T_v s + 1} \tag{2.116}$$

Steady-state accuracy can be achieved by this type of controller only for plants with integrators, because of $K(0) \neq \infty$. If we design it suitably, though, we can make it react stronger than a proportional controller to changes in the control error. Therefore, it might be the reasonable choice if accuracy is not that important or guaranteed by another integrator, and if the control rate, which can be achieved using a proportional controller, is not high enough.

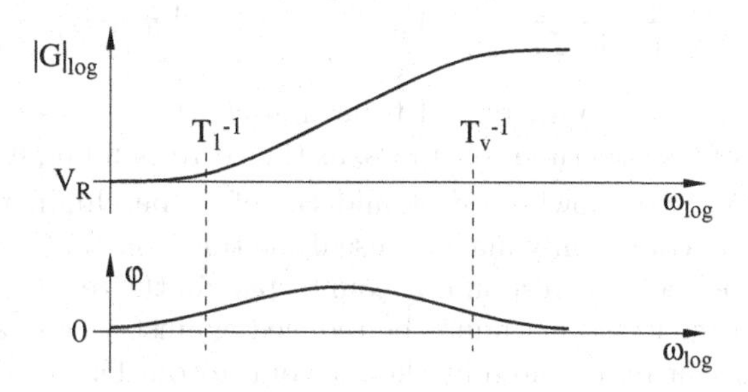

Fig. 2.39. Bode diagram of a PD controller

But, a PD controller can be used in a completely different way, too. If we examine the corresponding Bode diagram for a constellation with $T_1 > T_v$ (fig. 2.39), we find that adding a PD controller to a plant leads to a raise of the phase of the open-loop transfer function within an adjustable frequency range, as the phase of the open-loop function is just the sum of the phases of its single transfer elements. This way we can use a PD controller to bend the plant's Nyquist plot counterclockwise away from the point -1 for precisely this frequency range. The true control is then done by another controller.

We have to keep in mind, though, that a raise of the phase goes together with an increase of the gain for higher frequencies. This leads to a stretching of the Nyquist plot and therefore possibly to an approximation to the critical point. On the other hand, we can also choose $T_1 < T_v$. In this case, the gain is reduced, but for the price of lowering the phase. Whether using the PD controller as a *phase-correcting element* will be an advantage or not has to be decided for every individual case separately.

2.6.3 Controller Design

We have now introduced the different basic types of controllers, together with their characteristic traits. In the following, we present some methods which can be used to design their parameters. If the plant is not too complicated, even a few simple considerations might be sufficient to design a controller.

As an example of such considerations, we show the computation of a suitable PI controller for a second order lag. Assuming the second order lag to have two real poles, we can describe it by

$$G(s) = \frac{V}{(T_1 s + 1)(T_2 s + 1)} \tag{2.117}$$

We can now proportion the controller in a way that its lead can be reduced against a pole of the plant, so that just a first order lag with an integrator remains (cf. right-hand side Nyquist plot of fig. 2.36). This simplifies the open-loop transfer function to:

$$G(s)K(s) = \frac{V}{(T_1 s + 1)(T_2 s + 1)} V_R \frac{T_{pi} s + 1}{T_{pi} s} = V V_R \frac{1}{T_2 s (T_1 s + 1)} \tag{2.118}$$

where $T_{pi} = T_2 > T_1$. With regard to a high control rate, we have reduced the larger one of the two time constants, as the corresponding initial transient $e^{-\frac{t}{T_2}}$ passes by more slowly and should therefore be eliminated. Reducing time constants of course only makes sense if the time constant occurring in the transfer function has a corresponding counterpart in the real plant. Especially an equivalent time constant cannot be reduced against a pole, as in this case the initial transient represented by the pole of the transfer function only gives an approximation of the real initial transient.

After setting $T_{pi} = T_2$, we only have to determine the controller's remaining parameter V_R. In order to do this, we compute the transfer function of the closed loop:

$$T(s) = \frac{y(s)}{w(s)} = \frac{1}{\frac{T_1 T_2}{V V_R} s^2 + \frac{T_2}{V V_R} s + 1} \tag{2.119}$$

The closed-loop system is obviously a second order lag. We can adjust its damping ratio D using the parameter V_R. A comparison of its coefficients with the ones of equation (2.43) gives:

$$V_R = \frac{T_2}{4 V T_1 D^2} \tag{2.120}$$

Choosing a desired damping ratio D thus fixes the controller's second parameter. The closed loop behaves like an ordinary second order lag with a specified damping ratio.

In real applications, finding the right values for the controller parameters is—having simplified the plant's transfer function—often achieved by considerations like this, but a certain amount of intuition and experience in this field is required. The *root locus method* represents a more schematized way of doing this. For this method, we first have to determine the pole's positions of the closed-loop system with respect to the adjustable parameters of the controller. How the controller has to be adjusted we can directly derive from a configuration of the poles which we consider to be best. The question of which configuration will be the best still depends on the demands put on the controller, i.e. the special case of application. Here, a certain degree of freedom remains in the design process. Therefore, a certain amount of intuition is required for this method, too.

In contrast to this, the design process of a controller is fully schematized when using the adjustment rules by *Ziegler-Nichols*. By measuring the plant's step response we can evaluate certain characteristic data which we can then use to compute the parameters of the controller with well-know formulas. The control theory specific ideas underlying these formulas, however, are not that easy to comprehend and leave a user in the lurch of not knowing how to modify the controller parameters in case of bad results.

Another possibility is the optimization of a given quality measure (cf. (2.108)). Solving this extreme value problem automatically gives the controller parameters. The degree of freedom which is often overseen when using this method is the definition of the quality measure. Of course, the controller will be optimal only with respect to the given quality measure. If this is not choosen carefully, then no convenient control can be expected. Because of this, some intuition is required for this method, too.

All methods discussed so far share the fact that the structure of the controller has to be fixed, and only the parameters can be determined using the preferred method. And it is obvious that not every type of controller is suitable to stabilize any kind of plant. For example, a plant consisting of a series connection of two integrators can only be controlled by a PID controller, but never by a P-, I- or PI controller. Before we can compute the controller parameters we always have to analyze the plant and a find a suitable type of controller.

We do not need this step when using the so-called analytic methods. These methods not only determine the controller parameters, but also the controller structure, i.e. the transfer function of the entire controller. Here, of course, controller transfer functions can result which have no common features with a PID controller anymore. As an example, we present the *compensating controller* method. In this method, a model transfer function $M(s)$ has to be defined for the closed loop, that results from certain demands put to the closed-loop transfer function:

$$T(s) = \frac{G(s)K(s)}{1 + G(s)K(s)} \stackrel{!}{=} M(s) \tag{2.121}$$

It follows for the controller

$$K(s) = \frac{1}{G(s)} \frac{M(s)}{1 - M(s)} \tag{2.122}$$

The controller comprises the inverted transfer function of the plant, and by this means fully eliminates the plant's influence onto the open-loop transfer function $G(s)K(s)$. Of course we have to choose $M(s)$ in a way that a realizable controller results—one, where the degree of the numerator's polynomial is less than the degree of the denominator's polynomial. The fact that we usually cannot determine the poles and zeros of $G(s)$ exactly represents another problem. It may be the case that $G(s)$ and the factor $\frac{1}{G(s)}$ of the controller transfer function possibly do not compensate each other completely. As long as $G(s)$ has only poles and zeros with negative real part, this may be annoying, but it produces no problem. If worst comes to worst, (decreasing) oscillations may occur, which should have been eliminated. But if $G(s)$ has a pole with a positive real part and this pole is not entirely reduced by $K(s)$, then the open-loop transfer function $G(s)K(s)$ will also have an instable pole. Even if this does not necessarily lead to instability of the closed-loop system, an instable open-loop transfer function should still be avoided as far as possible. The feedback line of the controlled system might get undone—for example by a faulty sensor, thus producing an open loop. Now, for an open loop, the transfer characteristic of the entire system is only determined by the instable open-loop transfer function.

But the open-loop instability can be avoided easily by choosing $M(s)$ in a way that $1 - M(s)$ includes the zero that corresponds to an instable pole of $G(s)$. In doing so we can reduce the instable pole and zero on the right hand side of equation (2.122) against each other, and the controller will not have to compensate this pole at all.

In case $G(s)$ has a zero with positive real part, the same problem occurs. In order to prevent $K(s)$ from getting an instable pole, we have to choose a corresponding zero for $M(s)$. Summing up these ideas we can say that, when choosing a model transfer function $M(s)$, we must take the plant's special features into account from the very beginning. Therefore this method, too, has its degrees of freedom, which will be of strong influence to the controller design process.

2.6.4 Structure Expansion

Reference variable filter

We can achieve even better possibilities by leaving the structure given in figure 2.31 behind and inserting additional elements and connections into the control loop. A very simple method is to use a *reference variable filter* (fig. 2.40). With this device, the reference transfer function becomes

$$T(s) = \frac{y(s)}{w(s)} = F(s)\frac{G(s)K(s)}{1 + G(s)K(s)} \tag{2.123}$$

while the disturbance transfer function $S(s)$ remains unchanged, as the filter lies outside the closed loop. For the same reason, there will also be no effect on the stability of the system. This way we are now able to modify reference transfer function and disturbance transfer function independently of each other. First, we design the controller $K(s)$ in order to achieve an optimal disturbance response, and then the reference variable filter $F(s)$ for the reference transfer behaviour.

Optimal disturbance response normally means a fast correction of disturbances. For this, we have to consider stability, but a high damping is usually not required. Accordingly, in the design process we obtain a controller with high amplification and, as a result of this, a badly damped system. In contrast to this, a sufficient minimum damping is of great importance regarding the reference transfer behaviour. But right this can now be achieved by means of the reference variable filter. Using, for example, a low-pass filter, a change of the reference value will affect the closed-loop system only delayed, i.e. as a continuous curve, and therefore it does not activate the closed loop to oscillate, even though the damping ratio of the closed-loop system is small. Regarding the reference transfer behaviour, the system seems to be reasonably damped now. By adding a reference variable filter we can more easily find a compromise between the contrary demands of fast corrections of disturbances on the one hand and a sufficient damping of the reference transfer function on the other hand.

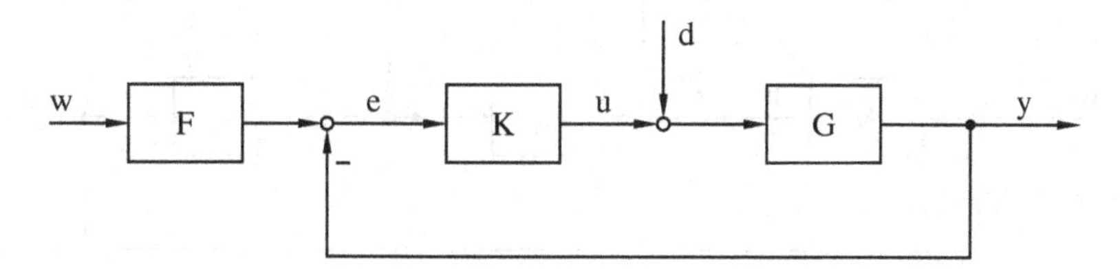

Fig. 2.40. Closed-loop system with reference variable filter

Disturbance Feedforward

Another simple expansion of structure is the *disturbance feedforward*, that also occurs outside the closed loop and has no influence on the stability. The corresponding structure is shown in figure 2.41. We assume that the disturbance affects the plant between the two parts $G_1(s)$ and $G_2(s)$. Our aim is to compensate the disturbance before it can influence the control variable. A prerequisite is of course that the disturbance variable can be measured. The simplest idea is to impress a signal of the same magnitude, but with opposite sign to the point where the disturbance is affecting. Unfortunately, this is usually not possible, as a disturbance often affects a plant at a point where no counteractions with an actuator can be taken. The course of a ship, for example, may be influenced by a current applied sideward to it, but no counterforces can be generated to compensate this effect at the same point. The only means that can be applied here is a counter steering of the rudder, i.e. using the actuating variable itself. Accordingly, a disturbance feedforward is usually applied directly at the controller output, allowing this additional information to be added to the actuating variable. If we want to achieve a total disturbance compensation at the point where the disturbance is affecting, the following equation has to be true:

$$F(s) = \frac{1}{G_1(s)} \tag{2.124}$$

As, in general, this function cannot be realized exactly, we have to make do with an imperfect approximation. But this plays no role for stability or control accuracy, because from the controller's point of view, impressing this quantity only represents another disturbance, which has to be corrected. Generally, a disturbance feedforward makes sense, if deviations between reference and control variables have to be kept as small as possible, so that the additional effort in measuring and calculation due to the disturbance feedforward is justified.

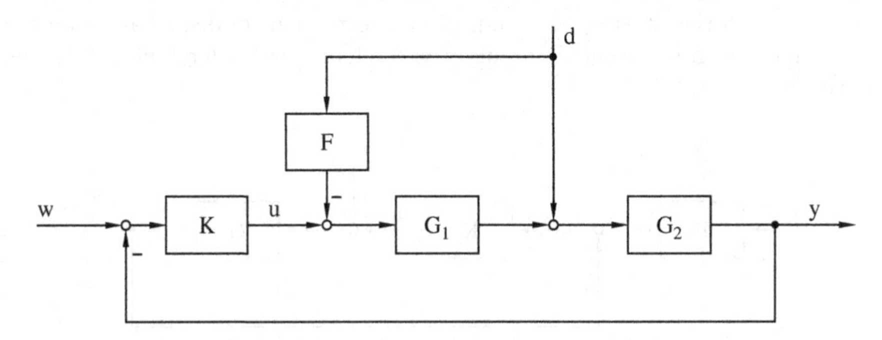

Fig. 2.41. Closed-loop system with disturbance feedforward

Complementary Feedback

Another possibility to improve an automatic control is the *complementary feedback*. In contrast to the two structure-expanding methods presented so far, in this case we add something to the closed loop and therefore influence the system stability (fig. 2.42). The idea is as follows: For a plant with low-pass behaviour, as usually given in real applications, the output signal occurs delayed compared to the internal quantities. If we can measure one or more of these internal signals, we can then—in case of a disturbance d_1—already provide the information about the forthcoming change in the control variable to the controller, even though the control variable itself is still unchanged. Accordingly, the controller is able to take countermeasures earlier, which leads to a better disturbance response. This advantage will of course not be available if the disturbance affects the plant behind the measuring point, which is the case for the disturbance d_2 in the given figure. In this case, we even have to keep in mind that the disturbance transfer function

$$\frac{y(s)}{d_2(s)} = \frac{G_2(s) + EKG_1G_2(s)}{1 + KG_1(s)(E(s) + 1)} \tag{2.125}$$

for an integrating controller type can only converge towards zero for $s \to 0$ (cf. (2.104)), if $E(s)$ compensates the pole of the controller at $s = 0$ in the term EKG_1G_2. For the sake of steady-state accuracy, $E(0) = 0$ has to be true for the function $E(s)$ in general.

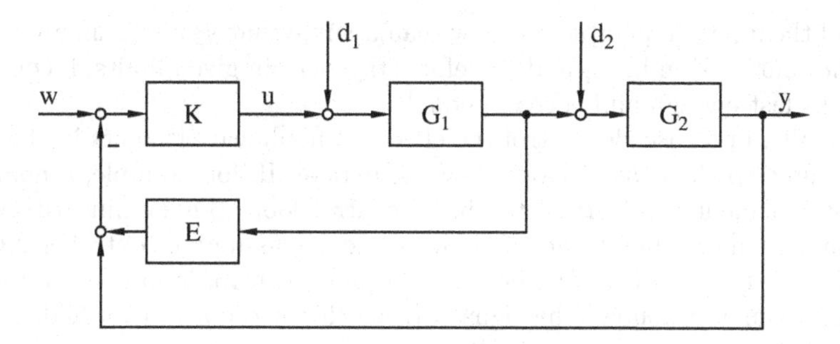

Fig. 2.42. Complementary feedback

Cascade Control

One special form of complementary feedback which is often used in real applications is the *cascade control*. In figure 2.43, we give an example of a system consisting of two loops. In principle, any number of loops can occur. A cascade control is a good choice if we can represent the plant as a series connection of different transfer elements with low-pass characteristic. The system is treated as a sequence of control loops fitted into each other. Each controller is responsible only for the part of the plant lying inside its feedback loop. In figure 2.43, controller 2 controls the variable y_2, and is given u_1—the output quantity of controller 1—as the reference variable. The closed inner loop on the other hand is part of the plant which controller 1 has to control. Its control variable is the system's output quantity, y_1.

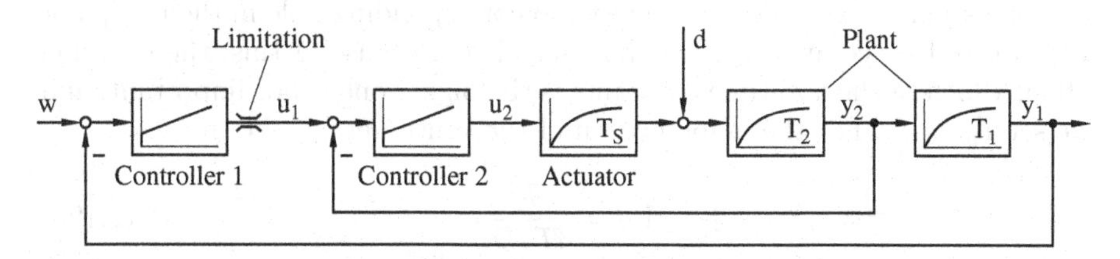

Fig. 2.43. Cascade control

One advantage is the fast correction of disturbances, just like for the complementary feedback. If, for example, a disturbance d occurs at the marked position in figure 2.43, controller 2 can already counteract as soon as a change in the value of y_2 is noticed. The deviation of the main control variable y_1 will then of course be smaller.

Another advantage is the possibility of limiting internal variables. In figure 2.43, such a limitation is marked at the output of controller 1. Regarding its functionality, this limitation corresponds to the nonlinear transfer element of figure 2.33. Although this limitation affects the variable u_1, it is intended to bound the internal quantity y_2, as we can assume that y_2 approximately takes up the course given by u_1 and therefore stays within given limits, if controller 2 works fast enough and correct enough.

Finally, in a cascade control the effects of nonlinear elements in the plant are limited to the control loops they are part of. If, for example, a nonlinear transfer element was part of the inner control loop, this nonlinearity would be almost unnotecible to the outer control loop—assuming controller 2 reacts sufficiently fast and exact—, because the inner control loop guarantees that y_2 approximately follows the signal u_1, which corresponds to a higher order lag and therefore linear transfer characteristic.

For practical applications, the most interesting advantage is how easy it is to put a cascaded arrangement into operation. First, we design controller 2 for the inner loop. Having done that, we can approximate the inner loop by a first order lag, assuming a sufficiently fast controller. Then, using this simplification, we can design controller 1 for the outer loop. As a basic outline, for a cascade control the controllers are designed successively from inner to outer loops, approximating the inner loop each time by a simple transfer element.

We illustrate this procedure using figure 2.43 as an example. Let a plant be given which consists of a series connection of three first order lags, where the first lag is an approximation for the dynamic behavior of the actuator. All three lags have a gain of 1, which makes calculations easier. The part of the plant controlled by controller 2 consists of two first order lags, a case we have already discussed. We choose a PI controller, whose time constant T_{pi} has to be set to the value of the bigger one of the plant's two time constants in order to compensate the corresponding pole. In our case, T_S is only an equivalent time constant anyway, i.e. there exists no corresponding pole in the real plant which could be compensated by he controller. Because of this, the question about which of the two time constants is the bigger one is not important, and we set $T_{pi} = T_2$. For the amplification of the controller we obtain

$$V_R = \frac{T_2}{4T_S D^2} \tag{2.126}$$

according to equation (2.120). The controller design is reduced to the definition of a suitable damping D. The inner loop appears to the outside as a second order lag with precisely this damping. If we select $D < 1$, then this second order lag will be able to oscillate, making control of the outer control loop more difficult. For this reason, we should consider a value $D \geq 1$. For all such values the inner loop will show an aperiodic transient behavior, which will be the slower, the larger D will be. In order to achieve an optimal control rate for aperiodic transient behavior, $D = 1$ is the right choice for the damping factor. We get the transfer function for the inner control loop

$$\frac{y_2(s)}{u_1(s)} = \frac{1}{4T_S^2 s^2 + 4T_S s + 1} \tag{2.127}$$

Approximating this transfer function according to equation (2.66) using a first order lag yields

$$\frac{y_2(s)}{u_1(s)} \approx \frac{1}{4T_S s + 1} \tag{2.128}$$

This way, the plant which controller 1 has to control will as well consist of two first order lags, and, neglecting the limitation, considerations in complete analogy to those already made will lead to $T_{pi} = T_1$ and

$$V_R = \frac{T_1}{16T_S D^2} \tag{2.129}$$

Static Feedforward

Besides those advantages mentioned, the cascade control has the disadvantage of a bad reference transfer behaviour, i.e. a bad response to a variation of the reference input. The reason is plain to see: If the reference value for the outer most control loop is set to a new value, this excitation first has to run through the entire cascade of controllers before it finally excites the plant itself and changes the output quantity. One way to correct this is a *feedforward*. It leads to an improvement in the reference signal transfer behaviour, and is therefore often used in combination with a cascade control. Figure 2.44 shows a so-called static feedforward. The reference transfer function is given by:

$$T(s) = \frac{y(s)}{w(s)} = \frac{G(s)(K(s) + V)}{G(s)K(s) + 1} \tag{2.130}$$

As the feedforwarding happens outside the closed loop, we can choose the factor V without having to worry about the stability of the system. We can also conclude this from the observation that V does not appear in the transfer function denominator. Steady-state accuracy is as well out of danger, because $\lim_{s\to 0} T(s) = 1$ will be true as long as the controller $K(s)$ contains an integrator, i.e. $\lim_{s\to 0} K(s) = \infty$.

We can explain the effect of a static feedforward like this: If the controller is of the PI-type, then in case of a change of the reference value Δw this step, multiplied by the proportional part P of the controller is passed on to the plant. At the same time, another step $V\Delta w$ reaches the plant via the

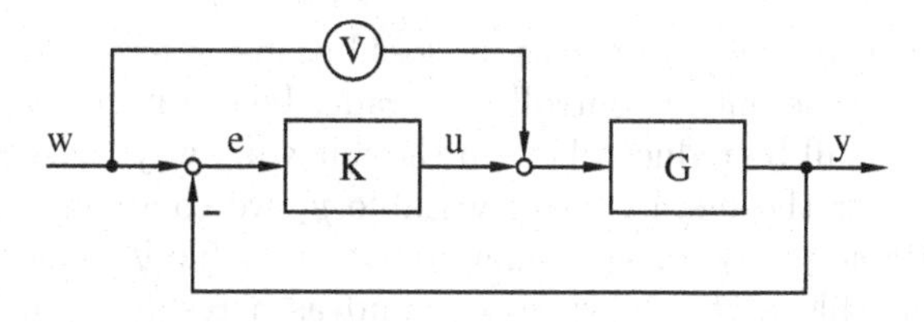

Fig. 2.44. Static feedforward

feedforward path. For a change in the reference signal, we can say that the feedforward path enlarges the controller's proportional part. The control signal y on the other hand, fed back to the controller, does not run through the feedforward path, i.e. further control actions depend only on the controller itself. The feedforward path remains constantly at the initial value, while the controller—if it contains an integrator—will keep on changing its actuating variable until the control error is vanished. During the initial phase, the excitation on the plant gets expanded by the feedforward, leading to an increase in the control rate, while during further phases the feedforward has no effect for the corrections of the control error, the control accuracy, and the stability. Increasing the control rate without endangering the stability is the reason, why static feedforwards can be found in many practical applications.

Dynamic Feedforward

In principle we can add static feedforwards to every controller of a cascade control. However, a more elegant way is provided by a dynamic feedforward, the so-called *reference variable generator* (RVG). The basic idea of feedforwarding—improving the reference transfer behaviour by means of an additional feedforward— remains the same. We explain the function principle of the RVG using figure 2.45.

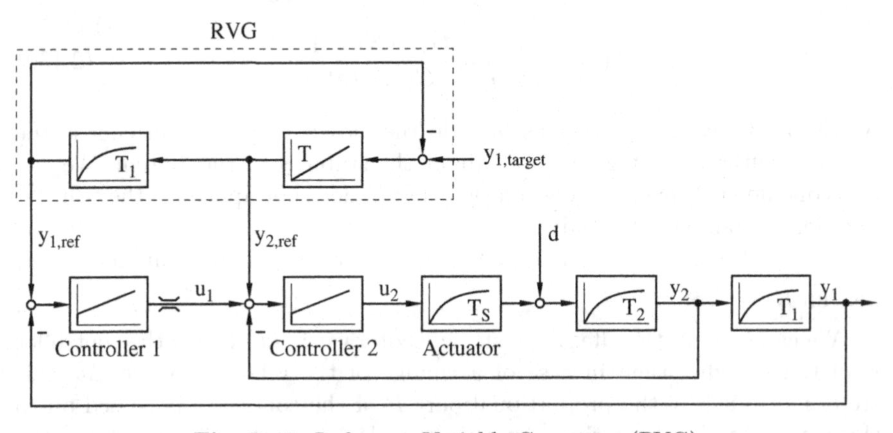

Fig. 2.45. Reference Variable Generator (RVG)

In a cascade control we have a time loss because a new reference value has to pass through the entire controller cascade, before it can affect the plant. This time loss shall be reduced here in the following way: Based on the target value $y_{1,target}$ for the main control variable y_1 we compute a corresponding reference variable $y_{2,ref}$ for the inner control variable y_2. Controller 2 forces its control variable y_2 to follow $y_{2,ref}$, and as a result, y_1 reaches $y_{1,target}$ theoretically without any action of controller 1.

The computation of a suitable reference variable $y_{2,ref}$ is done by a reference variable generator (RVG) using a model of the plant. The relationship between $y_{2,ref}$ and $y_{1,ref}$ has to be the same as between y_2 and y_1, that means here a first order lag with time constant T_1. We can see this transfer element in the plant and in the RVG as well.

If a new reference value $y_{1,target}$ for the main control variable is defined, the integrator output in the RVG remains changing until its input is equal to zero, i.e. $y_{1,ref} = y_{1,target}$. In this case $y_{2,ref}$ has also reached a value corresponding to $y_{1,target}$.

Besides this, the integrator in the RVG guarantees, that $y_{2,ref}$ has a continuous course, so that controller 2 can force the continuous variable y_2—as it is the output of a first order lag—to follow this course.

The entire RVG, i.e. the transfer behaviour between $y_{1,target}$ and $y_{1,ref}$, is equal to a second order lag, whose damping can be adjusted by the integrator time constant T. If T is large enough, the damping ratio is greater than one, so that the transient behaviour is aperiodic, and $y_{1,ref}$ as well as $y_{2,ref}$ change monotonously to their new values.

The question arises what we need controller 1 for. In real control systems, we have to take into account that the plant model inside the RVG might not be precise or that disturbances might affect the plant between y_2 and y_1 which cannot be recognized by controller 2. Therefore, the outer control loop with controller 1 is essential for final corrections and the overall control accuracy. As the reference variable $y_{1,ref}$ is computed by the RVG, too, it is guaranteed, that the reference variables for both controllers fit to each other and the controllers do not work against each other.

Like for the static feedforward, there arise no stability problems from the RVG—as long as its internal transfer function is stable—because the RVG works outside the closed-loop system. Therefore, the RVG can make a contribution to the system performance only in case of changes of the reference variable. But this is no drawback, as the fast correction of disturbances is already guaranteed by the cascade control structure.

Decoupling

In this chapter, we have so far neglected multi-input-multi-output(MIMO) systems , though in reality it is the common case that several quantities affect a system, and that also a set of output quantities is of interest, where each output variable possibly depends on more than one input variable. We say, that the variables are *coupled* to each other.

For some plants, we can define for each output variable exactly one input variable with a major influence on that output variable, while the influence of all other input variables is neglectible. In such a case, we can define SISO scalar transfer functions from each input to its output variable and design a controller for each of these partial systems like for a common SISO system. The influence of the other partial systems is ignored. The MIMO system is divided in several partial systems that are treated like SISO systems. The prerequisite of this method is the weak coupling between the partial systems.

Things are getting more complicated if the single quantities are coupled stronger. If, for example, one input variable of the MIMO plant is changed by a controller, then several output quantities could change, causing other controllers to react which might again affect the control variable of the first controller. This controller starts countermeasures causing again other countermeasures of other controllers. It is unlikely that the system will ever reach a stationary final state. In the worst case, the oscillations even can increase.

It is still possible, though, to design a separate control loop for every quantity if we can compensate the influence of other variables on this loop. Figure 2.46 shows the example of a double-input-double-output plant, where both actuating variables u_1 and u_2 affect each of the output quantities via the transfer elements G_{ij}. In order to establish that each of the two control loops can be proportioned on its own, we have to make sure that the quantities d_1 and d_2 are compensated. This can be done using the two signals a_1 and a_2, which are obtained from the actuating variables via the *decoupling elements* E_{ij}. As an example, in order to achieve a compensation of d_2 by a_2, the following equation has to be true:

$$G_{11}(s)a_2(s) \stackrel{!}{=} -d_2(s) \tag{2.131}$$

The decoupling therefore corresponds to a disturbance feedforward. From (2.131), letting $d_2(s) = G_{12}(s)u_2(s)$ and $a_2(s) = E_{12}(s)u_2(s)$, for the calculation of the decoupling element it follows that:

$$G_{11}(s)E_{12}(s)u_2(s) \stackrel{!}{=} -G_{12}(s)u_2(s) \qquad \Rightarrow \qquad E_{12}(s) = -\frac{G_{12}(s)}{G_{11}(s)} \tag{2.132}$$

And analogously for the second decoupling element:

$$E_{21}(s) = -\frac{G_{21}(s)}{G_{22}(s)} \tag{2.133}$$

As transfer functions occur in these formulae as a part of the denominator, restrictions regarding zeros with positive real part exist and also regarding the degree of the polynomials for the numerator and denominator. This means that such decoupling is in many cases only partly possible, or it may even be completely impossible.

The methods which we introduced in this chapter are excellently suitable for controlling linear single-input-single-output systems, and even some nonlinear systems (after linearization) or MIMO systems can be realized using these simple means. The vast majority of all industrially used controllers therefore consists of PID-based types with widely varying structural enlargements. In contrast to this, control research from the 60s on follows a different approach, which provides deeper insight into the system behavior and therefore allows a closed control design approach for MIMO systems of any degree.

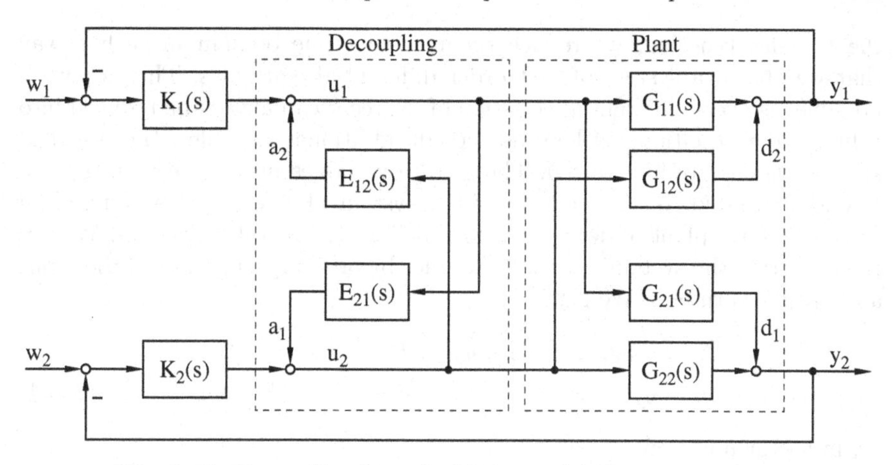

Fig. 2.46. Decoupling for a double-input-double-output system

It is the principle of state space control, which we will introduce in the next chapter.

2.7 State Space Description and State Space Control

2.7.1 Fundamentals

Definition of State Variables

We can characterize the methodology underlying the classical controller design procedures like this: The differential equations of the plant are transformed into the frequency domain, and then summed up to a transfer function for which—again in the frequency domain—a suitable controller has to be found. This procedure was developed in the middle of the 20th century, when control theory became an independent science. From the early 60s on, though, another—completely different—approach came up, which we introduce in this chapter. It is called the state space methodology and mostly due to Rudolf Kalman ([82]).

The important difference to the procedure used so far is the way in which internal variables of a system are treated. In the approaches mentioned so far we tried to eliminate internal variables and to develop a direct relation between the input and output variables with the transfer function. Exactly the contrary is the aim of the state space methodology. We focus on the system's internal variable, while the output variables receive only marginal attention. We show that examining the internal variables leads to a better understanding of the system's behavior.

The underlying principle is relatively simple. Instead of transforming the equations of the plant into the frequency domain and summing them up in the transfer function, we reduce them in the time domain in such a way that we obtain a system of first order differential equations. This is always possible, as every differential equation of degree k can always be reduced into k first order equations as long as enough additional variables, the so-called *state variables* are introduced. These state variables may correspond to real physical variable. In the end, we get a system of n first order differential equations for a plant of degree n, which defines the n state variables. We can then describe the system's output variables by ordinary functions of the input and output variables. We get:

$$\begin{aligned} \dot{x}_i &= f_i(x_1, x_2, ..., u_1, u_2, ..., t) \qquad i \in \{1, ..., n\} \\ y_j &= g_j(x_1, x_2, ..., u_1, u_2, ..., t) \qquad j \in \{1, ..., m\} \end{aligned} \tag{2.134}$$

or, in vector notation:

$$\begin{aligned} \dot{\mathbf{x}} &= \mathbf{f}(\mathbf{x}, \mathbf{u}, t) \\ \mathbf{y} &= \mathbf{g}(\mathbf{x}, \mathbf{u}, t) \end{aligned} \tag{2.135}$$

with $\mathbf{x} = [x_1, ..., x_n]^T$, $\mathbf{u} = [u_1, ...]^T$, $\mathbf{y} = [y_1, ..., y_m]^T$, $\mathbf{f} = [f_1, ..., f_n]^T$ and $\mathbf{g} = [g_1, ..., g_m]^T$.

These equations are called the system's *state space description*, where the u_i are the system's input variables, the x_i the state variables and the y_i the output variables. The f_i and g_i are initially any scalar functions of the state variables, the input variables and the time t. The output variables do not necessarily have to be different from the state variables. If an output variable is equal to a state variable, the corresponding function g_j is of course trivial: $y_j = g_j(\mathbf{x}, \mathbf{u}, t) = x_j$. In general, though, we have to assume that the number of output variables m is smaller than the number of state variables n.

Since $\mathbf{u}$ and $\mathbf{y}$ are defined as vectors, this approach obviously comprises MIMO systems. If necessary, we can find a direct relation between input and output variables from the state space description. We will give an example of this later on. A restriction to linear and time-invariant systems, as so far required by the application of the Laplace transform, can thus be avoided.

Characteristic Features of State Variables

In order to get a feeling of what a state variable represents, let us consider a simple example: Assume we are given a body of mass m, which is influenced by a force f_m (fig. 2.47). Effects caused by friction are neglected. The corresponding equations are:

$$\begin{aligned} f_m(t) &= m\, a(t) \\ a(t) &= \frac{dv(t)}{dt} \\ v(t) &= \frac{dl(t)}{dt} \end{aligned} \tag{2.136}$$

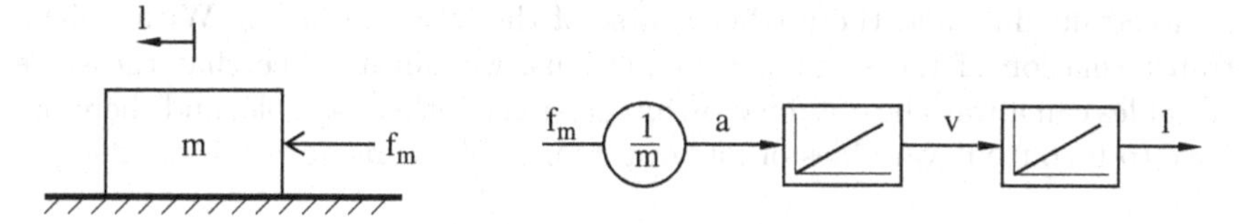

Fig. 2.47. Accelerated body

Starting at a certain time t_0, our task is to determine the body's position for a later time $t_1 > t_0$. A simple transformation of the given equations yields

$$
\begin{aligned}
a(t) &= \frac{1}{m} f_m(t) \\
v(t) &= \int_{t_0}^{t} a(\tau) d\tau + v(t_0) \\
l(t) &= \int_{t_0}^{t} v(\tau) d\tau + l(t_0)
\end{aligned}
\tag{2.137}
$$

and substituting into the last equation finally:

$$
l(t_1) = \int_{t_0}^{t_1} \left[\int_{t_0}^{\tau} \frac{1}{m} f_m(\sigma) d\sigma + v(t_0) \right] d\tau + l(t_0) \tag{2.138}
$$

We observe the following: The position $l(t_1)$ at a certain time can only be computed if the initial values $l(t_0)$ and $v(t_0)$ as well as the course of the system's input variable $f_m(t)$ are known for the interval $t \in [t_0, t_1]$. Earlier processes for $t < t_0$ play no role. From this it follows, that $l(t_0)$ and $v(t_0)$ obviously completely characterize the state of the system at time t_0. Knowing this state and the input variable $f_m(t)$, which affects the system from this time on, allows us to calculate any following state. $a(t_0)$ is not required, as we do not have to use integration for the computation of $a(t)$ from the input variable—in contrast to $v(t)$ and $l(t)$. Even more, we can eliminate the acceleration without further problems: It leads to a state space description of the system according to equation (2.134) with the state variables $x_1 = v$ and $x_2 = l$, the output variable $y = l$ and the input variable $u = f_m$:

$$
\begin{aligned}
\dot{x}_1 &= \frac{dv(t)}{dt} = \frac{1}{m} f_m(t) = f_1(u) \\
\dot{x}_2 &= \frac{dl(t)}{dt} = v(t) = f_2(x_1) \\
y &= l(t) = g(x_2)
\end{aligned}
\tag{2.139}
$$

Using the form of the two equations (2.135), we can also prove that the behavior of the system is clearly determined by the values of the state variables at a certain time and the further course of the input variables. With a little transformation of the state space equations, we can also see that the state variables can always be obtained by integration of other variable, and therefore have to be output variables of integrators in a block diagram (cf. fig. 2.47):

$$\mathbf{x}(t) = \int_{\tau=0}^{t} \mathbf{f}(\mathbf{x}(\tau), \mathbf{u}(\tau), \tau) d\tau + \mathbf{x}(0) \tag{2.140}$$

As in addition to this none of the arguments of the vector $\mathbf{f}$ is a derivative, the state variables are always the result of an integration over finite variable, and are therefore always continuous. Since the outputs of integrators can change only in a continuous manner, we can think of them as being a kind of a store, which might in many cases increase the comprehensibility of the state space description. For such store variables, we can choose continuous variables like the mass of a fluid in a tank or energy. The state variables then represent, for example, the energy content of the system.

If we sum up the state variables into a state vector $\mathbf{x} = [x_1, ..., x_n]^T$, this vector represents a point in the n-dimensional *state space*. Due to the continuity of the single components, these points form a trajectory in the course of time, a *state trajectory*. In figure 2.48, such a curve is given for the plant described above. Starting at the initial state $l(0) = v(0) = 0$, position and velocity increase for a constant positive acceleration. As during the initial phase the increase in velocity is greater than the one for the position, a parabolic curve results. This can be illustrated by the idea that velocity is needed first to obtain a following change in position. At time t_1 the force, or acceleration, respectively, switches to a negative value. The velocity decreases until the plant reaches the final position at time t_2. We can compute such a curve by eliminating the time from the equations describing the state variables, and describing v as a direct function of l.

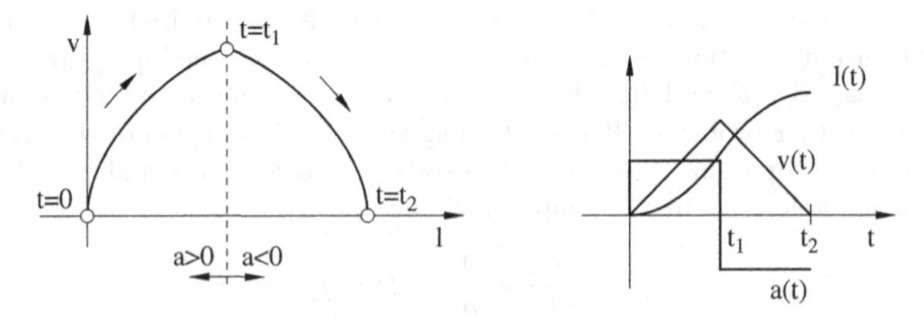

Fig. 2.48. Moving process with constant accelerations

State Space Description of Linear Systems

We can schematize the state space description even more, if we accept a limitation to linear and time-invariant systems without delays. This way the vector functions $\mathbf{f} = [f_1, ..., f_n]^T$ and $\mathbf{g} = [g_1, ..., g_m]^T$ become linear vector functions of the state variables and input variables. We can then write the equations (2.135) as:

$$\begin{aligned} \dot{\mathbf{x}} &= \mathbf{A}\mathbf{x} + \mathbf{B}\mathbf{u} \\ \mathbf{y} &= \mathbf{C}\mathbf{x} + \mathbf{D}\mathbf{u} \end{aligned} \tag{2.141}$$

$\mathbf{A}, \mathbf{B}, \mathbf{C}$ and $\mathbf{D}$ are matrices with constant coefficients. $\mathbf{A}$ is called the *system matrix*, $\mathbf{B}$ the *input matrix*, $\mathbf{C}$ the *output matrix* and $\mathbf{D}$ the *feed-through matrix*. Figure 2.49 illustrates these relations. The integrator in the figure represents a componentwise integration of the vector $\dot{\mathbf{x}}$.

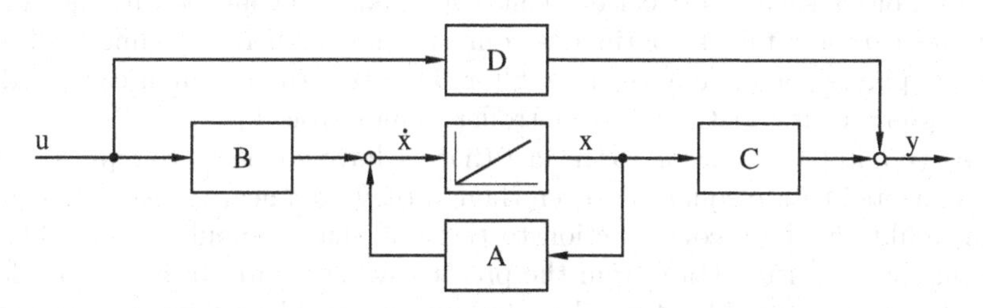

Fig. 2.49. State space description of a linear plant

For the SISO case, the system has only one input and one output variable, i.e. u and y are scalars, but still any number of state variables. $\mathbf{A}$ is for this reason always of degree $n \times n$, where n is the order of the plant. In contrast to this, $\mathbf{B}$ becomes an $n \times 1$-matrix for the single-input-single-output case, $\mathbf{C}$ a $1 \times n$-matrix and $\mathbf{D}$ a scalar.

In order to illustrate these differences, we extend our example from above by adding a spring and a velocity-dependent friction, which produces the system already shown in figure 2.4. The transformation of the equations (2.1) to (2.5) and the elimination of the acceleration gives the state equations

$$\begin{aligned} \frac{dv(t)}{dt} &= \frac{1}{m} f_m(t) - \frac{c_f}{m} v(t) - \frac{c_r}{m} l(t) \\ \frac{dl(t)}{dt} &= v(t) \end{aligned} \tag{2.142}$$

and the trivial output equation

$$l(t) = l(t) \tag{2.143}$$

describing the output. Here, it is useful to interpret the state variables as energy contents of the system. The position represents the displacement of the spring, and therefore as well the system's potential energy, while the velocity gives us a measure for the kinetic energy. Using matrix notation, the equations become

$$\dot{\mathbf{x}} = \begin{bmatrix} \dot{v} \\ \dot{l} \end{bmatrix} = \begin{bmatrix} -\frac{c_f}{m} & -\frac{c_r}{m} \\ 1 & 0 \end{bmatrix} \begin{bmatrix} v \\ l \end{bmatrix} + \begin{bmatrix} \frac{1}{m} \\ 0 \end{bmatrix} f_m = \mathbf{A}\mathbf{x} + \mathbf{B}\mathbf{u}$$

$$\mathbf{y} = l = \begin{bmatrix} 0 \; 1 \end{bmatrix} \begin{bmatrix} v \\ l \end{bmatrix} + 0 \, f_m = \mathbf{C}\mathbf{x} + \mathbf{D}\mathbf{u} \tag{2.144}$$

Because of $\mathbf{D} = \mathbf{0}$, in this example the output variable is a linear combination of the state variables and does not directly depend on the input variable. This way, the output variable as well as the state variables can change only in a continuous way, even for a step-like input variable. This holds for any type of plant where a low-pass filter, like an integrator or lag, forms part of the transmission path from the input to the output variable, because the output variable of such a transfer element—and hence the output variable of the system—can only take a continuous course. Since furthermore almost all real existing plants comprise a low-pass filter, $\mathbf{D} = \mathbf{0}$ is the common case, and is also a prerequisite for the some controller design procedures.

We should pay attention to the fact that we have no means to represent delay elements in state equations, even though they are linear transfer elements. This would also be a contradiction to the idea that it should be possible to compute every future state from the present system state: In order to calculate the output variable of a delay element we would need some knowledge about an older state because of its delay effect.

Normal Forms

In both examples used so far, the differential equations describing the plant were already of first order. Because of this, there was no need to introduce additional variables as state variables. They were obtained from the structure of the plant, and therefore had a real physical correspondence in form of the *velocity* and the *position*. However, it does not necessarily have to be this way. In principle, the state variables can be fixed according to any criteria, for example in order to give the system matrix $\mathbf{A}$ a special form. We can illustrate this by a simple example: Let a SISO system be given with the transfer function

$$G(s) = \frac{b_n s^n + b_{n-1} s^{n-1} + ... + b_0}{s^n + a_{n-1} s^{n-1} + ... + a_0} \tag{2.145}$$

First of all, we should try to develop a block diagram corresponding to this transfer function that only uses integrators and multiplications with constant factors (proportional elements). Since first multiplying a signal by a constant factor and then integrating produces exactly the same result as first integrating and then multiplying, there are obviously different possibilities to realize such a block diagram. Two of these possibilities are shown in figure 2.50.

They are two of the most frequently occurring forms, the so-called *controller canonical form* and the *observer canonical form*. These canonical forms have the advantage that the coefficients of the transfer function directly appear in the block diagram.

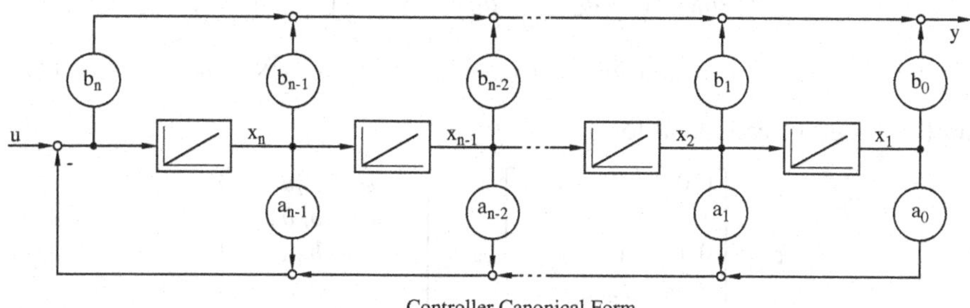

Controller Canonical Form

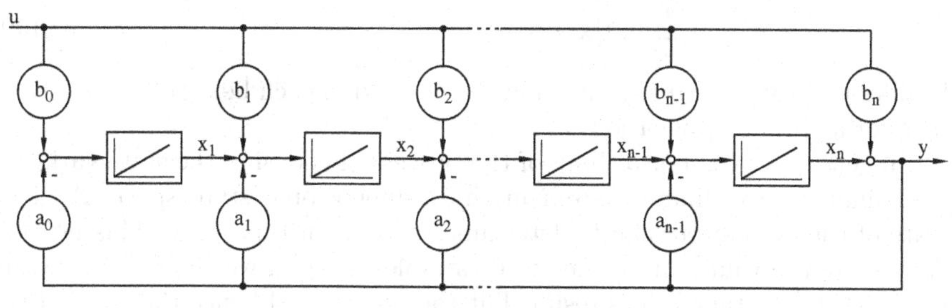

Observer Canonical Form

Fig. 2.50. Canonical forms of a rational transfer function

Both block diagrams contain n integrators each, and n also happens to be the system's degree. Hence, in order to give a state space description of the system, we need n state variables in addition to one input and one output variable. Since state variables are always obtained from integration of other variables, it seems reasonable to use the integrator output variables as state variables, as already shown in the block diagrams. For example, the equation describing the state variable x_n in observer canonical form is given by:

$$x_n(t) = \int_{\tau=0}^{t} b_{n-1}u(\tau) + x_{n-1}(\tau) - a_{n-1}(x_n(\tau) + b_n u(\tau))d\tau + x_n(0) \quad (2.146)$$

or, alternatively:

$$\dot{x}_n = b_{n-1}u + x_{n-1} - a_{n-1}(x_n + b_n u) \quad (2.147)$$

From the block diagrams, we get the system's complete state space description in controller canonical form

$$\dot{\mathbf{x}} = \begin{bmatrix} 0 & 1 & 0 & \cdots & 0 & 0 \\ 0 & 0 & 1 & \cdots & 0 & 0 \\ \cdots & \cdots & \cdots & \cdots & \cdots & \cdots \\ 0 & 0 & 0 & \cdots & 1 & 0 \\ 0 & 0 & 0 & \cdots & 0 & 1 \\ -a_0 & -a_1 & -a_2 & \cdots & -a_{n-2} & -a_{n-1} \end{bmatrix} \mathbf{x} + \begin{bmatrix} 0 \\ \cdots \\ \cdots \\ \cdots \\ 0 \\ 1 \end{bmatrix} u$$

$$y = [b_0 - b_n a_0, b_1 - b_n a_1, ..., b_{n-1} - b_n a_{n-1}]\mathbf{x} + b_n u \tag{2.148}$$

and in observer canonical form

$$\dot{\mathbf{x}} = \begin{bmatrix} 0\,0 \cdots 0 & -a_0 \\ 1\,0 \cdots 0 & -a_1 \\ 0\,1 \cdots 0 & -a_2 \\ \cdots\cdots & \cdots \\ 0\,0 \cdots 1 & -a_{n-1} \end{bmatrix} \mathbf{x} + \begin{bmatrix} b_0 - b_n a_0 \\ b_1 - b_n a_1 \\ b_2 - b_n a_2 \\ \cdots \\ b_{n-1} - b_n a_{n-1} \end{bmatrix} u$$

$$y = [0, ..., 0, 1]\mathbf{x} + b_n u \tag{2.149}$$

Besides these two normal forms many others exist, each being of some advantage for a special application.

In general, we can also think of the process of choosing the state variables as defining a coordinate system in the n-dimensional state space. A given state of the system at time t_1 determines a point in this space. This point is defined by the values of all the state variables at t_1. If we choose other state variables, then other values result, but the point—and hence the state of the system—always remains the same. Choosing different state variables merely corresponds to a change or a transformation respectively of the coordinate system.

Such a transformation can be described by a regular—and therefore as well invertible—transformation matrix $\mathbf{T}$. With this, we get $\mathbf{z} = \mathbf{T}^{-1}\mathbf{x}$ for the new state vector. Substituting $\mathbf{x} = \mathbf{Tz}$ and $\dot{\mathbf{x}} = \mathbf{T}\dot{\mathbf{z}}$ in (2.141) gives us the state space description for the new state variables:

$$\begin{aligned} \dot{\mathbf{z}} &= \mathbf{T}^{-1}\mathbf{ATz} + \mathbf{T}^{-1}\mathbf{Bu} \\ \mathbf{y} &= \mathbf{CTz} + \mathbf{Du} \end{aligned} \tag{2.150}$$

with the new system matrices

$$\mathbf{A}' = \mathbf{T}^{-1}\mathbf{AT} \qquad \mathbf{B}' = \mathbf{T}^{-1}\mathbf{B} \qquad \mathbf{C}' = \mathbf{CT} \qquad \mathbf{D}' = \mathbf{D} \tag{2.151}$$

Since we obtained the new system from the old one by a mere coordinate transformation, both representations must be fully equivalent.

Furthermore, the eigenvalues of the system matrix are defined by the zeros s_λ of the determinant

$$|s\mathbf{I} - \mathbf{A}| \tag{2.152}$$

Because of

$$|s\mathbf{I}-\mathbf{T}^{-1}\mathbf{A}\mathbf{T}| = |s\mathbf{T}^{-1}\mathbf{I}\mathbf{T}-\mathbf{T}^{-1}\mathbf{A}\mathbf{T}| = |\mathbf{T}^{-1}(s\mathbf{I}-\mathbf{A})\mathbf{T}| = |s\mathbf{I}-\mathbf{A}| \tag{2.153}$$

the eigenvalues of the new system matrix are the same as the eigenvalues of the old matrix. We will need this fact in the following derivations.

General Solution of a Linear State Equation

As another fundamental requirement for later calculations, we need the general solution of the state equation, which describes the direct relation between the system's input and output variable. The initial approximation for the solution is

$$\mathbf{x}(t) = e^{\mathbf{A}t}\mathbf{x}_0 + \int_0^t e^{\mathbf{A}(t-\tau)}\mathbf{B}\mathbf{u}(\tau)d\tau \tag{2.154}$$

with $\mathbf{x}_0 = \mathbf{x}(0)$ and the matrix-e-function

$$e^{\mathbf{A}t} := \mathbf{I} + \mathbf{A}t + \mathbf{A}^2\frac{t^2}{2!} + \mathbf{A}^3\frac{t^3}{3!} + ... = \sum_{k=0}^{\infty}\mathbf{A}^k\frac{t^k}{k!} \tag{2.155}$$

Now all that remains to be proven is that this approach actually meets the state equation

$$\dot{\mathbf{x}} = \mathbf{A}\mathbf{x} + \mathbf{B}\mathbf{u} \tag{2.156}$$

First, we can show that the series (2.155) is absolutely convergent for all matrices $\mathbf{A}$ and $|t| < \infty$. This allows differentiation by terms with respect to time and gives (in analogy to the scalar case):

$$\frac{d}{dt}e^{\mathbf{A}t} = \mathbf{A} + \mathbf{A}^2t + \mathbf{A}^3\frac{t^2}{2!} + \mathbf{A}^4\frac{t^3}{3!} + ... = \mathbf{A}e^{\mathbf{A}t} \tag{2.157}$$

Furthermore, we can make use of the Addition Theorem for the exponential function:

$$e^{\mathbf{A}t_1}e^{\mathbf{A}t_2} = e^{\mathbf{A}(t_1+t_2)} \tag{2.158}$$

Differentiation of the initial approximation (2.154) gives us

$$\mathbf{x}(t) = e^{\mathbf{A}t}\mathbf{x}_0 + e^{\mathbf{A}t}\int_0^t e^{-\mathbf{A}\tau}\mathbf{B}\mathbf{u}(\tau)d\tau$$

$$\begin{aligned}\dot{\mathbf{x}}(t) &= \mathbf{A}e^{\mathbf{A}t}\mathbf{x}_0 + \mathbf{A}e^{\mathbf{A}t}\int_0^t e^{-\mathbf{A}\tau}\mathbf{B}\mathbf{u}(\tau)d\tau + e^{\mathbf{A}t}e^{-\mathbf{A}t}\mathbf{B}\mathbf{u}(t) \\ &= \mathbf{A}\left[e^{\mathbf{A}t}\mathbf{x}_0 + \int_0^t e^{\mathbf{A}(t-\tau)}\mathbf{B}\mathbf{u}(\tau)d\tau\right] + \mathbf{B}\mathbf{u}(t) \\ &= \mathbf{A}\mathbf{x}(t) + \mathbf{B}\mathbf{u}(t)\end{aligned} \tag{2.159}$$

and thereby we can prove that (2.154) meets the state equation (2.156) and is thus a solution for this equation. Looking at it we can see that a certain state always depends on the initial state $\mathbf{x}_0$ and the course of the activation $\mathbf{u}(t)$ following it. If no external activation is given ($\mathbf{u}(t) = \mathbf{0}$), then the solution is reduced to the term $e^{\mathbf{A}t}\mathbf{x}_0$. The initial state $\mathbf{x}_0$ has to be multiplied by the matrix $e^{\mathbf{A}t}$ in order to produce the current state vector. For this reason, the matrix is also called the *transition matrix*.

We obtain the output variables from the state variables according to the output equation

$$\begin{aligned} \mathbf{y}(t) &= \mathbf{C}\mathbf{x}(t) + \mathbf{D}\mathbf{u}(t) \\ &= \mathbf{C}e^{\mathbf{A}t}\mathbf{x}_0 + \int_0^t \mathbf{C}e^{\mathbf{A}(t-\tau)}\mathbf{B}\mathbf{u}(\tau)d\tau + \mathbf{D}\mathbf{u}(t) \end{aligned} \tag{2.160}$$

2.7.2 Controllability and Observability

Using solution (2.154) we can can derive two system features which are very important for controlling linear and nonlinear plants as well, and therefore they are also very important for fuzzy controllers. For a thorough understanding of the two features, we first need to give a brief introduction to the concept of the so-called *state space control*. While for all types of controllers which we have introduced so far the plant's output variables were the variables to be controlled, in the case of a state space controller this will be the state variables. Now by controlling the state variables all the internal processes are subject to permanent control, which also makes the entire system much more manageable. Adjusting the output variables to given desired values should no longer be a problem, as they merely represent a linear combination of the state variables. But if we want to use a state space control, a system has to fulfill two criteria: *controllability* and *observability*. Both criteria which we are going to discuss now were introduced by Rudolf Kalman in 1960 ([82]).

Controllability is concerned with the problem whether the structure of a system permits to influence the state variables with the given actuating variables in a desired way. If this is the case, then the system is said to be controllable. Taking suitable influence on the state variables is again only possible if the course they take is known. Since usually only the system's output variables are available as measurable variables, we have to make sure that every state variable influences the output variables in a special way, in order to allow us to determine the course of the state variables from the measured output variables. If this holds, such a system is called observable.

We should therefore obviously examine and establish the system's controllability and observability before we start to design a controller. Both features depend only on the system structure, not on the type of control used. A plant which is not controllable or observable can therefore only be transfered into one which has these qualities by changing the configuration, and not by choosing another control algorithm. Concerning the controllability, these changes affect the type or number of the actuating variables, concerning the observability the type or number of the measured output variables. Controllability as well as observability are fundamental characteristics of a system—not a special problem of the state space control. Their systematic analysis, though, could only be established after the state space description had been developed.

Because of the importance of these two concepts, we ought to discuss them a little more in detail, starting with controllability. Two examples might help to make comprehension of this feature easier. The state space description for the first example is:

$$\begin{bmatrix} \dot{x}_1 \\ \dot{x}_2 \end{bmatrix} = \begin{bmatrix} a_{11} & 0 \\ 0 & a_{22} \end{bmatrix} \begin{bmatrix} x_1 \\ x_2 \end{bmatrix} + \begin{bmatrix} 0 \\ b_2 \end{bmatrix} u \tag{2.161}$$

Since obviously the only activation for the state variable x_1 is x_1 itself, taking influence via the actuating variable is not possible. The system is therefore not controllable. But a system is also said to be not controllable if its state variables cannot be influenced independently of each other. This is the case for the next example.

$$\begin{bmatrix} \dot{x}_1 \\ \dot{x}_2 \end{bmatrix} = \begin{bmatrix} a & 0 \\ 0 & a \end{bmatrix} \begin{bmatrix} x_1 \\ x_2 \end{bmatrix} + \begin{bmatrix} b \\ b \end{bmatrix} u \tag{2.162}$$

Here, not only the input variable u affects the two state variables in the same way, but also the internal feedback is the same for both variables. As a result of this, for example no transformation from the initial state $[x_1, x_2] = [1, 2]$ into the zero state $[0, 0]$ is possible for this system. A definition of controllability is therefore:

Definition 2.11 *A system is called (state-)controllable, if by means of a suitable course of its input variables a transformation from any initial state into the final state* $\mathbf{0}$ *is possible.*

The final state $\mathbf{0}$ imposes no special restrictions, since using a coordinate transformation we can make any point become the origin of our coordinate system. The question is whether or not it is already possible to determine the system's controllability by looking at its system matrices. With the following derivation, we will show that this is possible.

Our starting point is the general solution of the state equation:

$$\mathbf{x}(t) = e^{\mathbf{A}t}\mathbf{x}_0 + \int_0^t e^{\mathbf{A}(t-\tau)}\mathbf{B}\mathbf{u}(\tau)d\tau \tag{2.163}$$

After a finite time t_1, we expect the following equation to be true:

$$\mathbf{0} = \mathbf{x}(t_1) = e^{\mathbf{A}t_1}\mathbf{x}_0 + \int_0^{t_1} e^{\mathbf{A}(t_1-\tau)}\mathbf{B}\mathbf{u}(\tau)d\tau \tag{2.164}$$

From this, it follows that

$$-\mathbf{x}_0 = \int_0^{t_1} e^{-\mathbf{A}\tau}\mathbf{B}\mathbf{u}(\tau)d\tau \tag{2.165}$$

The system is controllable, if the matrices $\mathbf{A}$ and $\mathbf{B}$ have such a structure, so that for any $\mathbf{x}_0$ a course of the actuating variable $\mathbf{u}(\tau)$ exists, so that this equation holds.

The further computations shall give a condition for $\mathbf{A}$ and $\mathbf{B}$, that is easy to check. For this, we need the following theorem, that shall not be proven here:

Theorem 2.12 *Given an $n \times n$ matrix $\mathbf{A}$ and a function $\mathbf{F}$ of degree $p \geq n$ with*

$$\mathbf{F} = \mathbf{F}(\mathbf{A}^p, \mathbf{A}^{p-1}, \mathbf{A}^{p-2}, ..., \mathbf{A}) \tag{2.166}$$

then $\mathbf{F}$ can be replaced by the function $\mathbf{H}$ of degree $n-1$ with

$$\mathbf{F} = \mathbf{H}(\mathbf{A}^{n-1}, \mathbf{A}^{n-2}, ..., \mathbf{A}) \tag{2.167}$$

Concerning the matrix exponential function it follows that

$$e^{\mathbf{A}\tau} = \sum_{k=0}^{\infty} \mathbf{A}^k \frac{\tau^k}{k!} = \sum_{k=0}^{n-1} c_k(\tau)\mathbf{A}^k \tag{2.168}$$

Substituting this in equation (2.165) yields:

$$\begin{aligned}
-\mathbf{x}_0 &= \int_0^{t_1} \sum_{k=0}^{n-1} c_k(-\tau)\mathbf{A}^k\mathbf{B}\mathbf{u}(\tau)d\tau \\
&= \sum_{k=0}^{n-1} \mathbf{A}^k\mathbf{B} \int_0^{t_1} c_k(-\tau)\mathbf{u}(\tau)d\tau \\
&= \sum_{k=0}^{n-1} \mathbf{A}^k\mathbf{B}\mathbf{z}_k \qquad \text{where} \qquad \mathbf{z}_k = \int_0^{t_1} c_k(-\tau)\mathbf{u}(\tau)d\tau
\end{aligned} \tag{2.169}$$

Writing out the sum gives:

$$-\mathbf{x}_0 = \underbrace{[\mathbf{B}, \mathbf{A}\mathbf{B}, \mathbf{A}^2\mathbf{B}, ..., \mathbf{A}^{n-1}\mathbf{B}]}_{\mathbf{M}} \begin{bmatrix} \mathbf{z}_0 \\ \mathbf{z}_1 \\ \cdots \\ \mathbf{z}_{n-1} \end{bmatrix} \tag{2.170}$$

$\mathbf{M}$ is an $n \times (np)$ matrix where n is the number of state variables and p the number of actuating variables. $-\mathbf{x}_0$ is a linear combination of the np columns of $\mathbf{M}$. Obviously, for any $\mathbf{x}_0$ a solution can exist only if the column vectors of $\mathbf{M}$ span the entire n-dimensional space inside of which $\mathbf{x}_0$ can possibly lie:

Theorem 2.13 *A system is controllable if and only if the matrix*

$$\mathbf{M} = \left[\mathbf{B}, \mathbf{AB}, \mathbf{A}^2\mathbf{B}, ..., \mathbf{A}^{n-1}\mathbf{B}\right] \tag{2.171}$$

contains n linearly independent column vectors.

So far, we have only shown that controllability is a consequence of $\mathbf{M}$ having the highest possible rank n, while the theorem also states the counter-implication, but we will omit this proof here.

Controllability is sometimes also referred to as *state controllability*, in order to distinguish this term from the so-called *output controllability*, which refers to the possibility of influencing the output variables. As a further remark we should mention that other criteria for controllability exist besides this one due to Kalman, which can also be applied to linear plants only. For nonlinear plants—the favorite field of application for fuzzy controllers—such criteria so far have not been developed. Still, even if working in this field we should acquire a basic understanding of the concept of controllability as it will always be important to know whether we can influence a system in the desired way using the given actuating variables or not.

Our discussion of observability will not be as detailed as the one of controllability, due to its analogy to the latter. We can define observability like this:

Definition 2.14 *A system is observable if and only if any initial state vector $\mathbf{x}(t_0)$ can be reconstructed from the input and output variables $\mathbf{u}(t)$ and $\mathbf{y}(t)$, respectively, measured over a finite period of time $t \in [t_0, t_1]$.*

For practical applications, the definition would be more useful if it would deal with the possibility to compute the current state vector $\mathbf{x}(t_1)$ from the variables measured so far. This feature also exists and is called *reconstructability*. For linear time-invariant systems reconstructability is equivalent to observability. We also give the criterion of observability according to Kalman (without proof):

Theorem 2.15 *A system is observable if and only if the matrix*

$$\begin{bmatrix} \mathbf{C} \\ \mathbf{CA} \\ \mathbf{CA}^2 \\ \cdots \\ \mathbf{CA}^{n-1} \end{bmatrix} \tag{2.172}$$

contains n linearly independent row vectors.

2.7.3 Lyapunov's Definition of Stability for Linear Systems

An even more important feature than controllability and observability is the stability of a system. Now, that we have knowledge about the internal behaviour of a system, our earlier definitions of stability may seem somehow incomplete, as they are only based on the system's output variables in dependency to its input variables. For this reason, we present another definition originated by M.A. Lyapunov ([116]), which is limited to linear systems:

Definition 2.16 *A linear system is asymptotically stable if and only if its state variables converge to zero from any initial state without outer activation:*

$$\lim_{t\to\infty} \mathbf{x}(t) = \mathbf{0} \qquad \textit{with} \qquad \mathbf{u}(t) = \mathbf{0} \tag{2.173}$$

A stable system reaches a stable state on its own, independently of what the initial state is like. This definition will be generalized in chapter 2.8.4; furthermore, we will explain the difference between simple and asymptotic stability explained there, too. In the case of linear systems, stability is usually used in the sense of asymptotic stability. We can therefore omit any further differentiation.

In contrast to our earlier definitions 2.4 (finite step response) and 2.5 (bounded input - bounded output), where we related input to output variable in order to define the term stability, this time we use the system's internal processes following an initial state. Just like for controllability and observability the question arises whether or not stability can be determined by looking at the system matrices.

As the definition of stability according to Lyapunov assumes a system without any outer activation ($\mathbf{u} = \mathbf{0}$) and without respect to the output variables, the state equations (2.141) are reduced to a homogeneous differential equation in vector notation:

$$\dot{\mathbf{x}} = \mathbf{A}\mathbf{x} \tag{2.174}$$

For the definition of stability we now have to analyze on which conditions the solution to this equation converges to zero for any initial values. The further analysis will be easier if we use the Laplace transform of the equation instead of the equation itself. Because of the equation's linearity the Laplace transformation is allowed. In order to do so, we can transform the vectors component by component, in the same way as it is done for differentiation or integration. Using the Differentiation Theorem of the Laplace transform we get:

$$s\mathbf{x}(s) - \mathbf{x}_0 = \mathbf{A}\mathbf{x}(s) \tag{2.175}$$

or, alternatively:

$$\mathbf{x}(s) = (s\mathbf{I} - \mathbf{A})^{-1}\mathbf{x}_0 \tag{2.176}$$

Applying Cramer's Rule to the inverse matrix yields:

$$\mathbf{x}(s) = \frac{\mathbf{P}(s)}{|s\mathbf{I} - \mathbf{A}|} \mathbf{x}_0 \tag{2.177}$$

Here, $\mathbf{P}(s)$ is a matrix of polynomials, i.e. its single entries are polynomials in s. The notation (2.177) is of course only possible if the inverse exists, i.e. the determinant of $|s\mathbf{I} - \mathbf{A}|$ is different from zero. Now if s_i are the eigenvalues of $\mathbf{A}$, then

$$|s\mathbf{I} - \mathbf{A}| = \prod_{i=1}^{n} (s - s_i) \tag{2.178}$$

holds for the determinant, which we can use to expand every single element of $(s\mathbf{I} - \mathbf{A})^{-1}$ into partial fractions. Using matrix notation we obtain:

$$\mathbf{x}(s) = (s\mathbf{I} - \mathbf{A})^{-1}\mathbf{x}_0 = \frac{\mathbf{P}(s)}{\prod\limits_{i=1}^{n} (s - s_i)} \mathbf{x}_0 = \sum_{\mu=1}^{l} \sum_{\nu=1}^{r_\mu} \frac{\mathbf{M}_{\mu\nu}}{(s - s_\mu)^\nu} \mathbf{x}_0 \tag{2.179}$$

where l is the number of eigenvalues of the system and r_μ the multiplicity for the eigenvalue s_μ. $\mathbf{M}_{\mu\nu}$ is a matrix of constant coefficients. Finally, the retransformation into the time domain produces:

$$\mathbf{x}(t) = \sum_{\mu=1}^{l} e^{s_\mu t} \sum_{\nu=1}^{r_\mu} \frac{t^{\nu-1}}{(\nu-1)!} \mathbf{M}_{\mu\nu}\mathbf{x}_0 \tag{2.180}$$

We see, that every component of $\mathbf{x}(t)$ contains products of exponential functions and polynomials in t. In such a product, the exponential function is always the dominating term. If its real part is positive, then the product exceeds all limits. If it is negative, the exponential function converges so fast towards zero, that we can also neglect the contribution of the polynomial. Accordingly, the vector $\mathbf{x}(t)$ converges towards zero if and only if the real part of all coefficients s_μ is negative.

Theorem 2.17 *A linear time-invariant system is asymptotically stable in the sense of the definition according to Lyapunov if and only if all the eigenvalues of the system matrix* $\mathbf{A}$ *have a negative real part.*

The eigenvalues determine not only the system's stability, but also the form of the initial transients, as (2.180) shows. Depending on the size of the real parts, the system converges faster or slower to zero. In case of a pair of complex conjugated eigenvalues the system oscillates, just like for the scalar case at a complex conjugated pair of poles. The stability is independent of the state variables, as a change of the base $\mathbf{T}^{-1}\mathbf{A}\mathbf{T}$ of a matrix $\mathbf{A}$ does not affect the eigenvalues.

A question remains of how we can relate the new concept of stability to the two old definitions. In order to find an answer, we first transform the entire state space description (2.141) into the frequency range:

$$\begin{aligned} s\mathbf{x}(s) - \mathbf{x}_0 &= \mathbf{A}\mathbf{x}(s) + \mathbf{B}\mathbf{u}(s) \\ \mathbf{y}(s) &= \mathbf{C}\mathbf{x}(s) + \mathbf{D}\mathbf{u}(s) \end{aligned} \tag{2.181}$$

If $|s\mathbf{I} - \mathbf{A}| \neq \mathbf{0}$, we can form the inverse $(s\mathbf{I} - \mathbf{A})^{-1}$, and the following relation is true:

$$\mathbf{x}(s) = (s\mathbf{I} - \mathbf{A})^{-1}\mathbf{B}\mathbf{u}(s) + (s\mathbf{I} - \mathbf{A})^{-1}\mathbf{x}_0 \tag{2.182}$$

Substituting this into the initial output equation gives

$$\mathbf{y}(s) = \underbrace{(\mathbf{C}(s\mathbf{I} - \mathbf{A})^{-1}\mathbf{B} + \mathbf{D})}_{\mathbf{G}(s)}\mathbf{u}(s) + \mathbf{C}(s\mathbf{I} - \mathbf{A})^{-1}\mathbf{x}_0 \tag{2.183}$$

Looking at this equation, we find that we can interpret $\mathbf{G}(s)$ as the system's transfer matrix, which characterizes the transfer characteristic between the input and output of the system. In such an interpretation, the term depending on $\mathbf{x}_0$ represents the effect an initial disturbance has on the output variable. An element $G_{ik}(s)$ of $\mathbf{G}(s)$ can be regarded as a transfer function from the input variable u_k to the output variable y_i. For the SISO case, $\mathbf{G}(s)$ is reduced to an ordinary transfer function.

According to (2.177), we can represent the inverse $(s\mathbf{I} - \mathbf{A})^{-1}$ by a quotient of a polynomial-matrix and the determinant:

$$(s\mathbf{I} - \mathbf{A})^{-1} = \frac{\mathbf{P}(s)}{|s\mathbf{I} - \mathbf{A}|} \tag{2.184}$$

As the denominator in this fraction is not affected by the multiplication by the constant matrices $\mathbf{B}$ and $\mathbf{C}$, the determinant has to be the (common) denominator of all transfer functions $G_{ik}(s)$ in $\mathbf{G}(s)$. Accordingly, the determinants' zeros, i.e. the eigenvalues of the system matrix, form the poles of all scalar transfer functions $G_{ik}(s)$.

Now according to theorem 2.6 these poles determine the system's stability. If the system is stable in the sense of the definition according to Lyapunov (i.e. all eigenvalues of the system matrix have a negative real part), then the same has to be true for all poles of the transfer functions $G_{ik}(s)$, and therefore, the transfer functions are stable, too. But if all elements $G_{ik}(s)$ of $\mathbf{G}(s)$ are stable scalar transfer functions, then the entire system is stable in the sense of the definitions 2.4 and 2.5. Thus, the earlier mentioned transfer stability is a consequence of Lyapunov-stability.

But vice versa, this does not necessarily hold, as poles and zeros of the transfer functions may cancel out each other and therefore not all eigenvalues of the system matrix have to be poles of the transfer functions. If all poles have a negative real part, there might be some eigenvalues of the system matrix left that do not have. The concept of stability according to Lyapunov is therefore more general than our older definitions. This should not strike us, since if all internal processes of a system are stable, there cannot be an output variable with an instable course. On the other hand, a system might appear to be stable at the outside, while internal variables in fact are instable and they are just not noticed because they compensate each other or their depending output variables are not measured.

2.7.4 Design of a State Space Controller

Now, after this thorough discussion of the characteristics of a linear system given in state space description, let us describe the design of a linear state space controller. The idea of this representation is to provide an overview of the methods and possibilities of classical, linear control theory. We should then be able to decide whether the use of a fuzzy controller instead of a classical one will really be of use for the individual case.

Concept

In the following, we assume that $\mathbf{D} = \mathbf{0}$ and $\mathbf{B}$ and $\mathbf{C}$ both have the highest rank, i.e. the columns of $\mathbf{B}$ and the rows of $\mathbf{C}$ are linearly independent. If $\mathbf{B}$ was not of highest rank, then the vector of actuating variables would have more components than needed. Of course, this case can appear in real applications, but it only makes the theory more complicated and does not lead to an improvement of the automatic control. Similar comments can be made concerning $\mathbf{C}$. Here, the highest rank implies that all output variables are linearly independent. If one output variable was linearly dependent of the others, it would only be redundant information, unimportant for the controller design process. This latter case appears relatively often, especially in safety relevant applications. Redundant information, though, should be collected and summed up before being processed to the controller as a measured variable. We can therefore neglect this case in our presentation.

In addition to this $\mathbf{D}$ is usually equal to $\mathbf{0}$ for most of the real applications, that are normally low-pass plants. This can be illustrated by the fact that $\mathbf{D}$ represents the direct connection between input and output variable. For $\mathbf{D} \neq \mathbf{0}$, a step of an input variable also causes a step of at least one output variable. This is not possible for plants with a low-pass characteristic. Therefore this condition also imposes no special restrictions on the applicability of the design processes.

With the knowledge about the internal variables, we now have a simple and elegant way to influence the system, which is illustrated in figure 2.51. All we do is multiply the state vector by the constant matrix $\mathbf{F}$ (all the elements of the matrix are constants), and feed it back to the actuating variable.

The state space description of this system is given by:

$$\begin{aligned} \dot{\mathbf{x}} &= (\mathbf{A} + \mathbf{BF})\mathbf{x} + \mathbf{Bw} \\ \mathbf{y} &= \mathbf{Cx} \end{aligned} \tag{2.185}$$

We can see that using this automatic control yields a new system with the system matrix $(\mathbf{A} + \mathbf{BF})$. We now have to construct the matrix $\mathbf{F}$ in a way

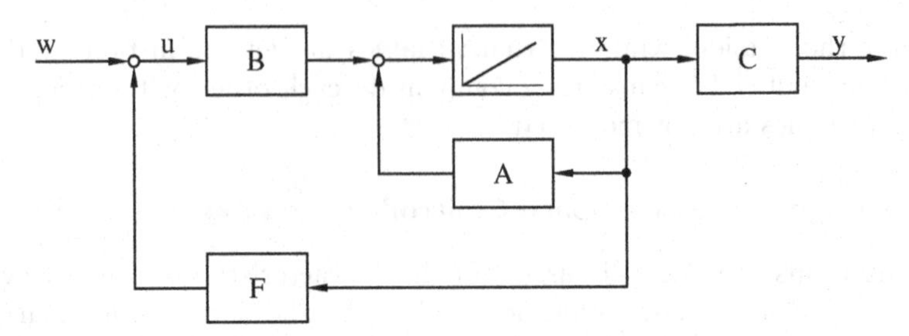

Fig. 2.51. The basic structure of a state space control

that the system matrix shows a suitable configuration of eigenvalues, i.e. that we obtain stability and sufficient damping. If we can achieve this, then the system will—for a constant input variable—always reach a stable steady state from the initial state. In addition to this, steady-state accuracy still has to be guaranteed, i.e. in this steady state the output variable $\mathbf{y}$ has to take the desired value $\mathbf{w}$. In the following paragraphs we discuss two procedures for the construction of the matrix $\mathbf{F}$ in more detail, as well as some methods which can be used to obtain steady-state accuracy.

The basic difference between a state space controller according to figure 2.51 and the PID-controller described earlier is that we need to know all the state variables if we want to use a state space control. This might be a problem for applications in the field, as usually not all of the state variables can be measured. Using so-called observers (which we will present later on) provides a remedy for this. If a plant is observable, an observer allows the computation of all state variables from the measured input and output variables.

Another difference compared to the PID-controller is the state space controller's lack of internal dynamics. The latter is described merely by a constant matrix $\mathbf{F}$, while for the former differential equations are used. This can be explained by the observation that the only source of information for a PID-controller are the output variables of the plant, while a state space controller has a permanent access to all internal information about the entire plant. The PID-controller has to compensate this deficiency of information by more complex internal calculations. In contrast to this, the state space controller corresponds to a simple, multidimensional proportional element.

Many procedures for the design of the matrix $\mathbf{F}$ have been developed. Just like the ones for the PID-controllers, each of these methods has certain advantages and disadvantages. Which procedure is the most suitable once again has to be decided for every individual case separately. The derivation of the methods is relatively complicated. For this reason we only present two of the most frequently used methods in the following sections.

Pole Placement

One standard procedure is the *pole placement*, in which the eigenvalues of the system matrix $(\mathbf{A}+\mathbf{BF})$ have to be given in order to allow computation of the matrix $\mathbf{F}$ describing the controller (*controller-matrix*). In order to keep things simple, we discuss this method for SISO plants only and neglect proof of the formulas (see [1] or [133] for details).

We need to know the plant's characteristic polynomial, i.e. the denominator of the transfer function

$$|s\mathbf{I}-\mathbf{A}| = s^n + q_{n-1}s^{n-1} + ... + q_1 s + q_0 \tag{2.186}$$

Then, we can choose the characteristic polynomial of the closed loop, i.e. its coefficients p_k, according to our own preferences. The only condition is that all of its zeros must have a negative real part so that the closed loop is stable:

$$|s\mathbf{I}-(\mathbf{A}+\mathbf{BF})| = s^n + p_{n-1}s^{n-1} + ... + p_1 s + p_0 \tag{2.187}$$

Both polynomials can be described by their coefficient vectors:

$$\begin{aligned} \mathbf{q} &= \left[q_{n-1}\ q_{n-2}\ \cdots\ q_0\right] \\ \mathbf{p} &= \left[p_{n-1}\ p_{n-2}\ \cdots\ p_0\right] \end{aligned} \tag{2.188}$$

Using

$$\mathbf{W} = \left[\mathbf{B}\ \mathbf{AB}\ \mathbf{A}^2\mathbf{B}\ \cdots\ \mathbf{A}^{n-1}\mathbf{B}\right] \begin{bmatrix} 1 & q_{n-1} & q_{n-2} & \cdots & q_1 \\ 0 & 1 & q_{n-1} & \cdots & q_2 \\ 0 & 0 & 1 & \cdots & q_3 \\ \cdots & \cdots & \cdots & \cdots & \cdots \\ 0 & 0 & 0 & \cdots & 1 \end{bmatrix} \tag{2.189}$$

we get the design equation for the controller:

$$\mathbf{F} = (\mathbf{q}-\mathbf{p})\mathbf{W}^{-1} \tag{2.190}$$

In this last expression we have to pay attention to the fact that for the SISO-case explained here the input matrix $\mathbf{B}$ is only a simple vector of dimension $n \times 1$. Because of this, $\mathbf{W}$ becomes a matrix of dimension $n \times n$. Existence of a solution, i.e. a controller $\mathbf{F}$, obviously depends on the invertability of $\mathbf{W}$. Now this matrix is a product of a triangular matrix, which is always invertible, and the controllability matrix, which is—for controllable systems—of rank n and therefore also invertible. Accordingly, it is not possible to compute a controller for non-controllable systems. The reason is plain to see: with the pole placement method we try to modify the system's eigenvalues, and in doing so we also manipulate the transient behavior of all state variables. This can, of course, only succeed if all state variables can in principle be influenced.

It may also suffice if we can influence at least those state variables which would be instable without control. This is called a *stabilizable* system:

Definition 2.18 *A system given by* $(\mathbf{A}, \mathbf{B})$ *is said to be stabilizable if a controller-matrix* $\mathbf{F}$ *exists, so that all eigenvalues of* $(\mathbf{A} + \mathbf{BF})$ *(the system matrix of the controlled system) have a negative real part.*

For such systems, a (slightly modified) version of the pole placement method is still possible.

It should be mentioned that for SISO plants, according to equation (2.190), exactly one controller—if one at all—exists which can be used to obtain a given configuration of eigenvalues for the closed loop. For MIMO systems, on the other hand, infinitely many controllers or solutions respectively exists if the plant is controllable. In this case further criteria have to be taken into account in order to decide for the optimal controller.

The following example illustrates the pole placement method and also the structure of a state space controller. Let a third order SISO plant be given in controller canonical form (fig. 2.50). According to (2.148), the matrices $\mathbf{A}$ and $\mathbf{B}$ are:

$$\mathbf{A} = \begin{bmatrix} 0 & 1 & 0 \\ 0 & 0 & 1 \\ -a_0 & -a_1 & -a_2 \end{bmatrix} \qquad \mathbf{B} = \begin{bmatrix} 0 \\ 0 \\ 1 \end{bmatrix} \tag{2.191}$$

Therefore, the controllability matrix is given by

$$\begin{bmatrix} \mathbf{B} \; \mathbf{AB} \; \mathbf{A}^2\mathbf{B} \end{bmatrix} = \begin{bmatrix} 0 & 0 & 1 \\ 0 & 1 & -a_2 \\ 1 & -a_2 & -a_1 + a_2^2 \end{bmatrix} \tag{2.192}$$

leaving $\mathbf{W}$ to be

$$\mathbf{W} = \begin{bmatrix} 0 & 0 & 1 \\ 0 & 1 & -a_2 \\ 1 & -a_2 & -a_1 + a_2^2 \end{bmatrix} \begin{bmatrix} 1 & a_2 & a_1 \\ 0 & 1 & a_2 \\ 0 & 0 & 1 \end{bmatrix} = \begin{bmatrix} 0 & 0 & 1 \\ 0 & 1 & 0 \\ 1 & 0 & 0 \end{bmatrix} = \mathbf{W}^{-1} \tag{2.193}$$

If we fix a characteristic polynomial $\mathbf{p} = \begin{bmatrix} p_{n-1} \; p_{n-2} \cdots p_0 \end{bmatrix}$ for the closed loop, we obtain

$$\begin{aligned} \mathbf{F} = (\mathbf{q} - \mathbf{p})\mathbf{W}^{-1} &= \left(\begin{bmatrix} a_2 \; a_1 \; a_0 \end{bmatrix} - \begin{bmatrix} p_2 \; p_1 \; p_0 \end{bmatrix}\right) \begin{bmatrix} 0 & 0 & 1 \\ 0 & 1 & 0 \\ 1 & 0 & 0 \end{bmatrix} \\ &= \begin{bmatrix} a_0 - p_0 \;\; a_1 - p_1 \;\; a_2 - p_2 \end{bmatrix} \end{aligned} \tag{2.194}$$

for the controller matrix. The block diagram 2.52 illustrates that, by adding the controller feedback, every single coefficient a_i of the plant is replaced by the coefficient p_i of our polynomial $\mathbf{p}$. This very simple procedure is a result of the special structure of the controller canonical form. (This is why it was given this name.)

The advantage of the pole placement method is its simplicity. Its disadvantage is that choosing the coefficients p_i requires a certain amount of intuition

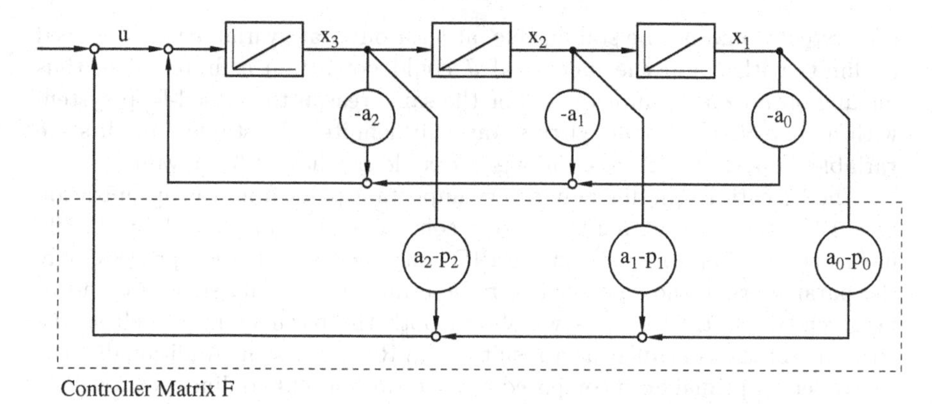

Fig. 2.52. State space control of a plant in controller canonical form

and expertise. It is difficult to keep track of which effects are caused by which eigenvalues, especially for MIMO-systems. Because of this, we usually need some tries to find a good controller for a given plant, i.e. to define suitable coefficients for the closed-loop eigenvalues.

Riccati Design

Another method is the design of an *optimal state space controller* ([135]). Here, our aim is to find the controller which minimizes the functional

$$J = \int_0^\infty \left(\mathbf{x}^T(t)\mathbf{Q}\mathbf{x}(t) + \mathbf{u}^T(t)\mathbf{R}\mathbf{u}(t)\right) dt \tag{2.195}$$

while converting the system from the initial state into a steady state. The functional will be small if the state vector converges rapidly to the zero vector, and if this convergence is achieved by using small values of the actuating variables. In principle, $\mathbf{Q}$ and $\mathbf{R}$ represent adjustable weights which we can use to manipulate the course of the actuating variables and the state variables. Both matrices have to be symmetrical. In addition to that, $\mathbf{Q}$ has to be positive semi-definite, and $\mathbf{R}$ positive definite. For a stabilizable system $(\mathbf{A}, \mathbf{B})$ the solution to this optimization problem is given by the controller matrix

$$\mathbf{F} = -\mathbf{R}^{-1}\mathbf{B}^T\mathbf{P} \tag{2.196}$$

with the (positive definite) solution $\mathbf{P}$ of the algebraic *Riccati equation*

$$\mathbf{A}^T\mathbf{P} + \mathbf{P}\mathbf{A} - \mathbf{P}\mathbf{B}\mathbf{R}^{-1}\mathbf{B}^T\mathbf{P} + \mathbf{Q} = \mathbf{0} \tag{2.197}$$

Because of this, the resulting controller is often called *Riccati controller*. The stabilizability of the plant is a requirement for the existence of a solution. If the system would not be stabilizable, at least one state variable would exceed all limits with t and the functional J could not attain a finite value, thus making optimization impossible. For the same reason the closed-loop system with a controller constructed this way will definitely be stable, i.e. all state variables converge to zero. Otherwise, J could not have a finite value.

The Riccati design differs in a very important point from the optimization of a PID-controller with regard to a control quality functional (as briefly mentioned in chapter 2.6.3). In the PID controller optimization process, only the parameters of the controller were determined, while its structure had to be given. In contrast to this we obtain both the parameters as well as the structure of the controller as a result of the Riccati design. Additionally, the controller is optimal even compared to time-variant and nonlinear controllers, as can be shown by calculus of variation.

Since solving a Riccati equation is a common problem for which suitable numerical algorithms are available ([105]), we can design an optimal controller using (2.196) automatically once we have fixed $\mathbf{A}, \mathbf{B}, \mathbf{R}$ and $\mathbf{Q}$. But we still need some intuition and experience in order to find suitable weight matrices $\mathbf{R}$ and $\mathbf{Q}$, as, in the end, the definition of the functional determines the controller. Finally, it should be pointed out that many other functionals exist which lead to entirely different controllers. The basic idea, however—minimizing the functional—is the same for all approaches.

2.7.5 Linear Observers

Having introduced the two most well-known methods for the design of state space controllers, we now discuss the observers which we have mentioned already. With the help of an observer the state vector $\mathbf{x}$ can be calculated from the vector of measured output variables $\mathbf{y}$. If we cannot directly measure the state variables needed by the controller and we have to compute them from the output vector (i.e. the measurable variables), then an observer is just as important as the controller itself. This fact is often ignored in the design of fuzzy controllers.

The easiest way to calculate the state vector from the output vector would be the direct one: $\mathbf{x} = \mathbf{C}^{-1}\mathbf{y}$. But as in most cases the number of output variables differs from the number of state variables, $\mathbf{C}$ is not a square matrix and therefore not invertible. Hence, this approach is beyond question.

D.G.Luenberger ([118],[119]) proposed estimating the state vector with the help of a model of the plant. The model variables are computed simultaneously to the real plant variables, with the same input variables are used for both systems (fig. 2.53). Then a state vector $\hat{\mathbf{x}}$ and an output vector $\hat{\mathbf{y}}$ are computed for the model, which possibly do not correspond exactly to the real variables $\mathbf{x}$ and $\mathbf{y}$. The differences between the real output variables and the ones of the model are therefore fed back to the model via a correcting matrix $\mathbf{H}$ to improve the model estimation.

We can read off the estimation error $\tilde{\mathbf{x}} = \mathbf{x} - \hat{\mathbf{x}}$ from the block diagram:

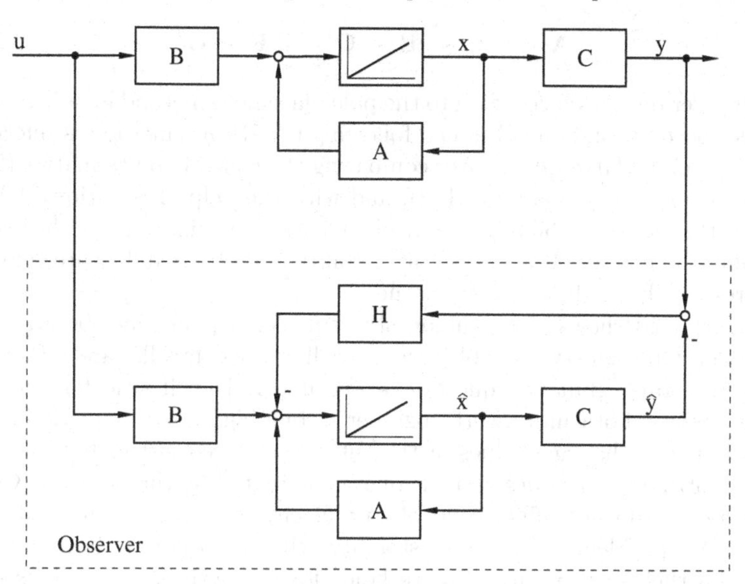

Fig. 2.53. Structure of an observer

$$\begin{aligned}\dot{\tilde{\mathbf{x}}} &= \dot{\mathbf{x}} - \dot{\hat{\mathbf{x}}} = \mathbf{Ax} + \mathbf{Bu} - [\mathbf{A}\hat{\mathbf{x}} + \mathbf{Bu} + \mathbf{H}\hat{\mathbf{y}} - \mathbf{Hy}] \\ &= \mathbf{Ax} + \mathbf{Bu} - [\mathbf{A}\hat{\mathbf{x}} + \mathbf{Bu} + \mathbf{HC}\hat{\mathbf{x}} - \mathbf{HCx}] \\ &= [\mathbf{A} + \mathbf{HC}](\mathbf{x} - \hat{\mathbf{x}}) = [\mathbf{A} + \mathbf{HC}]\tilde{\mathbf{x}} \end{aligned} \tag{2.198}$$

If all the eigenvalues of the matrix $(\mathbf{A} + \mathbf{HC})$ have a negative real part, the estimation error converges to zero for the steady state. Since $\mathbf{A}$ and $\mathbf{C}$ are already determined by the plant, we have to find a suitable correcting matrix $\mathbf{H}$. Obviously, when designing an observer, a problem similar to the one in the design of a state space controller occurs: In order to design the controller we had to find a matrix $\mathbf{F}$ which would make the closed-loop system matrix $(\mathbf{A} + \mathbf{BF})$ become stable. Now we need to determine a matrix $\mathbf{H}$ which does the same for $(\mathbf{A} + \mathbf{HC})$. The problem is not totally equivalent, since $\mathbf{C}$ and $\mathbf{H}$ changed order compared to $\mathbf{B}$ and $\mathbf{F}$. Nevertheless, we can still put down the design of an observer to the design of a state space controller.

Here, the characteristic polynomial of the matrix $(\mathbf{A} + \mathbf{HC})$ is a determinant, which is invariable to transposition:

$$|s\mathbf{I} - (\mathbf{A} + \mathbf{HC})| = |(s\mathbf{I} - (\mathbf{A} + \mathbf{HC}))^T| = |s\mathbf{I} - \mathbf{A}^T - \mathbf{C}^T\mathbf{H}^T| \tag{2.199}$$

If we compare this expression to the determinant which we obtained in the design of the controller,

$$|s\mathbf{I} - \mathbf{A} - \mathbf{BF}| \tag{2.200}$$

then we see that the controller design methods can be used here, too, if we make the following replacements:

$$\mathbf{A} \to \mathbf{A}^T \qquad \mathbf{B} \to \mathbf{C}^T \qquad \mathbf{F} \to \mathbf{H}^T \tag{2.201}$$

An observer designed according to the pole placement method is called a *Luenberger observer*, and one obtained following the Ricatti method is referred to by the name *Kalman filter*. After comparing the controllability matrix (2.171) to the observability matrix (2.172), and with the help of equation (2.201) it follows that the controllability of a plant is the criterion for whether we can compute the matrix **H** or not. Controllability and observability are said to be features which are dual to one another.

Figure 2.54 shows the common structure for a plant, an observer and a controller. On their own, the observer as well as the controller are—if designed properly—stable, but the question arises if this is still true for the entire system, where both units work together in one big closed loop. Here, it can be shown that the eigenvalues of the entire system consist of both the state space control's eigenvalues and the ones contributed by the observer. Control and observer do not affect each other's eigenvalues. They can be designed separately; problems concerning stability will not occur. If both units are stable on their own, then the same holds for the entire system. This fact is referred to by the term *Separation Theorem*. A prerequisite for this theorem is that the model exactly matches the plant. This is usually not the case for

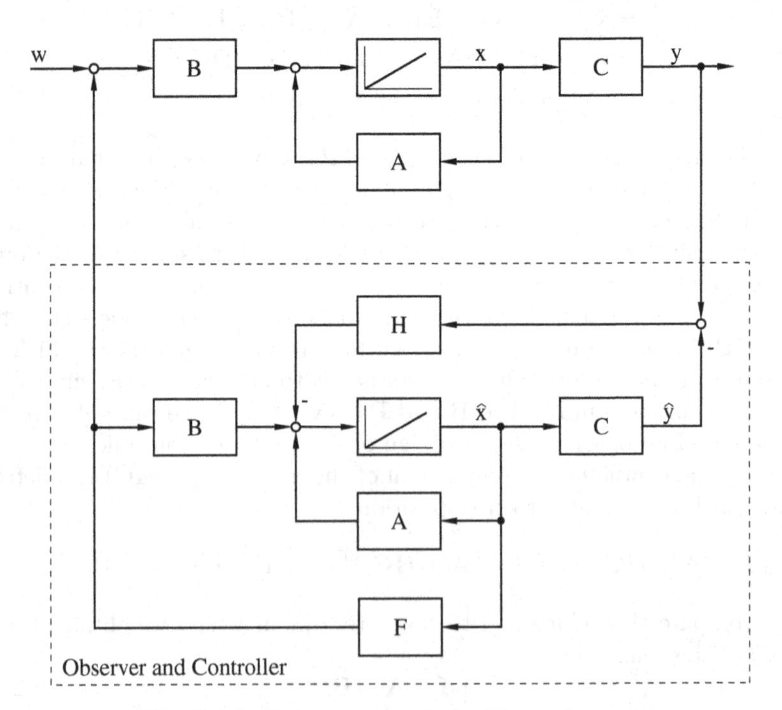

Fig. 2.54. State space control with an observer

applications in the field. But it was found out, that even for an unprecise model it is still possible to develop controller and observer approximately independently of each other without putting the stability of the entire system at risk.

2.7.6 Steady-State Accuracy of State Space Controllers

So far, our focus has always been on the system matrix of the closed-loop system, and thereby on the stability and the transient behavior. Accuracy, i.e. correspondence of reference values $\mathbf{w}$ and control variables $\mathbf{y}$, is a further demand on an automatic control, at least for the steady state. Unfortunately, to achieve accuracy in addition to stability by a suitable design of the controller matrix $\mathbf{F}$ is only possible in very few cases.

As a remedy, we can multiply the referecnce vector $\mathbf{w}$ with a matrix of constants $\mathbf{M}$. We can think of this matrix as being a multidimensional gain outside the closed loop, that can be adjusted in a way that for the steady and undisturbed state all output variables correspond to their reference variables. For stationary disturbances though, control errors still occur: Since $\mathbf{F}$ as well as $\mathbf{M}$ are constant matrices, only an actuating variable proportional to the reference values and state variables is produced. If this variable is insufficient, it will not be corrected. For a stead-state accurate control though, it would be necessary to re-adjust this variable until the control error was gone. Therefore, we need an additional integrator with the control error as input variable. Its output variable will be changing until the control error is zero. No steady state can occur before that.

Figure 2.55 illustrates this strategy. The point where the reference input affects the system remains unchanged, it is only drawn in a different way. $\mathbf{w}$ is now treated like a disturbance, and not like a reference vector as it was before. We add the components of the integral vector $\mathbf{e}$ over the control error $\mathbf{w} - \mathbf{y}$ as additional state variables to the system. These state variables have to be taken into account by the state space control. As long as the state space controller $\mathbf{F}$ of the expanded system is able to stabilize it, the system will get to a steady state sooner or later. The steady state is reached if all the variables remain unchanged. This can only happen if the input variables for all integrators are zero, as these variables would otherwise be added to the integrator output and cause a change of the output variable. Therefore, at the steady state the control variable $\mathbf{y}$ has to be equal to the reference variable $\mathbf{w}$, and control accuracy is reached. The state equation of the system for which we have to design the controller is now given by

$$\begin{bmatrix} \dot{\mathbf{x}} \\ \dot{\mathbf{e}} \end{bmatrix} = \begin{bmatrix} \mathbf{A} & \mathbf{0} \\ -\mathbf{C} & \mathbf{0} \end{bmatrix} \begin{bmatrix} \mathbf{x} \\ \mathbf{e} \end{bmatrix} + \begin{bmatrix} \mathbf{B} \\ \mathbf{0} \end{bmatrix} \mathbf{u} \tag{2.202}$$

In the methods which we introduced so far, we always had to know the complete state vector, and therefore most of the time an observer was needed.

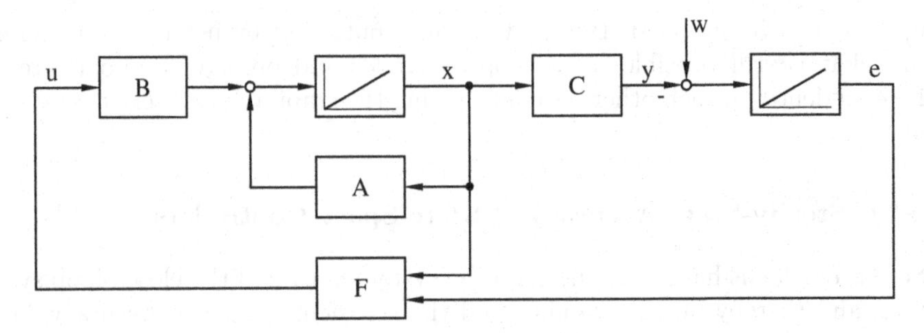

Fig. 2.55. Expanded state space control for steady-state accuracy

On the other hand, several approaches have been made to develop a state space control which is based on the measured output variables (fig. 2.56). In contrast to a state space controller, which is sometimes referred to by the term *state feedback*, the principle of this attempt is called *output feedback*. Here, the design gets more complicated because—compared to the state space control—less information is available to the controller, but we still expect it to produce a similar result.

One possibility is to design a common state space controller **F**, and then to compute the output feedback **R** in a way that the eigenvalues of the closed loop correspond as much as possible to the ones which we would have received from the (full) state space controller. We can also construct **R** with the help of a modified pole placement method or Riccati method. All these methods share that they are not as elegant and straight forward as the design methods for common state space controllers, and that the equations emerging in these processes can be solved only numerically, or even not at all (cf. [44]).

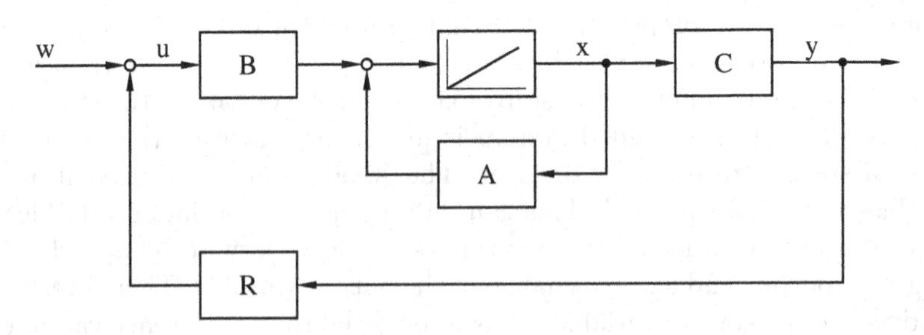

Fig. 2.56. State space control by output feedback

2.7.7 Norm-Optimal Controllers

The *norm-optimal controllers* ([37],[36],[132],[208],[209]) provide a totally different approach, but we have to make some preliminary remarks before we can explain them.

As already mentioned several times, the model of the plant which we use as the basis for the design of a controller usually does not exactly match reality. Errors may occur because, for clarity's sake, not all physical features are modeled, for example the dynamics of metering elements and actuators. Also the linearization of the model, which we require because the choosen controller design method can only be applied to linear plants, causes a deviation between the real and the modeled behavior of the plant. Finally, errors may be produced by changes of the plant in the course of time. The weight of a plane, for example, changes during a flight due to the consumption of fuel, or the buoyancy changes caused by the atmospheric pressure or the aircraft altitude. Compared to the other dynamic processes which occur during a flight, these changes happen very slowly, and because of this we do not treat it as separate influencing variables, but only as a change in the plant's parameters.

Of course we are interested in that the controller, whose design was based on a model, and which is now used to control a real plant, guarantees a stable system even if deviations between the model and the plant occur. A controller that satisfies this criterion is called a *robust controller*. The use of the term *robustness* only makes sense if we can quantify at the same time how big the deviation between plant and model can be before the system becomes instable. Without such a quantification, virtually any controller could be said to be robust. A PID-controller for a single-input-single-output plant, for example, is always designed in a way that the Nyquist plot of the open-loop transfer function passes the point -1 at a certain distance, in order to obtain a sufficient minimum damping. If the plant changes a little, then the Nyquist plot and, accordingly, the damping changes, too. The distance to the crucial point might shrink, but the system should still be stable, and the controller is robust. But still, a certain residual risk remains, as it is usually not possible to quantify exactly how much the plant may differ from the model before the Nyquist plot touches the point -1, or even passes it on the wrong side. We therefore say that a controller is *robust* only if the permissible variation between plant and model can be quantified.

Furthermore, norms can be defined for signal courses and transfer functions. In such a norm, a positive real number is assigned to a signal, a transfer function or a transfer matrix. This number indicates the "'magnitude"' of the element of interest. A signal's p-Norm, for instance, is given by

$$||u||_p := \left(\int_{-\infty}^{\infty} |u(t)|^p dt \right)^{\frac{1}{p}} \tag{2.203}$$

if the indefinite integral exists. For $p = 2$ we obtain the 2-norm

$$||u||_2 := \sqrt{\int_{-\infty}^{\infty} |u(t)|^2 dt} \tag{2.204}$$

that can be interpreted as the signal's content of energy. If, for example, $u(t)$ represents the voltage at a resistor R, then the power absorbed by this resistor is given by $P = u(t)i(t) = \frac{1}{R}u^2(t)$, and the energy by

$$\int_{-\infty}^{\infty} P(t)dt = \frac{1}{R}\int_{-\infty}^{\infty} u^2(t)dt = \frac{1}{R}||u||_2^2 \tag{2.205}$$

which means that the energy is proportional to the squared 2-norm of the voltage.

For $p \to \infty$ we get the ∞-norm

$$||u||_\infty = \sup_t |u(t)| \tag{2.206}$$

which defines the signal's maximum amplitude.

Obviously, regarding a certain norm several signals can have the same positive real number. Therefore, such a value will always represent a whole class of signals. In addition to this, calculations are easier for a scalar variable than for a signal course. A disadvantage of using the norm may be that we lose the details about the course of the signal with the time, but we usually do not need these details anyway. Especially for disturbance signals we will never be able to predict the course which they will take, but we can estimate their energy content or their maximum amplitude.

Norms for vectors of signals or transfer matrices are defined in a similar way. They can therefore also be applied to MIMO systems. Concerning the ∞-norm of a transfer matrix it can be shown that

$$||\mathbf{G}(j\omega)||_\infty = \sup_{||\mathbf{x}||_2 \neq 0} \frac{||\mathbf{y}||_2}{||\mathbf{x}||_2} \tag{2.207}$$

This way, the ∞-norm represents the largest possible factor with which the "energy" of the input signal $\mathbf{x}$ is transferred to the output signal $\mathbf{y} = \mathbf{Gx}$. Exact definitions and some further discussion concerning norms can be found in the appendix.

Our aim for the design process is now to construct a controller in a way that the norm of the transfer matrix is minimized, i. e. that the output signals with respect to the input signals get as small as possible regarding a certain norm. An example for such a configuration is shown in figure 2.57.

The vectors $\mathbf{m}$, which describes the noise of the measured values, and $\mathbf{z}$, which represents the disturbances that affect the plant, are defined as input signals. The output signals are the control variable $\mathbf{y}$ and actuating variable $\mathbf{u}$. Weighting each variable by *weight matrices* $\mathbf{W}_i$ is possible.

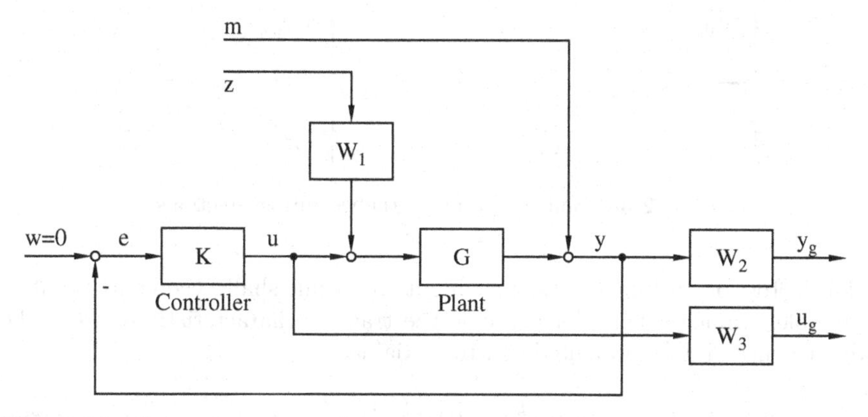

Fig. 2.57. Illustration of transfer matrices used for the computation of a norm-optimal control

If we want to minimize the transfer characteristic, then on the one hand the control variables $\mathbf{y}$ should attain values as small as possible and lie close to the reference value $\mathbf{w} = \mathbf{0}$, despite disturbances and noise. This guarantees implicitly also the stability of the system. On the other hand, the values of the actuating variable $\mathbf{u}$ shall also be minimized. That means, that the control goal $\mathbf{y} = \mathbf{w} = \mathbf{0}$ shall be reached with the smallest actuating variables possible, in order to protect the actuating unit.

However, the demands for a small control deviation and for small values of the actuating variables contradict each other, so that the norm of the entire transfer matrix cannot be infinitely small. Therefore, we have to make a compromise between the different demands by a suitable adjustment of the weighting functions.

As the weighting functions will be fixed, and the plant is given, the desired controller remains as the only part of the transfer matrix which is still undetermined. If the norm of this matrix is minimized numerically by a computer algorithm, then the corresponding controller will be generated more or less as a side effect. We refrain from presenting a derivation and a more detailed description of the algorithms involved in this process, as this would fill a book on its own ([132]). The important point is that these algorithms find the minimum using analytic methods instead of numerical search. It is therefore guaranteed that the optimum found by the algorithms is the true global optimum—not a local optimum, as often detected by searching procedures. Nowadays any control software libraries provide these algorithms, so the only thing which the control engineer has to do is to determine the transfer characteristic, that has to be minimized, by the weighting functions.

The easiest way to explain the role of the weighting functions is to treat the system shown in figure 2.57 as a SISO system, and to use the ∞-norm as the norm that has to be minimized. For this setup, the maximum amplifica-

Fig. 2.58. Bode diagrams of the weighting functions

tion factor for all frequencies from input to output shall become as small as possible. Among others, for example the transfer characteristic between the disturbance and the weighted control variable

$$\frac{y_g}{z} = W_2 \frac{GW_1}{1+GK} \tag{2.208}$$

gets minimized. The algorithm will find a controller for which this function will be as small as possible for all frequencies.

We set $W_1 = 1$ and choose W_2 to be a low-pass function—for example a first order lag with large values $|W_2|$ for low frequencies (fig. 2.58). As the transfer behaviour of the entire transfer function gets minimized for all frequencies, we obtain a residual transfer function $\frac{G}{1+GK}$, that takes very small values especially for low frequencies. But this function just describes the transfer characteristic between the disturbance and the control variable. Accordingly the control variable shows only little reaction to low-frequency disturbances. This again implies that the controller can correct disturbances of this type very well.

If we even use an integrator with infinite gain for dc signals for W_2, then necessarily such a controller has to result which sets the transfer characteristic $\frac{G}{1+GK}$ for dc signals to zero—otherwise the value for the norm of the entire transfer function could not be finite anymore. But if the transfer characteristic for dc disturbance signals is zero, i.e. if the control variable is equal to the reference value 0 even for steady disturbances, this means steady-state accuracy.

This can only be achieved by a controller for which $K(0) \to \infty$ holds, because only for this case the transfer characteristic for disturbance signals will be zero for $s = 0$. Accordingly, either the controller comprises an integral part, or it has an infinitely high constant gain, which is of course not realizable. For this reason, we treat the actuating variables—the controller output—also as output variables in the minimization process. As they too should attain the smallest values possible, the process will produce a controller with an integral part instead of one with infinite constant gain.

The actuating variable is considered by minimizing the transfer characteristic between the measurement noise and the weighted actuating variable

$$\frac{u_g}{m} = -\frac{W_3 K}{1+GK} \tag{2.209}$$

The reflections are similar as above: In contrast to W_2, we have to choose W_3 to show a high-pass filter-like behavior (fig. 2.58). As a result of this, the residual term $\frac{K}{1+GK}$, i.e. the transfer characteristic between measurement noise and actuating variable, gets small for high frequencies. This is desired, as the controller should in deed show as little reaction as possible to high-frequency disturbances of the measurings. Minimization of this function, however, has also another aspect to it, and this aspect requires some additional remarks.

We can express differences between the real plant and the model using an additive component. An example is given in the upper diagram in figure 2.59, where $\mathbf{G}_0$ represents the nominal model of the plant and $\mathbf{G} = \mathbf{G}_0 + \Delta\mathbf{G}$ the real plant. We can redraw the diagram; the result is shown in the lower part of the figure. If we choose a suitable controller $\mathbf{K}$ for the nominal model $\mathbf{G}_0$, then the inner loop will always be stable, but we have to analyze under which conditions also the outer circuit will be steady.

For the investigation of this problem we have to assume that the deviation $\Delta\mathbf{G}$ is stable. Fortunately, this is not a severe restriction, as in the application of this method $\Delta\mathbf{G}$ can be chosen freely. Thus the open loop—the series connection of inner loop and $\Delta\mathbf{G}$—will be stable.

Now we can make the very conservative statement, that the entire closed-loop system will be stable, if the open-loop gain is less than 1 for all frequencies (*Small Gain Theorem*). We want to prove this theorem only for SISO plants, which is relatively easy with the help of the Nyquist Theorem. The Small Gain Theorem implies here, that the gain of the open-loop transfer function always has to be less than 1. Accordingly, the Nyquist plot does not revolve around the critical point -1 and the phase shift regarding this point is zero. Now first, the transfer function cannot have any poles on the axis of imaginaries, as otherwise its gain would be infinite for small frequencies and hurt the prerequisite of the Small Gain Theorem. Besides this, the open-loop transfer function does not have any poles in the right half-plane, as all of its parts are stable. Summed up, all poles lie in the left half-plane. It follows from the Nyquist Theorem, that the necessary phase shift of the Nyquist plot regarding 1 has to be zero, and therefore, the closed-loop system is stable.

For a MIMO system, it follows from the Small Gain Theorem, that the ∞-norm of the transfer matrix of the open-loop system has to be less than 1. According to figure 2.59 this transfer matrix is $\Delta\mathbf{GK}(\mathbf{I} + \mathbf{G}_0\mathbf{K})^{-1}$. Our closed loop is therefore stable if

$$||\Delta\mathbf{GK}(\mathbf{I} + \mathbf{G}_0\mathbf{K})^{-1}||_\infty < 1 \tag{2.210}$$

holds.

If we write the transfer function given by equation (2.209) as a transfer matrix for multi-input-multi-output plants, then we can easily see that in comparison to (2.210) only the admissible model deviation $\Delta\mathbf{G}$ has to be replaced by the weighting function $\mathbf{W}_3$:

$$\mathbf{W}_3\mathbf{K}(\mathbf{I} + \mathbf{G}_0\mathbf{K})^{-1} \tag{2.211}$$

Therefore, after the computation of the controller we can use the matrix $\mathbf{W}_3$ to estimate the admissible model deviation: We have to multiply the weighting matrix $\mathbf{W}_3$ with a suitable factor, so that—with the modified matrix $\mathbf{W}_3'$—the norm $||\mathbf{W}_3'\mathbf{K}(\mathbf{I}+\mathbf{G}_0\mathbf{K})^{-1}||_\infty$ takes the value 1. For this case, $\mathbf{W}_3'$ is equal to the admissible deviation $\Delta\mathbf{G}$ between the model and plant for which the real system is still stable following the Small Gain Theorem (2.210). In this manner we also obtain a measure of the controller's robustness when designing an ∞-norm-optimal controller. This is of course an advantage for applications in the field.

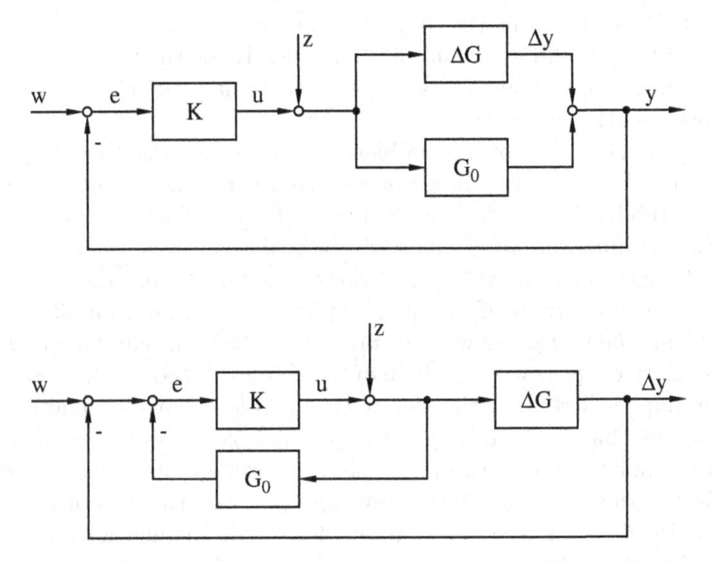

Fig. 2.59. Splitting a real plant up into a nominal model and its deviation

It is also possible to derive a controller in the opposite direction: We define the desired robustness $\Delta\mathbf{G}$ or $\mathbf{W}_3$ respectively first, and then we estimate the controller that minimizes the norm of the term (2.211). If it is smaller than one, then the stability of the closed-loop system is guaranteed even for the given deviations between the model and the real plant.

The biggest problem for this approach is the very conservative way of estimating the stability using the Small Gain Theorem, which we explain now for an example of a single-input-single-output system. For such a system, this theorem is equivalent to the demand that the gain of the Nyquist plot of the open-loop transfer function always has to be less then 1. But, according to the Nyquist Theorem much larger gains would be permissible, as long as the point -1 is not revolved. An unnecessarily small gain of the Nyquist plot or the open-loop transfer function respectively corresponds to an unnecessarily small controller gain, which then leads to an unnecessarily small control rate. In the worst case, this may even lead to the controller becoming unusable.

Another difficulty is choosing suitable weighting functions, which obviously have a crucial influence on the controller that is being constructed. Besides this, we have to pay attention to some minor boundary conditions when making our choice, if the method should produce a result. Finally, the plant has to be controllable as well as observable. But still, despite all these requirements, the norm-optimal controller represent a powerful tool for the design of controllers, especially when complex MIMO systems have to be controlled.

Even though the modern methods of controlling which we introduced in this chapter are superior to the classical PID-controller, their use in industrial applications increases only slowly. This is mainly a result of their complexity. An experienced engineer will find it much easier to adjust gain and integration time of a PI controller than to define the weighting matrices for the Riccati or a norm-optimal controller design. Nevertheless, it should be expected that the state space description, and together with it the modern methods of controller design, will gain acceptance. On the one hand, only these methods can cope with complex multi-input-multi-output systems and produce suitable controllers for these plants. On the other hand, it is possible to transfer a system given in state space description directly into a computer simulation. This will become increasingly important, as already today virtually every controller of higher quality gets tested in a simulation before being released.

Regarding fuzzy controllers, knowledge about the state space description and the corresponding design procedures is valuable for two reasons. First, we will see that a fuzzy controller is nothing more than a nonlinear state space controller and the theory which has been presented in this chapter can therefore be at least partly reused. On the other hand, in order to judge about the capability of fuzzy controllers the best methods available in classical control engineering should be used in comparisons, and not only PID controllers.

2.8 Nonlinear Systems

2.8.1 Features of Nonlinear Systems

In the preceding sections, we discussed only linear plants and the corresponding controllers, i.e. only systems, which can be described by linear differential equations. Almost every real-world system, though, contains one or more nonlinear transfer elements, like, for example, a fuzzy controller. Now already the occurrence of only one nonlinear relation converts a linear system of equations into a nonlinear one, and in order to be able to describe the latter we have to switch from using the simplified linear form of the state space description

$$\begin{aligned}\dot{\mathbf{x}} &= \mathbf{A}\mathbf{x} + \mathbf{B}\mathbf{u} \\ \mathbf{y} &= \mathbf{C}\mathbf{x} + \mathbf{D}\mathbf{u}\end{aligned} \tag{2.212}$$

back to the general form (see (2.135)):

$$\begin{aligned}\dot{\mathbf{x}} &= \mathbf{f}(\mathbf{x}, \mathbf{u}) \\ \mathbf{y} &= \mathbf{g}(\mathbf{x}, \mathbf{u})\end{aligned} \tag{2.213}$$

We could even try to cope with time-variant systems, which could be expressed by adding the parameter t to the functions $\mathbf{f}$ and $\mathbf{g}$, but we refrain from discussing these systems.

Many of the tools available in linear control theory are not applicable to nonlinear systems. We can, for example, no longer use Laplace transformation, which was only introduced for linear systems. The same holds for Nyquist plots, which only for linear systems describe how the amplitude and phase of a given sinusoidal input signal are changed on their way through the system to the output. Also the principle of superposition (theorem 2.1) is now inapplicable, i.e. if several input signals affect the system simultaneously, it is no longer possible first to compute the corresponding output signals separately and then add those up to obtain the overall response of the system. The proportional property does not hold for nonlinear systems either (again, see 2.1). Therefore, we cannot assume any longer that increasing the input signal by a certain factor will lead to a change of the output signal by the same factor. As a consequence of this it becomes useless to characterize a system by step responses, as steps of different heights as input signals may produce completely different output signals.

2.8.2 Application of Linear Methods to Nonlinear Systems

Linearization around a Set Point

For all these reasons, it will be of interest to the control engineer to find a way of representing a nonlinear system as a linear one, and then be allowed to apply the same methods to it which were developed for linear systems. One possibility to do this is the *linearization* of the system around a set point, i.e. the behavior of the nonlinear system is represented by a linear model, that describes the real behavior in the operating region around the set point as exactly as possible. For this linearization, we first have to expand the system's state space description into a Taylor's series around the given set point. For example, let $x = x_0$, $u = u_0$ and $f(x_0, u_0) = 0$ be the set point of the SISO system

$$\dot{x} = f(x, u) \tag{2.214}$$

The deviations from this set point are given by

$$\begin{aligned}\Delta x &= x - x_0 \\ \Delta u &= u - u_0 \\ \dot{\Delta x} &= \dot{x} = f(x, u)\end{aligned} \tag{2.215}$$

Expanding $f(x, u)$ into a Taylor's series around the operating point yields

$$f(x,u) = f(x_0,u_0) + \frac{\partial f}{\partial x}(x_0,u_0)\Delta x + \frac{\partial f}{\partial u}(x_0,u_0)\Delta u + r(x,u) \qquad (2.216)$$

The residual term $r(x,u)$ consists of the derivations of higher degree, which we will neglect here. From (2.215) and (2.216) we get with $f(x_0,u_0) = 0$

$$\dot{\Delta x} = \frac{\partial f}{\partial x}(x_0,u_0)\Delta x + \frac{\partial f}{\partial u}(x_0,u_0)\Delta u = a\Delta x + b\Delta u \qquad (2.217)$$

Thus, the deviation from the set point is given by a linear differential equation with coefficients a and b. The formula $y = g(x,u)$ can be linearized in a similar way. This produces a completely linear model for the operating region of the nonlinear system, to which we could now apply any method of linear control theory.

The procedure for the MIMO case is similar: Linearization of

$$\dot{\mathbf{x}} = \mathbf{f}(\mathbf{x},\mathbf{u}) \qquad (2.218)$$

gives

$$\dot{\Delta \mathbf{x}} = \mathbf{F}_\mathbf{x}(\mathbf{x}_0,\mathbf{u}_0)\Delta\mathbf{x} + \mathbf{F}_\mathbf{u}(\mathbf{x}_0,\mathbf{u}_0)\Delta\mathbf{u} = \mathbf{A}\Delta\mathbf{x} + \mathbf{B}\Delta\mathbf{u} \qquad (2.219)$$

where the single coefficients of the Jacobian matrices $\mathbf{F}_\mathbf{x}$ and $\mathbf{F}_\mathbf{u}$ are defined by

$$[\mathbf{F}_\mathbf{x}]_{i,j} = \frac{\partial f_i}{\partial x_j} \quad \text{and} \quad [\mathbf{F}_\mathbf{u}]_{i,j} = \frac{\partial f_i}{\partial u_j} \qquad (2.220)$$

where $\mathbf{f} = [f_1, ..., f_n]^T$.

Linearization is frequently used when tools and design methods of linear control theory shall be applied to nonlinear plants. The nonlinear system is linearized, and for this linear model a linear controller is developed. We have to keep in mind, though, that the deviations between model and real plant increase the further away the point of interest lies from the set point. The linear controller therefore has to be of sufficient robustness. Still, sometimes the nonlinearity can be of a kind that already for a relatively small distance to the set point the deviation between the real plant and the linearized model gets extremely large. As a result of this, the controller has to be of high robustness, which again might decrease the control rate. In a case like that, it will often not be possible to obtain a useful controller by linearization.

Exact Linearization

The nonlinearity of the plant may sometimes also be compensated by an inverse nonlinearity inside the controller, producing on the whole a purely linear system. Figure 2.60 shows a very simple example of this procedure. The sine function of the plant gets compensated by the controller's arc sine. Thus, in the end the system consists only of a linear plant with the sufficient PI-controller.

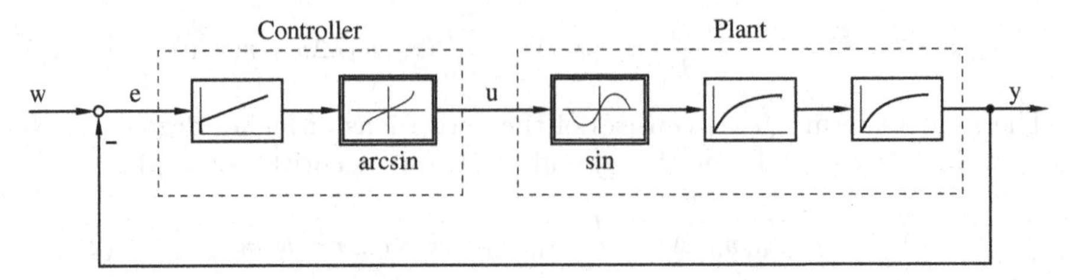

Fig. 2.60. A simple example of an exact linearization

This method is called *exact linearization* ([73], [177]). The procedure can be generalized to multi-input-multi-output systems of the form

$$\begin{aligned}\dot{\mathbf{x}} &= \mathbf{a}(\mathbf{x}) + \mathbf{B}(\mathbf{x})\mathbf{u} \\ \mathbf{y} &= \mathbf{c}(\mathbf{x})\end{aligned} \tag{2.221}$$

which are called *analytically linear systems* (ALS).

An example of such a system is a dc machine (cf. block diagram in fig. 2.61) with the input quantities *excitation flux* ϕ_e and *armature voltage* u_a and the states *armature current* i_a and *rotational speed* ω. We obtain the speed of an unloaded machine by integration over the *driving torque* T_a. The driving torque again is the product of armature current and excitation flux. The armature current is driven by the difference between armature voltage and induced counter voltage $\phi_e\omega$. All in all, the state space description is given by

$$\begin{pmatrix} \dot{i}_a \\ \dot{\omega} \end{pmatrix} = \begin{pmatrix} c_1 i_a \\ 0 \end{pmatrix} + \begin{pmatrix} c_2 & c_3\omega \\ 0 & c_4 i_a \end{pmatrix} \begin{pmatrix} u_a \\ \phi_e \end{pmatrix} \tag{2.222}$$

with the machine-dependent constant parameters c_i, which are not shown in the block diagram in order to maintain readability. This state space description is obviously of the form given in (2.221).

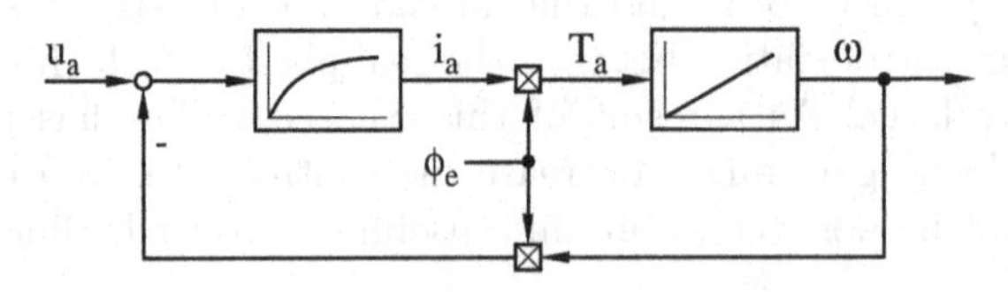

Fig. 2.61. Block diagram of a dc machine with external excitation

The description of this method for general multi-variable ALS would exceed the proportions of this book, so we only explain the corresponding variant for single-input-single-output systems. It should still be sufficient to illustrate the basic idea of exact linearization. To begin with, we can simplify the equation describing the system to

$$\dot{\mathbf{x}} = \mathbf{a}(\mathbf{x}) + \mathbf{b}(\mathbf{x})u$$
$$y = c(\mathbf{x}) \tag{2.223}$$

because we have only one input and one output variable now. Furthermore, we have the following definition for the so-called *relative system degree*:

Definition 2.19 *A single-input-single-output ALS as described in (2.223) is of relative degree d in a region R around the point* $\mathbf{x}_0$ *if it complies with the following conditions:*

1. *For all* $\mathbf{x} \in R$ *and* $k < d-1$*, it holds* $L_{\mathbf{b}}L_{\mathbf{a}}^{k}c(\mathbf{x}) = 0$
2. $L_{\mathbf{b}}L_{\mathbf{a}}^{d-1}c(\mathbf{x}) \neq 0$

$L_{\mathbf{a}}$ and $L_{\mathbf{b}}$ are the Lie derivatives as defined in the appendix. For linear SISO systems, the relative degree corresponds to the difference between the degree n of the denominator polynomial and the degree r of the numerator polynomial of the transfer function: $d = n - r$. For a linear SISO plant, a relative degree $d = n$ would therefore indicate that the numerator polynomial merely consists of a constant factor.

Among other requirements, we now have to make sure that the nonlinear ALS is of the maximum relative degree $d = n$. We can also apply this method to cases where $d < n$, but then the formulas involved get more complicated. In the first step, using the coordinate transformation

$$\mathbf{z}(\mathbf{x}) = \begin{pmatrix} z_1(\mathbf{x}) \\ z_2(\mathbf{x}) \\ \cdots \\ z_n(\mathbf{x}) \end{pmatrix} = \begin{pmatrix} c(\mathbf{x}) \\ L_{\mathbf{a}}c(\mathbf{x}) \\ \cdots \\ L_{\mathbf{a}}^{n-1}c(\mathbf{x}) \end{pmatrix} \tag{2.224}$$

we obtain from (2.223) the normal form (cf. fig. 2.62)

$$\dot{\mathbf{z}} = \begin{pmatrix} \dot{z}_1 \\ \dot{z}_2 \\ \cdots \\ \dot{z}_{n-1} \\ \dot{z}_n \end{pmatrix} = \begin{pmatrix} z_2 \\ z_3 \\ \cdots \\ z_n \\ f(\mathbf{z}) \end{pmatrix} + \begin{pmatrix} 0 \\ 0 \\ \cdots \\ 0 \\ g(\mathbf{z}) \end{pmatrix} u$$
$$y = \begin{pmatrix} 1 \; 0 \cdots 0 \end{pmatrix} \mathbf{z} \tag{2.225}$$

Here, f and g are nonlinear, scalar functions in $\mathbf{z}$. If we apply the equation

$$u = \frac{1}{g(\mathbf{z})}(-f(\mathbf{z}) + u^*) \tag{2.226}$$

to compute the actuating variable u from a quantity u^* (that we still have to define) as shown in fig. 2.62, we transform the system into

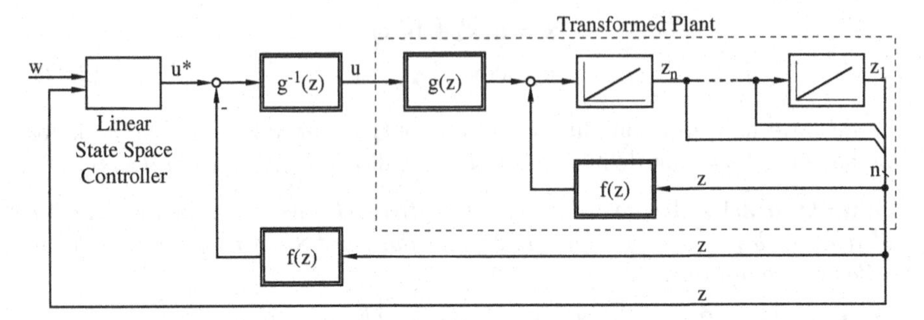

Fig. 2.62. Illustration of the basic idea of exact linearization

$$\dot{\mathbf{z}} = \begin{pmatrix} 0\,1 \cdots 0 \\ \cdots\cdots\cdots \\ 0\,0 \cdots 1 \\ 0\,0 \cdots 0 \end{pmatrix} \mathbf{z} + \begin{pmatrix} 0 \\ 0 \\ \cdots \\ 1 \end{pmatrix} u^*$$

$$y = \begin{pmatrix} 1\,0 \cdots 0 \end{pmatrix} \mathbf{z} \tag{2.227}$$

This is a purely linear, controllable and observable system, for which only a linear controller still has to be determined. $u^*(\mathbf{z})$ will be the actuating variable of this linear controller, which can be designed for the linear system (2.227) using any method for linear state space controllers. The output quantity u which actually affects the nonlinear plant, results from u^* according to (2.226).

Writing u depending on $\mathbf{x}$, we obtain

$$u(t) = \frac{1}{L_{\mathbf{b}} L_{\mathbf{a}}^{n-1} c(\mathbf{x}(t))} (-L_{\mathbf{a}}^n c(\mathbf{x}(t)) + u^*(\mathbf{z}(\mathbf{x}(t)))) \tag{2.228}$$

i. e. u depends on the plant's state variables, and thus we have a nonlinear state space controller. If the state variables cannot be measured, we need an observer to enable control. Under certain conditions, we can use the same procedure as for the design of the controller, i. e. application of a suitable transformation reduces the problem to a linear observer design problem. Instead of discussing this special task, we will describe the design idea of an observer for general nonlinear systems in section 2.8.11, as such a type of observer may be needed for fuzzy control.

Furthermore, it should be pointed out that two preconditions of this method are severe restrictions: First, f and g have to be known, and second, $g(\mathbf{z})$ (or $\mathbf{G}(\mathbf{z})$ in case of a MIMO system) needs to be invertible for all states $\mathbf{z}$. If these conditions are met, though, the exact linearization is a very elegant controller design method.

Adaptive Control

The *adaptive control* provides another way to make knowledge and methods of linear control theory applicable to nonlinear systems. A controller is said to be adaptive if it is continuously (or at least in certain time intervals) adjusted by a supervising system. This is done in order to achieve a better control performance and adaptation to a possibly changing plant behaviour (c.f. fig. 2.63). In this process the supervising system permanently receives measured values describing the current behavior of the plant. Whether the change in the plant's behavior is caused by a variation in time of the plant parameters or merely by changing to a different set point of a nonlinear time-invariant plant is not important. The change of the plant behavior just has to be slow compared to the other dynamic processes, i. e. its derivative with respect to time can be neglected compared to the other quantities. Otherwise, the changing parameter should be treated like a state variable.

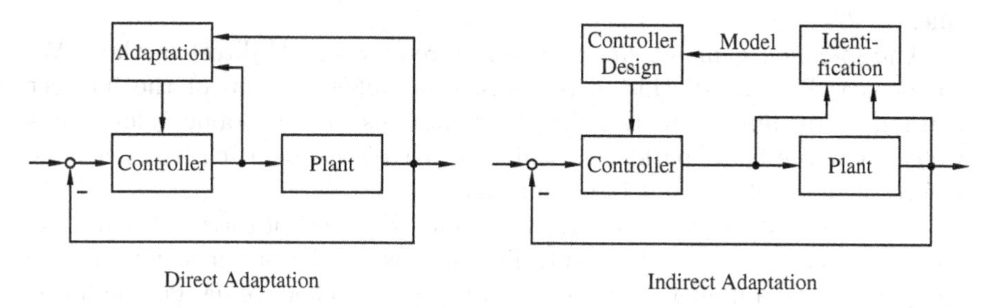

Fig. 2.63. Types of adaptive control

On the whole, there are two approaches: the *direct* and the *indirect adaptive methods*. With the indirect methods, in short intervals a new linear model of the plant is computed based on new measured values by an identification algorithm, while the system is in operation. The result of computing a linear model around a set point corresponds to a common linearization according to equation (2.219). The only difference is that in this case the calculation is done numerically by an identification process, while in the former mentioned linearization it is done in an analytical way. After an identification step is finished, a suitable linear controller corresponding to the plant model is also computed automatically according to a design procedure chosen in advance. All this is done while the system is still in full operation. Once the new controller is designed, it replaces the old one that has been working so far. It should be guaranteed that the actuating variable follows a continuous curve, in order to avoid oscillations and unnecessary activation of the plant with every change to a new controller.

The crucial point of this method is the identification process. The identification algorithm can get numerically instable and even produce an extremely faulty model of the plant, if the measured values used in this process are disadvantageously distributed, i. e. the pieces of information they contain are not independent of each other. This threat already exists because of the fact that newly measured values in a closed control loop can never be independent from the ones measured earlier. For example: an input signal of a plant first affects the plant itself and then the controller via the feedback loop. Finally, it reappears transformed at the plant's input. An additional check for plausibility of the identified model should therefore be included, if the indirect adaptive method is used.

The direct adaptive method is based on the same idea as the indirect approach, but in contrast to this it lacks the identification process. Instead of this, the controller is adjusted according to an adaptation rule based on the values which are currently being measured. This rule has to be set up beforehand. While in the indirect method the necessary information about the plant is obtained in every identification step, here the information about the plant has to be available in advance, as it has to be used in the adaptation rule.

The so-called *gain scheduling* is an extremely simple direct method. We could as well think of this method as a simplified version of the indirect method—only limited to time-invariant systems—as the same calculations which there are being performed while the system is operating here get done before the controller is put into operation.

First we need to choose several set points. For each of these set points we create a linear model of the plant. The models can be obtained by analytic computation or a numerical identification process. Then we design one linear controller for every set point, based on the linear models developed for these points. The controllers for the various set points may differ only in their parameters; the type has to be the same for all of them, i. e. it is not possible to construct a PI-controller for one set point and a PID-controller for another one.

Once the system is running, the adaptation algorithm only has to load the corresponding set of controller parameters every time a change to a different set point occurs. This way the online-identification can be skipped, which is why the gain scheduling is considered to be a direct adaptive method. In order to avoid step-like changes in the actuating variable, the parameters should not be replaced all at once when switching from the old controller to the new one, but in a continuous way. The ideal situation would be to have a continuous function of the controller parameters depending on the state of the plant, as the adaptation rule.

Unfortunately, for adaptive methods—although they seem very plausible and produce remarkable results—in the most cases no proof for stability exists. One of the exceptions are the so-called TSK controllers, that will be discussed in detail in sections 4.1.3, 4.2.2 and 5.1.

Cascade Control

The problems caused by a nonlinearity in the control loop can in many cases be compensated by a *cascade control* (multi-loop control, cf. fig. 2.64). If, for example, the speed of an electrical drive has to be controlled, then we need a static converter (which shows an extremely nonlinear behavior) as the actuator. For this reason we have to see the current as an internal control variable and construct an inner control loop consisting of a current controller and the static converter as the controlled plant. This controller, of course, has to be designed using methods of nonlinear control theory. But from outside, this inner control loop behaves approximately like a simple first order lag, with its output quantity i being proportional to the actuating variable u of the rotational speed controller. The nonlinear effects caused by the static converter are therefore limited to the inner control loop. We can therefore design the rotational speed controller of the outer circuit for an almost linear plant.

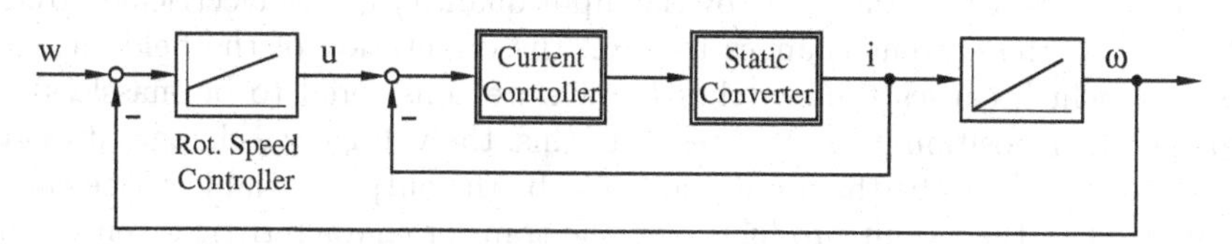

Fig. 2.64. Linearization with cascade control

In the methods introduced so far, we tried to limit or eliminate nonlinear effects, in order to allow application of linear methods to nonlinear systems. Still, many cases remain were we cannot use any of these methods, and therefore have to deal with the given nonlinearities explicitly. Nonlinearities need not be disadvantageous—in some cases they are even added to use their effects for the improvement of the controller. So nonlinearities occur not only unintentionally in the plant or actuator, but also deliberately inside the controller. Time-optimal controllers, which will be discussed later on, and of course fuzzy controllers are examples of this. A complete theory, which can be given for example for the linear state space control, does not exist for the nonlinear systems since there are simply too many different nonlinear phenomena.

2.8.3 Switching Elements

Ideal 2-Point Switching Control

We now focus on switching elements, which are frequently used in practical applications and provide a good example to convey some understanding of which effects can actually occur when coping with nonlinearities. The most simple type is the *ideal 2-point switching element* or bang-bang element, which we can as well see as an ideal switch. Figure 2.65 shows the corresponding characteristic curve. The value of the output variable is 1 for positive input values and -1 for negative ones. In order to avoid ambiguities, we define $y(0) = 1$ as the output for $x = 0$.

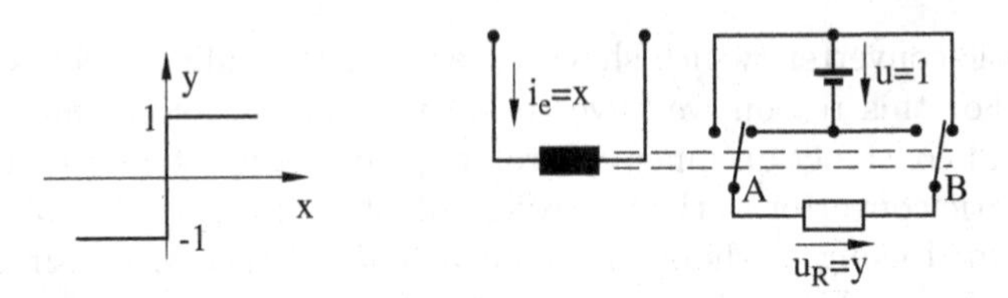

Fig. 2.65. Ideal 2-point switching controller

To the right of the characteristic curve a circuit is shown which could be used to produce this curve assuming that ideal conditions are given. The switches A and B are coupled to each other. They are driven by the field of the inductivity, which is controlled by the input quantity i_e, the electrical current. As soon as the current changes its sign, the orientation of the field changes as well, which causes the switches (which are considered to be massless) to change their position, too. As a result of this, the voltage u_R changes its sign. If we consider i_e to be the input and u_R to be the output variable of the entire system, then the circuit produces just the transfer characteristic given by the curve.

In practical applications, switching elements are frequently used as actuators. There, the output quantity may switch between 0 and 1 as well as between -1 and 1. A transistor or simple light switches represent such a type of a switch. These examples are already sufficient to answer our question of why such types of transfer elements are used at all, as they only make any theory more complicated: A switch is more simple and cheaper than a continuous transfer element. It is of course a disadvantage that its output quantity does not cover a continuous region, but we can compensate this disadvantage by pulse-width modulation, a method which we will describe later on.

An interesting fact is that 2-point switching elements can be used not only as actuators in a control loop, but also as controllers on their own. In principle, ideal 2-point switching controllers are the most simple type of controller one can think of. Figure 2.66 shows an example of an ideal 2-point switching controller with a plant consisting of a double integrator.

We can think of the two integrators as being an acceleration plant (cf. fig. 2.47). If we look at it this way, then the actuating variable u represents the body's acceleration a, while the control variable corresponds to the body's position l. The actuating variable u takes either the positive or the negative maximum value, and the body gets accelerated either into the positive or the negative direction, but always with maximum power.

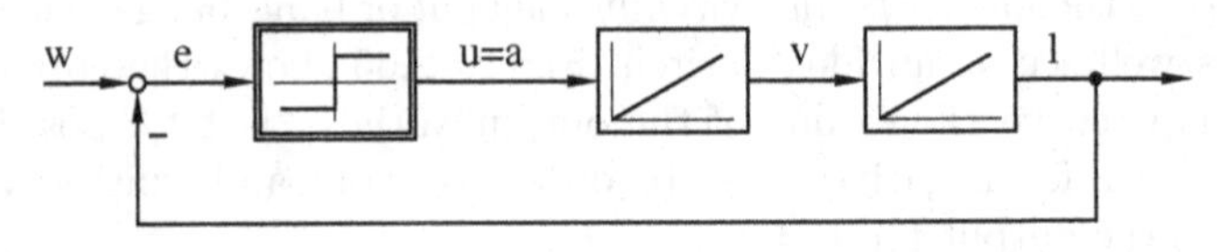

Fig. 2.66. System controlled by an ideal 2-point switching controller

The course of a control process for this plant with respect to time is given in figure 2.67. Let the actual value l of the position in the beginning be less than the reference value w. The control error e is therefore positive, which forces the controller to react with a maximum positive value for the regulated quantity u of the acceleration a respectively. The body gets accelerated into the direction of the reference value, but reaching it the velocity v is bigger than zero, the body overshoots the mark and we obtain a control error into the other direction. The controller responds with a maximum negative value for the actuating variable, which first slows the body down and then accelerates it into the opposite direction. The same process is repeated—this time only the other way round. The body obviously never reaches a steady state.

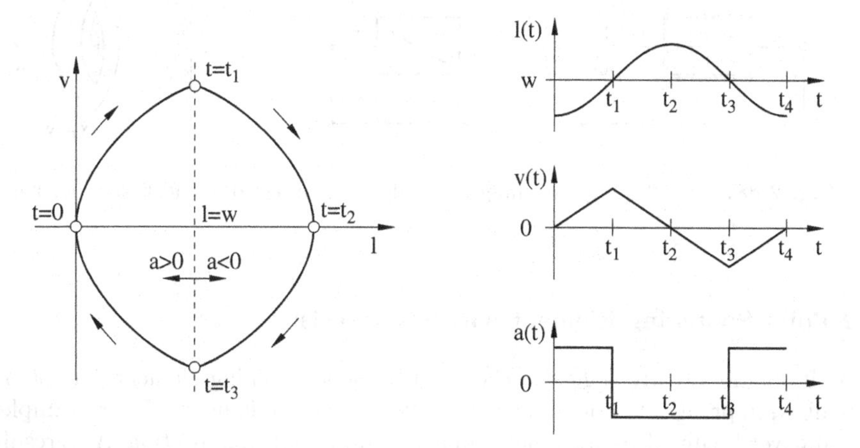

Fig. 2.67. State space trajectory of an ideal 2-point switching controller with double integrator and time-dependent courses of the control loop variables

We can as well describe this behavior in the state plane. As discussed earlier, the position l and the velocity v are state variables of the system. We can therefore also characterize the control process by a trajectory in the $v - l-$state plane. For this case of a steady oscillation, the trajectory is a closed curve, which is traversed over and over again, and the output quantity l produces oscillations around the reference value w.

The bigger the distance between the initial and the reference position gets, the longer it will take the body to reach this reference value, and also the higher its velocity will be when reaching this value. This will then enlarge the overshot into the other direction, resulting in an increase of the oscillation's amplitude as well as of the time interval between two passings through zero. Accordingly, the amplitude as well as the frequency of this *steady oscillation* depend on the initial conditions. A behavior like this cannot occur for linear systems. There, the frequency of an oscillation is always determined by the corresponding complex conjugated pair of poles of the transfer function. Only the amplitude depends on the initial conditions.

In contrast to this, we obtain decaying oscillations if the plant consists of a first order lag and an integrator, as shown in figure 2.68. Here, we can choose x_1 and x_2 as the state variables. The trajectory obviously converges towards the final value $(x_1, x_2) = (w, 0)$. The system always reaches this final state independently of the initial state and is therefore a stable one.

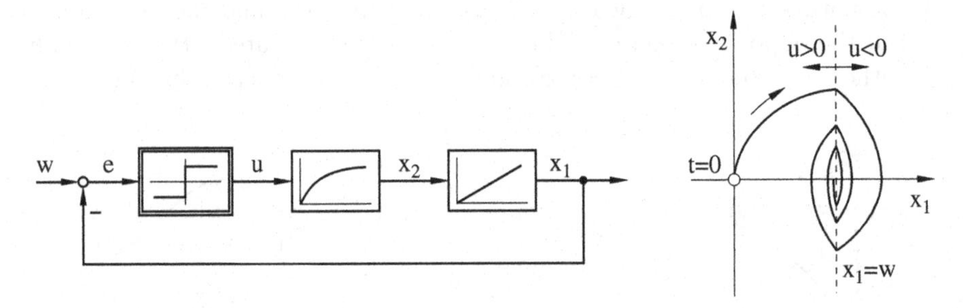

Fig. 2.68. Ideal 2-point switching controller with integrator and first-order lag

2-Point Switching Element with Hysteresis

As its name already suggests, the ideal 2-point switching element cannot be realized in practical applications. The two switches in figure 2.65, for example, will never be massless, and they will also produce sticking friction. As a result of this, they switch only after the field strength has reached a certain minimum value, and not, as assumed, already at the point where the field or the current i_e changes its sign. The transfer element therefore sticks to its old value for a while after the change of sign occurs. The output quantity changes only after the input variable exceeds a certain threshold. Around the origin, the characteristic curve of such a transfer element will therefore be ambiguous (fig. 2.69). Which branch of the curve will be valid depends on the previous state. In this respect, we can consider this transfer element to be a kind of 2-point switching element with memory. Such an effect is called *hysteresis*. Almost all switching elements in practical applications show a stronger or weaker hysteresis.

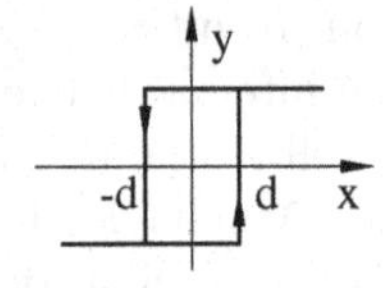

Fig. 2.69. Characteristic curve of a 2-point switching element with hysteresis

Figure 2.70 shows a 2-point switching element with hysteresis as a controller for a double integrator as the plant. The switching of the actuating variable or the acceleration a respectively, happens with some delay to the passing through zero of the control error e. This behavior can be represented in the state plane by a parallel shift of the switching line. For example: the controller switches from the positive to the negative value if $e = -d$ or $l = w + d$—and not for $e = 0$. Now this is precisely the case for the switching line which is shifted by d to the right. In a similar way, the switching line for the switch from a negative to a positive value is placed at $l = w - d$. Because of this delayed shifting the oscillation gets amplified, and the system becomes instable.

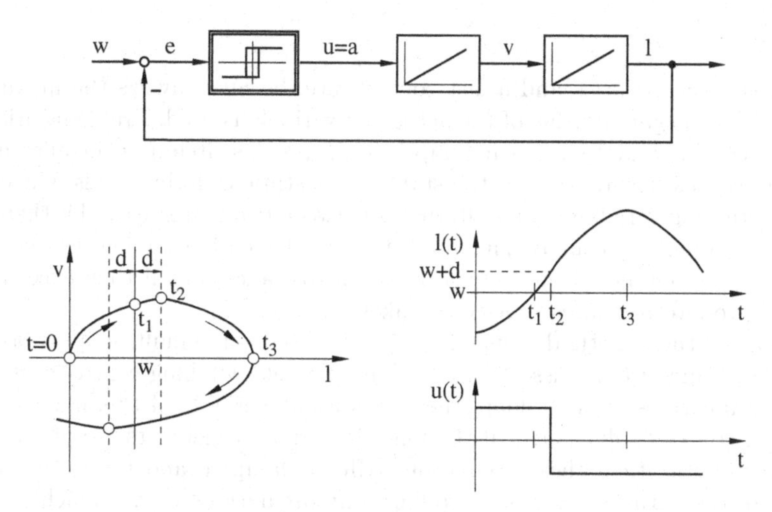

Fig. 2.70. 2-point switching controller with hysteresis and double integrator

On the other hand, if we link up the 2-point switching controller with a first order lag plus integrator, then—from a certain time on—the system will produce a steady oscillation, independently from the initial conditions. In figure 2.71 it can be seen clearly how the system ends up in the same oscillation, starting from two different initial states. Such an oscillation is called a *limit cycle*. In contrast to the steady oscillation discussed before, where the frequency and the amplitude depend on the initial state, for the limit cycle the frequency as well as the amplitude are determined by the system parameters, and they are totally independent of the initial state.

Limit cycles, too, can not occur in linear systems, as for linear systems the amplitude always depends on the initial state. We will define steady oscillations and limit cycles as a special type of nonlinear phenomena more precisely later on.

Although the stability of realizable 2-point switching controllers with hysteresis is already deteriorated compared to ideal 2-point switching controllers, they are frequently used in practical applications. Their advantage is a simple

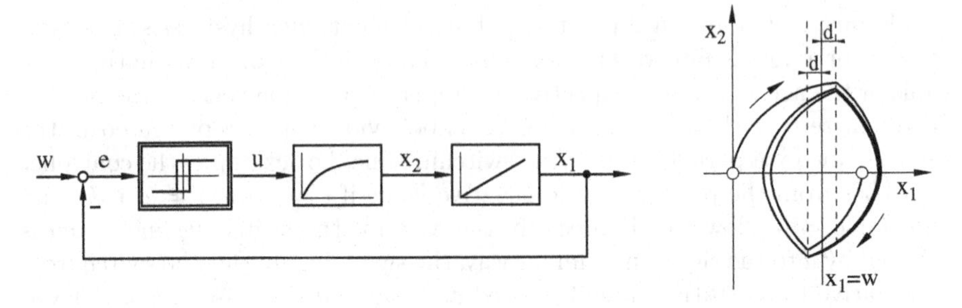

Fig. 2.71. 2-point switching controller with hysteresis and first order lag plus integrator

structure, a low price and a fast control rate, because always the maximum positive or negative value of the actuating variable is used. Problems arise by the fact that a system which comprises a 2-point switching controller never really settles down, even for the stable state still oscillating. This is because the actuating variable always alternates between two extremes and is therefore always to big or to small. The amplitudes of this oscillation can be accepted as long as they lie within a given range of tolerance, but if they exceed this range, countermeasures have to be taken.

One of these methods which is frequently used is the multi-loop or cascade control, which we discussed earlier. If the 2-point switching controller is used in the innermost control loop, then the output variable of this inner circuit will of course oscillate. But if the following transfer elements are of low-pass characteristic, then these oscillations will be damped, and the output variable of the entire system will produce only oscillations with a much smaller amplitude, which may possibly lie within the given range of tolerance.

3-Point Switching Element

Another possibility is using a controller with more than two possible output states. An example of this is the ideal 3-point switching element or the realizable 3-point switching element with hysteresis (fig. 2.72). We assume an integrator plus higher order lag following the switching controller as the plant. u will be different from zero as long as the control error e and correspondingly the difference $w-y$ lies outside the interval $[-\varepsilon, \varepsilon]$. As with $u \neq 0$ the integrator input is different from zero, its output variable y' changes, and therefore also y. Only for $u = 0$ the integrator stops at the value reached, the system settles down and takes a steady state.

$u = 0$ is equivalent to the fact, that the control deviation $e = w - y$ lies inside the interval $[-\varepsilon, \varepsilon]$. Therefore, the system achieves a steady state, if the output variable y lies within the range of tolerance $[-\varepsilon, \varepsilon]$ around the reference value w. This is sufficient for many applications. Steady-state accuracy ($w = y$), though, can not be reached.

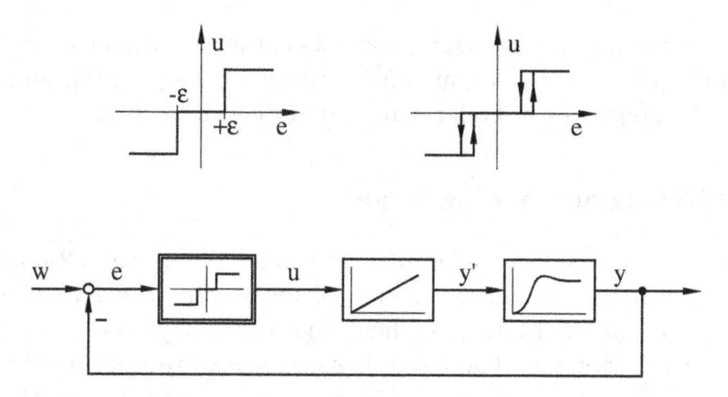

Fig. 2.72. 3-point switching controller with and without hysteresis

It should be mentioned that a 3-point switching controller is useful if it is followed by an integrator, as shown in figure cIn order to enable the system to settle down, a value of y' has to be the input to the last part of the plant, that corresponds to the entire output variable y lying within the range of tolerance $[-\varepsilon, \varepsilon]$ of w. Such a quantity y', though, can in most cases not be produced by the 3-point switching controller, as it comprises only three output values. For a 3-point switching controller without following integrator the control loop would produce the same steady oscillations as when using a 2-point switching controller. In contrast to this, with the help of an integrator we can produce the necessary value of y' exactly because of its continuous range for its output variable.

But the closed loop of fig. 2.72 does not reach a steady state in any case: If for example the integration time of the integrator is very short compared to the time constants of the following transfer elements, the output variable y cannot follow the integrator output y' fast enough. The integrator will cross the *right* range of values, i.e. the values that could produce output values for y inside the tolerance range, while the output variable y is still outside the tolerance range. When y reaches this tolerance range, y' has already left the *right* range of values. y will follow y' and also leave the tolerance range again. Therefore, if the parameters are choosen badly, we can obtain oscillations with a 3-point switching controller, too.

A big advantage for the 3-point switching controller is, that in technical systems the integrator stands relatively often at the beginning of the plant, so for the constellation 3-point switching controller plus integrator in many cases no additional integrator has to be added to the plant. One example is to control a boiler pressure with a valve whose opening cross section is adjusted by a motor. Closing the valve produces a pressure increase inside the boiler, and opening the valve a pressure decrease. The valve motor is steered by a switch with the three possibilities *open*, *close* and *stop*. If the signal value is *open*, the motor turns in the one direction, that opens the valve, and for *close* we get a turn in the other direction to close the valve. In this structure the switch is the 3-point switching controller, the transfer behavior from the switch to the valve cross section can be described by an integrator, and the dynamical behavior of the boiler can be described by a higher order lag.

Early Switching and Sliding Mode

As we have seen in the given examples, a system comprising a 2-point switching controller always oscillates. If in addition to this the 2-point switching controller also contains hysteresis, then the system may even become instable for the delayed switching. Therefore, it seems useful to improve the system's behavior by early switching. Accordingly, we would like the switching line of an ideal 2-point switching controller to be shifted to the left in the upper half of the state plane, and to the right in its lower half.

Alternatively, a counterclockwise revolving of the switching line would have the same effect. For the system given by figure 2.66 or 2.67, we can realize this revolving if we replace the equation $e = 0$ or $l = w$, respectively, which describes the switching line in state plane, by $l = w - kv$, where $k > 0$. This implies the following defining equation for the actuating variable:

$$u = \begin{cases} 1 & : \quad 0 < w - kv - l \\ -1 & : \quad 0 > w - kv - l \end{cases} \tag{2.229}$$

This behavior can obviously be achieved if we choose $w - kv - l$ instead of e or $w - l$ respectively as the input variable of the 2-point switching controller. The resulting structure, with an additional feedback loop, is shown in figure 2.73. Looking at the state trajectory, we can clearly see the improvement in the system's stability. The amplitudes of the oscillation continue to decrease until finally reaching the reference value $w = l$.

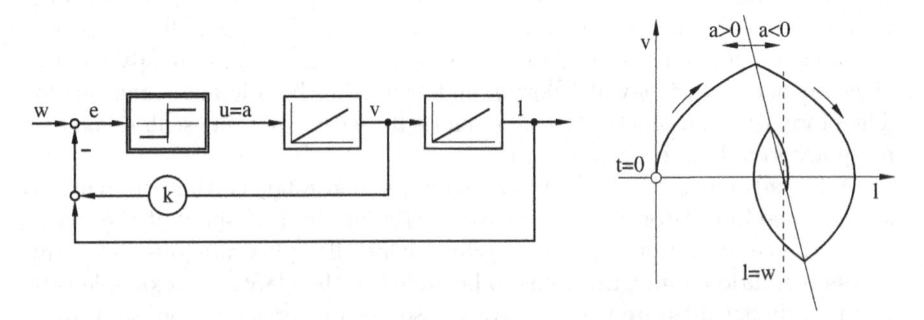

Fig. 2.73. 2-point switching controller with additional feedback loop

If we take a closer look at the transient behavior, it turns out that the system is not as ideal anymore as it seemed on first glance (fig. 2.74). For example: At some point the state trajectory approaches the inclined switching

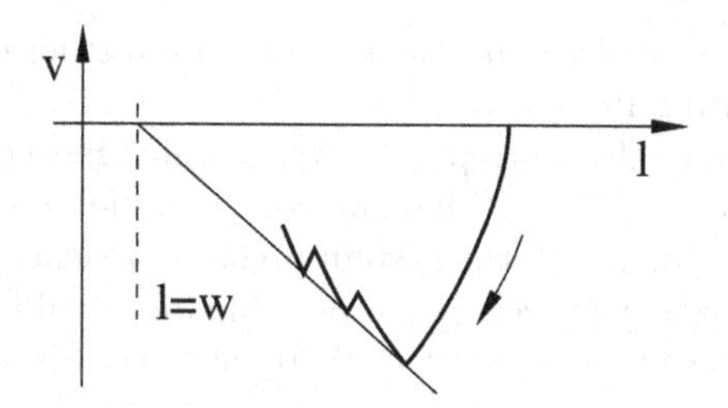

Fig. 2.74. Sliding mode for a switching element

line from the right with $u = -1$. When reaching this line, the controller switches to $u = 1$. Normally, the state of the system should now move off to the left of the switching line, but since the inclination of the state trajectory is bigger than the the one of the switching line, the system leaves to the right. As the condition $u = -1$ holds for this region, an instant switch back of the controller to -1 occurs and the system starts to approach the switching line again. In that way the system slides into the final state with a—theoretically—infinitely high switching frequency. Such a behavior is called *sliding mode*. An infinitely high switching frequency can of course never occur in reality, which is why this explanation should be regarded more as an illustration than as a proof. But indeed, we can obtain a very high switching frequency in such a structure shortly before the reference point is reached. An exact derivation of a sliding mode controller is given in section 2.8.10.

Concerning 2-point switching controllers with hysteresis, the system's behavior differs a little bit from this description, because of the delayed switching. The underlying principle, though, remains the same.

Using a 3-point switching controller has the advantage that the controller switches off as soon as the output variable gets sufficiently close to the reference value. Because of this, high-frequency shiftings in the final phase can be avoided.

Time-Optimal Control

So far, a system comprising a switching transfer element always overshot the mark when it had to adjust to a new reference value. It reached the reference state only after a switching process of several stages—if it could reach it at all. We now analyze this behavior shortly for a double integrator, i. e. the example of an accelerated body. Let the initial state be given by $(l, v) = (0, 0)$, and let the reference state be $(w, 0)$. In order to reach the reference state, first the body has to be accelerated. Overshooting would therefore mean that the initial phase of acceleration lasted too long, and that it was not possible to slow the body completely down when reaching the reference value. One improvement in the controller's behavior is therefore to start braking early enough. We obtain time-optimal behavior if the body is accelerated as long as possible, and braking starts just at the moment where any further delay would lead to the body overshooting its mark.

An examination of the state plane reveals which method we need in order to achieve this effect (fig. 2.75). First, we compute the switching curve. This is the state trajectory for which the system precisely reaches the reference point at the maximum negative acceleration possible for $v > 0$ (or at the maximum positive acceleration possible for $v < 0$). If the system's state lies below this switching curve (point 1), then the system can be accelerated positively until the curve is reached. Then acceleration is set to the maximum negative value, and the system follows the switching curve to the reference value. If on the other hand the state of the system lies above the curve (point 2), then it is not possible to slow the system down sufficiently and get it to a standstill at the reference point. Instead of this, the system will be at state 3 when the brake is complete. In order to get to the reference state from this position, the system first has to be accelerated further negatively until it reaches the lower branch of the switching curve. From there it can be transferred into the reference state at the maximum positive acceleration possible.

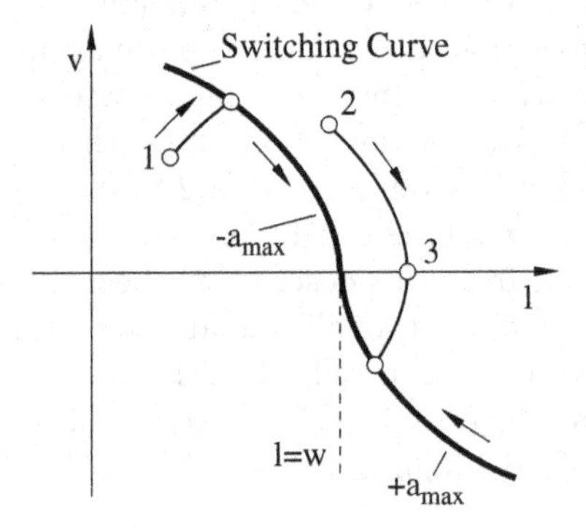

Fig. 2.75. Time-optimal process

A structure which allows such adjustments is called *time-optimal control*. It is shown in figure 2.76. Controller 1 is used to calculate the value of the switching curve v_S for the corresponding control error e. The switching curve in this block is printed the other way around because l appears with a negative sign in the control error. We can illustrate this by the following consideration: For a positive error for example, i. e. if the reference value is bigger than the actual value, the corresponding position in figure 2.75 will be to the left of the target. In this case the corresponding value of the switching curve therefore has to be positive—just as it is shown by the curve drawn inside the block for controller 1.

Then the calculated value v_S of the switching curve is compared to the actual velocity v. If the difference is positive, then the system lies below the

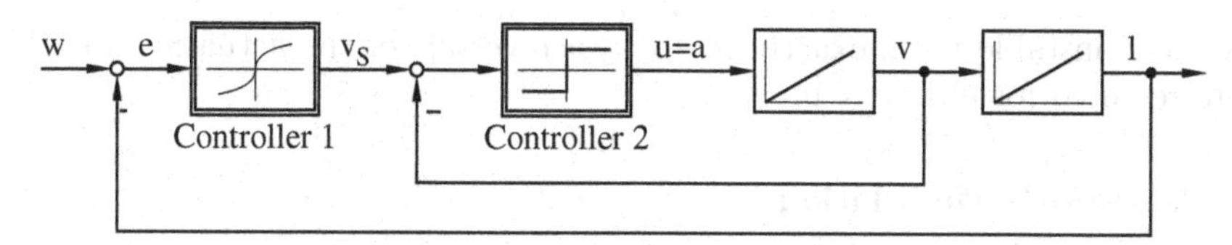

Fig. 2.76. Time-optimal control

switching curve. Accordingly, the 2-point switching controller 2 produces the maximum positive actuating variable possible. As soon as the difference $v_S - v$ becomes negative the controller switches to the maximum negative value and the system starts moving along the switching curve into the direction of the reference state.

After reaching this point, though, the behavior is undefined. Controller 2 will always produce an output different from zero, so the system cannot settle down. It will start to oscillate around the reference point with a—theoretically speaking—infinitely high switching frequency. Here a possibility for a shutdown should be kept in mind, which could, for example, be realized by replacing the 2-point switching controller by a 3-point switching controller.

With the help of calculus of variation it can be proven that the control processes which are obtained by using a controller of this structure will always be of optimum-time. This should also be clear from bare intuition: First the plant is accelerated with maximum positive or negative power (depending on the initial state) and then, at the last possible moment, again the maximum power is used to slow the plant down. Obviously such a behavior can be produced only with the help of a switching controller.

For a controller with hysteresis, the system cannot reach the reference point exactly following the switching curve, because of delayed switching. Instead of this, it will move slightly besides the curve and therefore exceed the reference position a bit. In the vicinity of the final position, oscillations of small amplitude and very high frequency will occur.

In addition to this, for applications in the field the problem arises that in order to allow calculation of the switching curve the plant has to be known exactly, while such a precise knowledge is usually not available. But still, even if the switching curve is not precisely calculated, we can obtain a relatively good control behavior.

An time-optimal control is also possible for systems of a higher degree. For a third order system, for example, the switching curve has to be replaced by a switching plane in state space. Using a maximum positive or negative actuating variable u of the controller, first this switching plane is reached. There, u has to change its sign. The system will move in the switching plane towards a switching curve, that lies in the plane. Then, the sign of u has to change again, so that the system converges towards the final state right on this switching curve. It can be shown that for a linear plant of degree n

without instable poles exactly $n - 1$ sign reversals of the actuating variable are required for a control process.

Pulse-Width Modulation

Finally, we should mention the pulse-width modulation (PWM), a method of high importance to applications in the field. The disadvantage of switching elements is that their output variable can take only a few, discrete values. On the one hand, this makes them useless for a high-quality control with steady-state accuracy. On the other hand, there is a vast interest in making them available to as many control applications as possible because of their low price. Here the PWM provides a method suitable for impressing quasi-continuous behavior on a switch, using an intelligent steering mechanism.

For PWM, a switching element gets switched with a high frequency following a special switching pattern, so that it produces a rectangular signal of high frequency with variable pulse width. If we use this signal as the actuating variable of a low-pass plant, then the signal's high frequency components are filtered out, and only the mean value of the oscillation affects the plant output.

Because of this, we can see the mean as the actuating variable. As on the other hand the mean depends continuously on the switching pattern, our only problem is to find the right pattern which we can use to impress a linear control behavior on the switching element. The *linearization by feedback* is a very comprehensible method which can be used to produce such a switching pattern and, accordingly, a quasi-linear controller (fig. 2.77).

The inner loop, consisting of the 2-point switching controller and the feedback function $G_R(s)$, will oscillate—given it is stable. To be precise, u will produce rectangular signals and y_R will oscillate around the input variable e. The smaller the width of the hysteresis of the 2-point switching controller gets, the higher the frequncy of the rectangular signal be and the less will y_R move away from e. If in addition to this the plant $G(s)$ comprises low-pass characteristics, then only the mean value of the rectangular signal will be of effect at the plant output y. The same holds for the influence which the actuating variable has on the internal feedback variable y_R. Because of this, we can restrict the analysis to mean values only. $\bar{y}_R(s) = G_R(s)\bar{u}(s)$ holds, and also—if the width of the hysteresis is sufficiently small—$e \approx \bar{y}_R$. This implies

$$\bar{u}(s) \approx \frac{1}{G_R(s)} e(s) \tag{2.230}$$

i. e. with regard to the mean values the overall transfer characteristic of the controller with internal feedback corresponds approximately to the reciprocal of its internal feedback transfer function. Depending on which function is chosen, we can approximate any desired linear control behavior.

Many other methods for pulse-width modulation are available besides the linearization by feedback. Often, optimized switching patterns are stored in

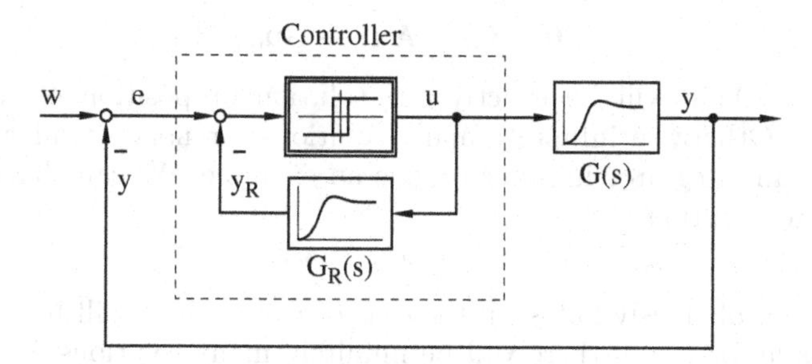

Fig. 2.77. Linearization achieved by internal feedback

lookup-tables and then, depending on the mean $\bar{u}$ that has to be produced (cf. [111]), the pattern needed is retrieved from it. The entire pulse-width modulation is done by one IC. The PWM is of high importance especially for the field of electrical drives control.

We have finished our discussion of different aspects of switching transfer elements. Our aim was to give a comprehensible explanation of the most important nonlinear effects, as well as to convey a basic understanding of the problems which occur when dealing with nonlinear control loops. The question of stability is of high importance to a control engineer. In order to be able to analyze the stability of a nonlinear system, a change from the more intuitive approach, which we have used so far, to exact descriptions of the problem is required. The following definitions and theorems are given only for systems which are continuous with respect to time, but they also hold for time-discrete systems in a similar manner. Later we will also give a discussion of the main criteria of stability for nonlinear systems, which can also be applied to fuzzy controllers.

2.8.4 Definition of Stability for Nonlinear Systems

Position of Rest

In order to be able to give an exact definition of stability for nonlinear systems, we first have to introduce the term *position of rest* ([45],[46],[56]):

Definition 2.20 *For a given constant vector* $\mathbf{u}_0$ *as the input, a dynamic system is at a position of rest—characterized by the state vector* $\mathbf{x}_R$*—if and only if the state variables remain constant, i. e. if*

$$\mathbf{0} = \dot{\mathbf{x}} = \mathbf{f}(\mathbf{x}_R, \mathbf{u}_0) \tag{2.231}$$

holds.

Without fixing a constant input vector, it would obviously be possible that the system never reaches a position of rest. For a linear system, we obtain its positions of rest from

$$\mathbf{0} = \dot{\mathbf{x}}_R = \mathbf{A}\mathbf{x}_R + \mathbf{B}\mathbf{u}_0 \tag{2.232}$$

If $|\mathbf{A}| \neq 0$, there will be exactly one solution or position of rest $\mathbf{x}_R = -\mathbf{A}^{-1}\mathbf{B}\mathbf{u}_0$. Otherwise infinitely many solutions, or none at all will result. Let us use an integrator in order to give an example. We can characterize it by the state equation

$$\dot{x} = 0x + 1u \tag{2.233}$$

Here, $|\mathbf{A}| = 0$ obviously holds. In the case of $u \neq 0$, there will be no position of rest, while for $u = 0$ there will be infinitely many solutions for the given equation. This result seems reasonable, as an integrator will continue to add or subtract if the input variable is different from zero, and it will stop instantly for $u = 0$ at its actual value.

While a linear system possesses no, one or infinitely many positions of rest, for nonlinear systems also a finite number of solutions other than 1 can occur. An example is given by the pendulum in figure 2.10. If the body is hung up on a stiff rod, then we get obviously positions of rest for $\alpha = 0$ and $\alpha = \pi$.

Lyapunov's Definition of Stability

There exists an important difference of a qualitative nature for the positions of rest of the pendulum just mentioned, which we can describe precisely using the definition of stability due to *Lyapunov* ([116]):

Definition 2.21 *A position of rest $\mathbf{x}_R$ is said to be stable for a given constant input value $\mathbf{u}_0$ if and only if for every $\varepsilon > 0$ a $\delta > 0$ can be found such that for any $|\mathbf{x}(0) - \mathbf{x}_R| < \delta$ the condition $|\mathbf{x}(t) - \mathbf{x}_R| < \varepsilon$ $(t \geq 0)$ is satisfied.*

In other words: A position of rest is stable if and only if, for $t > 0$, the state $\mathbf{x}(t)$ remains within an arbitrary small region (ε) around the position of rest as long as the initial state is situated sufficiently close (δ) to the position of rest (fig. 2.78).

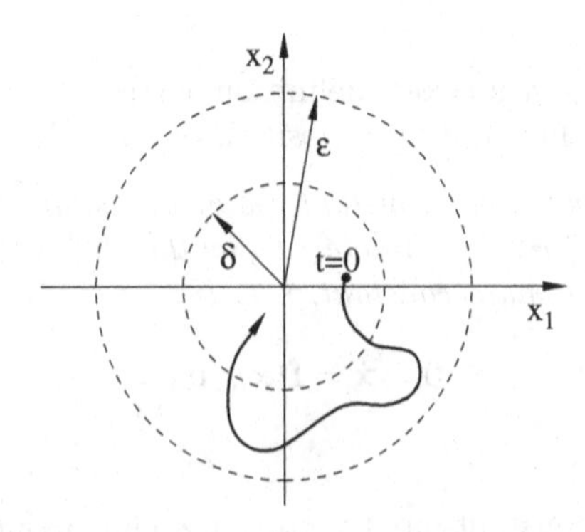

Fig. 2.78. Illustration to Lyapunov's definition of stability

According to this definition, the upper position of rest of the pendulum is instable, while the lower one is stable. If we displace the pendulum a little bit from the lower position of rest and then consider this position to be the initial one, then of course the pendulum will start to swing, but it will never move further away from the position of rest than the distance between this and the initial position. In this example, for any arbitrary region ε, that shall not be left by the system for $t > 0$, there exists a distance $\delta = \varepsilon$ the initial state has to lie in to meet this criterion.

The same obviously does not hold for the upper position of rest: Assume a demand that the pendulum should not leave a region ε of some degrees around the upper position of rest. The only initial state that meets this condition is the position of rest itself. If the initial state differs from this position of rest only a little bit, then the pendulum will swing down and leave the reference region. On the other hand, the definition demands that we can find a δ-region with $\delta > 0$ for the initial state for any ε in order to obtain a stable position of rest. As for the upper position of rest the state remains in that region only for $\delta = 0$, the condition for a stable position of rest is not fulfilled. This position of rest is therefore unstable.

Another simple example is given by the ideal 2-point switching controller with double integrator (fig. 2.67). The system is in steady oscillation around the position of rest $(l, v) = (w, 0)$, with the amplitude depending on the initial state. If we now demand that for $t > 0$ the system state has to lie within a certain ε-region around the position of rest, then all we need to do is pick the right initial state. Therefore, this system is stable according to Lyapunov's definition.

One should be aware of the fact that stability according to Lyapunov only guarantees that a certain region (which can be chosen freely) will never be left. For many applications however, this is not sufficient. There, in addition to this, we have the additional demand that the given position of rest also has to be reached sooner or later. This leads to the term of *asymptotic stability*:

Definition 2.22 *A position of rest* $\mathbf{x}_R$ *is said to be asymptotically stable if it is stable for any constant activation* $\mathbf{u}_0$*, and if it possesses a* β*-region where* $\lim_{t\to\infty} \mathbf{x}(t) = \mathbf{x}_R$ *for* $|\mathbf{x}(0) - \mathbf{x}_R| < \beta$*, i. e. there exists a region, where from any initial state inside this region the system converges towards the position of rest. The set of all initial states from which the system converges to* $\mathbf{x}_R$ *is called the position of rest's* domain of attraction*. If this domain comprises all initial states which can possibly occur under certain given technical restrictions, then the position of rest is said to be* asymptotically stable in the large*. If the domain consists of the entire state space, then the position of rest is said to be* globally asymptotically stable*.*

Once again we can use the example of the pendulum to illustrate this definition, especially its lower position of rest. In the case of an ideal pendulum and an initial displacement from the position of rest, the pendulum will swing *ad infinitum* and never reach the position of rest. The position of rest is stable in the sense of Lyapunov, but it is not asymptotically stable. But if we take, for example, the air resistance into consideration, then the amplitude of the oscillation will decrease continuously and the position of rest will be reached—in theory after an infinite period of time. For this case the position of rest is asymptotically stable, but not globally asymptotically stable, as there is one point in the state space from which no trajectory leads to the lower position of rest—the upper position of rest. But if we assume that the pendulum hangs on the ceiling and can therefore never reach the upper position of rest, then we can consider the system to be asymptotically stable in the large.

We can also apply Lyapunov's definition of stability to time-variant systems. As such systems can change within the flow of time, the definition given above has to be fulfilled for any initial time point, not only for $t = 0$. Accordingly, δ has to be a function depending on t as well as on ε. Finally, for the case where δ is a function depending only on ε even though the system is a time-variant one, this type of stability is called *constant stability*.

Stability of Trajectories

So far, our discussion of stability was only concerned with the positions of rest, but these concepts can also be applied to trajectories. In all the given definitions one only has to substitute 'trajectory' for 'position of rest'. Just like positions of rest, trajectories can be instable, stable or asymptotically stable.

For example, let us analyze the oscillation produced by a double integrator with ideal 2-point switching controller (fig. 2.67). For this oscillation amplitude and frequency—i.e. the course which the oscillation takes—depend on the initial state. A different initial state leads to a different cycle. If, for example, the initial state (for $t = 0$) is shifted slightly to the right in the state plane, then this implies a smaller initial displacement with regard to the position of rest and, accordingly, also a lower oscillation amplitude. The resulting trajectory will be similar to the original one, but closer to the position of rest than the first trajectory. Obviously, for any ε-region around the original trajectory we can find a corresponding δ-region around this trajectory inside of which an initial state has to lie so that its resulting oscillation trajectory lies within the ε-region of the original trajectory. This is precisely what stability according Lyapunov describes. This type of oscillation is called *steady oscillation*.

Asymptotic stability, though, is not given, as the oscillation resulting from the modified initial state will never converge into the original oscillation, and we could only speak of asymptotic stability if this condition was fulfilled. An example of this is the oscillation of a 2-point switching controller with hysteresis and integrator plus first order lag (fig. 2.71). From any initial state, the trajectory will sooner or later converge into the same given oscillation. Such an oscillation with asymptotical transient behavior is called a *limit cycle*. So, by the help of the Lyapunov stability definition we were able to define precisely the difference between steady oscillation and limit cycle.

Limit cycles need not necessarily be asymptotically stable—they can also be instable. An instable limit cycle is defined by the fact that trajectory starting at an initial state adjacent to the limit cycle moves off from the limit cycle's trajectory. A limit cycle can even be stable and instable at the same time. For a system of second order, the cycle divides the state plane into two regions, an inner and an outer one. It is therefore possible that for the inner region the trajectories from any initial state converge into the limit cycle, while outside they move away from it. Such a limit cycle is stable inside and instable outside. This is of course only possible in theory, as even a limit cycle which is instable in only one direction can never exist for a longer time. All we need is a minor disturbance that pushes the system's state to the outer side of the limit cycle, and the system leaves the cycle and will never reach the cycle again. Anyway, we should keep the possibility of limit cycles with different stability behavior in mind, as they will appear again later on, when we discuss the method of the describing function.

If we want to examine the stability of an oscillation, we should pay attention to the fact that the definition of stability according to Lyapunov was only applied to the trajectories in state space. If the course of the state variables with respect to time would be analyzed, then an entirely different picture would result. Let us consider the pendulum once more in order to illustrate this idea. If we neglect air resistance, it produces a steady oscillation whose parameters depend on the initial state. Starting at one initial state, we get a steady oscillation of a certain frequency and amplitude. Starting at a slightly bigger initial displacement, we achieve a similar steady oscillation with a slightly lower frequency and a slightly larger amplitude. If we sketch the trajectories for these two oscillations, they will be of the same basic form, and they will be adjacent to each other. The trajectory for the oscillation with the smaller amplitude will lie inside the other one. The oscillation is stable in the sense of the definition according to Lyapunov. Now, if we draw the course of the pendulum position for both cases with respect to time, then the curves diverge from each other because of the oscillations' different frequencies. If stability was defined with regard to these curves, then the oscillation would be considered to be instable.

For this reason, in [129] and [158] an oscillation is said to be asymptotically stable only if Lyapunov's definition of stability applies to the time dependent course of the state variables. If in contrast to this only stability with regard to the trajectories is given, this is referred to by the term *orbital stability*. For practical application, though, it is irrelevant to distinguish between these two features, as usually only the general form of the oscillation will be of interest, and not the explicit course of the state variables. Therefore, in this book we will keep on to judge the stability of oscillations only with respect to their trajectories.

Stability of Linear Systems

After this detailed discussion of Lyapunov's definition of stability for nonlinear systems, we want to discuss the link between this definition and the linear system stability definition. A linear system is—corresponding to def. 2.16—asymptotically stable if and only if its state variables converge towards zero from any initial state without external activation being necessary. The question is: Does this definition harmonize with the definitions 2.21 and 2.22?

The first thing we notice is that definition 2.16 deals with the stability of the system, while in definitions 2.21 and 2.22 we are talking about the stability of a single point of rest. To explain this, let us analyze a linear system with a constant activation $\mathbf{u}_0$:

$$\dot{\mathbf{x}} = \mathbf{A}\mathbf{x} + \mathbf{B}\mathbf{u}_0 \tag{2.234}$$

Any position of rest $\mathbf{x}_R$ which results from this activation meets the differential equation

$$\dot{\mathbf{x}}_R = \mathbf{A}\mathbf{x}_R + \mathbf{B}\mathbf{u}_0 \tag{2.235}$$

Subtracting (2.235) from (2.234) and substituting $\Delta\mathbf{x} = \mathbf{x} - \mathbf{x}_R$ yields

$$\dot{\Delta\mathbf{x}} = \mathbf{A}\Delta\mathbf{x} \tag{2.236}$$

Now, for the stability analysis of a point of rest we have to take into account only the distance between the state vector and the point of rest. Therefore, this investigation can be performed based on eq. (2.236). But, in this equation the input vector as well as the point of rest do not appear any more, so that the result of the stability analysis will be the same for any other point of rest and any other input vector. If one point of rest is asymptotically stable for one certain input vector, every point of rest for any input vector will be asymptotically stable as well. For that reason—for linear systems—we may talk about the stability of the entire system instead of the stability of one single point of rest. This is an important difference compared to nonlinear systems, where different positions of rest might correspond to completely different types of stability.

If one certain point of rest of a linear system is asymptotically stable for one certain input vector according to def. 2.21 and 2.22, this will hold for every other point of rest and especially for $\mathbf{x} = \mathbf{0}$ and $\mathbf{u} = \mathbf{0}$. Therefore, the prerequisites of def. 2.16 are fulfilled, and the system is also asymptotically stable according to that definition for linear systems. The same would hold for instability. Finally we can say that stability according to def. 2.16 for linear systems follows from defs. 2.21 and 2.22 for any point of rest of that system.

To prove the equivalence of the definitions, we also have to show the reverse way, that means, from def. 2.16 also defs. 2.21 and 2.22 have to follow. Because of the same stability characteristic of all points of rest of a linear system, we can prove this for the most simple case, i.e. for the position of rest given by $\mathbf{x} = \mathbf{0}$ and $\mathbf{u} = \mathbf{0}$, and broaden the result to the entire system.

From (2.234) we get with $\mathbf{u} = \mathbf{0}$

$$\dot{\mathbf{x}} = \mathbf{A}\mathbf{x} \tag{2.237}$$

Now, this system shall be asymptotically stable according to def. 2.16, i.e. all state variables converge to zero without any external input from any initial state. From theorem 2.17 it follows, that all eigenvalues of $\mathbf{A}$ have a negative real part. For that case also eventually occurring oscillations have a decreasing amplitude. Therefore, for any ε-region around the point of rest $\mathbf{x} = \mathbf{0}$, that shall not be left by the state vector for $t > 0$, we can find a δ-region, where the initial state has to be in: $\varepsilon = \delta$. With this result, the point of rest is stable according to def. 2.21. And the asymptotic stability according to def. 2.22 is guaranteed, because all state variables converge to zero. Here, an analogous proof for instability can be given, too, and the equivalence of the different stability definitions for linear systems is proven.

Besides this, the position of rest given by $\mathbf{x} = \mathbf{0}$ and $\mathbf{u} = \mathbf{0}$ is globally asymptotically stable, i.e from every initial state in the state space the trajectories converge to this zero state. For the proof it is sufficient to show, that no other points of rest exist. But this is an easy step, because if $\mathbf{A}$ possesses only eigenvalues with negative real parts, we have $|\mathbf{A}| \neq 0$, and this implies that equation (2.237) can have only one solution $\mathbf{x} = \mathbf{0}$ for $\dot{\mathbf{x}} = \mathbf{0}$.

Controllability and Observability

Besides stability, two other important characteristics of a system are controllability and observability. As already mentioned in section 2.7.2, we should make sure in advance, i. e. before designing the controller, that it is actually possible to influence the system in the desired way. This means we have to analyze the system for controllability. Furthermore, if we use state variables as input for the controller, then we should also check that we can compute them from the measurable output quantities. This corresponds to observability.

Two methods to check these features are available for nonlinear systems. The first is to linearize the model and then apply the criteria of controllability and observability for linear systems to it. Once again the problem arises that a linear model provides a sufficient approximation of the behavior of a nonlinear system only for a small region around the set point. The results of an analysis for controllability and observability are therefore limited to a small region of the state space.

The second method is to adjust the definitions and criteria in a suitable way for nonlinear systems. [177], for example, gives definitions of reachability and distinguishability of states, but sufficient criteria (cf. theorem 2.13 and 2.15) for nonlinear systems in general, which could also be applied easily, do not exist. Such criteria exist only for special classes of nonlinear systems, as, for instance, bilinear systems.

Definition of the Zero Position of Rest

In the following sections we will make the simplifying assumption that at the position of rest under consideration all quantities of the system are zero. If this does not hold, then the system has to be redefined. We can interpret this redefinition as a switch from the system's actual quantities $\mathbf{x}$—which stands not only for the state variables here—to their deviations $\Delta\mathbf{x} = \mathbf{x} - \mathbf{x}_R$ from the position of rest. This vector of all system variables fulfills the demand that all system variables are zero at the position of rest: $\Delta\mathbf{x} = \mathbf{0}$. We should emphasize that this step is an exact redefinition of the system—not a linearization around the working point.

For a linear system this redefinition is not necessary. Because the stability behavior of all positions of rest is equal there, we can always take the zero position of rest $\mathbf{x} = \mathbf{0}$ for our stability analysis. But for nonlinear systems, the redefinition is important and necessary. Figure 2.79 shows the required steps for the example of a standard closed-loop system consisting of a linear and a nonlinear part. The nonlinear part is given by the function $\mathbf{u} = \mathbf{f}(\mathbf{e}, \dot{\mathbf{e}})$, which depends on the error $\mathbf{e}$ and its derivative $\dot{\mathbf{e}}$.

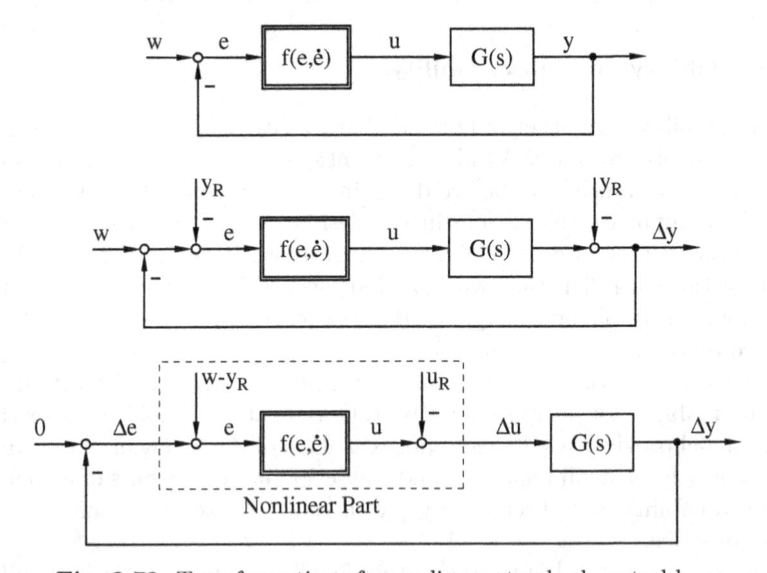

Fig. 2.79. Transformation of a nonlinear standard control loop

Let the position of rest under consideration be characterized by the vectors $\mathbf{w}, \mathbf{u}_R, \mathbf{y}_R$ and $\mathbf{e}_R = \mathbf{w} - \mathbf{y}_R$. If we want to use the deviations $\Delta\mathbf{u}, \Delta\mathbf{y}$ and $\Delta\mathbf{e}$ from the position of rest instead of the quantities $\mathbf{u}, \mathbf{y}$ and $\mathbf{e}$ themselves, then we have to proceed as follows (fig. 2.79): First, we have to subtract the variable $\mathbf{y}_R$ at two points of the closed loop. The effects of the two subtractions to the system cancel each other out because of the negative feedback, i.e. the systems remains unchanged.

In the second step, the subtraction of $\mathbf{y}_R$ at the output of the linear part shall be replaced by the subtraction of a vector $\mathbf{u}_R$ at the input of the linear part, where $\mathbf{u}_R$ represents the position of rest's vector of actuating variables. The dc signal $\mathbf{u}_R$ can be easily computed from the dc signal $\mathbf{y}_R$ with $\mathbf{y}_R = \mathbf{G}(s=0)\mathbf{u}_R$. $\Delta\mathbf{u} = \mathbf{u} - \mathbf{u}_R$ represents the deviation from the position of rest. If we furthermore sum up the input quantities $\mathbf{w}$ and $\mathbf{y}_R$ at the input of the nonlinear part, then we can write the new nonlinear transfer characteristic as

$$\Delta\mathbf{u} = \bar{\mathbf{f}}(\Delta\mathbf{e}, \Delta\dot{\mathbf{e}}) = \mathbf{f}(\Delta\mathbf{e} + \mathbf{w} - \mathbf{y}_R, \frac{d}{dt}(\Delta\mathbf{e} + \mathbf{w} - \mathbf{y}_R)) - \mathbf{u}_R \tag{2.238}$$

The resulting, new, entire system meets the condition that all of its quantities will be zero at the position of rest. In the following, we will assume, that before any stability analysis the system is redefined, so that the position of rest under consideration is always the zero position.

2.8.5 Lyapunov's Direct Method

We are now able to discuss the question of how to determine the stability of the position of rest for any given system. Following definitions 2.21 and 2.22 strictly, we would have to solve the nonlinear differential equations for any possible initial state and then to define the required ε- and δ-regions. This is obviously impossible because of infinitely many possible initial states. Therefore, we need methods or criteria, that enable us to estimate the stability behavior of a given position of rest without complicated computations.

For a second order system, an analysis can be made in the state plane, just like we have shown that previously for the nonlinear switching elements. For systems of a higher degree, though, other criteria are required, which we present in the following sections.

The first of these criteria was developed by Lyapunov himself and is referred to as the *direct method*. This method is based on the following underlying indea: We define a scalar *Lyapunov function* $V(\mathbf{x})$, depending on the state vector, which has to be zero at the origin of the state plane—that was defined as the position of rest—and must be increasing with the distance to origin. We can think of V as being a generalized distance to the position of rest. As an example, the contour lines of such a function in the state plane are given in figure 2.80. Furthermore, a state vector—and therefore also the function V—follows a certain curve in the course of time, determined by the system's state equation. If we can show that the derivative of the function V with respect to time is negative for any state vector $\mathbf{x}$, then this implies that starting at any initial state the state trajectory passes all contour lines of V from outside to inside, and that therefore the state vector converges towards zero. Hence, in this case the system is asymptotically stable.

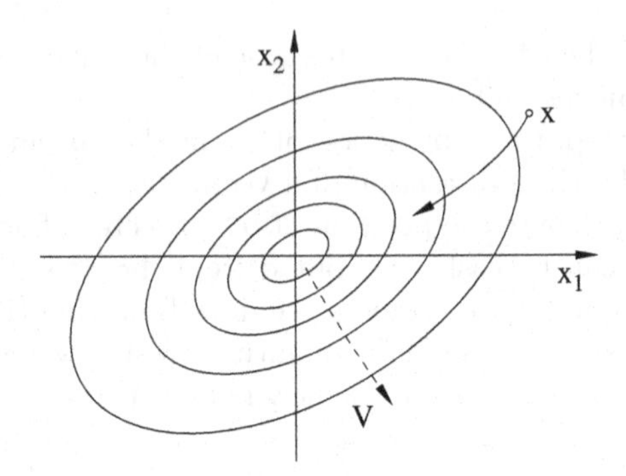

Fig. 2.80. Illustration of Lyapunov's direct method

Theorem 2.23 *Let $\mathbf{x} = \mathbf{0}$ be the position of rest of the dynamic system $\dot{\mathbf{x}} = \mathbf{f}(\mathbf{x})$. Let $V(\mathbf{x})$ be a positive definite function (i. e. $V(\mathbf{x}) > 0$ for $\mathbf{x} \neq \mathbf{0}$ and $V(\mathbf{x}) = 0$ for $\mathbf{x} = \mathbf{0}$), where $V(\mathbf{x})$ and its first partial derivatives are continuous functions within a certain region around the position of rest. Let furthermore its derivative with respect to time,*

$$\dot{V} = \sum_{i=1}^{n} \frac{\partial V}{\partial x_i} \dot{x}_i = \sum_{i=1}^{n} \frac{\partial V}{\partial x_i} f_i \tag{2.239}$$

be negatively definite for this area. Then the position of rest is asymptotically stable, and the region forms its domain of attraction.

Let G be an area lying inside the region for which $V < c$ holds ($c > 0$) and which is bounded by $V = c$. If in addition to this G is limited and comprises the position of rest, then G belongs to the position of rest's domain of attraction.

If the domain of attraction comprises the entire state space and if furthermore $V(\mathbf{x}) \to \infty$ follows from increasing distance from the position of rest $|\mathbf{x}| = \sqrt{x_1^2 + ... + x_n^2} \to \infty$, then the position of rest is globally asymptotically stable.

If $\dot{V}$ is negatively semi-definite ($\dot{V}(\mathbf{x}) \leq 0$), then only the simple stability can be guaranteed. But if the set of points with $\dot{V} = 0$ comprises no other trajectory except $\mathbf{x} = \mathbf{0}$, then we have asymptotic stability here, too.

A proof of this theorem can be found for example in [104]. Because of our considerations before, we can neglect comments on the first part of the theorem, but some additional remarks on the last three paragraphs may be helpful.

Generally, any consideration concerning the domain of attraction is easier if we think of V in terms of contour lines in state space. The second paragraph is illustrated in figure 2.81. In this example, $\dot{V}$ is negatively definite only in between the two dotted lines; outside of it the function is positively definite. It is because of this distribution that a trajectory as sketched in the figure is possible. As long as the trajectory is running between the dotted lines it crosses the contour lines of V from the outside to the inside, and in the rest of the state space from the inside to the outside. It is therefore not guaranteed that the position of rest's domain of attraction is the entire area with $\dot{V} < 0$ in between the dotted lines. Only the region G bounded by the contour line H is for sure part of it. As the entire line lies in the area for which $\dot{V} < 0$ holds, it is impossible to cross this line from the inside to the outside. Accordingly, we always have to use a closed contour line to describe the boundaries of a domain of attraction, and this is precisely the demand which is made in the second paragraph.

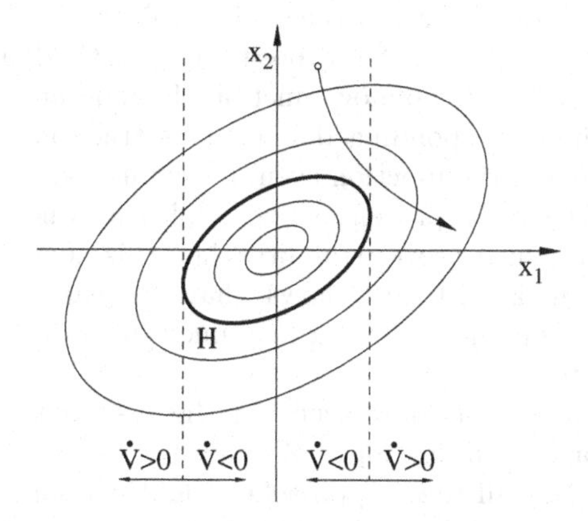

Fig. 2.81. Domain of attraction of a stable position of rest

Thinking of V in terms of contour lines also helps to explain the demand for unlimited growth of V at infinite distances from the origin: If V did not continue to grow with increasing distance to the origin, then some contour lines would exist which would stretch out into infinity, but which would never be closed. Two adjacent contour lines could then keep an infinite distance to each other at infinity. It would then require—for $\dot{V} < 0$—an infinite period of time for the state curve to cross the contour line with the smaller value for V after having crossed the contour line with the bigger value. The point in state space would spend an infinite period of time between the two contour lines, possibly at an infinite distance from the origin, and the system would not be stable even for $\dot{V} < 0$.

The last paragraph of the theorem is easier to comprehend. If V, and correspondingly the distance to the origin remain constant ($\dot{V} = 0$) instead of getting smaller in the course of time, then obviously only the simple stability is given. If, on the other hand, the points in state space for which $\dot{V} = 0$ holds do not form a continuous trajectory, then in between these points the trajectory will contain states with $\dot{V} < 0$ (assuming it did not reach the origin already). V will therefore be not a strictly monotonously decreasing function, but a monotonously decreasing one. The origin will be reached sooner or later, and the system is therefore asymptotically stable.

We can also use the direct method to prove the instability of a position of rest. Analogously to theorem 2.23, this time it has to be shown that $\dot{V}$ is positively definite. For the proof of stability as well as for the proof of instability of a position of rest there exist many variations of theorem 2.23 for different boundary conditions ([56]), among these also the so-called Lyapunov's theorems of instability. Even for time-variant systems (e. g. [18], [158]), and for systems with external activations ([104]) some theorems do exist. The basic idea, the use of a Lyapunov function, remains the same in all cases, and therefore we neglect more detailed explanations.

Instead, let us briefly discuss the main problem of the direct method: Obviously, the size and form of the position of rest's domain of attraction depend crucially on the Lyapunov function. If another Lyapunov function is chosen, then the corresponding domain of attraction can be completely different. The choice of the function even determines whether the stability of the position of rest can be proven or not. And if it is not possible to find a Lyapunov function which proves stability, then this does not imply that the position of rest is instable. It just shows, that our search was not successful. To show instability, one has to find a Lyapunov function V whose derivative is positively definite.

For that reason, several approaches for the systematic search for a Lyapunov function have been developed through the years ([18], [45], [46], [52], [56], [158], [164]). But all these approaches could not solve the crucial problem, that no statement about the stability of a position of rest is possible if no Lyapunov function could be found.

Finally, in the late 90s a solution for that problem was developed for the extensive class of so-called TSK systems (see chap. 4.1.3), using LMI algorithms (see appendix G). With the help of these algorithms it is possible now, to find an answer for the general question of the existence of a Lyapunov function with negatively defined derivative for the given system. And if a Lyapunov function exists, we know that the corresponding position of rest is stable. We will introduce this method in section 4.2.2.

Here, we will illustrate an application of the direct method, using the spring-mass system of figure 2.4. It is not very reasonable to use the direct method for stability analysis here—an examination of the eigenvalues of the system matrix would be a much easier, but on the other hand it provides a very comprehensible example, and the calculations involved are not too complicated.

The system is not activated from outside. According to (2.144), the state equation is given by

$$\begin{bmatrix} \dot{v} \\ \dot{l} \end{bmatrix} = \begin{bmatrix} -\frac{c_f}{m} & -\frac{c_r}{m} \\ 1 & 0 \end{bmatrix} \begin{bmatrix} v \\ l \end{bmatrix} \tag{2.240}$$

The system's overall contents of energy is given by the sum of the kinetic energy of the moving body and the potential energy which is stored in the spring. Friction causes a loss of energy to the system, it will therefore finally stop to oscillate. As the energy is constantly decreasing, it is therefore reasonable to use the system's contents of energy as a Lyapunov function to prove the stability of the position of rest $(v, l) = (0, 0)$:

$$V = E = E_{\text{kin}} + E_{\text{pot}} = \frac{1}{2}mv^2 + \int_0^l f_r dx = \frac{1}{2}mv^2 + \int_0^l c_r x dx$$
$$= \frac{1}{2}mv^2 + \frac{1}{2}c_r l^2 \tag{2.241}$$

V and its derivatives are continuous functions. V is positive in the entire state space except for the origin $(v, l) = (0, 0)$. It exceeds all limits for $|\mathbf{x}| = |(v, l)^T| \to \infty$. Therefore, it meets all the requirements of theorem 2.23 for global stability. We only need to analyze whether $\dot{V}$ is negatively definite. The derivation of V with respect to time is given by

$$\begin{aligned} \dot{V} &= mv\dot{v} + c_r l\dot{l} \\ &= mv[-\frac{c_f}{m}v - \frac{c_r}{m}l] + c_r lv \\ &= -v^2 c_f \end{aligned} \tag{2.242}$$

making use of the state equation (2.240). The function is negatively semi-definite, as it is zero not only at the origin but at all states where $v = 0$. This is easy to explain. A loss of energy—and, correspondingly, a decreasing V—is caused by friction. This happens only for a velocity different from zero, but at the points of maximum displacement the velocity and therefore also the friction and $\dot{V}$ is equal to zero. So far, we can only guarantee global stability, but no asymptotic stability. An examination of the points in state space for which $\dot{V}$ is equal to zero reveals that they do not form a closed trajectory (with the exception of the origin). A state $(v = 0, l \neq 0)$ indicates a maximum displacement of the spring as well as a maximum value for the amplitude of the oscillation, but the spring immediately starts to accelerate the body, and the system attains a state with $v \neq 0$ and $\dot{V} < 0$. All in all, V is therefore a monotonously decreasing function and thus the global asymptotic stability of the system is proven according to the fourth paragraph of theorem 2.23.

2.8.6 Describing Function

Now that we have concluded the discussion on the direct method, we describe a totally different approach, the *describing function method*. In this method, which we will present here for SISO systems only, it is assumed that the system's output variable y oscillates around the position of rest $y = 0$. This leads, of course, to oscillations of the error e and actuating variable u of the system. We do not discuss the cause of these oscillations. The analysis of the oscillation allows us to draw conclusions concerning the system's stability with regard to the position of rest in question. The assumed oscillation can be a steady oscillation or a limit cycle, which we will not distinguish here.

One requirement is that the control loop can be separated into a linear and a nonlinear part, just like for the standard control loop given in figure 2.79. We assume that all the dynamic parts of the system, like, for example, integrators, delay elements etc. are contained in the linear part, while the nonlinear part is *instantaneously acting*:

$$u(t) = f(e, \mathrm{sgn}(\dot{e})) \tag{2.243}$$

which means that—in principle—the output variable u of the nonlinear part can be computed from the current input variable e without knowledge of former values of u or e. Regarding, for example, characteristic curve elements it is possible to compute the output variable $u = f(e)$ directly from the current value of the input variable e. Accordingly, these elements are instantaneously acting. Transfer elements with hysteresis are also considered to be instantaneously acting, even though in this case some knowledge of the history is needed as it would otherwise not be possible to decide which branch of the hysteresis loop is currently traversed. The term $\mathrm{sgn}(\dot{e})$ is used to express this history.

In addition to this, the characteristic curve of the nonlinear part has to be a monotonically increasing odd function (point symmetry to the origin). This is for example the case for the switching elements, which we introduced in the beginning of this section. In contrast to this, the transfer function of the linear part needs to be of strong low-pass characteristic. We illustrate the meaning of this feature in the following derivation. Even this requirement is usually fulfilled in many cases of real-life applications, and for this reason the method of the describing function can be widely used.

We start the derivation with the assumption that the output of the system is a sinusoidal oscillation $y(t) = -A\sin(\omega t)$, where our task is to determine the amplitude A and the frequency ω. The negative sign is not important here, but later it will simplify the computations. Since we redefined the system according to figure 2.79 before applying the new method, the reference signal w is equal to zero, and the input to the nonlinear part is $e(t) = A\sin(\omega t)$. Accordingly, we obtain a periodic output signal of the nonlinear part which can be described by a Fourier series with the fundamental frequency ω. Because of the point symmetry to the origin of the characteristic curve of the nonlinear part, the signal has no dc content:

$$u(t) = \sum_{k=1}^{\infty} A_k \cos k\omega t + B_k \sin k\omega t$$

$$\text{with} \qquad A_k = \frac{2}{T}\int_0^T u(t)\cos(k\omega t)dt$$

$$B_k = \frac{2}{T}\int_0^T u(t)\sin(k\omega t)dt$$
$$T = \frac{2\pi}{\omega} \tag{2.244}$$

This signal forms the input quantity to the linear part. According to theorem 2.3, every harmonic component at the input of a linear transfer element causes an oscillation of the same frequency at the output. If the low-pass characteristic of the linear part is sufficiently strong, then all *higher harmonics* (oscillations with frequencies higher than the basic frequency ω) are filtered out of the signal and only the *first harmonic* (basic frequency signal) remains. Whether or not the low-pass characteristic is sufficiently strong is difficult to describe formally. As a rule of thumb we can say that the degree of the denominator's polynomial of the transfer function should exceed the one of the numerator's polynomial by at least 2. In some cases, though, a difference of only 1 can already be sufficient. In any case one should check that at the end of the procedure, when all the parameters of the oscillation—and therefore also ω—have been calculated, all the signal's parts of higher frequencies $2\omega, 3\omega, ...$ can be sufficiently suppressed by the linear part. Otherwise a major requirement of this method is neglected and the entire calculation becomes invalid.

The basic frequency which remains at the output of the linear part represents the signal $y(t) = -A\sin(\omega t)$, which was chosen initially. All other parts of the oscillation which were produced at the output of the nonlinear part could not pass the linear part. But only components of the signal which are able to pass all parts of the control loop can contribute to a constant or even to an increasing oscillation of the entire system and therefore endanger the system's stability. For the stability analysis, we can therefore neglect all higher harmonics which occur at the output of the nonlinear part. It remains

$$u(t) = A_1\cos\omega t + B_1\sin\omega t = C_1\sin(\omega t + \varphi_1) \tag{2.245}$$

as the result, where $C_1 = \sqrt{A_1^2 + B_1^2}$ and $\varphi_1 = \arctan\frac{A_1}{B_1}$. $u(t)$ is therefore the result of a multiplication of the input signal $e(t) = A\sin(\omega t)$ by the factor $\frac{C_1}{A}$ and a phase shift of $-\varphi_1$, which corresponds precisely to the behavior of a purely linear delay element with a constant factor (cf. (2.38)). Hence, we can define a quasi-linear transfer function corresponding to the delay element which describes the behavior of the nonlinear part. Such a function is called *describing function*:

$$\frac{u}{e} = N(A,\omega) = \frac{C_1(A,\omega)}{A}e^{j\varphi_1(A,\omega)} \tag{2.246}$$

It has to be pointed out that this type of linearization has nothing in common with the linearization around the set point (equation (2.219)). According to the definition of A_1 and B_1, C_1 and φ_1 depend on the amplitude A as well as on the frequency ω of the input signal. It is possible to show that for instantaneously acting nonlinearities the dependency on ω is dropped and therefore the parameters of the describing function rely only on the amplitude of the input signal:

$$N(A) = \frac{C_1(A)}{A} e^{j\varphi_1(A)} \tag{2.247}$$

This is a very important difference between a quasi-linear and a genuine linear transfer function, as the delay time and the gain of the latter depend on the frequency of the input signal only. Furthermore, the describing function describes the transfer characteristic of the nonlinear element only with regard to the basic frequency. We can therefore use the describing function like a linear transfer function only if it is guaranteed that the input signal of the nonlinear part is actually $e(t) = A\sin(\omega t)$. We will never be able to use it for example in the calculation of a step response.

In our case, the requirements are fulfilled, and therefore the describing function may be used like a linear transfer function. The system's open-loop transfer function now consists of the describing function and the transfer function of the linear part: $N(A)G(j\omega)$. In order to obtain a constant oscillation, we have to make sure that the output signal y appears unchanged at the output after traversing the entire closed loop. We get the resulting condition for such an oscillation by

$$y = -N(A)G(j\omega)y \tag{2.248}$$

or, alternatively

$$-1 = N(A)G(j\omega) \tag{2.249}$$

If we separate the real part of this complex equation from the imaginary part, we obtain two equations for the two unknown quantities—the amplitude A and the frequency ω to be precise. If a solution to (2.249) exists, then a corresponding oscillation is possible for the system. This oscillation may be a steady oscillation or a limit cycle. It is also possible that several solutions exist, which implies that several different oscillations could be obtained. If no solution exists, then no *harmonic oscillation* (an oscillation consisting of harmonics) can occur in the closed loop. Non-harmonic oscillations, though, are still possible, but usually unlikely. As mentioned earlier, one should check for every possible oscillation whether its components of higher frequencies are sufficiently extracted by the low-pass filter of the linear part, as this is an essential requirement of the complete algorithm.

We can obtain a suitable description of the stability characteristic of a possible oscillation in a graphical approach, together with the help of the Nyquist criterion (theorem 2.9). This criterion determines the required phase shift of the open-loop transfer function's nyquist plot regarding the point -1. In equation (2.249), the left hand side represents this crucial point and the right hand side is the open-loop transfer function. Now, if we rewrite it as

$$-\frac{1}{N(A)} = G(j\omega) \tag{2.250}$$

then another interpretation gets possible: Now, the open-loop transfer function consists only of the linear part, while the crucial point gets extended to a curve $-\frac{1}{N(A)}$, depending on the amplitude A.

In this case, we first have to compute or to measure the describing function $N(A)$ of the nonlinear part. Next, we draw the curve $-\frac{1}{N(A)}$ in the complex plane. Then we have to measure or calculate the frequency response $G(j\omega)$ of the linear part and draws its Nyquist plot, again in the complex plane. Every intersection of the two curves represents a solution to equation (2.250) and represents a possible oscillation, whose stability behavior can be obtained from the Nyquist criterion. We will give some examples for this now.

In the first example, the nonlinear part consists of an ideal 2-point switching controller. First, we compute its describing function. The parameters C_1 and φ_1 of the describing function result from the coefficients A_1 and B_1, which therefore need to be calculated first. The output $u(t)$ for a sinusoidal input signal is a rectangular signal (fig. 2.82). For $T = \frac{2\pi}{\omega}$, we obtain:

$$\begin{aligned} B_1 &= \frac{2}{T}\int_0^T u(t)\sin(\omega t)dt \\ &= \frac{2}{T}2K\int_0^{\frac{T}{2}} \sin(\omega t)dt = \frac{4K}{\pi} \\ A_1 &= \frac{2}{T}\int_0^T u(t)\cos(\omega t)dt = 0 \end{aligned} \tag{2.251}$$

and from this

$$\begin{aligned} C_1 &= \sqrt{A_1^2 + B_1^2} = B_1 = \frac{4K}{\pi} \\ \varphi_1 &= \arctan\frac{A_1}{B_1} = \arctan 0 = 0 \end{aligned} \tag{2.252}$$

or

$$N(A) = \frac{C_1(A)}{A}e^{j\varphi_1(A)} = \frac{4K}{A\pi} \tag{2.253}$$

as the describing function. The phase lag $-\varphi_1$ of the describing function is therefore zero and the gain $\frac{C_1}{A} = \frac{4K}{A\pi}$. This should also seem reasonable, as the rectangular signal at the output of the 2-point switching controller is in phase balance to the sinusoidal oscillation at the input end and so the phase delay has to be zero. Furthermore the amplitude of the rectangular signal has got the same value for any input signal. But since the gain is defined to be the ratio between the output amplitude and the input amplitude, it has to be of inverse proportionality to the amplitude of the input signal.

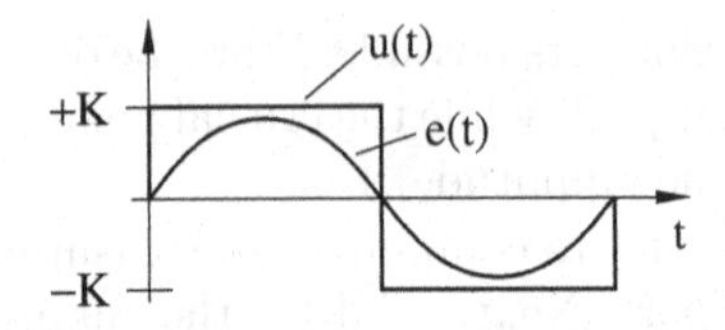

Fig. 2.82. Input signal and output signal of an ideal 2-point switching element

Tables of other describing functions can be found in, for example, [18], [45], [46] and [198]. The most thorough information on describing functions is given in [50], which can be regarded as a standard for this topic.

If the describing function is known, we can proceed to the analysis for stability. First we sketch the graph of the function $-\frac{1}{N(A)}$ in the complex plane, depending on the amplitude, in the complex plane. According to (2.253), for the 2-point switching controller we obtain $-\frac{1}{N(A)} = -\frac{A\pi}{4K}$, i.e. a curve lying on the axis of reals that continues to move away from the origin in the negative direction for increasing values of A. We then add the Nyquist plot of the linear part to this representation. From the resulting points of intersection and the position of the curves to each other we can now draw conclusions about the stability of the system. Figure 2.83 shows some examples for the case that the nonlinear part of a standard control loop (fig. 2.79) consists of an ideal 2-point switching controller.

For the example in the left upper corner, the linear part consists of an integrator and a first order lag with the transfer function

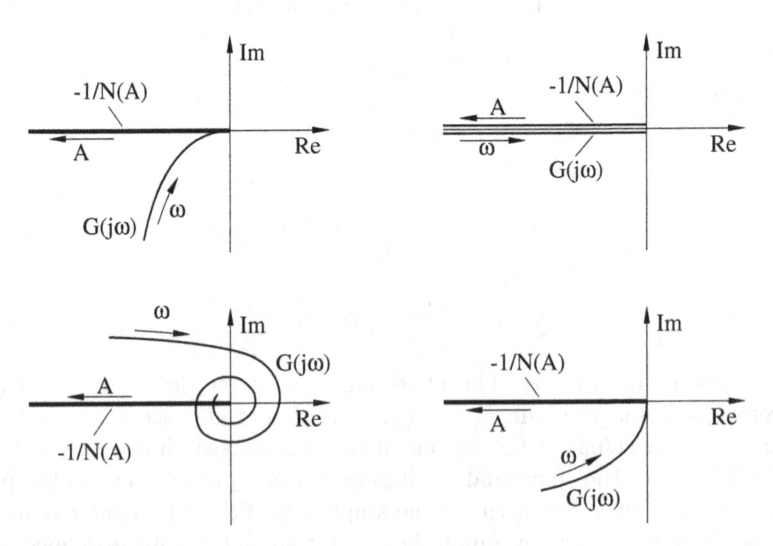

Fig. 2.83. Stability analysis with describing function for systems containing a 2-point switching controller

$$G(s) = \frac{1}{s(Ts+1)} \tag{2.254}$$

Thus, the entire system corresponds to the system given in figure 2.68. The only intersection of the Nyquist plot and the curve of the describing function is at the origin, i.e. for $A = 0$, $\omega = \infty$. Correspondingly, this is also the only solution for equation (2.249). This means that for this system only an oscillation with amplitude $A = 0$—i.e. no oscillation at all—is possible.

With regard to the interpretation of equation (2.250) we could also regard the curve $-\frac{1}{N(A)}$ as the amplitude-depending crucial point of the Nyquist criterion. In this case $G(s)$ represents the open-loop transfer function. According to the Nyquist criterion, its Nyquist plot needs a phase shift by a certain degree with respect to this point in order to produce a stable system. As $G(s)$ comprises an integrator, the necessary phase shift is equal to $+\frac{\pi}{2}$. This value is just what we are given if we neglect the intersection at the origin. Regarding any other point of the curve $-\frac{1}{N(A)}$ the phase shift has exactly this value. This is easy to see, if draw a vector from the curve $-\frac{1}{N(A)}$ to the Nyquist plot of the linear part and check its phase shift with increasing ω. Therefore, this system is stable.

For the example in the upper right corner, the linear part consists of a double integrator, and we obtain the system shown in figure 2.66. The curves of the describing function and the Nyquist plot of the linear part lie exactly one on top of the other. We get infinitely many intersection points and, correspondingly, infinitely many solutions to equation (2.249). Here, the Nyquist plot of the linear part gets larger values of ω the closer it gets to the origin, while the values of A for the curve of the describing function get the bigger, the further it moves away from the origin. For an intersection, and accordingly for a possible solution or oscillation, we can state that the amplitude will be the bigger, the smaller the frequency gets. This corresponds to the results we achieved earlier for this system. The resulting steady oscillation depends on the initial state of the system. This can be seen in figure 2.67. The bigger the amplitude gets, the slower the oscillation or the lower the frequency will be.

The linear part for the example in the lower left corner consists of a double integrator with delay time element. Since the Nyquist plot of the linear part spirals into the origin, we get infinitely many intersection points of the two curves. The question is which oscillation we will actually obtain. The following explanation is not totally precise, but provides a good illustration and finally produces the correct result. First, we assume that the system's state can be characterized by one intersection, and that the system is oscillating. If some minor disturbance occurs and, possibly, the amplitude is diminished a bit, then the system will move a little to the right, following the curve of the describing function. But this point is a crucial point, like any other point of the curve $-\frac{1}{N(A)}$. The phase shift of the Nyquist plot around this point will surely be negative, while it should be $+\pi$ according to the Nyquist criterion, because of the two integrators in the linear part. For this reason, the system is instable regarding this point, and the oscillation increases. The system moves back to the left to the intersection point. Therefore, the oscillation is stable concerning a reduction of the amplitude.

For an augmentation of the amplitude, caused by another disturbance, the system moves further to the left following again the curve of the describing function. The point which is reached is again a crucial one. The phase shift of the Nyquist plot is once more definitely negative because of its spiral form, i.e., according to the Nyquist criterion, we get another case of instability. Accordingly, the oscillation increases and moves on to the left to the next intersection point further away from the origin.

The same considerations hold for all intersection points, i.e. all limit cycles are, regarding the state plane, stable inside and instable outside. Because of this, the system will continue for every disturbance to move from intersection to intersection, further and further away from the origin. This corresponds to continuously growing values for the amplitude of the oscillation, and the system is therefore instable.

In the last example, we discuss the example in figure 2.73 where we added a feedback loop to the 2-point switching controller with double integrator. Our first task is to transform the system into a standard control loop (fig. 2.79). In order to do so, we define the 2-point switching element to be the nonlinear part and regard everything else as the linear part of the control loop. Next, we have to compute the transfer function for the linear part, which we obtain by relating the output variable u to the input variable e of the nonlinear part. In the case of the standard control loop, this relation is given by $e = -G(s)u$. For the system at hand, we have

$$e(s) = -u(s)(\frac{k}{s} + \frac{1}{s^2}) \tag{2.255}$$

according to figure 2.73, and therefore

$$G(s) = -\frac{e(s)}{u(s)} = \frac{ks+1}{s^2} \tag{2.256}$$

The Nyquist plot of this function is given in the lower right corner of figure 2.83. Like in the first example, the only intersection of the two curves is in the origin at $A = 0$, which means, that no harmonic oscillation is possible for this system. The phase shift of the Nyquist plot with regard to the crucial point, i.e. to the curve of the describing function is π. Because of the two integrators, this is (according to the Nyquist criterion) precisely the value required to get a stable entire system. We should still be cautious concerning this result, as the difference between the degree of the denominator's and the numerator's polynomial is only 1. It is therefore questionable whether the low-pass-filter-effect of the linear part—which is a requirement—will be sufficient here or not. In this case, though, we can accept the result as it is confirmed by the consideration we made regarding figure 2.73.

The curve of the describing function looks a bit different for a 2-point switching element with hysteresis. Because of the hysteresis, the switching from one output value to the other is delayed here compared to the ideal 2-point switching controller (and therefore also compared to a sinusoidal oscillation input signal). For this reason, the phase lag φ is different from zero, and the describing function is no longer a purely real-valued function. We skip about the calculation of this function; the shape of the curve $-\frac{1}{N(A)}$ can be seen in figure 2.84.

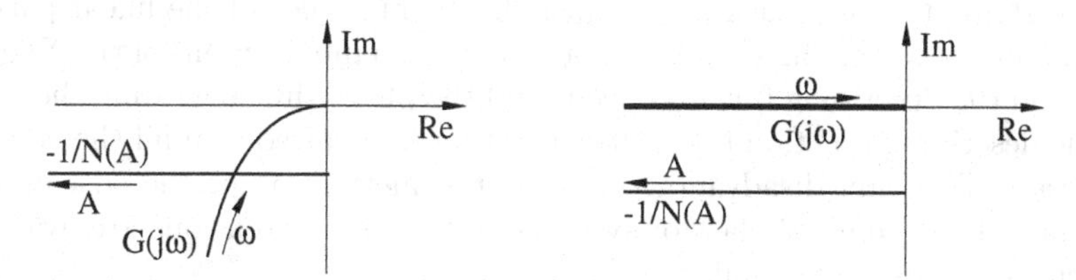

Fig. 2.84. Describing function for a 2-point switching controller with hysteresis

For the example on the left hand side of figure 2.84, the linear part once again consists of a first order lag with an integrator (cf. fig. 2.71). The curves intersect at an amplitude other than zero. We should therefore expect the case of a limit cycle or steady oscillation, but we still have to analyze the stability of this oscillation. Again, we make use of an inexact but more comprehensible explanation. We assume that the system is currently in a state corresponding to the intersection of the two curves. Then some disturbance occurs that reduces the amplitude of the oscillation a bit. The system's state can now be characterized by a point slightly to the right of the intersection, lying on the curve of the describing function. The phase shift of the Nyquist plot with respect to this point is approximately $-\pi$, while the phase shift required for stability according to the Nyquist criterion should be $+\frac{\pi}{2}$ because of the integrator of the linear transfer function. We have another case of instability: the oscillation rises, the amplitude increases and the system moves back to the point that corresponds to the intersection of the two curves.

If—caused by another disturbance—the system moves along the describing function to the left, then the phase shift is approximately $+\frac{\pi}{2}$. In this case, the system is stable. The oscillation settles down and the system moves back to the intersection. As the overall result, we find that it is not possible for the system to move away from the intersection. Accordingly this oscillation is a stable limit cycle.

In the example illustrated on the right hand side, the linear part consists of a double integrator. According to the Nyquist criterion, the phase shift of the Nyquist plot with respect to the crucial point should be $+\pi$, but actually it attains negative values lying between $-\frac{\pi}{2}$ and $-\pi$, regarding the entire curve of the describing function. This system is therefore instable, which is confirmed by the considerations concerning figure 2.70.

After all these examples, we are now convinced that for the experienced user the describing function method provides a simple and comprehensible method in the stability analysis. It is easy to obtain the required information, as the describing function as well as the Nyquist plot can be measured in case a representation by formulas should be too difficult or even impossible. In addition to this, the graphical representation provides such an excellent description that it can also be used to develop possibilities for the stabilization of a system. The only task is to adjust the Nyquist plot of the linear part by adding linear correcting elements in a way that no intersections of the Nyquist plot and the describing function occur and the phase shift regarding the curve of the describing function takes the value that is required to fulfil the Nyquist criterion. The only disadvantage is that the method as presented can only be applied to a limited class of systems, but several extensions are available which we discuss in the following.

One important restriction on this method is the demand for a sufficiently strong low-pass characteristic of the linear part. In [50], the suggestion is made to truncate the expansion into a Fourier series after a term of higher degree if the low-pass characteristic of the describing function is insufficient. But in doing so, we lose one big advantage of this method—the representation of nonlinearities by a linear transfer function which leads to the easy handling of this method.

[50] also presents a method of how to compute describing functions for signal shapes other than harmonic oscillations, e.g. Gaussian noise or dc signals. In those cases, we can no longer obtain the describing function from the first term of an expansion into a Fourier series. Instead of this, we define a linear transfer element with initially unknown parameters. We then compute these parameters in a way that the mean square error between the output signal of this linear transfer element and the one of the given nonlinear element is minimized with respect to a given input signal. If we assume the linear transfer element to be a delay element with adjustable gain, then this approach will—for a sinusoidal input signal—produce the same result as the one with the expansion into a Fourier series.

[45] discusses the possibility of how the describing function method can be used if the control loop consists not only of one linear and one nonlinear part like the standard control loop, but of several nonlinear parts which are separated from one another by linear parts.

Figure 2.85 shows a simple example. On condition that the linear parts are of sufficient low-pass characteristic, we can represent e_1 and e_2 by their first harmonics:

$$\begin{aligned} e_1 &= A_1 \sin \omega t \\ e_2 &= A_2 \sin(\omega t + \varphi_2) \end{aligned} \tag{2.257}$$

or, alternatively, as complex-valued pointers:

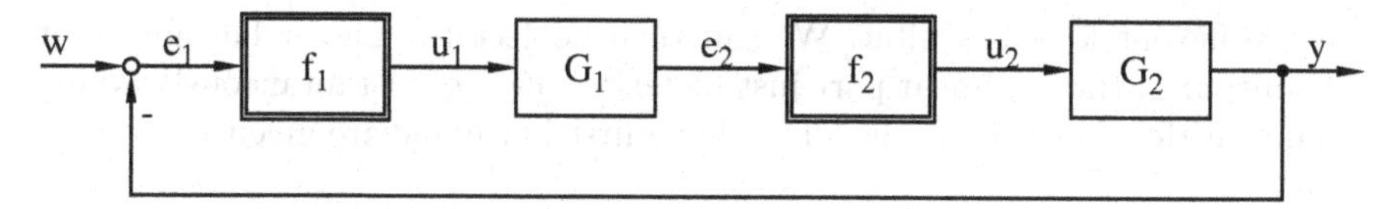

Fig. 2.85. Expanded control loop for the describing function method

$$\begin{aligned} e_1 &= A_1 e^{j\omega t} \\ e_2 &= A_2 e^{j(\omega t + \varphi_2)} \end{aligned} \tag{2.258}$$

Next, we substitute quasi-linear describing functions $N_1(A_1, \omega)$ und $N_2(A_2, \omega)$ for the nonlinear parts. Accordingly, we get

$$-1 = N_1(A_1, \omega) G_1(\omega) N_2(A_2, \omega) G_2(\omega) \tag{2.259}$$

as the equation describing the oscillatory balance. We can split this equation up into the imaginary part and the real part. This leaves us with two equations but three unknown quantities A_1, A_2 and ω. Since describing functions can be treated like linear transfer functions, we can use the relation between the input signals of the nonlinear elements to get

$$\begin{aligned} e_2 &= N_1(A_1, \omega) G_1(\omega) e_1 \\ A_2 e^{j(\omega t + \varphi_2)} &= N_1(A_1, \omega) G_1(\omega) A_1 e^{j\omega t} \\ A_2 e^{j\varphi_2} &= N_1(A_1, \omega) G_1(\omega) A_1 \end{aligned} \tag{2.260}$$

Examination of the gains provides us with the required third equation:

$$A_2 = |N_1(A_1, \omega)| \, |G_1(\omega)| A_1 \tag{2.261}$$

Unfortunately, a graphical solution for this set of equations is possible only for a small number of cases, but numerical solutions can still be found.

For higher importance to practical applications is the possibility to modify this method in a way that it can be applied to nonlinearities which are not instantaneously acting, but which have certain internal dynamics. In this case, the output variable u of the nonlinear part depends not only on the input signal e (or alternatively on the output y of the linear part) but also on its derivatives: $u = f(e, \dot{e}, ...)$. This dependency occurs as well for nonlinear parts without internal dynamics if the input to the nonlinear part consists not only of the control error or the output of the plant, but also of its derivatives. This constellation occurs for example for fuzzy controllers.

Let us now assume that the nonlinear part is defined by the first order transfer characteristic $u = f(e, \dot{e})$, instead of simply $u = f(e)$. Furthermore, we assume this function to be an odd one, $f(-e, -\dot{e}) = -f(e, \dot{e})$, and that it is monotonically increasing in e for any $\dot{e} > 0$. Also, the linear part has to be a sufficient low-pass filter. We can then neglect the higher harmonics at the output of the nonlinear part just like in the case of instantaneously acting nonlinearities. Now the coefficients of the first harmonic are given by

$$
\begin{aligned}
A_1 &= \frac{2}{T} \int_0^T f(e, \dot{e}) \cos(\omega t) dt \\
&= \frac{2}{T} \int_0^T f(A \sin(\omega t), A\omega \cos(\omega t)) \cos(\omega t) dt \\
B_1 &= \frac{2}{T} \int_0^T f(e, \dot{e}) \sin(\omega t) dt \\
&= \frac{2}{T} \int_0^T f(A \sin(\omega t), A\omega \cos(\omega t)) \sin(\omega t) dt
\end{aligned} \tag{2.262}
$$

where $T = \frac{2\pi}{\omega}$. Using the same formulas as for the instantaneously acting nonlinearities, we get another describing function, $N(A, \omega)$, but this time one which depends on the amplitude A as well as on the frequency ω. Accordingly, we obtain a family of curves $-\frac{1}{N(A,\omega)}$ instead of only one curve $-\frac{1}{N(A)}$ as the graphical representation for the describing function, with ω as its parameter. This means that for every frequency ω_1, we get one amplitude-depending curve $-\frac{1}{N(A,\omega_1)}$.

In an analysis for stability, we sketch this family of curves as well as the Nyquist plot of the linear part in the complex plane, and interpret the family as the crucial point of the Nyquist criterion. Once again, we can use the relation between the Nyquist plot and the family of curves to draw conclusions concerning the stability of the system. As an example, figure 2.86 shows the Nyquist plot of a third order lag and a family of curves which could result for a fuzzy controller.

If any of the intersections represents an oscillation of the system, it will surely be the case of a stable limit cycle. In order to give an explanation,

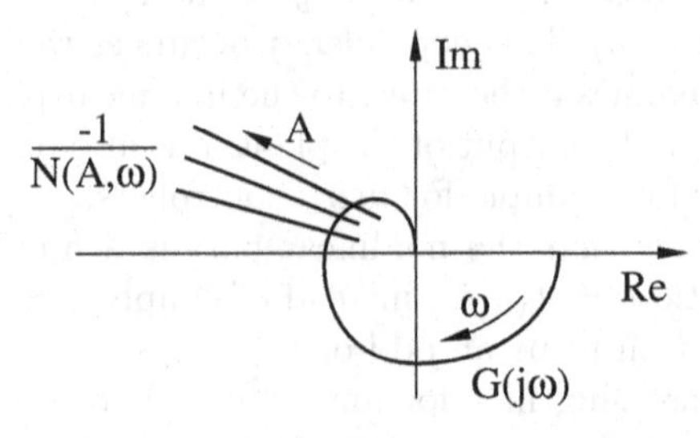

Fig. 2.86. Using a frequency-depending describing function for stability analysis

we assume that the system currently performing an oscillation represented by one intersection. If the amplitude is diminished because of a disturbance, then the system moves into the direction of the origin, following the curve of the describing function it is lying on. It reaches a point which is enclosed by the Nyquist plot. The phase shift of the Nyquist plot around this point is negative, but it should be zero for a stable system with a third-order lag according to the Nyquist criterion. The current state is therefore instable, the oscillation grows and the system moves back to the intersection. If on the other hand—starting at the intersection—the amplitude increases because of a disturbance, then the system moves away from the origin, again following its curve of the family of curves. The phase shift of the Nyquist plot around the point which now represents the state of the system is zero, the system is stable, the oscillation settles down and the systems moves back to the intersection.

Having a family of curves, we have to take into account that not every intersection represents a possible oscillation. Until now, where we had only one curve of the describing function, in every intersection the amplitude of the oscillation was given by the curve of the describing function and the frequency by the Nyquist plot of the linear part, and every intersection represented a solution of eq. (2.250). Now, for every curve of the family of curves of the describing function one frequency is fixed. An intersection with the Nyquist plot can therefore represent a possible oscillation or solution respectively only, if this fixed frequency is equal to the frequency given by the Nyquist plot in that point. Only in that case this frequency is the frequency of a possible oscillation, and the amplitude is given by the describing function.

Therefore, a graphical solution under these conditions is pure chance, so that the only useful way to obtain the amplitude and frequency of the oscillation is a numerical solution of equation (2.250)

$$G(j\omega) = -\frac{1}{N(A,\omega)} \tag{2.263}$$

Generally, we can also determine the describing function $N(A,\omega)$ always in a numerical way. This will be sensible if no analytical description of the nonlinear part is available, for example in the case of a fuzzy controller. In order to do so, we pick a pair of values (A_1,ω_1) and feed-forward the corresponding sine wave at the input of the nonlinear part. This will produce a periodic oscillation at its output, but not necessarily a sinusoidal one. However, we can use the least square method to approximate this oscillation by a sinusoidal oscillation. A comparison between the approximating output oscillation and the input oscillation yields the gain V and the phase lag $-\varphi$ for the nonlinear part and the values (A_1,ω_1), which then again leads to the complex value $N(A_1,\omega_1)$ for the describing function. This way, we can obtain a pointwise description of the describing function.

[195] even discusses an extension of this method for MIMO systems, but since the prerequisites for this extension cannot be verified in practical

applications, any analysis for stability using this method will be fairly insecure. We therefore refrain from discussing this topic.

2.8.7 Popov Criterion

We can now turn to another method which is based on the criterion of stability according to Popov. This is, in contrast to the method of the describing function, an exact criterion, but, for special cases, it may lead to very conservative results as it is a sufficient but not a necessary criterion. This means that there might be stable systems whose stability cannot be proven with this criterion. On the other hand, though, it is very easy to apply. Again, we require that the system can be split up into an instantaneously acting nonlinear part and a linear part.

The method shall be described for SISO systems first. The characteristic curve of the nonlinear part and the Nyquist plot of the linear part have to be known. Let us first give an additional definition in order to keep the formulation of the criterion as simple as possible (cf. fig. 2.87):

Definition 2.24 *A characteristic curve $f(e)$ runs through a sector $[k_1, k_2]$ if*

$$k_1 \leq \frac{f(e)}{e} \leq k_2 \qquad \textit{and} \qquad f(0) = 0 \tag{2.264}$$

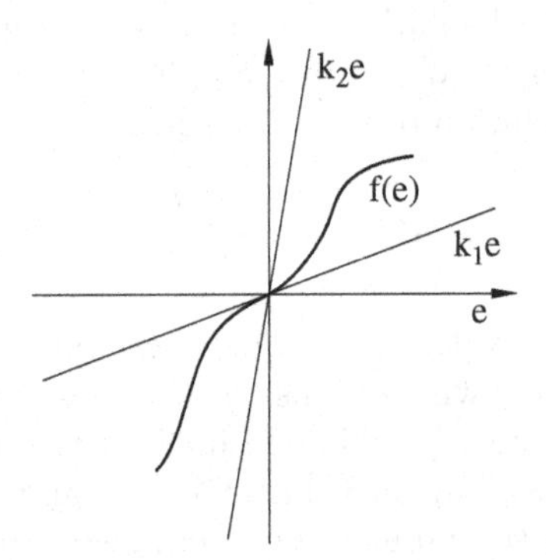

Fig. 2.87. Sector of a characteristic curve

We are now set to present the Popov criterion. It can be proven using the direct method of Lyapunov, but we refrain from giving this proof (cf. [2]):

Theorem 2.25 *(Popov criterion) Let a closed loop be given consisting of a linear and a nonlinear part. Let the transfer function $G(s)$ be a purely rational function. All the poles of the function have a negative real part, and the gain is $V_l = 1$. Furthermore, let the degree of the polynomial of the denominator be* *bigger than the degree of the one of the numerator. Let the nonlinear part be given by a well-defined and piecewise continuous characteristic curve* $u = f(e)$. *If under these conditions the inequality*

$$Re((1 + jq\omega)G(j\omega)) > -\frac{1}{k} \qquad (2.265)$$

holds with $0 < k \leq \infty$ *and any finite* q *for all frequencies* $0 \leq \omega < \infty$, *then the closed-loop system has a globally asymptotically stable position of rest* $u = y = 0$ *for any characteristic curve running through the sector* $[0, k]$. *This is also referred to be* absolutely stable *inside the sector* $[0, k]$. *If the right-hand side of the inequality can be set to zero (i.e.* $k = \infty$*), then the sector becomes* $[0, \infty]$.

If the function $G(s)$ *has also one pole lying on the axis of imaginaries (in addition to the poles with negative real part), then additionally it has to be proven that the linear transfer function* $\frac{\varepsilon G(s)}{1+\varepsilon G(s)}$ *is stable for at least one* $\varepsilon > 0$ (limit stability*). In this case, the permissible sector for the nonlinear characteristic curve is given by* $(0, k]$.

If the linear transfer function has several poles on the axis of imaginaries, then in addition to the limit stability another restriction is given in the form that only finite values of k are allowed and the sector of absolute stability is reduced to $[\varepsilon, k]$ *Furthermore, no frequency must exist which meets the following system of equations:*

$$Re(G(j\omega)) = -\frac{1}{k} \quad \text{and} \quad \omega Im(G(j\omega)) = 0 \qquad (2.266)$$

But now, (2.265) may also be an equality.

One may think that the requirement of a gain $V_l = 1$ for the linear part may be quite a strong one, but in fact it is not, as we can cover a factor $V_l \neq 1$ by the nonlinear part. Instead of $u = f(e)$, we then get a characteristic curve $\tilde{u} = V_l f(e) = \tilde{f}(e)$.

Let us also give some explanation on the other prerequisites for the linear transfer function. If a characteristic curve actually lies on the border $k_1 = 0$ of the sector, as mentioned in the theorem, it means that the actuating variable u and the input quantity of the linear part are always equal to zero. Now for this case, the linear part of the system is without outer stimulation—left to its own, so to say. If we then demand for asymptotic stability of the entire system, it can only be achieved by a linear part which reaches a position of rest from any initial state without any influences from outside. This is the reason for the demand of a negative real part of all poles (cf. theorem 2.17) of the theorem.

If the linear transfer function also has purely imaginary poles (for example contributed by an integrator), then the linear part of the system would not reach the position of rest without outer activation. For this reason we have to exclude 0 as the lower limit of the sector. But this restriction for the sector is still not sufficient. We also have to show that the closed loop can be stabilized in principle by the characteristic curve $f(e) = \varepsilon e$. Accordingly, we have to replace the nonlinear part by a linear gain ε, and then we have to prove stability for this new, purely linear system

$$\frac{\varepsilon G(s)}{1 + \varepsilon G(s)} \tag{2.267}$$

We refer to this additional property by the term *limit stability*. It is not difficult to prove, as it is a purely linear problem.

We should finally discuss the additional restrictions made in the last paragraph of the theorem. The reduction to finite values of k means that, for example, an ideal 2-point switching element does not meet the requirements for the application of this theorem, as the inclination of its characteristic curve at the origin is infinitely high. And the condition that the system of equations (2.266) must not be fulfilled by any frequency corresponds to the demand that the Popov locus, which we are going to introduce soon, must not run through the point $(-\frac{1}{k}, 0)$.

We can extend the theorem for a linear part containing a delay time. All we need to do is to add the restriction that the nonlinear characteristic curve is continuous (instead of just being piecewise continuous) and that q cannot be chosen arbitrarily, but has to be bigger than zero.

Several other special cases are described in [2]. For practical applications, however, they are not that important. One should be aware of the fact that the Popov criterion makes no statements about characteristic curves which leave a sector. We cannot use the Popov criterion to prove instability.

The question remains of how we have to proceed if we want to apply the criterion to a given problem. Let, for example, a nonlinear characteristic curve and the Nyquist plot of the linear part be given. The latter also fulfills the prerequisites of the theorem. The question is whether the closed loop is stable or not. In order to find out, we have to determine the permissible sector $[0, k]$ using inequality (2.265), and check if the given characteristic curve lies inside this sector or not. We first fix an arbitrary value for q and compute the corresponding value k from inequality (2.265) so that this inequality then holds for all ω. If the characteristic curve lies inside the sector defined by k, then the stability of the system is proven.

A problem occurs if this sector does not contain the given characteristic curve. Since theorem 2.25 is only a sufficient criterion of stability, we cannot come to a decision for this case. The next question would then be whether a q exists which would produce a larger sector. A similar question arises if the nonlinear characteristic curve was left to be determined during a controller design process. In this case, we would want to obtain a sector for the curve which is as large as possible. The fundamental aim is therefore to find the q that maximizes k.

A very elegant graphical solution exists to this problem. In order to apply it, we first have to rewrite the Popov inequality (2.265) as

$$\mathrm{Re}(G(j\omega)) - q\omega\mathrm{Im}(G(j\omega)) > -\frac{1}{k} \tag{2.268}$$

Then we define a new Nyquist plot $\tilde{G}(j\omega) = \tilde{x} + j\tilde{y}$ with the real part $\tilde{x} = \mathrm{Re}(G(j\omega))$ and the imaginary part $\tilde{y} = \omega\mathrm{Im}(G(j\omega))$. This is the so-called *Popov locus*. The Popov inequality becomes

$$\tilde{x} - q\tilde{y} > -\frac{1}{k} \tag{2.269}$$

or, rewritten for $\tilde{x}$:

$$\tilde{x} > q\tilde{y} - \frac{1}{k} \tag{2.270}$$

This inequality has to be true for all values of ω, i.e. for every point of the Popov locus. The borderline case of this inequality is

$$\tilde{x}_G = q\tilde{y} - \frac{1}{k} \tag{2.271}$$

or, alternatively,

$$\tilde{y} = \frac{1}{q}(\tilde{x}_G + \frac{1}{k}) \tag{2.272}$$

which is a line with the inclination $\frac{1}{q}$ and the $\tilde{x}$-intercept $-\frac{1}{k}$. This limit line gives the corresponding real part $\tilde{x}_G$ for any imaginary part $\tilde{y}$ of the Popov locus. On the other hand, the real part $\tilde{x}$ of the Popov locus according to equation (2.270) has to be bigger than the real part determined by the limit line. Accordingly, inequality (2.270) (and thereby (2.265), too) is true for all values of ω only if the Popov locus runs to the right of the limit line, i.e. in the region of larger real parts (fig. 2.88).

The procedure to determine the maximum border k of a sector is illustrated in figure 2.88. Using the (measured or computed) Nyquist plot of the linear part $G(j\omega)$, we first have to sketch the Popov locus. Then, we have to draw the limit line. Its inclination $\frac{1}{q}$ can be chosen arbitrarily, since according to

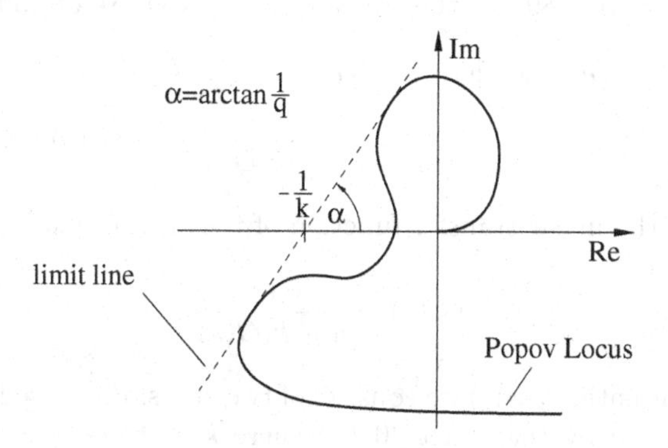

Fig. 2.88. Graphical approach to determine the maximum sektor

theorem 2.25, q can have any value. It just has to run to the left of the Popov locus, in order to meet inequality (2.265). The intersection $-\frac{1}{k}$ between the limit line and the axis of reals determines the upper border k of the sector. The further this intersection is situated to the right, the bigger k gets. We obviously obtain the maximum k if the limit line is almost a tangent to the Popov locus, as it is shown in fig. 2.88. Of course, it must not be a genuine tangent, as this would require inequality (2.265) to permit equality, too. In practical applications, however, we can generously skip about this difference since we would never obtain exact values from the measuring and drawing.

For practical applications it is also interesting to make use of a sector transformation. Theorem 2.25 is based on 0 or ε as the lower limit to the sector. If the characteristic curve runs through a sector $[k_1, k_2]$ (where $k_1 < 0$), then a direct use of this theorem is impossible. In such a case, we first have to replace the characteristic curve $u = f(e)$ by the transformed curve $u_t = f_t(e) = f(e) - k_1 e$ as illustrated in figure 2.89. The transformed characteristic curve then runs through a sector $[0, k]$, where $k = k_2 - k_1$. We can also think of this action as adding a proportional element with gain $-k_1$ to the closed loop system, parallel to the nonlinearity. In this interpretation the nonlinearity $f(e)$ together with the proportional element form the transformed nonlinearity $f_t(e)$.

Such a modification of the closed-loop system would of course distort the result of the stability analysis. This is why we have to provide a compensation for the adjustment obtained by the transformation of the characteristic curve before we can start with an analysis. The easiest thing to do is to add another proportional element with gain k_1 in parallel to the transformed nonlinearity $f_t(e)$. It just compensates the first proportional element. In the analysis for stability, this element is treated as a part of the linear system. Now the question is how the transfer function of the modified linear part of the system looks like. For an untransformed system we had the relation

$$\frac{e}{u} = -G(s) \tag{2.273}$$

According to figure 2.89, for the transformed system we obtain

$$\begin{aligned} (u_t + k_1 e)G(s) &= -e \\ \frac{e}{u_t} &= -\frac{G(s)}{1 + k_1 G(s)} = -G_t(s) \end{aligned} \tag{2.274}$$

and hence for the linear transfer function of the transformed system:

$$G_t(s) = \frac{G(s)}{1 + k_1 G(s)} \tag{2.275}$$

Sector-transformation therefore consists of two steps: First, we have to replace the sector $[k_1, k_2]$ by the sector $[0, k]$, where $k = k_2 - k_1$, and second, the transfer function $G(s)$ of the linear part is replaced by $G_t(s)$ according to

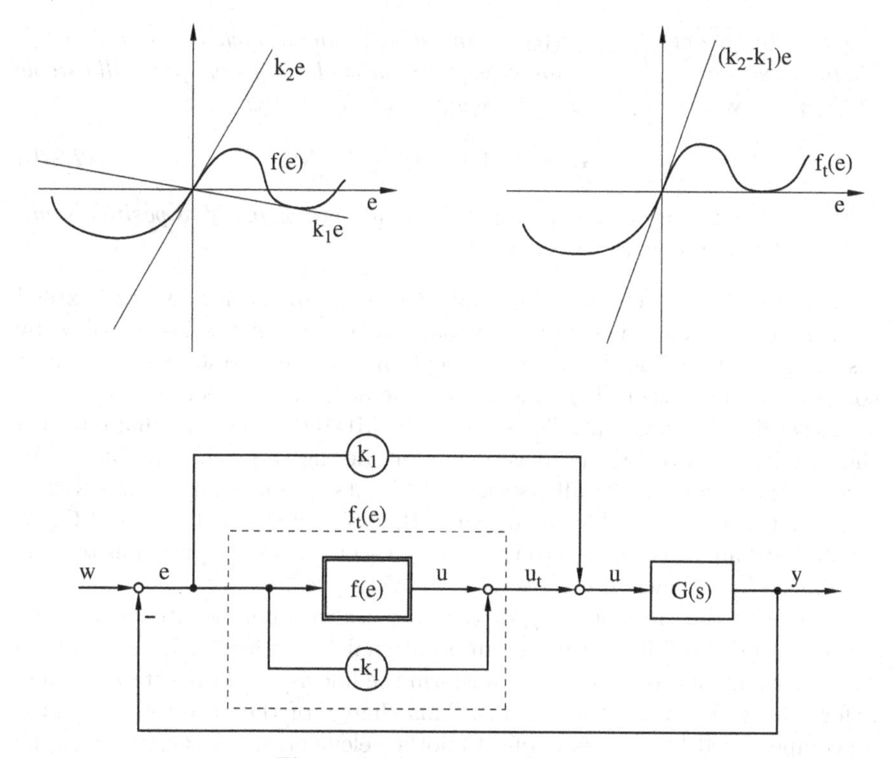

Fig. 2.89. Sector transformation

(2.275). We can then apply the Popov criterion to the transformed system, provided that G_t meets the criterion's requirements for the linear part of the system. If we can prove stability of the transformed system, then the original system is also stable.

It should be mentioned that such a sector transformation can be advantageous also for $k_1 > 0$. If we know, for example, that the characteristic curve runs through the sector $[k_1, k_2]$ $(k_1 > 0)$, then the sector-transformation lowers the upper limit of the sector $k = k_2 - k_1$, and thereby also diminishes the sector $[0, k]$ for which we have to prove stability. The demands put on the linear part of the system will then be less severe.

Finally, let us discuss how the method can be extended for multi-input-multi-output systems. The corresponding version of the Popov criterion is:

Theorem 2.26 *(Popov criterion for MIMO Systems) Let a standard closed-loop system be given, consisting of a linear and a nonlinear part. The linear part is defined by the linear transfer matrix* $\mathbf{G}(s)$ *and the nonlinear part by the vector* $\mathbf{f}$*. The vectors* $\mathbf{e}$*,* $\mathbf{u}$ *and* $\mathbf{y}$ *have the same dimension. The single transfer functions* $G_{ij}(s)$ *of the transfer matrix may have poles with a negative real part only. The single coefficients of* $\mathbf{f}$ *consist of piecewise continuous, well-defined characteristic curves which depend on the corresponding components of the* *input vector* $\mathbf{e}$ *only* ($u_i = f_i(e_i)$)*, and which run through the sector* $[0, k_i]$*. Under these conditions the closed-loop system is globally asymptotically stable at the point* $\mathbf{w} = \mathbf{u} = \mathbf{y} = \mathbf{0}$ *if the square transfer matrix*

$$\mathbf{G}_p(s) = (\mathbf{I} + s\mathbf{Q})\mathbf{G}(s) + \mathbf{V} \tag{2.276}$$

is strictly positive real. $\mathbf{Q}$ *is any real diagonal matrix and* $\mathbf{V}$ *a positive semi-definite diagonal matrix, where* $v_{ii} = \frac{1}{k_i} \geq 0$.

If the vectors $\mathbf{u}$ and $\mathbf{y}$ do not have the same dimension, we can extend them by artificial additional coefficients. This extension can be described by inserting constant linear transfer elements into the closed-loop system, that compensate each other. This method will be described in section 2.8.9.

According to theorem 2.26, stability of MIMO systems is guaranteed if the matrix $\mathbf{G}_p$ is strictly positive real. Considering theorem F.1, this means among other things that all its elements $[\mathbf{G}_p(s)]_{ij}$ must have poles with a negative real part only. This holds, since the poles of the coefficients of $\mathbf{G}_p(s)$ remain unchanged compared to the ones of $\mathbf{G}(s)$, and—as a prerequisite—the functions $G_{ij}(s)$ have only poles with a negative real part.

If some of the functions $G_{ij}(s)$ contain poles with non-negative real parts, we have the possibility to insert an additional linear feedback element into the linear part of the system, to transform the linear part into a stable system before the stability analysis begins. This change of the entire system must be compensated by the insertion of another element at another place of the system, to keep the overall transfer characteristic of the closed loop unchanged. This method will be described in section 2.8.9, too.

For the further discussions we will now assume, that the vectors $\mathbf{u}$ and $\mathbf{y}$ are of same dimension and that the linear part is stable, i.e. the functions $G_{ij}(s)$ contain only poles with negative real parts. Now we have to check, if the matrix $\mathbf{G}_p(s)$ is positive real or if the matrix $\mathbf{H}_p(j\omega)$ (that is constructed from $\mathbf{G}_p(s)$ and that corresponds to the matrix $\mathbf{H}$ of theorem F.1) has only positive eigenvalues for all frequencies respectively.

If the characteristic curves of the nonlinear part are given together with the boundaries of the sectors k_i, and therefore also the elements of $\mathbf{V}$ are fixed, we can construct $\mathbf{G}_p$ and $\mathbf{H}_p$ using one chosen matrix $\mathbf{Q}$ and hope that all eigenvalues of $\mathbf{H}_p(j\omega)$ are positive for all frequencies. If this is not the case, then the same question as for the SISO case remains: Does such a matrix $\mathbf{Q}$ exist at all, for which $\mathbf{H}_p(j\omega)$ gets positive eigenvalues? And how can this matrix be found? It is therefore more sensible to determine the free parameters in a way that sectors result which are as large as possible—just what we did in the SISO case.

First, we have to observe that the permissible sectors for the nonlinear characteristic curves are determined by the elements of the diagonal matrix $\mathbf{V}$. The smaller the v_{ii}, the larger the corresponding sectors. We therefore set the matrix $\mathbf{V}$ initially to zero. This means that the upper bound of all sectors is the maximum value ∞. Next, we construct the matrix

$$\mathbf{H}_p(j\omega) = \frac{1}{2}(\mathbf{G}_p(j\omega) + \bar{\mathbf{G}}_p^T(j\omega))$$
$$= \frac{1}{2}((\mathbf{I} + j\omega\mathbf{Q})\mathbf{G}(j\omega) + \bar{\mathbf{G}}^T(j\omega)(\mathbf{I} - j\omega\mathbf{Q})) \qquad (2.277)$$

according to theorem F.1, using an arbitrary matrix $\mathbf{Q}$. By means of numerical methods we then optimize $\mathbf{Q}$ in a way that the smallest eigenvalue of $\mathbf{H}_p(j\omega)$ gets as large as possible for all frequencies. We stop the optimization process prematurely if this value exceeds zero for all frequencies. For this case, $\mathbf{G}_p$ is strictly positive real with $\mathbf{V} = \mathbf{0}$, and the permissible sectors for all characteristic curves are $[0, \infty]$. If, at the end of the optimization process, the smallest occurring eigenvalue has a value $\mu < 0$ for some frequency, we have to pick $\mathbf{V} = |\mu|\mathbf{I}$, as for this choice we get

$$\mathbf{H}_p(j\omega) = \frac{1}{2}((\mathbf{I} + j\omega\mathbf{Q})\mathbf{G}(j\omega) + \bar{\mathbf{G}}^T(j\omega)(\mathbf{I} - j\omega\mathbf{Q})) + |\mu|\mathbf{I} \qquad (2.278)$$

for $\mathbf{H}_p(j\omega)$ instead of (2.277). This results in all the eigenvalues of the matrix being shifted to the right by $|\mu|$, so even the smallest eigenvalue then lies in the positive region. $\mathbf{H}_p(j\omega)$ is then positive definite for all frequencies, i.e. $\mathbf{G}_p$ with $\mathbf{V} = |\mu|\mathbf{I}$ is strictly positive real. The boundaries for all sectors are then given by $[0, \frac{1}{|\mu|}]$.

Of course, we could also choose a different $\mathbf{V}$, which would probably produce higher upper bounds than $\frac{1}{|\mu|}$ for some of the characteristic curves, but for different elements in the diagonal of $\mathbf{V}$, we would no longer be able to predict the effects on the eigenvalues of $\mathbf{H}_p(j\omega)$ that easily.

A sector transformation is also possible for the MIMO case. For SISO systems, we obtained this transformation by adding two proportional elements with gains $-k_1$ and k_1 to the nonlinear part and the linear part, respectively, so that they compensated each other in their overall effects (fig. 2.89). In the case of a MIMO system, this proportional element is replaced by a constant diagonal matrix $\mathbf{D}$. We get

$$f_i'(e_i) = f_i(e_i) - d_{ii}e_i \qquad (2.279)$$

for the single components of the new nonlinear transfer characteristic $\mathbf{f}'$, and for the linear part (c.f. fig. 2.89 and eq. (2.275))

$$\mathbf{G}' = [\mathbf{I} + \mathbf{G}\mathbf{D}]^{-1}\mathbf{G} \qquad (2.280)$$

Finally, it should be mentioned that the Popov criterion can be applied easily (at least for the SISO case) and is therefore suitable for practical applications. It is easy to obtain the necessary information about the system. For the linear part of the system, the measured frequency response is sufficient (or, for MIMO systems, the different frequency responses $G_{ij}(j\omega)$), while for the nonlinear part only the sector where the characteristic curve runs through has to be known. A disadvantage might be that the Popov criterion as a sufficient but not necessary criterion produces merely conservative results, i.e. in many cases stability cannot be proven with the Popov criterion even though the system is stable.

In the MIMO case there is one requirement that makes the Popov criterion rather unsuitable for practical applications: It is the fact, that every component of the nonlinear part must depend only on its corresponding component of the input vector $u_i = f_i(e_i)$. It follows, that for example a nonlinear (fuzzy) MIMO-controller has to be a parallel connection of SISO controllers. But such a controller structure is unsuitable for real MIMO plants, where one input variable usually affects several output variables, and a controller therefore should have similar internal connections. Obviously, it is more convenient to use the Hyperstability criterion for MIMO plants as introduced in section 2.8.9. And this is also the reason, why we discuss the before mentioned vector extension (to get vectors of same dimension) and the stabilization of the linear part in that section.

Last, but not least, we should mention the famous *Assumption of Aisermann*: If the nonlinear part $u = f(e)$ of the system is replaced by a proportional element with gain k_1, and the entire resulting system is stable, and the same is also true for another gain $k_2 > k_1$, then the system is also stable for any nonlinear characteristic curve running through the sector $[k_1, k_2]$. Even though this assumption seems plausible, it is not generally true. A refutation is given in [56], and counter-examples exist already for systems of second order. It is not possible to predict a system's stability by comparing the nonlinear characteristic curve to linear ones. Unfortunately, this happens frequently in real applications, which is why we give this warning.

2.8.8 Circle Criterion

The next criterion of stability which we are going to discuss is called the *circle criterion*. It is based on exactly the same prerequisites as the Popov criterion. Again, the system has to consist of a linear part and a nonlinear part, but this time the transfer characteristic of the nonlinear part need not be represented by a static characteristic curve. In the case of a SISO system with static characteristic curve, we can derive this method from the Popov inequality (2.265) by setting the free parameter q to zero, performing a few transformations and interpreting the result graphically (cf. [45]). The derivation which we present here, however, has the advantage of being expendable to MIMO systems in a straight way. It is based on the use of norms (cf. [18]).

We mentioned norms already when discussing norm-optimal state space controllers. A further, more thorough explanation is given in the appendix. For example, the norm of a transfer matrix can be interpreted as a kind of maximum amplification factor between the input and the output vector of the system. According to equation (C.17), the ∞-norm of a linear transfer matrix $\mathbf{G}$, with $\mathbf{y} = \mathbf{G}\mathbf{u}$, is given by

$$||\mathbf{G}(j\omega)||_\infty = \sup_\omega \sup_{\mathbf{u}\neq 0} \frac{|\mathbf{G}(j\omega)\mathbf{u}|}{|\mathbf{u}|} \tag{2.281}$$

and the ∞-norm of a nonlinear transfer function with $\mathbf{f}(\mathbf{e}, \dot{\mathbf{e}}, ...) = \mathbf{u}$ is, according to equation (C.20),

$$||\mathbf{f}||_\infty = \sup_{\mathbf{e}\neq 0} \frac{|\mathbf{u}|}{|\mathbf{e}|} \tag{2.282}$$

where $\mathbf{e}, \mathbf{u}$ and $\mathbf{y}$ represent the quantities of the closed-loop system according to figure 2.79. Regarding SISO systems, this becomes (cf. equation (C.19)):

$$\begin{aligned} ||G(j\omega)||_\infty &= \sup_\omega |G(j\omega)| \\ ||f||_\infty &= \sup_{e\neq 0} \frac{|u|}{|e|} \end{aligned} \tag{2.283}$$

According to these definitions, the output quantity of a closed-loop system consisting of a linear and a nonlinear part is given by $\mathbf{y} = \mathbf{G}(j\omega)\mathbf{f}(\mathbf{e}, \dot{\mathbf{e}}, ...)$. If $\mathbf{f}$ would be a linear transfer matrix $\mathbf{F}$, we could write $\mathbf{y} = \mathbf{GFe}$, and $\mathbf{GF}$ would be the matrix of the open-loop transfer function. But, as $\mathbf{f}$ is defined only as a nonlinear vector function of $\mathbf{e}$, here the matrix of the open-loop transfer function cannot be given directly. We can just, according to the definitions of the different norms, estimate the maximum transfer factor from $|\mathbf{e}|$ to $|\mathbf{y}|$. It is given by the product of the individual norms $||\mathbf{G}||\ ||\mathbf{f}||$.

Furthermore, the Small Gain Theorem, which we also already discussed in the section on norm-optimal state space controllers, is still valid. It states that the closed loop consisting of the linear and the nonlinear part is definitely stable if the maximum transfer factor from $|\mathbf{e}|$ to $|\mathbf{y}|$ is less than one and the linear part on its own is stable. The first condition is easy to understand, because it guarantees that $|\mathbf{y}| < |\mathbf{e}|$. With this, the feedback feeds a vector $\mathbf{y}$ back into the control loop, that is smaller than $\mathbf{e}$. This diminished quantity is then further reduced by running through $\mathbf{f}$ and $\mathbf{G}$, and so on, so that $\mathbf{e}$, $\mathbf{u}$ and $\mathbf{y}$ are decreased continuously and converge to zero. Therefore, the entire system is stable.

On the other hand, we have to consider the case that the output quantity $\mathbf{u}$ of the nonlinear part is constantly zero. Here, the output quantity $\mathbf{y} = \mathbf{Gu}$ and, correspondingly, the transfer factor from $|\mathbf{e}|$ to $|\mathbf{y}|$ are zero for any input vectors $\mathbf{e}$, too, and therefore the first requirement of the Small Gain Theorem is fulfilled. But under these circumstances the linear part would receive no outer activation, so in addition it has to be guaranteed that the linear part reaches a steady state from any initial state without any outer stimulation. This means that the linear part has to be stable, which is ensured by the second requirement of the Small Gain Theorem.

Together with the Small Gain Theorem and the estimate for the maximum transfer factor from $|\mathbf{e}|$ to $|\mathbf{y}|$ as mentioned above, the stability of $\mathbf{G}$ together with the condition

$$||\mathbf{G}(j\omega)|| \; ||\mathbf{f}|| < 1 \tag{2.284}$$

is considered to be a sufficient condition for the stability of the closed-loop control system consisting of a nonlinear and a linear part. If we choose the ∞-norm for all the norms involved, we obtain

$$||\mathbf{G}(j\omega)||_\infty \; ||\mathbf{f}||_\infty < 1 \tag{2.285}$$

for a MIMO system, and for a SISO system we get

$$\sup_\omega |G(j\omega)| \; \sup_{e \neq 0} \frac{|u|}{|e|} < 1 \tag{2.286}$$

using (2.283). The norm of the linear part is the maximum distance between the Nyquist plot and the origin, while the norm of the nonlinear part corresponds to the maximum absolute value of the amplification factor between the input and the output of the nonlinear transfer element.

If we can find upper and lower bounds of the sector for the nonlinear transfer function, as for example given for the characteristic curve in figure 2.87, then this amplification factor will definitely be less than the maximum absolute value of a sector's boundary

$$\sup_{e \neq 0} \frac{|u|}{|e|} \leq \max\{|k_1|, |k_2|\} \tag{2.287}$$

Substituting this into equation (2.286) gives a new, more restrictive condition for the stability of the closed loop:

$$\sup_\omega |G(j\omega)| \max\{|k_1|, |k_2|\} < 1 \tag{2.288}$$

or, alternatively,

$$\sup_\omega |G(j\omega)| < \frac{1}{\max\{|k_1|, |k_2|\}} \tag{2.289}$$

The system is therefore stable if the distance between the Nyquist plot of the stable, linear part and the origin is always less than the reciprocal value of the maximum absolute value of a boundary of a sector. This means that only the boundary with the larger value is the deciding factor. But then again, we should be allowed to modify the other boundary in a way that $|k_1| = |k_2|$ and $k_1 < 0 < k_2$ hold, without affecting the result of the inequality. In doing so, we enlarge the permissible sector for the nonlinear transfer characteristic without restricting the stability condition for the linear part.

We can come up with the same consideration if the nonlinear transfer characteristic is given and it is bounded by a sector $[k_1, k_2]$, where $|k_1| \neq |k_2|$. Using a sector-transformation from $[k_1, k_2]$ to $[-k_d, k_d]$ with $k_d = \frac{1}{2}|k_2 - k_1|$ (fig. 2.90), the right-hand side of inequality (2.289) becomes $\frac{1}{k_d}$. It has increased because of $k_d < \max\{|k_1|, |k_2|\}$, and therefore the demand on the

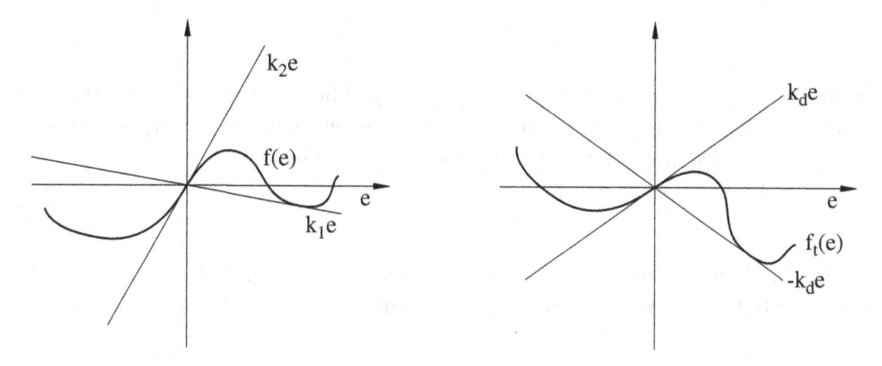

Fig. 2.90. Sector transformation for the circle criterion

linear part has become less severe. The resulting formula shall be derived in the following.

Just like in the case of the Popov criterion, we perform this transformation by inserting additional proportional elements (cf. fig. 2.89), but this time we have to rotate the sector by the mean value $k_m = \frac{1}{2}(k_1 + k_2)$ instead of the lower boundary k_1. With $k_d = \frac{1}{2}|k_2 - k_1|$, the nonlinear transfer characteristic is now bounded by the symmetrical sector $[-k_d, k_d]$, and the linear part of the system changes to (cf. (2.275))

$$G_t(s) = \frac{G(s)}{1 + k_m G(s)} \tag{2.290}$$

Equation (2.289) becomes

$$|G_t(j\omega)| < \frac{1}{k_d} \tag{2.291}$$

for all ω. Attention should be paid to the fact that the linear part of the Small Gain Theorem is now $G_t(s)$ instead of $G(s)$, so now G_t has to be a stable transfer function, while G gets free from any demands at the moment. Furthermore, we can now compute k_d and k_m only if $k_2 < \infty$, thus $k_2 = \infty$ has to be excluded. Substituting G_t leads to

$$k_d|G(j\omega)| < |1 + k_m G(j\omega)| \tag{2.292}$$

Next, we square this inequality; the squares of the absolute values are replaced by the products of the complex quantities and their complex conjugated values:

$$0 < (k_m^2 - k_d^2)G(j\omega)\bar{G}(j\omega) + k_m(G(j\omega) + \bar{G}(j\omega)) + 1 \tag{2.293}$$

With $k_m^2 - k_d^2 = k_1 k_2$, we obtain:

$$0 < k_1 k_2 G(j\omega)\bar{G}(j\omega) + k_m(G(j\omega) + \bar{G}(j\omega)) + 1 \tag{2.294}$$

Now we have to distinguish between several cases, depending on the signs of k_1 and k_2. In the first case, we assume $k_1 k_2 > 0$, i.e. both boundaries have the same sign. Here, we can rewrite the inequality to

$$|G(j\omega) - m| > r \tag{2.295}$$

letting $r = \frac{1}{2}|\frac{1}{k_1} - \frac{1}{k_2}|$ and $m = -\frac{1}{2}(\frac{1}{k_1} + \frac{1}{k_2})$. The Nyquist plot therefore has to lie outside a circle with radius r around the center m (fig. 2.91, upper left). For $k_1 < 0 < k_2$, with the same abbreviations, we get:

$$|G(j\omega) - m| < r \tag{2.296}$$

In this case, the Nyquist plot has to lie inside the circle (fig. 2.91, upper right). For $k_1 = 0, k_2 > 0$, we lose the first summand in (2.294), leaving

$$\text{Re}(G(j\omega)) > -\frac{1}{k_2} \tag{2.297}$$

The Nyquist plot should therefore run to the right of the line defined by $-\frac{1}{k_2}$ (fig. 2.91, lower left). Similarly, for $k_1 < 0, k_2 = 0$, we obtain a line through $-\frac{1}{k_1}$, and the Nyquist plot should run to the left of it (fig. 2.91, lower right).

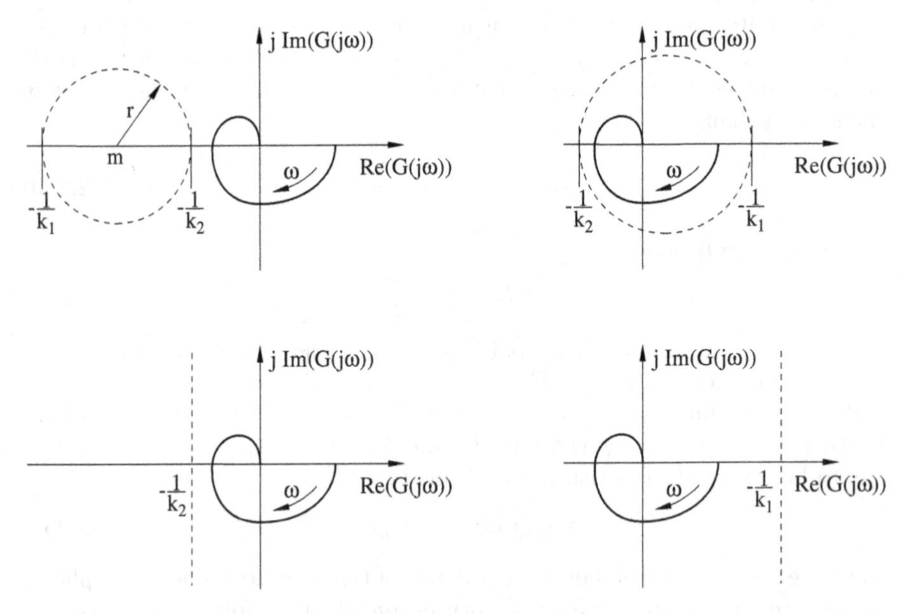

Fig. 2.91. Illustrations to the circle criterion

For the last three cases another problem occurs, as, in principle, because of $0 \in [k_1, k_2]$ each of them permits a characteristic curve $f(e) = 0$. This would leave the linear part of the system to its own, as we mentioned already in our discussion of the Popov criterion and the Small Gain Theorem. Stability of the entire system can therefore be achieved only if the linear part of the system is stable on its own. Accordingly, we get another demand in addition to the one that G_t is stable and all its poles have a negative real part: The same must hold for G, too.

For any constellation of boundaries k_1, k_2, we can determine a region $V(k_1, k_2)$ where the Nyquist plot of the linear part must not run through if the entire system should be stable. If we think of a line as being a circuit with infinite radius, then this forbidden region is always of a circular shape. This is where the name circle criterion stems from:

Theorem 2.27 *(Circle criterion) Let a closed-loop system be given, consisting of a linear and a nonlinear part. Let the nonlinear transfer characteristic be bounded by a sector $[k_1, k_2]$, where $k_2 < \infty$. If k_1 and k_2 are of opposite signs, or if one of the boundaries is equal to zero, then the transfer function of the linear part $G(s)$ has to be stable. The function $G_t(s) = \frac{G(s)}{1+k_m G(s)}$, where $k_m = \frac{1}{2}(k_1 + k_2)$, has to be stable in any case, and the degree of the numerator of G_t has to be lower than the one of its denominator. If, for all $\omega > 0$, under these circumstances the Nyquist plot $G(j\omega)$ runs outside the forbidden region $V(k_1, k_2)$, which is determined by k_1 and k_2, then the closed-loop system has a globally asymptotically stable position of rest at $u = y = 0$.*

It is interesting to investigate the relations between the circle criterion, the Popov criterion and the method of harmonic balance. As already mentioned, in the case of static nonlinearities we can derive the circle criterion from the Popov criterion by setting the free parameter q of the Popov inequality (2.265) to zero, performing some transformations and interpreting the result graphically (cf. [45]). But to "sacrifice" a free parameter means to make a sufficient criterion of stability become more restrictive, which means that the circle criterion is even more conservative than the Popov criterion. From that it follows, that we might not be able to prove stability with the circle criterion, but with the Popov criterion. Instability, however, can not be proven with any of the two criteria, as both are sufficient but not necessary ones.

Similar considerations hold for the relation between the circle criterion and the method of harmonic balance. Here, it can be proven that the curve of the describing function $-\frac{1}{N(A)}$, computed for a given characteristic curve, always lies entirely inside the forbidden region $V(k_1, k_2)$ which is determined by the circle criterion ([18]). If we can then prove stability using the circle criterion—i. e. if we can show that the Nyquist plot of the linear part runs outside the forbidden area—then we could as well prove stability using the method of harmonic balance, as the linear Nyquist plot and the curve of the describing function could not intersect. The conclusion in opposite direction, however, does not hold. If the Nyquist plot does not intersect the curve of the describing function, this does not guarantee that it remains entirely outside the larger, forbidden region of the circle criterion.

The circle criterion is therefore the most conservative of the three criteria. But on the other hand, as it is the only one that can also be used for dynamic nonlinearities, it is the one with the widest range of application (if we refrain from considering some special cases which are covered by the Popov criterion). Besides that, it is the easiest to apply. The boundaries of a sector of a nonlinear transfer element can be determined easily, and the Nyquist plot of the linear part of the system can be obtained by measuring. When concerned with any applications, one should therefore first try to prove stability using the circle criterion. The other criteria only have to be used if this attempt is unsuccessful.

We can easily adjust the circle criterion, which we derived from the Small Gain Theorem, to MIMO systems. The resulting method for checking a system's stability, however, is not so easy to apply as the one for the SISO system. Again, we start with inequality (2.285):

$$||\mathbf{G}(j\omega)||_\infty \; ||\mathbf{f}||_\infty < 1 \tag{2.298}$$

The norm of the nonlinear part of the system is determined following equation (2.282):

$$||\mathbf{f}||_\infty = \sup_{\mathbf{e}\neq\mathbf{0}} \frac{|\mathbf{u}|}{|\mathbf{e}|} \tag{2.299}$$

If $u_i = f_i(e_i)$ is true for every component of the vector $\mathbf{u}$, then a relatively simple but nonetheless precise estimation of $||\mathbf{f}||_\infty$ is possible. We assume each of these nonlinear functions runs within a sector $[k_{i1}, k_{i2}]$. Then we first have to add a diagonal matrix $\mathbf{M}$ in parallel to the nonlinearity according to figure 2.92, since we want the sector for each single component to be symmetrical. This means,

$$m_{ii} = -\frac{1}{2}(k_{i1} + k_{i2}) \tag{2.300}$$

has to be true for the elements of $\mathbf{M}$. Next, we insert a diagonal matrix $\mathbf{H}$. The transformed characteristic curves and their symmetrical boundaries get multiplied by the components of $\mathbf{H}$. If we choose

$$h_{ii} = \frac{2}{|k_{i2} - k_{i1}|} \tag{2.301}$$

then all characteristic curves of the new nonlinearity $\mathbf{f}'$ lie within the sector $[-1, 1]$. For every i, the ratio $\frac{|u_i'|}{|e_i|}$ is therefore at most equal to 1, which is why we can estimate the nonlinearity's norm by $||\mathbf{f}'||_\infty \leq 1$ according to (2.299).

The extension of the nonlinearity by $\mathbf{M}$ and $\mathbf{H}$ needs some equivalent actions outside of $\mathbf{f}'$, i.e. in the linear part of the closed loop, that compensate the effects of $\mathbf{M}$ and $\mathbf{H}$. Otherwise the stability analysis would be performed with a modified system, and the stability results would be worthless for the original system. In fig. 2.92 it is shown, how $\mathbf{H}$ (in the nonlinear part) is compensated by a series connection of $\mathbf{H}^{-1}$ (in the linear part), and how $\mathbf{M}$ (with a positive sign in the nonlinear part) is compensated by a parallel connection of $\mathbf{M}$ (with a negative sign in the linear part—outside of $\mathbf{f}'$). Therefore, the both systems in fig. 2.92 are equivalent to each other.

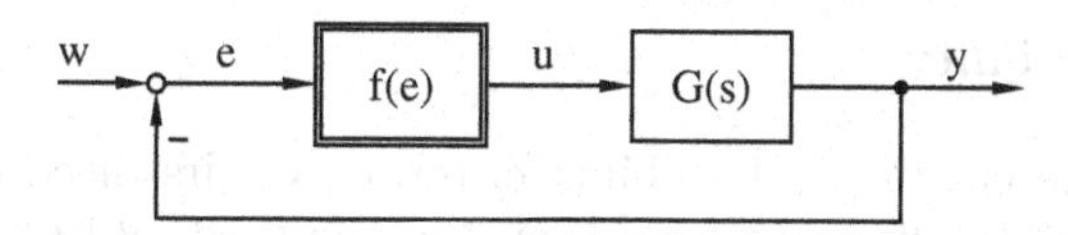

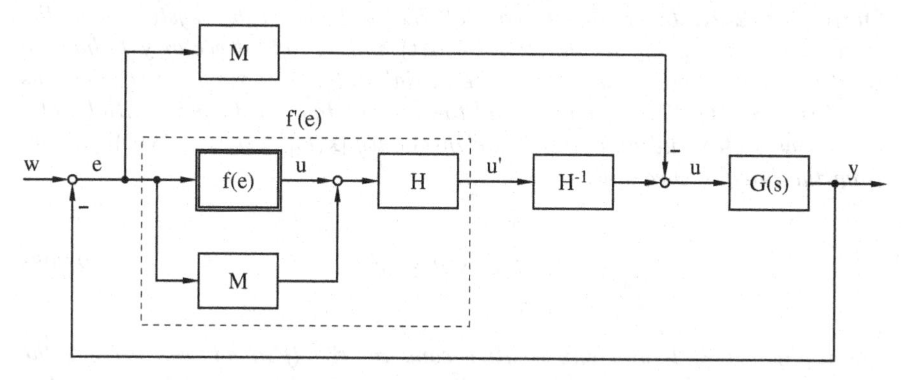

Fig. 2.92. Sector transformation for a MIMO system

For the lower system of fig. 2.92, we get for the linear transfer characteristic from $\mathbf{u}'$ to $\mathbf{e}$

$$\mathbf{G}' = (\mathbf{I} - \mathbf{GM})^{-1}\mathbf{GH}^{-1} \tag{2.302}$$

and for the stability condition (2.298)

$$||\mathbf{G}'(j\omega)||_\infty \; ||\mathbf{f}'||_\infty < 1 \tag{2.303}$$

With $||\mathbf{f}'||_\infty \leq 1$ this becomes

$$||(\mathbf{I} - \mathbf{GM})^{-1}\mathbf{GH}^{-1}||_\infty < 1 \tag{2.304}$$

This inequality certainly holds if

$$||(\mathbf{I} - \mathbf{GM})^{-1}||_\infty \; ||\mathbf{GH}^{-1}||_\infty < 1 \tag{2.305}$$

is true. From (C.15) it follows that

$$||(\mathbf{I} - \mathbf{GM})^{-1}||_\infty = \frac{1}{||\mathbf{I} - \mathbf{GM}||_\infty} \tag{2.306}$$

so that we finally get the stability condition

$$||\mathbf{GH}^{-1}||_\infty < ||\mathbf{I} - \mathbf{GM}||_\infty \tag{2.307}$$

Since the computation of the ∞-norm is nowadays included in any control-engineering specific software tool, this condition can usually be checked by merely clicking on a button. If we need an algebraic solution, we can estimate the ∞-norm by other norms which are easier to compute (cf. [18]). Of course, such an estimation can be very rough. Finally, we have to prove the stability of $\mathbf{G}$ and $\mathbf{G}'$, just like for the SISO case; but since this is a genuine linear problem, it should by fairly easy.

2.8.9 Hyperstability

To be able to present the next stability criterion, we first need to introduce a new, more strict stability definition than the one used by Lyapunov. This is the *hyperstability* ([162], [163]):

Definition 2.28 *Let a linear, controllable and observable system with the state vector* $\mathbf{x}$ *be given. Its input vector* $\mathbf{u}(t)$ *and its output vector* $\mathbf{y}(t)$ *have to be of the same dimension.* $\mathbf{x}(0)$ *is the initial state. Accordingly,* $\mathbf{y}(t)$ *depends on* $\mathbf{x}(0)$ *and* $\mathbf{u}(t)$*. The system is said to be hyperstable if, for any initial state, any input vector and any* $\beta_0 > 0$ *the inequality* $|\mathbf{x}(t)| \leq \beta_0 + \beta_1 |\mathbf{x}(0)|$ *follows from the integral inequality*

$$\int_0^T \mathbf{u}^T \mathbf{y} dt \leq \beta_0^2 \tag{2.308}$$

for any $0 < t < T$ *and any positive constant* β_1*. If in addition to this the state vector converges towards zero,* $\lim_{t\to\infty} \mathbf{x}(t) = 0$*, then the system is said to be asymptotically hyperstable.*

The underlying idea of this definition is: If the product of the input and output quantities of a hyperstable system is bounded in a certain sense, then the state variables are limited, too. The condition that the input and the output quantity are of the same dimension is required because otherwise the product $\mathbf{u}^T\mathbf{y}$ could not be computed.

It is interesting to compare this definition to the ones which we have already introduced so far. The first two of them, def. 2.4 (finite step response) and def. 2.5 (BIBO stability), deal with the reaction of a system output to a certain input signal, while the definition according to Lyapunov 2.21 focuses on the system's internal behavior (state variables) without external stimulation in reaction to a certain initial state. In contrast to this, the hyperstability criterion takes the initial state as well as outer activations of the system into account.

An (asymptotically) hyperstable system is obviously also (asymptotically) stable in the sense of Lyapunov, as inequality (2.308) is certainly true for an input vector $\mathbf{u}(t) = \mathbf{0}$, i.e. for a system without any external stimulation. In this case, for a hyperstable system the state vector will be limited by $|\mathbf{x}(t)| \leq \beta_0 + \beta_1 |\mathbf{x}(0)|$, so the system is also stable according to Lyapunov. And the additional restriction for asymptotical stability is the same in both cases. A linear system which is stable in the sense of Lyapunov, however, is also stable according to the definitions 2.4 and 2.5, as we have already demonstrated. Thus, hyperstability is the strictest definition of stability we have introduced so far.

This is also illustrated by the fact that, for example, a feedback loop consisting of two hyperstable linear systems $\mathbf{H}_1$ and $\mathbf{H}_2$, as shown in figure 2.93, is again a hyperstable system with the input quantity $\mathbf{w}$ and the output quantity $\mathbf{y}$, which can be proven. A feedback loop consisting of two Lyapunov-stable systems on the other hand does not necessarily result in another Lyapunov-stable closed-loop system.

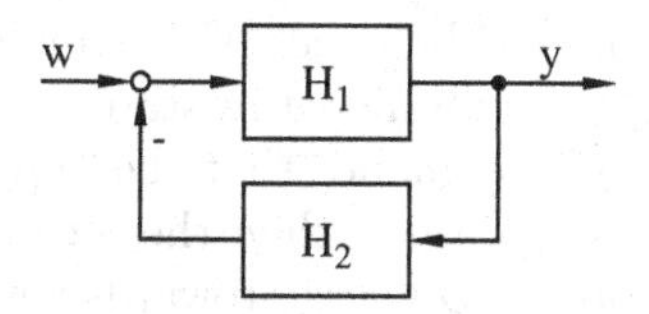

Fig. 2.93. Hyperstable feedback loop consisting of two hyperstable systems

From (2.308), we see that hyperstability is in some way an extension of the absolute stability which was mentioned in the Popov criterion. Absolute stability according to theorem 2.25, for example in the sector $[0, \infty)$, requires that the characteristic curve $f(e)$ lies inside this sector. We can express this condition as well by demanding $f(e)e \geq 0$. For $f(e) = u$ and $e = -y$, this becomes $uy \leq 0$. The relation to (2.308) is plain to see.

We can also come up with an interpretation of hyperstability in terms of energy. For example, we can think of u and y as representing the electric current and voltage at the input of an electrical circuit, and of the state variable x as a storehouse of energy, e.g. the potential drop at a capacitor. The product of u and y then is equivalent to the electric power supplied, and the integral of this product to the electric energy. If the latter is limited according to (2.308), then the energy stored inside (and hence also x) also has to be bounded in a hyperstable circuit. This is always true for passive electrical circuits, which are therefore always hyperstable. However, if a circuit contains active elements, like, for instance, amplifiers, then hyperstability is not necessarily given.

We should mention that our definition is a simplified and quite restricted version of the general definition (cf. [152], [163]), which refers to nonlinear, time-variant systems, and the absolute values are replaced by generalized functions. For such general systems, however, there result no practically applicable criteria of stability.

Instead, let us now develop a criterion of stability for the standard control loop as given in figure 2.79, based on our definition for linear systems. We assume that the position of rest under consideration is given by $\mathbf{w} = \mathbf{y} = \mathbf{0}$, otherwise we have to redefine the system suitably. If the nonlinear part meets the inequality

$$\int_0^T \mathbf{u}^T \mathbf{e}\, dt \geq -\beta_0^2 \tag{2.309}$$

for all $T > 0$, then the input and output quantities of the linear part $\mathbf{u}$ and $\mathbf{y}$ obviously comply with condition (2.308) given in definition 2.28, because of $\mathbf{e} = -\mathbf{y}$. If we can show additionally, that the linear part of the system is hyperstable, it is guaranteed that its state variables remain limited—independently from the internal behavior of the nonlinear part of the system. In case of asymptotical hyperstability, the state variables even converge towards zero. If we furthermore demand, that the nonlinear part of the system comprises no internal state variables, then there can be no state variables in the entire system which do not converge towards zero. Now this means, that for the position of rest $\mathbf{x} = \mathbf{0}$ the entire system is asymptotically stable in the sense of Lyapunov. We get:

Theorem 2.29 *The nonlinear standard control loop has got the (asymptotically) stable position of rest $\mathbf{x} = \mathbf{0}$ for the reference value $\mathbf{w} = \mathbf{0}$, if the linear part of the system is (asymptotically) hyperstable and the static, nonlinear part of the system complies with the integral inequality*

$$\int_0^T \mathbf{u}^T \mathbf{e}\, dt \geq -\beta_0^2 \tag{2.310}$$

for all $T > 0$.

Now, what do we have to do if the nonlinear part is not static but also comprises internal state variables? As already mentioned, internal actions of the nonlinear part have no effect on the boundedness of the state variables of the linear part, as long as inequality (2.310) is met. Therefore, in order to guarantee asymptotical stability of a standard control loop with a dynamic nonlinear part, we only have to show—in addition to the conditions put up in theorem 2.29—that the state variables of the nonlinear part converge towards zero. Unfortunately, no universally valid criterion exists which is also easy to apply. In many cases, however, if the structure of the nonlinear part is not too complicated, we can estimate the course of the state variables more of less by hand. And if the state variables of the nonlinear part are, from a technical point of view, without importance, then we can even do without such an analysis. Then, of course, we can no longer talk of asymptotical stability of the entire system, only of that the states of the linear part converge towards zero for the given reference value.

The question arises of how to proceed in practical applications. Already the demand that the vectors $\mathbf{u}$ and $\mathbf{y}$ be of the same dimension may often cause the first problem, as this is, in many cases, not given initially. Usually the nonlinear part (e.g. a fuzzy controller) will have more input than output quantities. In order to achieve same dimension here, auxiliary output quantities have to be added to the nonlinear part of the system, which are set to a constant zero. Accordingly, the number of input quantities of the linear part as well as its transfer matrix have to be adjusted, too.

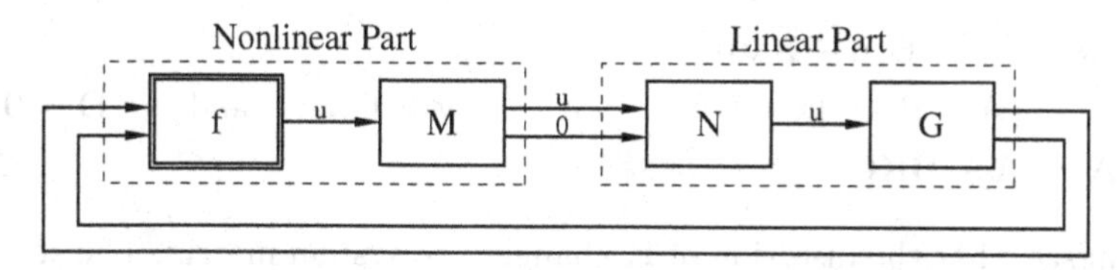

Fig. 2.94. Inserting additional matrices in order to obtain the same dimension of the input and output vectors

The definition of additional output quantities is equivalent to the insertion of two matrices **M** and **N** into the closed loop (fig. 2.94). To keep the systems' transfer behaviour unchanged, **NM** = **I** has to be true. In figure 2.94, the matrices **N** and **M** are:

$$\mathbf{N} = (1\ 0) \qquad \mathbf{M} = \begin{pmatrix} 1 \\ 0 \end{pmatrix} \tag{2.311}$$

The output quantity u of the nonlinear system therefore turns into the output vector $\mathbf{u} = [u, 0]^T$, and the linear transfer matrix

$$\mathbf{G}(s) = \begin{pmatrix} G_1(s) \\ G_2(s) \end{pmatrix} \tag{2.312}$$

into the quadratic transfer matrix

$$\mathbf{G}(s)\mathbf{N} = \begin{pmatrix} G_1(s)\ 0 \\ G_2(s)\ 0 \end{pmatrix} \tag{2.313}$$

Now both parts of the system have the same number of input and output quantities. In the following sections, we refrain from giving an explicit representation of these matrices **M** and **N**, i.e. we assume that **f** and **G** are already suitably adjusted parts of the system.

Let us now analyze the linear part of the system for hyperstability. To do this, we need the following theorem, which we provide without proof:

Theorem 2.30 *A linear, time-invariant, controllable and observable system is asymptotically hyperstable if and only if its transfer function $G(s)$ (or its quadratic transfer matrix $\mathbf{G}(s)$) is strictly positive real (c.f. appendix F).*

As mentioned in the discussion of theorem F.1, a prerequisite of hyperstability of the linear part is its stability. If this is not the case, we can try to obtain a stable system by applying some transformation. For this purpose, we insert a feedback-matrix $\mathbf{K}$ for the linear part as shown in figure 2.95. The matrix $\mathbf{D}$ is initially zero. Because of the feedback, we then get a new linear system $\mathbf{G}' = (\mathbf{I} + \mathbf{G}\mathbf{K})^{-1}\mathbf{G}$. If we choose $\mathbf{K}$ suitably and if certain prerequisites are met, then all eigenvalues of $\mathbf{G}'$ have a negative real part.

In the state space description the effect of $\mathbf{K}$ gets more clear:

$$\begin{aligned} \dot{\mathbf{x}} &= \mathbf{A}\mathbf{x} + \mathbf{B}(\mathbf{u} - \mathbf{K}\mathbf{y}) \\ &= (\mathbf{A} - \mathbf{B}\mathbf{K}\mathbf{C})\mathbf{x} + \mathbf{B}\mathbf{u} \qquad \text{with} \qquad \mathbf{y} = \mathbf{C}\mathbf{x} \qquad \text{and} \qquad \mathbf{D} = \mathbf{0} \\ \mathbf{A}' &= \mathbf{A} - \mathbf{B}\mathbf{K}\mathbf{C} \end{aligned} \tag{2.314}$$

We can see, that the insertion of $\mathbf{K}$ changes the system matrix. The stabilization can only succeed, if $\mathbf{A}$, $\mathbf{B}$, and $\mathbf{C}$ have a certain structure.

We have assumed, that in the original system there is no direct link from the actuating variable to the output variable ($\mathbf{D} = \mathbf{0}$). But in principle, the system can also be stabilized for $\mathbf{D} \neq \mathbf{0}$. The equations get just more complicated.

Of course, we have to provide a compensation for the modification of the entire system which results from inserting $\mathbf{K}$. Otherwise the stability analysis would be taken out on a different control-loop. It is easy to see that the insertion of $\mathbf{K}$ adds the term $-\mathbf{K}\mathbf{y}$ to the initial input quantity $\mathbf{u}$ of the linear part. If we also add a matrix $\mathbf{K}$ in parallel to the nonlinear part $\mathbf{f}$, then we already obtain the desired compensation, because of $-\mathbf{K}\mathbf{e} = +\mathbf{K}\mathbf{y}$. The given extended system (with $\mathbf{D} = \mathbf{0}$) is equivalent to the original one. In our stability analysis, we therefore have to replace the linear part $\mathbf{G}$ by $\mathbf{G}'$ and the nonlinear part $\mathbf{f}$ by $\mathbf{f}'$. This transformation of the system can be compared to the sector transformation of the Popov criterion or of the circle criterion. We adjust a given standard control loop until it complies with the preconditions of the stability analysis criterion. By proving stability of the transformed system, we also prove the stability of the original system.

Now, we have a stable, linear system $\mathbf{G}'(s)$. Next, according to theorem F.1 we need to check whether the matrix

$$\mathbf{H}'(j\omega) = \frac{1}{2}(\mathbf{G}'(j\omega) + \bar{\mathbf{G}}'^T(j\omega)) \tag{2.315}$$

has only positive eigenvalues for all frequencies ω. This can be done in a numerical way. The required steps are similar to the ones which were taken in the Popov criterion for MIMO systems. Because of the frequency-dependent $\mathbf{H}'$, we get a frequency-dependent curve for every eigenvalue. If all these curves for the eigenvalues are positive, then the linear system $\mathbf{G}'$ is strictly positive real.

Otherwise, another transformation (fig. 2.95) of the system is required. The objective of this transformation is, to make the linear part of the system positive real by parallel connection of a diagonal matrix with elements as small as possible. The smaller its elements are, the larger sectors we get for the transfer characteristic of the nonlinear part, as we will show later.

First, we need to find the smallest value $d < 0$ of all eigenvalues of $\mathbf{H}'$, with respect to ω. Adding a matrix $\mathbf{D} = |d|\mathbf{I}$ to $\mathbf{G}'$ produces the system $\mathbf{G}'' = \mathbf{G}' + \mathbf{D}$ with

$$\begin{aligned}\mathbf{H}''(j\omega) &= \frac{1}{2}(\mathbf{G}''(j\omega) + \bar{\mathbf{G}}''^T(j\omega)) = \frac{1}{2}(\mathbf{G}'(j\omega) + \mathbf{D} + \bar{\mathbf{G}}'^T(j\omega) + \mathbf{D}) \\ &= \mathbf{H}'(j\omega) + |d|\mathbf{I} \end{aligned} \tag{2.316}$$

as the matrix assigned to it. Compared to the eigenvalues of $\mathbf{H}'$, the ones of $\mathbf{H}''$ are obviously shifted to the right by $|d|$. They are therefore all positive. As furthermore $\mathbf{G}''$ has the same, stable poles as $\mathbf{G}'$, the extended system becomes strictly positive real.

It is also possible to use a positive semi-definite diagonal matrix $\mathbf{D}$ for the extension, where the elements of $\mathbf{D}$ are not all equal to one another. In this case, however, it is not possible to state a direct relation between these elements and the shift of the eigenvalues of $\mathbf{H}'$. This can make finding matrix $\mathbf{D}$ more complicated and time-consuming. But to get larger sectors for the nonlinear transfer characteristics, a different choice for these elements may sometimes be necessary.

The diagonal matrix $\mathbf{D}$, that makes the linear part strictly positive real, can be computed in a different way, based on the state space description of the system according to theorem F.2: First, we choose a matrix $\mathbf{L}$ with $grad(\mathbf{L}) = n$. With $\mathbf{L}$ and the system matrix $\mathbf{A}'$ we can compute a matrix $\mathbf{P}$ from the Lyapunov equality (F.3).

As $\mathbf{A}'$ is the system matrix of a stable system $\mathbf{G}'$, all eigenvalues of $\mathbf{A}'$ are negative. Furthermore, we already mentioned in appendix F, it follows from $grad(\mathbf{L}) = n$, that $\mathbf{L}\mathbf{L}^T$ is a symmetrical, positive definite matrix. It follows from theorem D.1, that $\mathbf{P}$ is positive definite and fulfills the prerequisite of theorem F.2.

Because $\mathbf{L}$ is regular, it is also invertible, and $\mathbf{V}$ results from equation (F.4) by

$$\mathbf{V} = \mathbf{L}^{-1}(\mathbf{C}'^T - \mathbf{P}\mathbf{B}') \tag{2.317}$$

As $\mathbf{D}$ shall be a diagonal matrix, it is symmetrical ($\mathbf{D} = \mathbf{D}^T$). Therefore, we can transform equation (F.5) into

$$\mathbf{D} = \frac{1}{2}\mathbf{V}^T\mathbf{V} \tag{2.318}$$

According to figure 2.95 this matrix is added in parallel to the stable linear part $\mathbf{G}'$. Because we used equations (F.3) - (F.5) for the computation of $\mathbf{D}$, the resulting system $\mathbf{G}''$ meets the prerequisites of theorem F.2 and is therefore strictly positive real.

In contrast to the method mentioned before, it is not guaranteed here, that the diagonal elements of $\mathbf{D}$ are as small as possible. Therefore, regarding to the further steps, that will be described in the following, the first method is the better one.

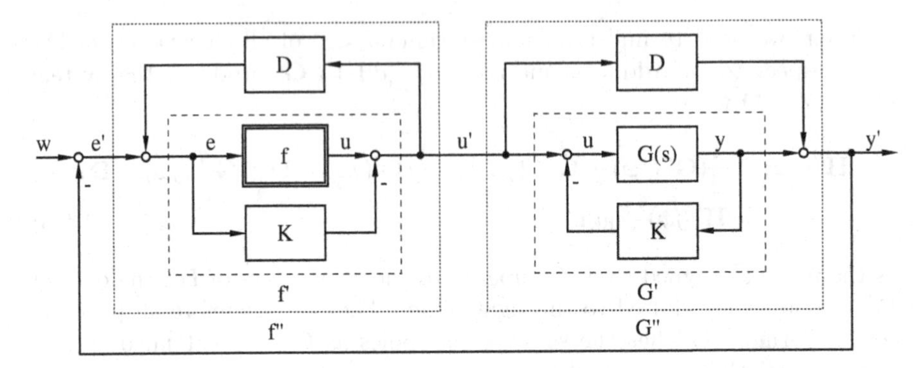

Fig. 2.95. Extending the linear part of the system in order to guarantee hyperstability

Now, we have to provide a compensation for the modification which we achieved by adding $\mathbf{D}$ to the linear part. This time, the effect of inserting $\mathbf{D}$ is that $-\mathbf{Du}'$ is added to the original input quantity of the nonlinear part $\mathbf{e}$. A feedback loop for the nonlinear part $\mathbf{f}'$ containing $\mathbf{D}$ compensates this effect. The system shown in figure 2.95 is therefore equivalent to the original system. Altogether, for our stability analysis we replace the linear part $\mathbf{G}$ by $\mathbf{G}''$ and the nonlinear part $\mathbf{f}$ by $\mathbf{f}''$. If we can prove that the transformed system is stable, then so is its original.

Before we discuss the last step of the stability analysis, let us make a list of all transformation steps discussed so far:

- Adding of two matrices $\mathbf{N}$ and $\mathbf{M}$, to get same dimension of the vectors $\mathbf{u}$ and $\mathbf{e}$ or $\mathbf{y}$ respectively.
- Adding of a matrix $\mathbf{K}$ to stabilize the linear part.
- Adding of a matrix $\mathbf{D}$ to make the linear part positive real.

After the transformations, the linear part $\mathbf{G}''$ of the transformed system is for sure asymptotically hyperstable. All we have to do to prove the stability of the entire system is to show that the nonlinear part of the extended system $\mathbf{f}''$ complies with the inequality

$$\int_0^T \mathbf{u}'^T \mathbf{e}' dt \geq -\beta_0^2 \tag{2.319}$$

or, alternatively,

$$\int_0^T [\mathbf{f}(\mathbf{e}) - \mathbf{Ke}]^T [\mathbf{e} - \mathbf{D}(\mathbf{f}(\mathbf{e}) - \mathbf{Ke})] dt \geq -\beta_0^2 \tag{2.320}$$

It is definitely sufficient, if the i-th components of both vectors of the integrand are of the same sign. This leads to the sector condition

$$
\begin{aligned}
0 \leq \frac{f_i(\mathbf{e}) - \mathbf{k}_i^T \mathbf{e}}{e_i} \leq \frac{1}{d_{ii}} & \qquad \text{if } e_i \neq 0 \\
f_i(\mathbf{e}) - \mathbf{k}_i^T \mathbf{e} = 0 & \qquad \text{if } e_i = 0
\end{aligned}
\tag{2.321}
$$

for all i where $d_{ii} > 0$ and $\mathbf{K} = [\mathbf{k}_1, \mathbf{k}_2, ...]^T$. $\mathbf{k}_i$ is the i-th row of $\mathbf{K}$. For $d_{ii} = 0$, the upper sector bound becomes ∞. It is easy to see, that the permissible sector is the bigger, the smaller the value we chose for d_{ii}, and how important it is to choose $\mathbf{D}$ in a way that its elements are as small as possible.

If the linear system is stable from the very beginning, we can skip over the transformation with matrix $\mathbf{K}$, and the conditions for the sectors change to

$$
\begin{aligned}
0 \leq \frac{f_i(\mathbf{e})}{e_i} \leq \frac{1}{d_{ii}} & \qquad \text{if } e_i \neq 0 \\
f_i(\mathbf{e}) = 0 & \qquad \text{if } e_i = 0
\end{aligned}
\tag{2.322}
$$

Together with the demand that the linear part is strictly positive real, these conditions correspond to the demands of the Popov criterion for MIMO systems. This is not surprising, as equation (2.276) of the Popov criterion with $\mathbf{Q} = \mathbf{0}$ can be interpreted in a way, that a stable, linear system $\mathbf{G}$ has to be extended by a diagonal matrix $\mathbf{V}$ to make it positive real. But exactly this is, what we have done in this chapter.

Let us therefore illustrate the differences between the two criteria: In contrast to the criterion of hyperstability mentioned just above, the Popov criterion contains a matrix $\mathbf{Q}$, which we can choose in a way that in the end sectors as large as possible result for the nonlinear characteristic curves. In so far, the Popov criterion is an extension of the criterion of hyperstability. On the other hand, the Popov criterion can be applied to time-invariant, static characteristic curves only. In addition to this, any curve may depend only on a single component of the input vector ($u_i = f_i(e_i)$). In contrast to this, the nonlinear part of the criterion of hyperstability only has to comply with the integral inequality (2.310). Internal dynamics as well as any dependencies on the input quantities are permitted.

In practical applications, the conditions (2.321) and (2.322) can be checked analytically only in very simple cases. Normally, a numerical check is needed: A sufficient large and representative set of error vectors $\mathbf{e}$ must be chosen, and for every single vector of this set the stability conditions have to be checked. But if a numerical analysis is needed anyway, it is even better to check the integrand of (2.320) instead of the conservative conditions (2.321) and (2.322). If this term is positive for all error vectors $\mathbf{e}$, then condition (2.320) will definitely be fulfilled.

Let us demonstrate an application of the criterion of hyperstability by giving a very simple example: Let a SISO system be given whose nonlinear part consists of a multiplication of $e(t)$ by a time-dependent amplification $k(t)$ (fig. 2.96, top). Let the transfer function $G(s)$ of the linear part be strictly positive real and thus asymptotically hyperstable. In order to prove asymptotic stability of the entire system for the zero position $w = u = y = 0$, all we have to do is examine inequality (2.310). For $u = ke$, we get:

$$\int_0^T u(t)e(t)dt = \int_0^T k(t)e^2(t)dt \geq -\beta_0^2 \tag{2.323}$$

This inequality is certainly true if the integrand is positive, i.e. $k(t) \geq 0$.

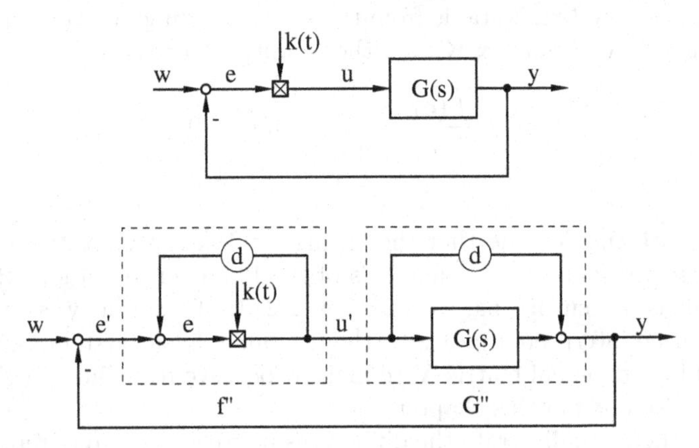

Fig. 2.96. Application example for the hyperstability criterion

Now let us assume that the linear transfer function in our example is asymptotically stable, but no longer hyperstable. All of its poles therefore still have a negative real part, but the function is not strictly positive real. According to theorem F.1, this means that the values of the real part of the frequency response $G(j\omega)$ can now be negative, too, i.e. a part of the Nyquist plot lies in the left-hand side of the complex plane. Figure 2.97 and the lower part of figure 2.96 illustrate how we have to proceed in such a case. We shift the Nyquist plot until it lies completely to the right of the axis of imaginaries. This can be achieved by adding a proportional element in parallel to the linear part. At the same time, this modification of the entire system has to be compensated by adding a feedback loop to the nonlinear part. Finally, we obtain a transformed system with the linear part $G''(s)$ and the nonlinear part f''. $G''(s)$ is strictly positive real and therefore asymptotically hyperstable. All we still have to do is examine the integral inequality for the nonlinear part. For

$$u'(t) = \frac{k(t)}{1 - dk(t)} e'(t) \tag{2.324}$$

we get

$$\int_0^T u'(t)e'(t)dt = \int_0^T \frac{k(t)}{1 - dk(t)} e'^2(t)dt \geq -\beta_0^2 \tag{2.325}$$

This inequality is certainly true for $\frac{k(t)}{1-dk(t)} \geq 0$, or for $0 \leq k(t) < \frac{1}{d}$. The demand on $k(t)$ is now more restrictive than in the first setup, as $G(s)$ is not strictly positive real here.

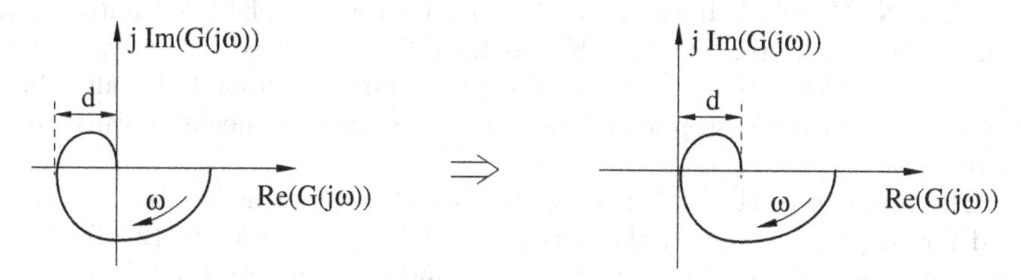

Fig. 2.97. Shifting the Nyquist plot in order to obtain hyperstability

At the end of this chapter, we want to introduce a version of the hyperstability criterion (see [145]) that is based on theorem F.2 in the appendix. In this version, the possibly necessary extensions of the linear part of the system will be used to get additional degrees of freedom for the stability analysis and thus make it less conservative.

In contrast to the former version, here the first step is the insertion of a stabilization matrix **K**, the second step is the extension **N** to get the same number of input and output variables of the two parts of the system (**M** results directly from $\mathbf{NM} = \mathbf{I}$), and the third step is the insertion of a diagonal matrix **D** to make the linear part positive real. **D** will be computed according to the second variant of the former version, that means **D** results with the equations (F.3), (2.317) and (2.318) from an independently selected, regular matrix **L**.

K, **N** and **L** can be interpreted as parameters of the stability analysis, that can be chosen independently regarding some secondary conditions.

Regarding **N** and **M** and taking into account the changed order of the system extensions we now get

$$\int_0^T (\mathbf{M}(\mathbf{f}(\mathbf{e}) - \mathbf{K}\mathbf{e}))^T (\mathbf{e} - \mathbf{D}\mathbf{M}(\mathbf{f}(\mathbf{e}) - \mathbf{K}\mathbf{e}))dt \geq -\beta_0^2 \tag{2.326}$$

instead of equation (2.320) as the condition for stability. This inequality can be easily understood, if we insert the matrices **M** and **N** in fig. 2.95, that are taken into consideration explicitly in this version. **M** is inserted directly after the subtraction of **u** and **Ke**, and **N** is inserted directly before the subtraction of $\mathbf{u}'$ and **Ky**. These positions result from the order of the system extensions.

The condition is fulfilled, if the integrand is positive for any **e**:

$$(\mathbf{M}(\mathbf{f}(\mathbf{e}) - \mathbf{K}\mathbf{e}))^T (\mathbf{e} - \mathbf{D}\mathbf{M}(\mathbf{f}(\mathbf{e}) - \mathbf{K}\mathbf{e})) \geq 0 \tag{2.327}$$

Now an interative loop is run through: The matrices **N**, **K** and **L** will be fixed independently in the first step, just meeting the secondary conditions **L** to be regular, $\mathbf{M}$ existing with $\mathbf{NM} = \mathbf{I}$ and $\mathbf{K}$ stabilizing the linear system. From $\mathbf{L}$ we get the matrix $\mathbf{D}$ with (F.3), (2.317) and (2.318). Then we check inequality (2.327) for a suitable set of vectors $\mathbf{e}$. If it is fulfilled for all of these vectors, the entire system is stable. Otherwise, in the next iterative run other matrices $\mathbf{N}$, $\mathbf{K}$ and $\mathbf{L}$ have to be chosen, and the complete computation is done again. With every run, $\mathbf{N}$, $\mathbf{K}$ and $\mathbf{L}$ can be optimized in a way, so that the left-hand side of inequality (2.327) gets as large as possible for all values of $\mathbf{e}$. The algorithm is stopped as soon as it does not take negative values any more.

The problem of the optimization is, that there does not exist a gradient field for the dependency of the left-hand side of the inequality (2.327) from the coefficients of the three matrices $\mathbf{N}$, $\mathbf{K}$ and $\mathbf{L}$ to be optimized. Therefore, the search for the optimal coefficients cannot be systematic, but only with the help of an evolutionary algorithm. Nevertheless, such an optimization is useful, as the result of the stability analysis with optimized matrices will be less conservative than with non-optimized matrices, even if the optimization does not provide the absolute optimum.

2.8.10 Sliding Mode Controllers

After this introduction to some of the stability criteria for nonlinear systems, we now discuss a design method for SISO system controllers. As we will see later on, this method can also be used for an analysis of the stability of fuzzy controllers. We are concerned with the *sliding mode controller* ([155]). It requires that the state space model of the plant conforms to

$$x^{(n)}(t) = f(\mathbf{x}(t)) + u(t) + d(t) \tag{2.328}$$

where $\mathbf{x} = (x, \dot{x}, ..., x^{(n-1)})^T$ is the state vector, $u(t)$ the actuating variable and $d(t)$ an unknown disturbance load. For the linear SISO case, such a model is equivalent to the controller canonical form (fig. 2.50).

The reference value x_d does not necessarily have to be a constant value, which is why we introduce a vector $\mathbf{x}_d = (x_d, \dot{x}_d, ..., x_d^{(n-1)})^T$ which contains the reference values as its components. Accordingly, we also have to replace the output error $e = x_d - x$ by the error vector $\mathbf{e} = \mathbf{x}_d - \mathbf{x} = (e, \dot{e}, ..., e^{(n-1)})^T$. The controller's aim is to achieve $\mathbf{e} = \mathbf{0}$, i.e. the output error and all of its derivatives should be zero.

However, we do not try to reach this aim directly, but by establishing another goal first, which is characterized by the differential equation

$$0 = q(\mathbf{e}) = (\frac{\partial}{\partial t} + \lambda)^{n-1} e \tag{2.329}$$

$$= e^{(n-1)} + \binom{n-1}{1} \lambda e^{(n-2)} + \binom{n-1}{2} \lambda^2 e^{(n-3)} + ... + \lambda^{n-1} e$$

$$= e^{(n-1)} + g_\lambda(\mathbf{e}) \tag{2.330}$$

where $\lambda > 0$. If the error vector meets this differential equation, it will always converge towards zero, from any initial state. If we try to reach $q = 0$, then the initial goal is achieved automatically.

This is easy to understand: Equation (2.329) corresponds to the differential equation of a series connection of $n-1$ first order lags (fig. 2.98). If the input quantity q becomes zero, then so do e and its $n-1$ derivatives.

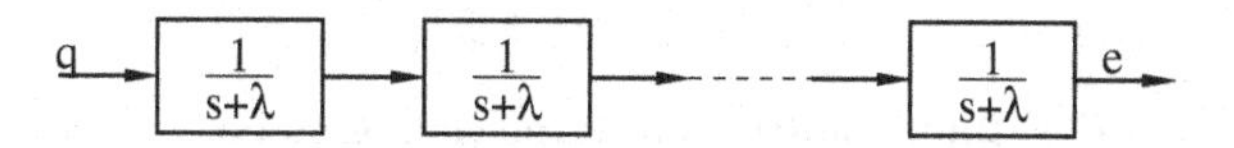

Fig. 2.98. Graphic interpretation of q

Next, we would like to replace again the goal $q(\mathbf{e}) = 0$ by another condition. In fact, the function $q^2(\mathbf{e})$ will be positive everywhere except for $q(\mathbf{e}) = 0$, the control goal. If we can guarantee that the derivative of this function always meets the condition

$$\frac{\partial}{\partial t}(q^2(\mathbf{e})) < -2\eta|q(\mathbf{e})| \tag{2.331}$$

(where $\eta \geq 0$), then $q^2(\mathbf{e})$ will converge towards $q^2(\mathbf{e}) = 0$ for any initial value, which obviously also establishes $q(\mathbf{e}) = 0$. We can therefore consider q^2 to be a Lyapunov function. Keeping inequality (2.331) leads sooner or later to equation (2.330) being true. And as a result of this, as already mentioned, the initial control goal $\mathbf{e} = \mathbf{0}$ will be fulfilled, too.

The question arises what the advantage of a twice repeated goal replacement might be. First, we can rewrite equation (2.331) in simpler terms, as calculation of the derivate yields

$$q\dot{q} < -\eta|q| \tag{2.332}$$

and thereby

$$\dot{q}\ \mathrm{sgn}(q) < -\eta \tag{2.333}$$

Using this as a formulation of the control goal, we can now answer the above question. While the initial task for the controller—to adjust the original system to a given vector of reference values—was of degree n (because of the $n-1$ derivations of $\mathbf{e}$), then the degree of the task given by the new goal (2.333) is obviously only 1, because the interesting variable occurs only with its first derivative.

A geometrical interpretation of the different goals is also interesting. The condition $q(\mathbf{e}) = 0$ defines a hypersurface in the n-dimensional space spanned by $\mathbf{e}$. Complying with inequality (2.331) or (2.333), the system is forced to approach this hypersurface, and once it has reached the plane, it can never leave it again. Then the system slides down the hypersurface into the point $\mathbf{e} = \mathbf{0}$. $q(\mathbf{e}) = 0$ is therefore also called a *sliding surface*. Figure 2.99 gives an illustration of this for $n = 2$. Because of $0 = q(\mathbf{e}) = \dot{e} + \lambda e$, in this example

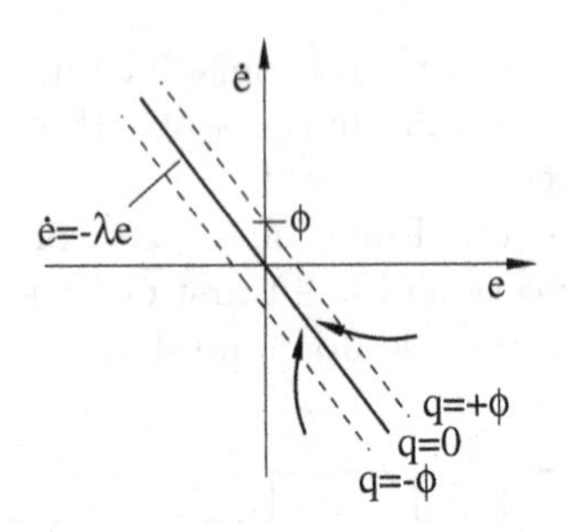

Fig. 2.99. Illustration to the sliding mode control

the hypersurface consists only of a line through the origin of the $e - \dot{e}-$plane. We will explain the other lines and the variable Φ later on.

The question is now which features the actuating variable must have in order to achieve that inequality (2.333) is always true. First, from (2.330) we get:

$$\begin{aligned} q &= e^{(n-1)} + g_\lambda(\mathbf{e}) \\ \dot{q} &= e^{(n)} + g_\lambda(\dot{\mathbf{e}}) = x_d^{(n)} - x^{(n)} + g_\lambda(\dot{\mathbf{e}}) \end{aligned} \tag{2.334}$$

and together with (2.328):

$$\dot{q} = g_\lambda(\dot{\mathbf{e}}) - f(\mathbf{x}) - u - d + x_d^{(n)} \tag{2.335}$$

From this, for inequality (2.333) it follows that

$$(g_\lambda(\dot{\mathbf{e}}) - f(\mathbf{x}) - u - d + x_d^{(n)}) \operatorname{sgn}(q) < -\eta \tag{2.336}$$

The function f can be split up into

$$f = f_0 + \Delta f \tag{2.337}$$

Here, f_0 denotes the nominal model of the plant, i.e. the part of the plant's model which we know is correct, while Δf represents the uncertainty of the model. If we then choose

$$u = -f_0(\mathbf{x}) + g_\lambda(\dot{\mathbf{e}}) + x_d^{(n)} + U\operatorname{sgn}(q) \tag{2.338}$$

for the actuating variable with a constant value U, that has to be computed later, inequality (2.336) becomes:

$$(-\Delta f(\mathbf{x}) - d) \operatorname{sgn}(q) - U < -\eta \tag{2.339}$$

It shall be possible to estimate the uncertainty of the model Δf and the disturbance load d by upper bounds:

$$|\Delta f| < F \qquad |d| < D \tag{2.340}$$

Hence, inequality (2.339), and therefore also (2.331) will be certainly true if

$$U = F + D + \eta \tag{2.341}$$

holds, and from (2.338) follows for the actuating variable

$$u = -f_0(\mathbf{x}) + g_\lambda(\dot{\mathbf{e}}) + x_d^{(n)} + (F + D + \eta)\mathrm{sgn}(q) \tag{2.342}$$

This defines the sliding mode controller. We can think of the first three summands as being an inverse model of the plant, while the last summand is mainly a contribution of the model's uncertainties and external disturbances. Furthermore, we can see that for the design of such a controller, on the one hand the plant has to be representable in the way of (2.328) and on the other hand the corresponding model of the plant f_0 has to be known. An uncertainty of the model of Δf is permissible, though we have to estimate an upper bound for it by F. If f_0 is completely unknown, we have to set $f_0 = 0$ and choose a sufficient large value for F. We also have to be able to estimate an upper bound D for the maximum amplitude of the external disturbance. Furthermore, it must be possible to measure the state vector $\mathbf{x}$. From $\mathbf{x}$ and the vector of the reference values $\mathbf{x}_d$ (which is known anyway), we can immediately compute $x_d^{(n)}$, the error vector $\mathbf{e} = \mathbf{x}_d - \mathbf{x}$ and hence also $g_\lambda(\dot{\mathbf{e}})$ and $q(\mathbf{e})$.

However, the estimation of the error vector $\mathbf{e}$ contains one problem: Normally, only the error $e = x_d - x$ is measured at the plant output, but we also need its n derivatives to compute the error vector $\mathbf{e}$. It is not possible to estimate these derivatives by simple discrete derivations, as the unavoidable measurement noise of e would affect the higher derivatives so intensively, that these could not be used for control any more. Therefore, $\dot{\mathbf{e}}$ can be estimated only by using a nonlinear observer. But this observer requires a relatively precise plant model f_0.

Finally, we have to choose values for the parameters λ and η. η determines the velocity with which the system approaches the hypersurface, according to inequality (2.331). The larger the value for η, the faster the system approaches the hypersurface. From equation 2.342, however, it follows that a faster approach also requires a larger value for the actuating variable. Technical aspects therefore have to be considered when fixing η.

λ defines the hypersurface, according to equation (2.330). If the system has already reached the hypersurface, this hypersurface determines the dynamic behavior of the system. The error converges the faster towards zero, the larger a value is chosen for λ. Again, technical aspects of the system have to be considered just like for η. However, once λ has been set to a certain value, this parameter completely determines the system's behavior on the hypersurface, independently of the plant parameters, their changes or external disturbances. This feature, however, characterizes a robust control, as for a robust control the desired behavior can be guaranteed even if the parameters of the plant change. And a measure of robustness, i.e. of the permissible difference between the real and the nominal plant, is given by F.

In principle, $q(\mathbf{e})$ can be defined by a general polynomial

$$q(\mathbf{e}) = \sum_{i=0}^{n-2} c_i e^{(i)} + e^{(n-1)} \tag{2.343}$$

instead of (2.329). The coefficients c_i have to be defined in a way, that all zeros of the polynomial

$$c(s) = s^{n-1} + c_{n-2}s^{n-2} + ... + c_1 s + c_0 \tag{2.344}$$

have a negative real part. This guarantees stability of the corresponding linear transfer element, and from this it follows, that e with $q(\mathbf{e}) = 0$ converges to zero. In contrast to only one free parameter λ (as we had before) we now have $n-1$ free parameters, that can be used to adapt the shape of the hyper surface better to the requirements of the plant and the technical restrictions. However, this adaptation is not trivial.

An undesired feature of the sliding mode controller according to equation (2.342) is the discontinuous course of the actuating variable at every sign reversal of q. The step increases with the plant's uncertainty F and the disturbance's estimated bound D. As a means to reduce the hight of this step, η can be set to zero, as this only affects the control rate. We could therefore achieve a continuous course of the actuating variable—for an exact model and an undisturbed plant. However, as this is never the case in any real application, we have to try other means in order to avoid this discontinuity. One possibility is to replace the sign function by

$$h(q) = \begin{cases} \frac{1}{\Phi} q & : |q| < \Phi \\ \text{sgn}(q) : |q| \geq \Phi \end{cases} \tag{2.345}$$

(fig. 2.100), but the sign function is needed to maintain inequality (2.331) in (2.342). If we replace it by $h(q)$, this inequality may no longer be true for $|q| < \Phi$. As a result of this, the behavior of the system we aimed to achieve by using the sliding mode controller can no longer be guaranteed for such values of q. The new controller only ensures that the system enters and remains in a zone $|q| < \Phi$ around the hypersurface $q = 0$ (fig. 2.99). It approaches the target, but with uncertainties of the model and disturbances being present it will never reach it exactly. On the other hand, however, it will also never again leave this zone, which we can therefore consider to be the region of tolerance for the controller. The larger Φ gets, i.e. the smoother the course of the actuating variable, the larger the region of tolerance has to be.

2.8.11 Nonlinear Observers

As the closing section of this chapter, let us turn to a completely different type of problem. As we will see, a fuzzy controller is in many cases nothing

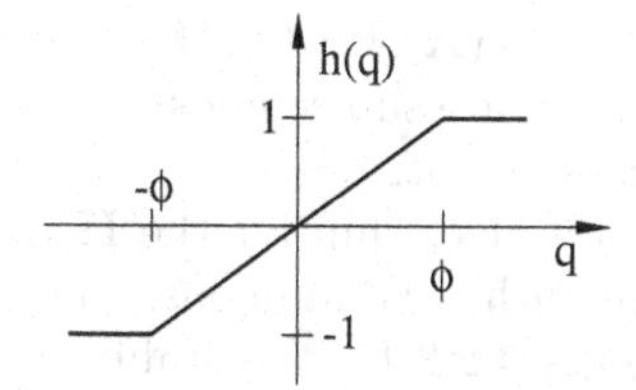

Fig. 2.100. An alternative to the sign function

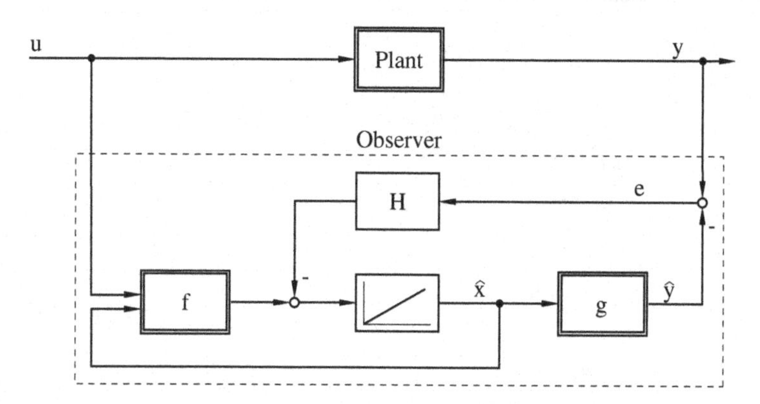

Fig. 2.101. Nonlinear observer

else but a nonlinear state space controller. Accordingly, the same problem as for linear state space controllers results: The course of the plant's states has to be known. If they can not be measured directly, then the use of an observer will be necessary.

The underlying idea is the same as for the linear observer. The same actuating variables $\mathbf{u}$ are impressed on both the plant itself and the nonlinear model of the system, and the difference $\mathbf{e}$ between the output of the model and the output of the plant is fed back as a correction term to the model, multiplied by a correcting matrix $\mathbf{H}$ (fig. 2.101). As for linear observers, it is a requirement that the system has no direct access from the actuating variable to the output quantity, i.e. $\mathbf{g}$ is a function depending on $\mathbf{x}$ only, and not on $(\mathbf{x}, \mathbf{u})$ as in equation (2.213).

The question is how we can determine the matrix $\mathbf{H}$. For general nonlinear systems there exists no algorithm, that could guarantee the stability of the observer and the exactness of the observed values. But the situation is different for TSK systems (see chap. 4.1.3), that have the same underlying idea as the former mentioned gain scheduling method. A TSK system consists of linear partial systems, whose output variables are superimposed depending on the current set point, to get the output variable of the entire system.

For a TSK observer the plant model is linearized at different set points, so that we get a linear model with the system matrices $(\mathbf{A}_i, \mathbf{B}_i, \mathbf{C}_i)$ for each set point. For every linear model we have to compute an observer correction matrix $\mathbf{H}_i$, so that the matrices $(\mathbf{A}_i, \mathbf{B}_i, \mathbf{C}_i, \mathbf{H}_i)$ form the linear observer at set point i. These linear observers are superimposed to a nonlinear TSK observer, depending on the current set point.

The precise computation algorithm for the $\mathbf{H}_i$ matrices shall be omitted here, as the formulas are rather voluminous, but it is very similar to the algorithms given in chapters 4.2.2 for the stability analysis of TSK systems and 5.1 for the design of TSK controllers. In these chapters we also refer to literature that deals with the design of TSK observers.

3

Design of Fuzzy Controllers

While in classical control theory it is always the objective to use analytical methods for the description of the system behavior and the following design of the controller, fuzzy systems are suitable for the modelling of vague knowledge, for example about a technical process or an existing (human) controller. From this principal difference there result the completely different methodologies to solve a given control problem.

In classical control engineering, as the first step we construct a model of the plant. As the second step, we design a suitable controller based on this model. This attempt can be said to be model-based. In contrast to this, the design of a fuzzy controller is rule-based. Here, no model is constructed. Instead, we design the controller directly, in some sense by intuition, while, of course, the design still requires some understanding of the plant's behavior.

Therefore, the fuzzy controller presents itself mainly for systems, where no model exists or where the model has such an inconvenient nonlinear structure, that a classical controller design is impossible. If a fuzzy controller is also suitable for other systems, shall be discussed at the end of this chapter.

3.1 Mamdani Controllers

The first model of a fuzzy controller we introduce here was developed in 1975 by Mamdani [120] on the basis of the more general ideas of Zadeh published in [211, 212, 213].

The *Mamdani controller* is based on a finite set $\mathcal{R}$ of if-then-rules $R \in \mathcal{R}$ of the form

$$R: \quad \text{If } x_1 \text{ is } \mu_R^{(1)} \text{ and } \dots \text{ and } x_n \text{ is } \mu_R^{(n)} \text{ then } y \text{ is } \mu_R. \tag{3.1}$$

$x_1, \dots, x_n$ are input variables of the controller and y is the output variable. Usually, the fuzzy sets $\mu_R^{(i)}$ or μ_R stand for linguistic values, that is, for vague

K. Michels et al.: *Fuzzy-Control: Fundamentals, Stability and Design of Fuzzy Controllers*, StudFuzz **200**, 235–257 (2006)
www.springerlink.com

concepts like "about null", "of average height" or "negative small" which, for their part, are represented by fuzzy sets. In order to simplify the notation, we do not distinguish between membership functions and linguistic values that they represent.

How to precisely interpret the rules is essential for understanding the Mamdani controller. Although the rules are formulated in terms of if-then statements, they should not be understood as logical implications, but in the sense of a piecewise defined function. If the rulebase $\mathcal{R}$ consists of the rules $R_1, \ldots, R_r$, we should understand it as a piecewise definition of a fuzzy function, that is

$$f(x_1, \ldots, x_n) \approx \begin{cases} \mu_{R_1} & \text{if } x_1 \approx \mu_{R_1}^{(1)} \text{ and } \ldots \text{ and } x_n \approx \mu_{R_1}^{(n)} \\ \vdots & \\ \mu_{R_r} & \text{if } x_1 \approx \mu_{R_r}^{(1)} \text{ and } \ldots \text{ and } x_n \approx \mu_{R_r}^{(n)}. \end{cases} \tag{3.2}$$

This is similar to the pointwise specification of a crisp function defined over a product space of finite sets in the form

$$f(x_1, \ldots, x_n) \approx \begin{cases} y_1 & \text{if } x_1 = x_1^{(1)} \text{ and } \ldots \text{ and } x_n = x_1^{(n)}, \\ \vdots & \\ y_r & \text{if } x_1 = x_r^{(1)} \text{ and } \ldots \text{ and } x_n = x_r^{(n)}. \end{cases} \tag{3.3}$$

We obtain the graph of this function by

$$\text{graph}(f) = \bigcup_{i=1}^{r} \left(\hat{\pi}_1(\{x_i^{(1)}\}) \cap \ldots \cap \hat{\pi}_n(\{x_i^{(n)}\}) \cap \hat{\pi}_Y(\{y_i\}) \right). \tag{3.4}$$

The "fuzzification" of this equation, by using the minimum for the intersection and the maximum (supremum) for the union, the fuzzy graph of the function described by the rule set $\mathcal{R}$ is the fuzzy set

$$\mu_{\mathcal{R}} : X_1 \times \ldots \times X_n \times Y \to [0,1],$$

$$(x_1, \ldots, x_n, y) \mapsto \sup_{R \in \mathcal{R}} \{\min\{\mu_R^{(1)}(x_1), \ldots, \mu_R^{(n)}(x_n), \mu_R(y)\}$$

or

$$\mu_{\mathcal{R}} : X_1 \times \ldots \times X_n \times Y \to [0,1],$$

$$(x_1, \ldots, x_n, y) \mapsto \max_{i \in \{1, \ldots, r\}} \{\min\{\mu_{R_i}^{(1)}(x_1), \ldots, \mu_{R_i}^{(n)}(x_n), \mu_{R_i}(y)\}$$

in the case of a finite rulebase $\mathcal{R} = \{R_1, \ldots, R_r\}$.

If a concrete input vector $(a_1, \ldots, a_n)$ for the input variables $x_1, \ldots, x_n$ is given, we obtain the fuzzy set

$$\mu_{\mathcal{R},a_1,\ldots,a_n}^{\text{output}} : Y \to [0,1], \qquad y \mapsto \mu_{\mathcal{R}}(a_1,\ldots,a_n,y)$$

as the fuzzy "output value".

The fuzzy set $\mu_{\mathcal{R}}$ can be interpreted as a fuzzy relation over the sets $X_1 \times \ldots \times X_n$ and Y. Therefore, the fuzzy set $\mu_{\mathcal{R},a_1,\ldots,a_n}^{\text{output}}$ corresponds to the image of the one-element set $\{(a_1,\ldots,a_n)\}$ or its characteristic function under the fuzzy relation $\mu_{\mathcal{R}}$. So in principle, instead of a sharp input vector we could also use a fuzzy set as input. For this reason, it is very common to call the procedure of feeding concrete input values to a fuzzy controller as *fuzzification*, i.e. the input vector $(a_1,\ldots,a_n)$ is transformed into a fuzzy set which is nothing else than the representation of a one-element set in term of its characteristic function.

Fuzzification can also be interpreted in another way. In the section on fuzzy relations we have seen that we can obtain the image of a fuzzy set under a fuzzy relation computing the cylindrical extension of the fuzzy set, intersecting the cylindrical extension with the fuzzy relation and projecting the result to the output space. In this sense, we can understand the cylindrical extension of the measured tuple or the corresponding characteristic function as a the fuzzification procedure which is necessary for intersecting it with the fuzzy relation.

Figure 3.1 shows this procedure. In order to make a graphical representation possible, we consider only one input and output variable. Three rules are shown in the figure:

If x is A_i, then y is B_i. $\qquad (i = 1, 2, 3)$

The fuzzy relation $\mu_{\mathcal{R}}$ is represented by the three pyramids. If the input value x is given, the cylindrical extension of $\{x\}$ defines a plane cutting through the pyramids. The projection of this cutting plane onto the y-axis (pointing from the front to the back) yields the fuzzy set $\mu_{\mathcal{R},x}^{\text{output}}$ which describes the desired output value (in fuzzy terms).

We can illustrate the computation of the output value by the following scheme. Figure 3.2 shows two rules of a Mamdani controller with two input

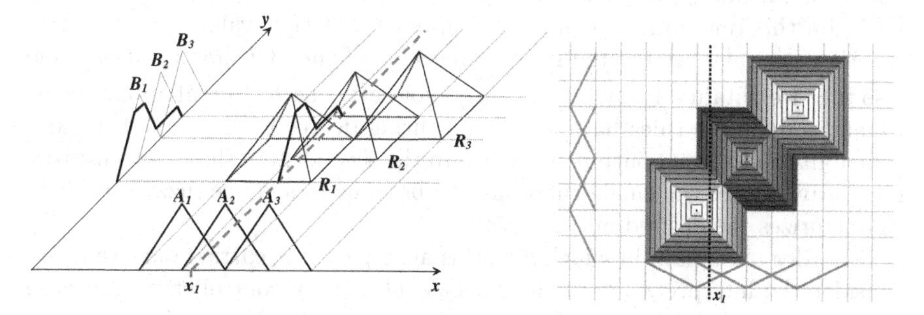

Fig. 3.1. The projection of the input value x_1 to the output axis y

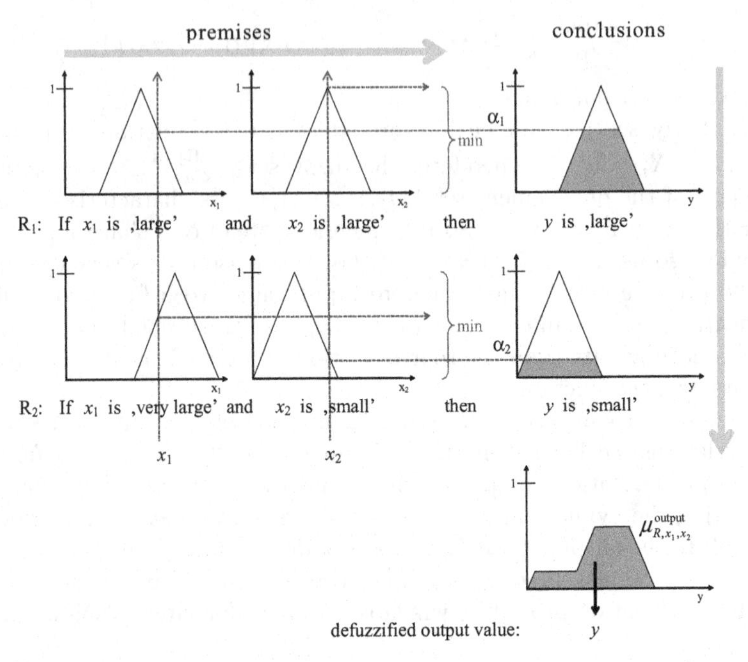

Fig. 3.2. Illustration of the Mamdani controller computation scheme

and one output variable. At first, we consider only one of the two rules, say the rule R. The degree to which the premise is satisfied for the given input values is determined taking the minimum of the membership degrees to the corresponding fuzzy sets. Then the fuzzy set in the conclusion of the rule is "cut" off at this level. This means, the membership degree to the actual output fuzzy set is the minimum of the membership degree to the fuzzy set in the conclusion part of the rule and the firing degree of the rule.

If the firing degree of the rule is 1, so we obtain exactly the fuzzy set in the conclusion part as a result, i.e. $\mu_R = \mu_{R,a_1,\ldots,a_n}^{\text{output}}$. When the rule is not applicable, meaning that the firing degree is 0, we obtain $\mu_{R,a_1,\ldots,a_n}^{\text{output}} = 0$, i.e. based on this rule nothing can be said about the output value.

All other rules are treated in the same way – figure 3.2 shows just one rule – so that we obtain a fuzzy set $\mu_{R,a_1,\ldots,a_n}^{\text{output}}$ for each rule R. In real applications, usually only a few rules have a non-zero firing degree at the same time and contribute to the final output fuzzy set. In the second step, these output fuzzy sets derived from the single rules have to be joint to a single fuzzy set which characterizes the desired output value.

In order to explain in which way this aggregation is carried out, we again consider the interpretation of the rulebase of a fuzzy controller in the sense of a piecewise definition of a fuzzy function (cf. (3.2)). For a crisp piecewise defined function the different r cases have to be disjoint or, if they overlap, they must provide the same output, because otherwise the function value is not well defined. We assume that each single case describes a "function value" in the form of a set for every input value: If a considered input value satisfies the condition of the corresponding case, then we obtain a one-element set with the specified function value. Otherwise, we obtain the empty set for this case. With this interpretation the actual function value or the one-element set which contains this function value is given by the union of the sets resulting from the single cases.

For this reason the (disjoint) union of the fuzzy sets $\mu_{R,a_1,\ldots,a_n}^{\text{output}}$ resulting from the single rules should be computed for the overall fuzzy output. In order to compute the union of fuzzy sets we use a t-conorm, for instance the maximum, so that we obtain

$$\mu_{\mathcal{R},a_1,\ldots,a_n}^{\text{output}} = \max_{R\in\mathcal{R}}\{\mu_{R,a_1,\ldots,a_n}^{\text{output}}\} \tag{3.5}$$

as the final output fuzzy set based on the rulebase $\mathcal{R}$, when the input vector $(a_1,\ldots,a_n)$ is given. In this way, the two rules shown in Figure 3.2 lead to the output fuzzy set which is also shown there.

In order to obtain a single crisp output value, the output fuzzy set has to be *defuzzified*. A large variety of heuristic defuzzification strategies has been proposed in the literature. However, without specifying more precisely how the fuzzy sets and rules should be interpreted, defuzzification remains a matter of pure heuristics. Therefore, we restrict our considerations to one defuzzification strategy and will discuss the issue of *defuzzification* in more detail after the introduction of conjunctive rule systems.

In order to understand the fundamental idea of defuzzification applied in the case of Mamdani controllers, we consider the output fuzzy set determined in figure 3.2 once again. The fuzzy sets in the conclusion parts of the two rules are interpreted as vague or imprecise values. In the same sense, the resulting output fuzzy set is a vague or imprecise description of the desired output value. Intuitively, the output fuzzy set in figure 3.2 can be understood in the sense that we should favour a larger output value, but should also consider a smaller output value with a lower preference. This interpretation is justified by the fact that the first rule, voting for a larger output value, fires with a higher degree than the second rule that points to a lower output value. Therefore, we should choose an output value that is large, but not too large, i.e. a compromise of the proposed outputs of the two rules, however, putting a higher emphasis on the first rule.

A defuzzification strategy which satisfies this criterion is the centre of gravity method (COG), or the centre of area method (COA). The output value of this method is the centre of gravity (or, to be precise, its projection to the output axis) of the area under the output fuzzy set, i.e.

$$\text{COA}(\mu_{\mathcal{R},a_1,\ldots,a_n}^{\text{output}}) = \frac{\int_Y \mu_{\mathcal{R},a_1,\ldots,a_n}^{\text{output}} \cdot y\,dy}{\int_Y \mu_{\mathcal{R},a_1,\ldots,a_n}^{\text{output}}\,dy}. \tag{3.6}$$

This method requires the implicit condition that the functions are integrable $\mu^{\text{output}}_{\mathcal{R},a_1,\ldots,a_n}$ and $\mu^{\text{output}}_{\mathcal{R},a_1,\ldots,a_n} \cdot y$ which is always satisfied, provided the membership functions appearing in the conclusion parts of the rules are chosen in a reasonable way, for instance when they are continuous..

3.1.1 Remarks on Fuzzy Controller Design

When choosing the fuzzy sets for the input variables, one should make sure that the domain of each input variable is completely covered, this means, for every possible value there exists at least one fuzzy set to which it has a non-zero membership degree. Otherwise, the fuzzy controller cannot determine an output value for this input value.

The fuzzy sets should represent vague or imprecise values or ranges. Therefore, convex fuzzy sets are preferable. Triangular and trapezoidal membership functions are especially suitable, because they have a simple parametric representation and determining the membership degrees can achieved with low computational effort. In ranges where the controller has to react very sensitively to small changes of an input value we should choose very narrow fuzzy sets in order to distinguish between the values well enough. We should, however, take into account that the number of possible rules grows very fast with the number of fuzzy sets. If we have k_i fuzzy sets for the ith input variable, then a complete rulebase that assigns to all possible combinations of input fuzzy sets an output fuzzy set, contains $k_1 \cdot \ldots \cdot k_n$ rules, when we have n input variables. If we have four input values with only five fuzzy sets for each of them, we obtain already 625 rules.

Concerning the choice of the fuzzy sets for the output variable similar constraints as for the input variables should be considered. They should be convex and in the ranges, where a very exact output value is important, narrow fuzzy sets should be used. Additionally, the choice of fuzzy sets for the output value strongly depends on the defuzzification strategy. It should be noted that the defuzzification of asymmetrical triangular membership functions of the form $\Lambda_{x_0-a,x_0,x_0+b}$ with $a \neq b$ does not always correspond to what we might expect. If only one single rule fires with the degree 1 and all others with degree 0, we obtain the corresponding fuzzy set in the conclusion part of this rule as the output fuzzy the set of the controller before defuzzification. If this is an asymmetrical triangular membership function $\Lambda_{x_0-a,x_0,x_0+b}$, we have $\text{COA}(\Lambda_{x_0-a,x_0,x_0+b}) \neq x_0$, i.e. not the point where the triangular membership function reaches the membership degree 1.

Another problem of the centre of gravity method is that it can never return a boundary value of the interval of the output variable. This means, the minimum and maximum value of the output domain are not accessible for the fuzzy controller. A possible solution of this problem is to extend the fuzzy sets beyond the interval limits for the output variable. However, in this case we have to ensure that the controller will never yield an output value outside the range of permitted output values.

For the design of the rulebase, completeness in an important issue. We have to make sure that for any possible input vector there is at least one rule that fires. This does not mean that for every combination of fuzzy sets of input values we have to formulate a rule with these fuzzy sets in the premise part. On the one hand, a sufficient overlapping of the fuzzy sets guarantees that there will still be rules firing, even if we have not specified a rule for all possible combinations of input fuzzy sets. On the other hand, there might be combinations of input values that correspond to a state which the system can or must never reach. For these cases it is not necessary to specify rules. We should also avoid contradicting rules. Rules with identical premise parts and different conclusion parts should be avoided.

The Mamdani controller we have introduced here is also called max-min controller because of equation (3.5) for computing the output fuzzy set $\mu_{\mathcal{R},a_1,\ldots,a_n}^{\text{output}}$. Maximum and minimum were used to calculate the union and the intersection, respectively, in equation (3.4).

Of course, also other t-norms and t-conorms can be considered instead of minimum or maximum. In applications the product t-norm is often preferred and the bounded sum $s(\alpha, \beta) = \min\{\alpha + \beta, 1\}$ is sometimes used as the corresponding t-conorm. The disadvantage of minimum and maximum is the idempotency property. The output fuzzy set $\mu_{R,a_1,\ldots,a_n}^{\text{output}}$ of a rule R depends only on the input variable for which the minimum membership degree to the corresponding fuzzy set in the premise is obtained. A change of another input variable will not affect the output of this rule, unless the change is large enough to let the membership degree of this input value to its corresponding fuzzy set drop below the membership degree of the other input values.

If the fuzzy sets $\mu_{R,a_1,\ldots,a_n}^{\text{output}}$ of different rules support a certain output value to some degree, the aggregation based on the maximum will only take the largest of these membership degrees into account. It might be desirable that such an output value obtains a higher support than another one that is supported to the same degree, but only by a single rule. In this case a t-conorm like the bounded sum should be preferred to the maximum.

In principle, we could also compute the firing degree of a rule and the influence of this degree to the fuzzy set in the conclusion part of this rule based on different t-norms. Some approaches even choose an individual t-norm for each rule.

Sometimes, even t-conorms are used to compute the firing degree of a rule. In this case, the corresponding rule must read as

$$\begin{aligned} R: \quad & \text{If } x_1 \text{ is } \mu_R^{(1)} \text{ or } \ldots \text{ or } x_n \text{ is } \mu_R^{(n)} \\ & \text{then } y \text{ is } \mu_R. \end{aligned}$$

In the sense of our interpretation of the rules as a piecewise definition of a fuzzy function this rule can be replaced by the following n rules.

$$R_i: \quad \text{If } x_i \text{ is } \mu_R^{(i)}$$
$$\text{then } y \text{ is } \mu_R. \quad (i = 1, \ldots, n)$$

In some commercial programs weighted rules are allowed. The resulting output fuzzy set of a rule is then multiplied using by the assigned weight. Such weights increase the number of the free parameters of a fuzzy controller. But the same effect can be achieved directly by an adequate choice of the fuzzy sets in the premise or the conclusion part of the rule without any weights. In most cases, weights make it more difficult to interpret a fuzzy controller.

The fundamental idea of the Mamdani controller as a piecewise definition of a fuzzy function requires implicitly that the premises parts of the rules represent disjoint fuzzy situations or cases. At this point we do not to formalize this concept of disjointness for fuzzy sets exactly. However, when this assumption is ignored, this can lead to an undesired behaviour of the fuzzy controller. For instance, refining the control actions cannot be achieved by merely adding rules without changing the already existing fuzzy sets. As an extreme example we consider the rule

$$\text{If } x \text{ is } I_X \text{ then } y \text{ is } I_Y,$$

where the fuzzy sets I_X and I_Y are the characteristic functions of the input and output domain, respectively, yielding membership degree 1 for any value. The rule can be read as

$$\text{If } x \text{ is anything then } y \text{ is anything.}$$

With the rule, the output fuzzy set will always be constantly 1, the characteristic function of the whole output domain. It does not depend on or change with other rules that we add to the rulebase. We will discuss this problem again, when we introduce conjunctive rule systems.

Another problem of the disjoint fuzzy cases is illustrated in figure 3.3 which shows an output fuzzy set where the defuzzification is difficult.

Should we interpolate between the two fuzzy values represented by the triangles like it would be done by the centre of gravity method? That would mean that a defuzzification yields a value whose membership degree to the output fuzzy set is 0 which surely does not fit to our intuition. Or do the two triangles represent two alternative output values from which we have to choose one? For example, the illustrated fuzzy set could be the output fuzzy set of a controller which is supposed to drive a car around an obstacle. Then

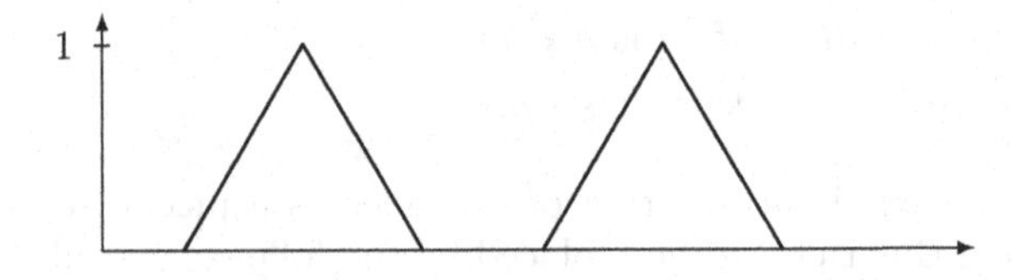

Fig. 3.3. Output fuzzy set obtained from two non-overlapping fuzzy sets

the fuzzy set says that we have to evade the obstacle either to the left or to the right, but not keep on driving straight ahead. The latter solution would be proposed by the centre of gravity method, leading to the effect that the car will bump into the obstacle. The interpretation of two alternative outputs is contradictory to the underlying philosophy of the Mamdani controller as a piecewise definition of a fuzzy function, because, in this case, the function is not well-defined, since two fuzzy values are assigned at the same time to only one input.

3.1.2 Defuzzification Methods

In recent years numerous defuzzification methods were proposed which were developed more or less intuitively on the basis of the fact that a fuzzy set, but no further information is given. But what is missing is a systematic approach which is based on a more rigorous interpretation of the fuzzy set which has to be defuzzified.

A general defuzzification strategy has to fulfil two tasks at the same time. It has turn a fuzzy set into a set and it has to choose one (fuzzy) value from a set of (fuzzy) values. It is not obvious in which order these two steps should be carried out. For example, the fuzzy set from figure 3.3 could be defuzzified by first choosing one of the two fuzzy values it represents, i.e. by choosing one of the two triangles. In the second step, the corresponding triangle, representing only a single fuzzy value must be turned into a suitable crisp value. We could also exchange the order of these two steps. Then we would first turn the fuzzy set into a crisp set that in this case would contain two elements. Afterwards we have to pick one of the values from this crisp set. These considerations are not taken into account in the axiomatic approach for defuzzification in [173], nor by most defuzzification methods which implicitly assume that the fuzzy set which is to be defuzzified represents just a single vague value and not a set of vague values.

The underlying semantics or interpretation of the fuzzy controller or fuzzy system is also essential for the choice of the defuzzification strategy. In the following section we will explain in further detail that the Mamdani controller is based on a interpolation philosophy. Other approaches do not fit into this interpolation scheme, as we will see in the section on conjunctive rule systems.

At this point we explain some other defuzzification strategies and their properties in order to better understand the issue of defuzzification.

The *mean of maxima* method (MOM) is a very simple defuzzification strategy which chooses as output value the mean values of the values with maximum membership degree to the output fuzzy set. This method is rarely applied in practice because for symmetrical fuzzy sets it leads to a discontinuous controller behaviour. Given the input values, the output value depends only on the output fuzzy set which belongs to the rule with the highest firing degree – provided that there are not two or more rule which by chance fire to the same degree and which have different fuzzy sets in their conclusion parts.

When we use symmetrical fuzzy sets in the conclusion parts of the rules in the context of the mean of maxima method, the output value will always be on of the centres of the fuzzy sets, except in the rare case when two or more rules fire with the same maximum degree. Therefore, the rule with the highest firing degree will determine the constant output until another rule takes over. Then the controller output will jump directly or with one intermediate step to the output of the rule that has now the maximum firing degree.

Both the centre of area method as well as MOM will result in the possibly non-desired mean value in the defuzzification problem shown in figure 3.3.

In [81] a method is proposed, how to avoid this effect of COA and MOM. There, they always choose the right-most value of the values with maximum membership degree to the output fuzzy set. (Alternatively one can always take the left-most one.) According to [81] the authors hold a patent on this method. But similar to MOM, it can also lead to discontinuous changes of the output value.

The centre of gravity method requires relatively high computational costs and does not have the interpolation properties we would expect. To illustrate this, we consider a Mamdani controller with the following rulebase:

If x is 'about 0' then y is 'about 0'
If x is 'about 1' then y is 'about 1'
If x is 'about 2' then y is 'about 2'
If x is 'about 3' then y is 'about 3'
If x is 'about 4' then y is 'about 4'

The terms 'about 0',..., 'about 4' are represented by fuzzy sets in the form of symmetrical triangular membership functions of width three, i.e. by $\Lambda_{-1,0,1}, \Lambda_{0,1,2}, \Lambda_{1,2,3}, \Lambda_{2,3,4}$ and $\Lambda_{3,4,5}$, respectively. It seems that these rules describe the function $y = x$, i.e a straight line. Applying the centre of gravity method, the resulting control function matches the straight line only at the values 0, 0.5, 1, 1.5,..., 3.5 and 4, at all other points the control function differs a little bit from the simple straight line as shown in figure 3.4.

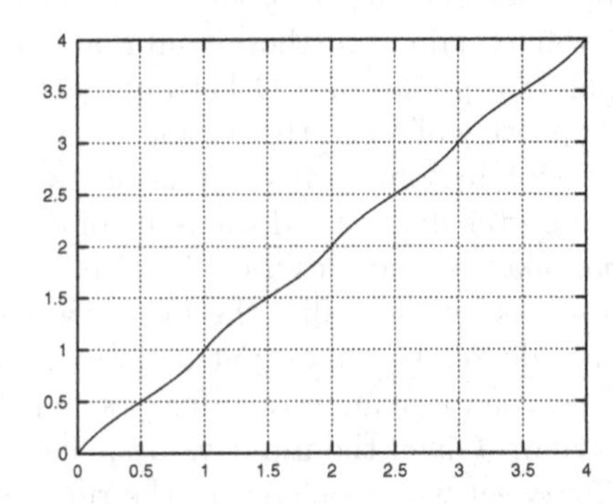

Fig. 3.4. Interpolation of a straight line using the centre of gravity method

These and other undesirable effects, which, for example, can appear using asymmetrical membership functions in the conclusion, can be avoided by using rules with a crisp value in the conclusion part. For describing the input values we still use fuzzy sets, but the outputs are specified not in terms of fuzzy sets, but as a single value for each rule. In this case, the defuzzification is also very simple. We take the weighted mean of the rule outputs, i.e. each rule assigns its firing degree as weight to its crisp output value in the conclusion part. In this way, we obtain

$$y = \frac{\sum_R \mu_{R,a_1,\ldots,a_n}^{\text{output}} \cdot y_R}{\sum_R \mu_{R,a_1,\ldots,a_n}^{\text{output}}} \tag{3.7}$$

as the output of the fuzzy controller. The rules have the form

$$R: \quad \text{If } x_1 \text{ is } \mu_R^{(1)} \text{ and } \ldots \text{ and } x_n \text{ is } \mu_R^{(n)} \text{ then } y \text{ is } y_R$$

with crisp output values y_R. $a_1, \ldots, a_n$ are the measured input values for the input variables $x_1, \ldots, x_n$ and $\mu_{R,a_1,\ldots,a_n}^{\text{output}}$ denotes as before already the firing degree of rule R for these input values.

3.2 Takagi-Sugeno-Kang Controllers

Takagi-Sugeno or Takagi-Sugeno-Kang controllers (TS or TSK models) [181, 186] use rules of the form

$$R: \text{ If } x_1 \text{ is } \mu_R^{(1)} \text{ and } \ldots \text{ and } x_n \text{ is } \mu_R^{(n)} \text{ then } y = f_R(x_1, \ldots, x_n). \tag{3.8}$$

In the same manner as in the case of the Mamdani controller (3.1) the input values in the rules are described by fuzzy sets. However, using a TSK model, the conclusion part of a single rule consists no longer of a fuzzy set, but determines a function with the input variables as arguments. The basic idea is that the corresponding function is a good local control function for the fuzzy region that is described by the premise part of the rule. For instance, if we use linear functions, the desired input/output behaviour of the controller is described locally (in fuzzy regions) by linear models. At the boundaries between single fuzzy regions, we have to interpolate in a suitable way between the corresponding local models. This is done by

$$y = \frac{\sum_R \mu_{R,a_1,\ldots,a_n} \cdot f_R(x_1, \ldots, x_n)}{\sum_R \mu_{R,a_1,\ldots,a_n}}. \tag{3.9}$$

Here $a_1, \ldots, a_n$ are the measured input values for the input variables $x_1, \ldots, x_n$, and $\mu_{R,a_1,\ldots,a_n}$ denotes the firing degree of rule R which results for these input values.

A special case of the TSK models is the variant of the Mamdani controller where the fuzzy sets in the conclusion parts of the rules are replaced by constantly values and the output value is calculated by equation (3.7). In this case, the functions f_R are constant.

In order to maintain the interpretability of a TSK controller in terms of local models f_R for fuzzy regions, a strong overlap of these regions should be avoided, since otherwise th interpolation formula (3.9) can completely blur the single models and mix them together into one complex model, that might have a good control behaviour, but looses interpretability completely. As an example we consider the following rules:

If x is 'very small' then $y = x$
If x is 'small' then $y = 1$
If x is 'large' then $y = x - 2$
If x is 'very large' then $y = 3$

First, the terms 'very small', 'small', 'large' and 'very large' are modelled by the four non-overlapping fuzzy sets in figure 3.5. In this case, the four functions or local models $y = x$, $y = 1$, $y = x - 2$ and $y = 3$ defined in the rules are reproduced exactly as shown in figure 3.5. If we choose only slightly overlapping fuzzy sets, the TSK model yields the control function in figure 3.6. Finally, figure 3.7 shows the result of the TSK model which uses fuzzy sets that overlap even more.

We can see that the TSK model can lead to slight overshoots as shown in figure 3.6, even if the fuzzy sets overlap only slightly. If we choose an overlap of the fuzzy sets typical for Mamdani controllers, the single local model are not visible anymore as figure 3.7 shows.

A suitable way to avoid this undesirable effect is to use trapezoidal membership functions instead of triangular ones, when working with TSK models. When we choose trapezoidal membership function in such a way that an overlap occurs only at the edges of the trapezoidal functions, the corresponding local models are reproduced exactly in the regions where membership degree is 1.

3.3 Logic-based Controllers

In this section we discuss the consequences resulting from an interpretation of the rules of a fuzzy controller in the sense of logical implications. In example 1.23 we have already seen how logical inference can be modelled on the basis of fuzzy relations. This concept is now applied to fuzzy control. In order to simplify the notation, we first restrict our considerations to fuzzy controllers with only one input and one output variable. The rules have the form

$$\text{If } x \text{ is } \mu \text{ then } y \text{ is } \nu.$$

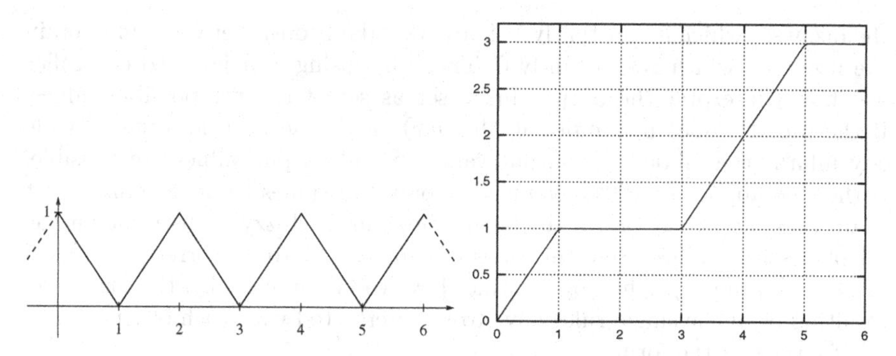

Fig. 3.5. Four non-overlapping fuzzy sets: exact reproduction of the local models

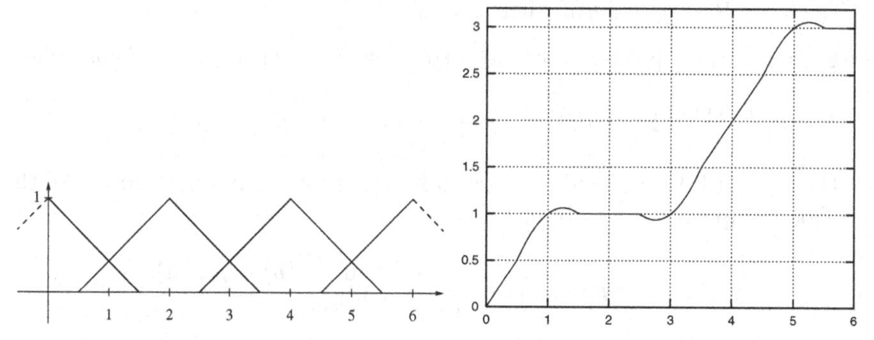

Fig. 3.6. Four slightly overlapping fuzzy sets: slight mixture of the local models

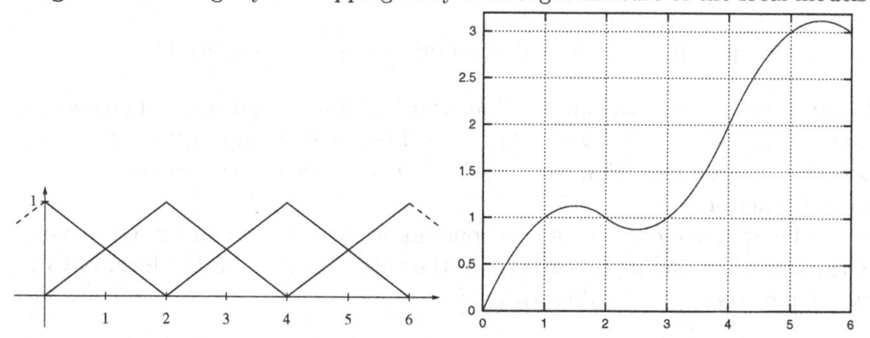

Fig. 3.7. Four strongly overlapping fuzzy sets: almost complete mix of the local models

With one single rule of this form and a given input value x we obtain a output fuzzy set according the computation scheme from example 1.23. When the input value x has a membership degree of 1 to the fuzzy set μ, the resulting output fuzzy matches exactly the fuzzy set ν, just as in the case of the Mamdani controller. However, in contrast to the Mamdani controller, the output fuzzy set becomes larger, the less the premise part of the rule is satisfied, i.e. the lower the value $\mu(x)$ is. In the extreme case $\mu(x) = 0$ we obtain as output the fuzzy set which is constantly 1. For a Mamdani controller we would obtain the fuzzy set which is constantly 0. Therefore, using a logic-based controller we should interpret the output fuzzy set as set of the yet possible values. If the premise does not apply at all ($\mu(x) = 0$), the rule does not provide any information about the output value and all output values are possible. If the rule applies to 100% ($\mu(x) = 1$), only the values from the (fuzzy) set ν are possible. Therefore, a single rule provides a (fuzzy) constraint on the set of possible values. Since all rules are considered to be correct (true), all fuzzy constraints specified by the rules have to be satisfied, i.e. the fuzzy sets resulting from the single rules have to be intersected with each other.

If r rules of the form

$$R_i\colon \quad \text{If } x \text{ is } \mu_{R_i} \text{ then } y \text{ is } \nu_{R_i}. \qquad\qquad (i = 1, \dots, r)$$

are given and the input is $x = a$, the output fuzzy set of a logic-based controller is

$$\mu_{\mathcal{R},a}^{\text{out, logic}} : Y \to [0,1], \quad y \mapsto \min_{i \in \{1,\dots,r\}} \{[\![a \in \mu_{R_i} \to y \in \nu_{R_i}]\!]\}.$$

Here, we still have to choose a truth function for the implication $\to$. With the Gödel implication we obtain

$$[\![a \in \mu_{R_i} \to y \in \nu_{R_i}]\!] = \begin{cases} \nu_{R_i}(y) & \text{if } \nu_{R_i}(y) < \mu_{R_i}(a) \\ 1 & \text{else,} \end{cases}$$

and the Łukasiewicz implication leads to

$$[\![a \in \mu_{R_i} \to y \in \nu_{R_i}]\!] = \min\{1 - \nu_{R_i}(y) + \mu_{R_i}(a), 1\}.$$

In contrast to the Gödel implication, which can lead to discontinuous output fuzzy sets, the output fuzzy sets of the Łukasiewicz implication are always continuous, provided that the involved fuzzy sets are continuous (as real-valued functions).

So far we have only considered one input variable. When we have to deal with more than one input variable in the rules, i.e. rules in the form of (3.1), we have to replace the value $\mu_{R_i}(a)$ by

$$[\![a_1 \in \mu_{R_i}^{(1)} \wedge \ldots \wedge a_n \in \mu_{R_i}^{(n)}]\!]$$

for the input vector $(a_1, \dots, a_n)$. For the conjunction we have to choose a suitable t-norm as its truth function, for example the minimum, the Łukasiewicz t-norm or the algebraic product.

In the case of the Mamdani controller, where the rules represent fuzzy points, is makes no sense to use rules of the form

$$\text{If } x_1 \text{ is } \mu_1 \text{ or } x_2 \text{ is } \mu_2 \text{ then } y \text{ is } \nu.$$

But if we use a logic-based controller, we can determine an arbitrary logic expression with predicates (fuzzy sets) over the input variables as premise, so that it is reasonable to have also rules with disjunction or negation for logic-based controllers [91]. We only have to specify suitable truth functions for the logical connectives.

We want to emphasize an essential difference between Mamdani controllers and logic-based ones. Each rule of a logic-based controller is a constraint for the control function [96], therefore the choice of very narrow fuzzy sets for the output and (strongly) overlapping fuzzy sets in the input can lead to contradictory constraints and the controller yields the empty fuzzy set (constantly 0) as output. While specifying the fuzzy sets this fact should be taken into account by preferring narrower fuzzy sets for the input variables and wider ones for the output variables.

Increasing the number of rules for the Mamdani controller leads, in general, to a more fuzzy output, because the output fuzzy set is the union of the output fuzzy sets resulting from the single rules. In the extreme case, the trivial but empty rule

If x is *anything* then y is *anything*,

where *anything* is modelled by the fuzzy set which is constantly 1, causes that the output fuzzy set to be constantly 1. This is independent of other rules occurring in the rulebase of the Mamdani controller. In a logic-based controller this true but useless rule has no effect and does not destroy the control function completely.

3.4 Mamdani Controllers and Similarity Relations

When we introduced the Mamdani controllers we have already seen that the fuzzy rules there represent fuzzy points on the graph of the control or transfer function which is described by the controller. Based on the concept of similarity relations as discussed in section 1.7, fuzzy sets, such as the ones appearing in the Mamdani controller, can be interpreted as fuzzy points. Here we discuss this interpretation of the Mamdani controller and its consequences in detail.

3.4.1 Interpretation of a Controller

At first, we consider a given Mamdani controller. Let us assume that the fuzzy sets defined for the input and output domains satisfy the conditions of theorem 1.33 or even better of theorem 1.34. In this case, we can derive similarity relations, so that the fuzzy sets can be interpreted as extensional hulls of single points.

Example 3.1 For a Mamdani controller with two input variables x and y and one output variable z we use the left fuzzy partition from figure 3.8 for the input variables and the right one for the output variable. The rulebase consists of four rules.

R_1: If x is *small* and y is *small* then z is *positive*
R_2: If x is *medium* and y is *small* then z is *null*
R_3: If x is *medium* and y is *big* then z is *zero*
R_4: If x is *big* and y is *big* then z is *negative*

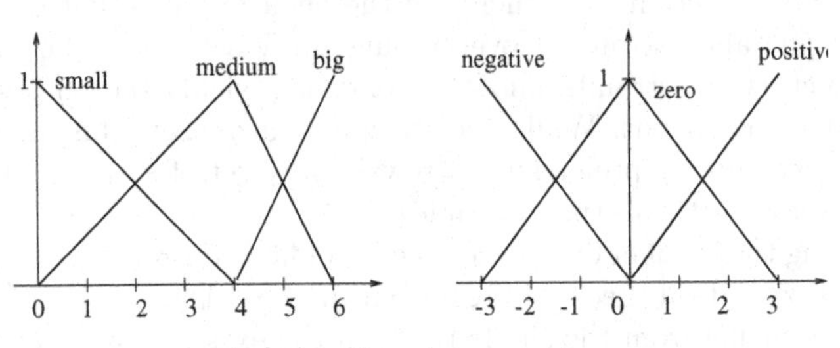

Fig. 3.8. Two fuzzy partitions

These fuzzy partitions satisfy the conditions of theorem 1.34, so that suitable scaling functions can be found. For the left fuzzy partition in figure 3.8 the scaling function is

$$c_1 : [0,6] \to [0,\infty), \qquad x \mapsto \begin{cases} 0.25 & \text{if } 0 \le x < 4 \\ 0.5 & \text{if } 4 \le x \le 6, \end{cases}$$

for the right fuzzy partition it is

$$c_2 : [-3,3] \to [0,\infty), \qquad x \mapsto \frac{1}{3}.$$

The fuzzy sets *small, medium, big, negative, zero* and *positive* correspond to the extensional hulls of the points 0, 4, 6, -3, 0 and 3, respectively, if the similarity relations induced by the above scaling functions are considered.

Then the four rules say that the graph of the function described by the controller should pass through the corresponding points (0,0,3), (4,0,0), (4,6,0) and (6,6,-3). □

The interpretation on the basis of similarity relations in the example above specifies four points on the graph of the control function as well as additional information encoded in the similarity relations. The construction of the whole control function is therefore an interpolation task. We want to find a function which passes through the given points and maps similar values to similar values in the sense of the similarity relations.

If we, for example, want to find a suitable output value for the input (1,1), we can see that (0,0) is the most similar input pair for which we know an output value, namely the value 3 according to rule R_1. The similarity degree of 1 to 0 is nothing but the membership degree of the value 1 to the extensional hull of 0, that is, to the fuzzy set *small*, which is 0.75. The input (1,1) is also similar to the input pair (4,0) in rule R_2, however much less similar than to

(0,0). The similarity degree of 1 to 4 is 0.25, the similarity of 1 to 0 is again 0.75. Thus the output value for (1,1) should be quite similar to the output value 3 for the input (0,0) (rule R_1), but also a little bit similar to the output value 0 of the input (4,0) (rule R_2).

So far we have left the question open, how the two similarity degrees, which we obtain for the two components x and y of the input pair, should be aggregated. A t-norm is a suitable choice in this case, for instance the minimum. As an example, let us determine, how well the output value 2 fits to the input (1,1). In order to answer this question, we have to compute the similarity degrees to the points defined by the four rules. Each of these similarity degrees is simply the minimum of the membership degrees of the three components 1, 1 and 2 to the three fuzzy in the corresponding rule.

In this way, for the point induced by rule R_1 we obtain a similarity degree of $2/3 = \min\{0.75, 0.75, 2/3\}$. For R_2 the result is $0.25 = \min\{0.25, 0.75, 2/3\}$. For the two rules R_3 and R_4 the similarity degree is 0, because already the considered input values do not fit to these rules at all. The similarity degree according to the four given points or rules corresponds to the best possible value, that is 2/3. Repeating the calculation for any value z in the input/output tuple $(1, 1, z)$, we obtain a function

$$\mu : [-3, 3] \to [0, 1]$$

for the given input (1,1). This function can be interpreted as a fuzzy set over the output range. When we compare this computation scheme with the calculations carried for the Mamdani controller, we obtain exactly the output fuzzy set (3.5) of the corresponding Mamdani controller.

3.4.2 Construction of a Controller

Instead of determining the scaling functions or similarity relations and the corresponding interpolation points indirectly from a Mamdani controller we can also specify them directly and then derive the corresponding Mamdani controller. The advantage of this procedure is on the one hand that we can no longer specify arbitrary fuzzy sets, but only fuzzy sets which are consistent in a certain way. And on the other hand the interpretation of the scaling functions and especially of the interpolation points that have to be specified is very simple. The scaling functions can be interpreted in the sense of example 1.31. In ranges where the control is very sensitive to a change of the value, we should distinguish very exactly between the single values. So we should choose a larger scaling factor. For ranges where the exact values are less important for the control, a small scaling factor is sufficient. This leads to very narrow fuzzy sets in ranges where very precise control actions have to be taken, while larger fuzzy sets are used in regions where a rough control is sufficient. In this way we can explain, why fuzzy sets near the operating point of a controller are usually chosen to be very narrow in contrast to the other ranges. The operating point requires very exact control actions in most cases. In contrast to this, if the process is far from the operating point, first of all rough and strong actions have to be carried out to force the process closer to its operating point.

Using scaling function it is also obvious which additional hidden assumptions are used for the design of a Mamdani controller. The fuzzy partitions are defined for the single domains and are then used in the rules. In the sense of scaling functions this means that the scaling functions for the different domains are assumed to be independent from each other. The similarity of the values in a domain does not depend on the values in other domains. In order to illustrate this issue, we consider a simple PD controller which uses as input variable the error – the difference from the reference variable – and the change of the error. For a very small error value it is obviously very important for the controller whether the change of the error is slightly positive or slightly negative. Therefore we would choose a larger scaling factor near zero in the domain of the change of error, resulting in narrow fuzzy sets. On the other hand, if the error value is large, it is not very important whether the change of error tends to be slightly positive or slightly negative. First of all we have to reduce the error. This fact speaks in favour for a small scaling factor near zero in the domain of the change of error, i.e. we should use wider fuzzy sets. In order to solve this problem, there are three possibilities:

1. We specify a similarity relation in the product space of the domains error and change of error which models the dependence described above. But this seems to be quite difficult, because the similarity relation in the product space cannot be specified using scaling functions.
2. We choose a high scaling factor near zero in the domain of the change of error. However, in this case, it might be necessary to specify many almost identical rules for the case of a large error value. The rules differ only in the fuzzy sets for the change of the error, like

   If error is *big* and change is *positive small* then y is *negative*.
   If error is *big* and change is *zero* then y is *big*.
   If error is *big* and change is *negative small* then y is *negative*.

3. We define rules that do not use all input values, for instance

   If error is *big* then y is *negative*.

The interpretation of the Mamdani controller in the sense of similarity relations also explains why it is very convenient that adjacent sets of a fuzzy partition meet each other at the level of 0.5. A fuzzy set is a vague value which will be used for the interpolation points. If a certain value has been specified, this value provides also some information about similar values, where similar should be interpreted in the sense of the corresponding similarity relation. So as long as the similarity degree does not drop to zero, there is still some information available. Therefore, once the similarity degree is zero, new information is needed, i.e. a new interpolation point must be introduced. Following this principle, the fuzzy sets will exactly overlap at the level of 0.5. Of course, the interpolation points could be chosen closer, provided sufficiently detailed knowledge about the process is available. This would lead to strongly overlapping fuzzy sets. However, this does not make sense, when we desire a representation of the expert's knowledge on the process that is as compact as possible. Therefore, new interpolation points are only introduced, when they are needed.

Even if a Mamdani controller does not satisfy the conditions of one of the theorems 1.33 or 1.34, calculate the similarity relations from theorem 1.32 will still provide important information. The corresponding similarity relations always exist and make the fuzzy sets extensional, even though they might not be interpreted as extensional hulls of points. The following results are taken from [94, 95].

1. The output of a Mamdani controller does not change if we use instead of a crisp input value its extensional hull as input.
2. The output fuzzy set of a Mamdani controller is always extensional.

This means that we cannot overcome the indistinguishability or vagueness that is inherently coded in the fuzzy partitions.

3.5 Fuzzy Control versus Classical Control

In this chapter we have seen, that the output variables of a fuzzy controller after the defuzzification depend as sharply from the input variables as the output variables of a classical (analytical) controller. Therefore, the fuzzy controller is a nonlinear characteristic surface without internal dynamics, and it can be interpreted as some kind of nonlinear state space controller. In a control loop, linear dynamical transfer elements for differentiation or integration can be added before or after such a characteristic surface.

But the same interpretation with characteristic surface and additional linear dynamical transfer elements holds for any classically designed (analytical) controller as well, so that there cannot be any difference in the behavior of a fuzzy or a classical controller. The only difference lies in the way, how both controller types are designed and described. And therefore, only about these two aspects it makes sense to discuss the advantages and disadvantages of fuzzy control compared with classical control.

Having presented these fundamental traits, let us now discuss the advantages and disadvantages of fuzzy controllers, and the range of application resulting from this. First of all it should be mentioned that the classical way of designing a controller offers certain advantages. We can carry out the construction of the plant model and the design of the controller based on it in a systematic way. Stability and eventually desired damping is guaranteed implicitly. Uncertainties and inaccuracies of the model, which arise from linearization, can be taken into account by robust controllers, especially norm-optimal controllers, which means that even in these cases stability of the controller can be assured.

In contrast to this, the design of a fuzzy controller is usually based on heuristics. This, of course, affects the time needed for construction, so that the design for complex plants may even become impossible. Furthermore, no guarantee for stability of the closed-loop system can be given. However, we should think of these aspects in relative terms, as a systematic way of controller design as well as the stability analysis have become subjects of intense research, which started in the late 80s, and some useful attempts at these fields are already available (cf. chapter 4 and 5).

Even the need for a huge computation time, which is an argument against fuzzy control, can be partly invalidated. Normally, for a fuzzy controller in every time-step all fuzzy rules have to be worked out, the output fuzzy set has to be estimated and defuzzificated. These large-scale computations normally exceed the power of available processors, especially if the time interval for the computation has to be small (because of the plant's dynamics) and the controller must be cheap. To solve this problem, for a sufficient number of input values the output values of the controller can be computed a priori and saved in a characteristic surface. In the run, we just have to take the saved values and make an interpolation to get the controller output. This controller output does not take exactly the same value as at the original controller, but the errors can be neglected if the resolution of the characteristic surface is high enough.

Another aspect to discuss is robustness. Because of their underlying fuzziness, fuzzy controllers are said to have a high degree of robustness. But, as we already mentioned above, the transfer characteristic of a fuzzy controller is just as clearly defined as the one of a conventional controller. Accordingly, there is nothing mysterious about its robustness, and it can be subject to discussion just like the robustness of the classical controller. Here, we only wish to emphasize the following: As already explained in chapter 2.7.7, using the term *robustness* is reasonable only if one can also quantify to what extent the real plant may differ from the nominal plant before the closed loop becomes instable. The unquantified attribute *robust* can be applied to almost any controller, and is therefore futile. Regarding fuzzy controllers, which are designed for a plant to which no model exists, such a quantification is impossible. And even if such a model is available, robustness can normally be demonstrated only by performing several runs of a simulation, with different values for the parameters of the plant. In this case, stability is therefore shown by giving a few selected examples, which is of course no proof of stability and robustness. On the other hand, in classical, linear control theory, methods are now available (cf. [133]) where the expected range of uncertainty can be defined for any single plant parameter. The controller which results from this method can thus guarantee closed-loop stability for any possible constellation of the plant.

Finally, we have the clearness of a fuzzy controller. Of course, a fuzzy rule is more clear and—especially for non-specialists—easier to understand than the transfer function of a PI controller or even the coefficient matrix of a state space controller. But if the plant and the controller reach a certain level of complexity, the fuzzy controller consists of up to several hundred rules, and not of only a few. In such a case, each single rule is still clear, but it is impossible to overview the complete rule base. The effect of a change of a certain input variable can only be predicted, if the complete rule base is simulated. On the other hand, at a coefficient matrix of a state space controller it can quite easily be seen even for high-order systems, if the output variables increase or decrease if one certain input variable changes.

We can therefore summarize for the useful application of a fuzzy controller: If a model of the plant exists, in differential or difference equations, and if this model can be used to design a controller by classical control methods (if necessary after some linearization), then one should also try these methods. Using a fuzzy controller might make sense if

- no model of the plant exists in the form of differential or difference equations.
- because of nonlinearities, the structure of the plant makes use of classical methods impossible
- the aims which the control should achieve are given in a vague way, as for example the demand that the automatic gear of a motor vehicle should change gears *smoothly*.
- the plant and the required control strategy are so simple that the design of a fuzzy controller takes less time than the construction of a model of the plant and the design of a classical controller.

In addition to this, it is also possible to use a fuzzy controller on a higher level instead of putting it into the closed loop as a genuine controller. For example, it can then be used to compute suitable desired values (trajectory planning), to adapt a classical controller to changing plant parameters, to supervise the entire process or to diagnose errors. There are no limits to the user's imagination. As here the fuzzy controller is not a part of the closed loop, there exist no control engineering specific problems, like, for instance, the stability problem. In these applications, the focus is usually more on the aspect of technical safety, like, for example, safety to failure, or whether the fuzzy controller actually covers all possible cases or not.

Although especially for this kind of applications the advantages of fuzzy systems take really big effect, we will not discuss them in this book. As these structures are on a higher level than the underlying control loop, they have to be application-specific, so that their representation would exceed the limits of this book. But in the different technical journals one can find examples from many fields of technology.

Let us finish this chapter with a discussion of some practical aspects. One of these aspects is how we can provide the input quantities for the fuzzy controller. If we can measure these quantities directly, it will be no problem. We can also compute simple differences without any difficulty—for example in order to obtain the error difference $\Delta e(k) = e(k) - e(k-1)$ of two following error values e. But already forming the second difference $\Delta^2 e(k) = \Delta e(k) - \Delta e(k-1) = e(k) - 2e(k-1) + e(k-2)$ can become a serious problem. This difference can be computed easily from the last measured values of the error, but normally it will not provide a useful result. The reason is that a calculation of the first difference corresponds to a high-pass filter of first order, and a calculation of the second difference even corresponds to a high-pass filter of second order. But a high-pass filter amplifies signals of high frequency, and therefore especially of measurement noise. For this reason, a second difference will represent a useful result only if the measured signal is almost free from noise.

A more promising approach is the use of an observer, as described in chapter 2.7.5 or 2.8.11. Here, no differences or differentiations are involved, only integration. An amplification of the measurement noise is therefore not possible. Unfortunately, an observer requires a fairly precise model of the plant, in the form of difference equations or differential equations. And if such a model is already available, the question arises whether a classical controller design would be more reasonable than a fuzzy one.

The same question occurs if the design and the optimization of a fuzzy controller is based on simulation—as a simulation, which permits useful conclusions, also requires a precise model of the plant.

4

Stability Analysis of Fuzzy Controllers

In principle, for the stability analysis of a fuzzy controller any method can be used which is suitable for the analysis of nonlinear systems. We presented the most well-known procedures in Chapter 2.8. None of these is better or worse than any other; which method is the best one to use depends only on the prerequisites. There, the structure of the system and the type of information describing the plant are usually the decisive points. In order to be able to come up with a reasonable decision for an individual case, we need to know exactly what the preconditions of any of those methods are.

For those methods, that are already presented in detail in chapter 2.8, in this chapter we present only the special features of these procedures which arise because of their application to fuzzy control systems. For the sake of simplicity, these methods will be presented in the same order here, by adding chapter 4.6 about norm-based stability analysis and two attempts about direct analysis in the state space (4.9). We will not describe methods which require a completely fuzzified system, i.e. methods, where the behavior is entirely given by fuzzy relational equations (cf. [22],[76],[83]), as these attempts are more of theoretical than of practical importance.

4.1 Preliminaries

4.1.1 Structure of the Closed-Loop System

Before starting with any analysis, it is important to realize the following fact: The controller consists not only of a genuine fuzzy component including fuzzification, rule base, inference mechanism and defuzzification. Providing the input and output quantities to this component also requires some computation. As an example, figure 4.1 shows a controller with the input quantity e, the output quantity u and the linear plant $G(s)$. One of the fuzzy rules could be given by, for example:

K. Michels et al.: *Fuzzy-Control: Fundamentals, Stability and Design of Fuzzy Controllers*, StudFuzz **200**, 257–307 (2006)
www.springerlink.com

$$\text{IF } e = \ldots \text{ AND } \dot{e} = \ldots \text{ AND } \int e = \ldots \text{ THEN } \dot{u} = \ldots \tag{4.1}$$

The fuzzy component therefore requires the input quantities *error* e, *derivative of the error* $\dot{e}$ and *integral of the error* $\int e$. Furthermore, it returns only the derivative $\dot{u}$ of the actuating variable u, instead of the variable itself. In addition to the fuzzy component, the controller contains a differentiation (or a discrete difference) and two integrations (or discrete summations). In order to facilitate naming, from now on we refer to the genuine fuzzy component (without internal dynamics) by the term *fuzzy controller*, and to the transfer element which contains this fuzzy controller, the differentiations and integrations by *controller*.

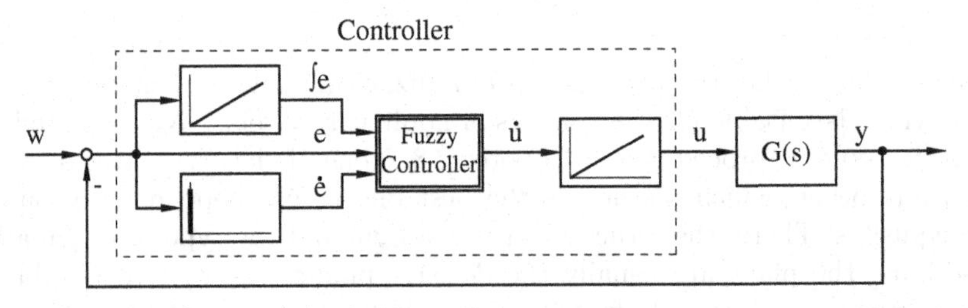

Fig. 4.1. Elements of a system with fuzzy controller

The structure of the system shown in figure 4.1 is not really suitable for stability analysis. On the one hand the segmentation does not correspond to the standard segmentation into a linear and a nonlinear part (fig. 2.79), which some methods require, on the other hand the block diagram contains a differentiation. Differentiating elements should generally not occur in block diagrams, as they interchange cause and effect. For example: We can describe the relation between acceleration and velocity by a differentiating element with velocity as its input quantity, or by an integrator with acceleration as input quantity. Now, acceleration is proportional to the driving force, and the force obviously causes the motion. In contrast to the differentiating block, the integrator "uses the cause" as input quantity and is therefore certainly the better way to represent this relation.

For this reason, we restructure the block diagram according to figure 4.2. The integrator at the output of the fuzzy controller remains unchanged. The linear plant $G(s)$ is split up into an integrator and a remaining transfer function $G'(s)$. With this segmentation we obtain the derivative of the output quantity $\dot{y}$, and accordingly the derivative of the error $\dot{e}$, without any differentiating element. Instead, we just need a subtraction of $\dot{y}$ and the derivative of the reference value $\dot{w}$. And as an analysis of stability is usually made at the set point zero, we get $\dot{w} = 0$ and can even omit this subtraction.

As another result of our restructuring—besides that we have avoided the differentiation—we have split up the circuit into a linear and a nonlinear part,

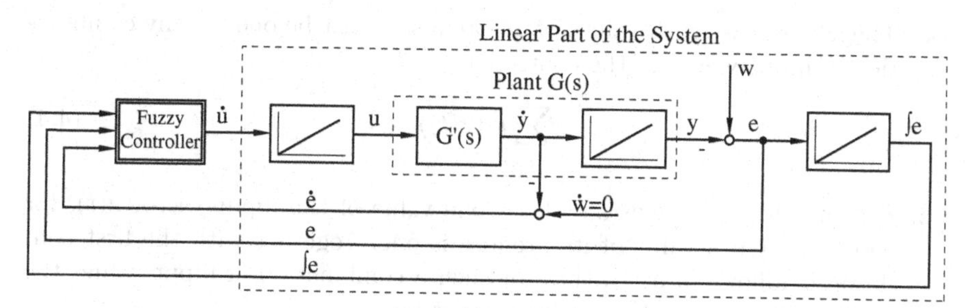

Fig. 4.2. Block diagram, restructured for stability analysis

which is a prerequisite for some analysis methods. Of course, this segmentation will be more difficult for plants which contain nonlinear parts. In such cases we can only try to subsume the nonlinear parts of the plant under the fuzzy controller. However, if this is not possible we can still use one of the other methods which do not require this segmentation.

4.1.2 Analytic Approximation of a Fuzzy Controller

In addition to restructuring of the system, some stability criteria demand that the transfer characteristic of the fuzzy controller is given in a certain, analytical form. To make the connection between the conventional and the analytical form clear, we want to develop the analytical form from the conventional form here.

The output vector $\mathbf{u} = [u_1, ..., u_m]^T$ of the controller shall be given by rules of the form

$$\text{If ... then } u_1 = \mu_{\mathbf{u}_{i,1}} \text{ and } u_2 = \mu_{\mathbf{u}_{i,2}} \text{ and ...} \tag{4.2}$$

or

$$\text{If ... then } \mathbf{u} = \mu_{\mathbf{u}_i} \tag{4.3}$$

with $\mu_{\mathbf{u}_i} = [\mu_{\mathbf{u}_{i,1}}, ..., \mu_{\mathbf{u}_{i,m}}]^T$ is a vector of fuzzy sets, and i is the number of the fuzzy rule.

The premise of each rule assigns a truth value to any possible combination of input values. With this value the rule will be activated. Therefore, each premise can be replaced equivalently by an analytical, multi-dimensional probability function. The premise of the i-th rule will be replaced by the function $k_i(\mathbf{z}) \in [0, 1]$, where $\mathbf{z}(t)$ is the vector of input variables. These input variables do not necessarily have to be state variables, but for some methods, this is required. For the input vector $\mathbf{z}(t)$, $k_i(\mathbf{z}(t))$ specifies the truth value, that the premise of rule i would have delivered, i.e. the value, with that the rule will be activated.

For all further considerations, the premises of the fuzzy controller shall meet the following two conditions:

1. The sum of the truth values of all premises must be one for any combination of input values of the controller:

$$\sum_i k_i(\mathbf{z}(t)) = 1 \tag{4.4}$$

2. For any rule there exists at least one value of the input vector $\mathbf{z}(t)$, for which the truth value of its premise is one. Together with the first condition it follows directly, that for such a combination of input values the truth values of all other premises must be zero.

These conditions are not restrictive, and they are normally met by practical fuzzy controllers. They ensure, that the fuzzy controller is defined for the entire range of input values, i.e. that for any combination of input values the output value of the controller can be computed. Besides that, the fuzzy sets in the premises must define normal fuzzy sets, i.e. they must contain at least one element with the truth value one.

After we have replaced the premises by analytical truth functions $k_i(\mathbf{z}(t))$, let us make a similar step for the conclusions and the defuzzification of the fuzzy fules. But while for the premises the analytical form was equivalent to the original form, the conclusions with defuzzification can only be approximated by an analytical function.

First, we have to estimate a suitable output vector $\mathbf{u}_i$ for each fuzzy rule, that will replace the vector of fuzzy sets $\mu_{\mathbf{u}_i}$ in the conclusion. The overall output quantity $\mathbf{u}$ of the controller can then be computed as an overlapping of these vectors $\mathbf{u}_i$ by

$$\mathbf{u} = \sum_i k_i(\mathbf{z}(t))\mathbf{u}_i \tag{4.5}$$

where $k_i(\mathbf{z}(t))$ are the truth functions of the premises.

At this formula we can see, that the vectors $\mathbf{u}_i$ have to be estimated dependent on the shape of the fuzzy sets $\mu_{\mathbf{u}_{i,1}}$ and the defuzzification method, to get a good approximation of the original controller transfer characteristic. We want to omit the relating computation steps here. They result from a comparison of the output quantity of the original controller (4.3) with the output quantity of the analytical form (4.5) for certain, relevant values of the input vector $\mathbf{z}(t)$. This way, we have approximated the original fuzzy controller by an analytical controller.

The vectors $\mathbf{z}(t)$, for that the truth function k_i takes the value one and all other truth functions the value zero, will be denoted as supporting points, so that the vector $\mathbf{u}_i$ ist the output vector of the controller at the i-th supporting point. Generally, with this definition the output vector of the controller is the weighed mean value of the output vectors at the supporting points. The weighting factors $k_i(\mathbf{z}(t))$ depend on the actual value of the input variables.

Such a fuzzy controller is therefore a genuine characteristic surface, as shown in fig. 4.3 for a fuzzy controller with two input variables z_1 and z_2 and one output variable u. In a characteristic surface, for certain supporting points the output values are given. If an input vector $(z_1, z_2)^T$ does not lie exactly on a supporting point, we have to interpolate between the neighbouring supporting points to get the output value. How this has to be performed is shown in the right drawing of the figure. We provide, that the distance between each two supporting points is one, what can be achieved easily by a suitable norming of the input quantities. Then, the output vector is computed by

$$u(\mathbf{z}) = (1-a)[(1-b)u_1 + bu_2] + a[(1-b)u_3 + bu_4] \tag{4.6}$$

with the input vector $\mathbf{z}$.

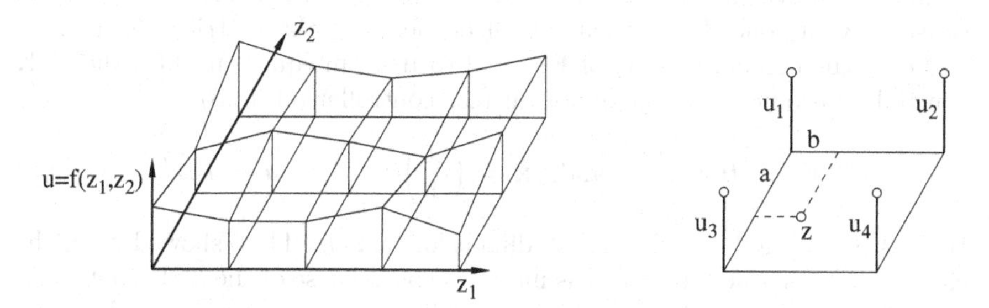

Fig. 4.3. Characteristic surface fuzzy controller

4.1.3 Takagi-Sugeno-Kang Systems

The fuzzy controller can also from the start be available as *Takagi-Sugeno-Kang controller (TSK controller)* (cf. chapter 3.2). It distinguishes itself by that the output variable of a rule is not determined by fuzzy sets $\mu_{\mathbf{u}_i}$ but by a function of arbitrary system variables $\mathbf{x}$ according to equation (3.8). But in connection with the stability analysis, here we only want to regard the TSK controllers where these functions are linear because otherwise we cannot derive any usable criteria:

$$\text{If ... then } \mathbf{u} = \mathbf{F}_i\mathbf{x} \tag{4.7}$$

$\mathbf{F}_i$ is a constant coefficient matrix. If we here also describe the premises with corresponding truth functions $k_i(\mathbf{z})$, we obtain the output variable of the TSK controller from the overlapping of the output variables of the single rules to

$$\mathbf{u} = \sum_i k_i(\mathbf{z})\mathbf{F}_i\mathbf{x}. \tag{4.8}$$

In this formula we can see that the control input variables $\mathbf{x}$ are not necessarily the same variables which determine the control premises ($\mathbf{z}$).

Here we can also determine supporting points, provided that the truth functions describe normal fuzzy sets and the condition (4.4) is satisfied. If the system is at the supporting point i, the transfer characteristic of the TSK controller is given by the purely linear relation $\mathbf{u} = \mathbf{F}_i\mathbf{x}$ which can even represent internal dynamics because the input vector $\mathbf{x}$ can also obtain previous values of $\mathbf{u}$. That was not possible for the earlier discussed characteristic surface.

So at a supporting point the TSK controller is like a linear transfer element. Therefore, a TSK controller can be understood as parallel connection of linear transfer elements where the mean results from its output values and is weighed depending on the input value. But thus the TSK controller corresponds just with the gain scheduling controller described in chapter 2.8.2.

For the TSK controller with the input variables $\mathbf{x} = [x_1, ..., x_n]^T$ we can define an additional artificial input variable x_{n+1} which always shows the constant value one. If we then set all elements of the matrices $\mathbf{F}_i$ to zero and only the column $(n+1)$ of $\mathbf{F}_i$ equal to $\mathbf{u}_i$ from equation (4.5), the TSK controller becomes a characteristic surface controller (cf. (4.5)):

$$\mathbf{F}_i = \begin{pmatrix} \mathbf{0} \; \mathbf{u}_i \end{pmatrix} \qquad \text{and} \quad \mathbf{x}' = \begin{pmatrix} \mathbf{x} \\ 1 \end{pmatrix} \qquad\qquad \mathbf{u}_i = \mathbf{F}_i\mathbf{x}' \tag{4.9}$$

Here $\mathbf{0}$ is the null matrix of the dimension $n \times n$. That showed that the characteristic surface controller is merely a special case of the TSK controller.

TSK systems can also be used for modelling given plant transfer elements. With the help of such a TSK model we can approximate any linear or non-linear transfer characteristic with or without internal dynamics with sufficient exactness, depending on the number of rules or supporting points, provided that it contains no hysteresis or delay element. The TSK model consists of several linear models, whose output variables can be overlaid with changing weight factors depending on the actual input values. For the state space model of a plant we get, for example,

$$\dot{\mathbf{x}}(t) = \sum_i k_i(\mathbf{z}(t)) \left[\mathbf{A}_i\mathbf{x}(t) + \mathbf{B}_i\mathbf{u}(t)\right] \tag{4.10}$$

In practical applications, we get such a model by chosing different supporting points as operating points, like in indirectly adaptive methods, and then carrying out a classical identification of the plant for each operating point. Thus, we get a linear model of the plant transfer characteristic at this operating or supporting point. The same is done at all of the supporting points and, thus, the TSK model of the plant is determined. We have to pay attention to the fact that we assume the same system structure for all of the operating points, that is, especially the same number of state and input variables, so that the matrices $\mathbf{A}_i$ und $\mathbf{B}_i$ show the same dimension at all of the operating points.

If a classical non-linear model of the plant already exists but because of its complexity a controller design is not possible, it can be very sensible to convert this model into a TSK model and design a TSK controller based on the TSK model. In [191] a suitable approach is presented. Before discussing this approach more detailed, we want to make some fundamental considerations.

Starting point of the considerations is a general non-linear model of the system:

$$\dot{\mathbf{x}} = \mathbf{f}(\mathbf{x}, \mathbf{u}) \tag{4.11}$$

Expanding the right-hand side for a given operating point $(\mathbf{x}_0, \mathbf{u}_0)$ into a Taylor series and stopping after the first term results in (cf. equation (2.216))

$$\dot{\mathbf{x}} = \mathbf{f}(\mathbf{x}_0, \mathbf{u}_0) + \frac{\partial \mathbf{f}}{\partial \mathbf{x}}(\mathbf{x}_0, \mathbf{u}_0)(\mathbf{x} - \mathbf{x}_0) + \frac{\partial \mathbf{f}}{\partial \mathbf{u}}(\mathbf{x}_0, \mathbf{u}_0)(\mathbf{u} - \mathbf{u}_0) \tag{4.12}$$

Obviously, this equation contains a constant part $\mathbf{f}(\mathbf{x}_0, \mathbf{u}_0)$ which is not necessarily equal to zero. In contrast, it will be probably non-zero at the most operating points if we carry out such a Taylor series expansion at several operating points.

The Taylor series expansion of a nonlinear function for an operating point will therefore normally not lead to linear partial models of the form $\dot{\mathbf{x}} = \mathbf{A}_i\mathbf{x} + \mathbf{B}_i\mathbf{u}$ but to affine partial models of the form

$$\dot{\mathbf{x}} = \mathbf{A}_i\mathbf{x} + \mathbf{B}_i\mathbf{u} + \mathbf{a}_i, \tag{4.13}$$

that is, linear models with constant parts. Unfortunately, for the controller design and also for stability analysis based on TSK models we need linear partial models.

Remedial action is taken by the already discussed approach in [191] with which the general nonlinear model can be transferred to a linear model at any operating point. The approach holds only for systems without external stimulation, but we present it here in order to give at least an idea of the necessary steps. Starting point is the general equation $\dot{\mathbf{x}} = \mathbf{f}(\mathbf{x})$ which is to be approximated for an operating point $\mathbf{x}_0$ by the linear equation $\dot{\mathbf{x}} = \mathbf{A}\mathbf{x}$. So, it has to hold in the environment of the operating point:

$$\mathbf{f}(\mathbf{x}) \approx \mathbf{A}\mathbf{x} \tag{4.14}$$

$$\mathbf{f}(\mathbf{x}_0) = \mathbf{A}\mathbf{x}_0. \tag{4.15}$$

We want to compute the coefficients of the matrix $\mathbf{A}$. For every row $\mathbf{a}_i^T$ of $\mathbf{A}$ it follows from (4.14) and (4.15):

$$f_i(\mathbf{x}) \approx \mathbf{a}_i^T\mathbf{x} \tag{4.16}$$

$$f_i(\mathbf{x}_0) = \mathbf{a}_i^T\mathbf{x}_0 \tag{4.17}$$

The Taylor series expansion of the left-hand side of equation (4.16) with a stop after the first term results in

$$f_i(\mathbf{x}_0) + \left(\frac{\partial f_i}{\partial \mathbf{x}}(\mathbf{x}_0)\right)^T(\mathbf{x} - \mathbf{x}_0) \approx \mathbf{a}_i^T\mathbf{x} \tag{4.18}$$

Inserting (4.17) into (4.18) gives

$$(\frac{\partial f_i}{\partial \mathbf{x}}(\mathbf{x}_0))^T(\mathbf{x}-\mathbf{x}_0) \approx \mathbf{a}_i^T(\mathbf{x}-\mathbf{x}_0) \tag{4.19}$$

Therefore, the coefficients of $\mathbf{a}_i$ have to be determined in a way that, firstly, $\mathbf{a}_i$ corresponds as exactly as possible to $\frac{\partial f_i}{\partial \mathbf{x}}(\mathbf{x}_0)$, i.e.

$$\frac{1}{2}\int_{-\infty}^{+\infty}(\frac{\partial f_i}{\partial \mathbf{x}}(\mathbf{x}_0)-\mathbf{a}_i)^T(\frac{\partial f_i}{\partial \mathbf{x}}(\mathbf{x}_0)-\mathbf{a}_i)dt \tag{4.20}$$

gets as small as possible, and secondly, the auxiliary condition $f_i(\mathbf{x}_0) = \mathbf{a}_i^T\mathbf{x}_0$ is satisfied. According to the theory of calculus of variations (cf. [23]) for solving the problem at first we form the Lagrange function:

$$L = \frac{1}{2}(\frac{\partial f_i}{\partial \mathbf{x}}(\mathbf{x}_0)-\mathbf{a}_i)^T(\frac{\partial f_i}{\partial \mathbf{x}}(\mathbf{x}_0)-\mathbf{a}_i) + \lambda(\mathbf{a}_i^T\mathbf{x}_0 - f_i(\mathbf{x}_0)) \tag{4.21}$$

The Euler-Lagrange equation results in

$$0 = \frac{\partial L}{\partial \mathbf{a}_i} = \mathbf{a}_i - \frac{\partial f_i}{\partial \mathbf{x}}(\mathbf{x}_0) + \lambda\mathbf{x}_0 \tag{4.22}$$

Multiplication of this equation with $\mathbf{x}_0^T$, inserting of equation (4.17) and resolving after λ leads to

$$\lambda = \frac{\mathbf{x}_0^T\frac{\partial f_i}{\partial \mathbf{x}}(\mathbf{x}_0) - f_i(\mathbf{x}_0)}{||\mathbf{x}_0||^2} \tag{4.23}$$

Let $\mathbf{x}_0 \neq \mathbf{0}$ be assumed. This term for λ is inserted in (4.22) and gives us the required calculation formula for $\mathbf{a}_i$:

$$\mathbf{a}_i = \frac{\partial f_i}{\partial \mathbf{x}}(\mathbf{x}_0) - \frac{\mathbf{x}_0^T\frac{\partial f_i}{\partial \mathbf{x}}(\mathbf{x}_0) - f_i(\mathbf{x}_0)}{||\mathbf{x}_0||^2}\mathbf{x}_0 \tag{4.24}$$

With this formula we can calculate for every operating point the rows of the matrix $\mathbf{A}$, which holds for that operating point, from the nonlinear function $\mathbf{f}(\mathbf{x})$. So we obtain a linear plant model for every operating point. These linear partial models can then be united to one TSK model of the entire plant.

We would like to emphasize particulary that TSK plant models cannot be used for a classical controller design. The overlapping of the single linear models causes errors in the higher derivatives of the models, such as negative amplification which can have disastrous effects on the controller design. In [143] such a case is illustrated.

The design of a linear state space controller with the coefficient matrix $\mathbf{F}_i$, however, is suitable for every linear partial plant model $(\mathbf{A}_i, \mathbf{B}_i)$. The single controllers are then combined to a TSK controller which shows the same premises (supporting points) as the TSK plant model. We get a TSK controller according to equation (4.8). But while there the vector $\mathbf{x}$ was a vector of arbitrary system variables, it is now the state vector. How far a controller designed in this way can really stabilize the system is discussed in chapter 4.2.2.

Inserting equation (4.8) of the TSK controller into (4.10) results in the state equation of a closed-loop system, which is not externally stimulated, in the form of a TSK model

$$\dot{\mathbf{x}}(t) = \sum_i \sum_j k_i(\mathbf{z}(t)) k_j(\mathbf{z}(t)) \left[\mathbf{A}_i + \mathbf{B}_i \mathbf{F}_j\right] \mathbf{x}(t) \tag{4.25}$$

$$\dot{\mathbf{x}}(t) = \sum_i \sum_j k_i(\mathbf{z}(t)) k_j(\mathbf{z}(t)) \mathbf{G}_{ij} \mathbf{x}(t) \tag{4.26}$$

with $\mathbf{G}_{ij} = \mathbf{A}_i + \mathbf{B}_i \mathbf{F}_j$. After reindexing and uniting with $\mathbf{A}_{g,l} = \mathbf{G}_{ij}$ and $k_l(\mathbf{z}(t)) = k_i(\mathbf{z}(t)) k_j(\mathbf{z}(t))$ we obtain

$$\dot{\mathbf{x}}(t) = \sum_l k_l(\mathbf{z}(t)) \mathbf{A}_{g,l} \mathbf{x}(t). \tag{4.27}$$

The index g at the system matrix shall illustrate that it is a model of a closed-loop system and not a model of the plant.

Taking into consideration an external stimulation results in the general form of the TSK model of a closed-loop with external stimulation (cf. [189]):

$$\dot{\mathbf{x}}(t) = \sum_i k_i(\mathbf{z}(t)) \left[\mathbf{A}_{g,i} \mathbf{x}(t) + \mathbf{B}_{g,i} \mathbf{w}(t)\right] \tag{4.28}$$

Because, in general, fuzzy controllers are realized with a microprocessor, we might also need the discrete form of the state equation:

$$\mathbf{x}(k+1) = \sum_i k_i(\mathbf{z}(k)) \left[\mathbf{A}_{g,i} \mathbf{x}(k) + \mathbf{B}_{g,i} \mathbf{w}(k)\right] \tag{4.29}$$

$\mathbf{x}(k)$ is the state vector of the system and $\mathbf{w}(k)$ the stimulation at time $t = kT$, where T is the sampling interval of the system. We have to pay attention to the fact that the matrices $\mathbf{A}_{g,i}$ and $\mathbf{B}_{g,i}$ are not identical to the ones in equation (4.28).

The TSK controller as a state space controller causes the question how to provide the state variables as input variables of the controller because in many cases the state variables cannot be measured directly. Like for classical state space controllers an observer is necessary. And because as well the plant model as the controller exist in the form of TSK systems it is sensible that the observer is also a TSK system.

It is obvious to choose the same supporting points for the observer like for plant model and controller. For every supporting point, at first, a linear observer is defined, according to figure 2.53 definiert:

$$\dot{\hat{\mathbf{x}}}(t) = \mathbf{A}_i \hat{\mathbf{x}}(t) + \mathbf{B}_i \mathbf{u}(t) + \mathbf{H}_i (\mathbf{y}(t) - \hat{\mathbf{y}}(t)) \tag{4.30}$$

$\mathbf{A}_i$ and $\mathbf{B}_i$ correspond to the matrices of the plant model of equation (4.10). $\hat{\mathbf{x}}(t)$ is the estimate of the state vector, and $\hat{\mathbf{y}}(t)$ the output vector of the observer which is the estimate of the real output vector $\mathbf{y}(t)$. Finally, $\mathbf{H}_i$ is the correction matrix with which the difference between real and assessed output vector is multipicated and fed back to the observer.

Then the single observers are united to one TSK observer:

$$\dot{\hat{\mathbf{x}}}(t) = \sum_i k_i(\mathbf{z}(t)) \left[\mathbf{A}_i\hat{\mathbf{x}}(t) + \mathbf{B}_i\mathbf{u}(t) + \mathbf{H}_i(\mathbf{y}(t) - \hat{\mathbf{y}}(t))\right] \tag{4.31}$$

Special attention should be paid to the input variables of the premises $\mathbf{z}$ of such an observer. Since if state variables are used as input variables of the premises, that can only be the estimates of the state variables. In all methods where it is about TSK observers we have to pay special attention to this fact and deal with it in the appropriate way.

4.2 Lyapunov's Direct Method

After these preliminary remarks, we are now set to turn to a discussion of the different methods for the stability analysis of fuzzy controllers. We start with the direct method of Lyapunov. As the prerequisite for this method the state equations of the plant have to be known, and the input variables of the fuzzy controller have to be state variables. However, we do not require that the system can be split up into a linear and a nonlinear part. This method can also be used for MIMO systems.

4.2.1 Application to Conventional Fuzzy Controllers

First of all, let us assume that we are looking at a fuzzy controller which contains a fuzzifying and a defuzzifying component as well as a rule base. Its transfer characteristic is therefore already determined, but it will be very difficult or even impossible to express it in an analytic form. Let us also assume that the reference value $\mathbf{w}$ of the closed circuit is zero. We can achieve this by redefining the system according to figure 2.79. We can then use $\mathbf{u} = \mathbf{r}(\mathbf{x})$, an undetermined function, as an initial approximation of the fuzzy controller, where $\mathbf{x}$ represents the plant's state vector. In contrast to this, we must know the differential equations describing the plant: $\dot{\mathbf{x}} = \mathbf{f}(\mathbf{x}, \mathbf{u})$. The state equation of the closed loop is then given by:

$$\dot{\mathbf{x}} = \mathbf{f}(\mathbf{x}, \mathbf{r}(\mathbf{x})) \tag{4.32}$$

Next, we have to find a positive definite Lyapunov function $V(\mathbf{x})$ such that its derivative $\dot{V}(\mathbf{x})$ is negative definite for all vectors $\mathbf{x}$ inside the region of the state space we are interested in. If such a function exists, then the position of rest under consideration is asymptotically stable according to theorem 2.23, and a part of the region, which is bounded by a closed contour line of V, is its domain of attraction.

The most simple approach is to fix a positive definite matrix $\mathbf{P}$ and define the Lyapunov function by $V = \mathbf{x}^T\mathbf{P}\mathbf{x}$. Since $\mathbf{P}$ is positive definite, the function can be guaranteed to be positive definite, too. Using (4.32), the derivative of this function is

$$\begin{aligned}\dot{V}(\mathbf{x}) &= \dot{\mathbf{x}}^T\mathbf{P}\mathbf{x} + \mathbf{x}^T\mathbf{P}\dot{\mathbf{x}} \\ &= \mathbf{f}^T(\mathbf{x}, \mathbf{r}(\mathbf{x}))\mathbf{P}\mathbf{x} + \mathbf{x}^T\mathbf{P}\mathbf{f}(\mathbf{x}, \mathbf{r}(\mathbf{x}))\end{aligned} \tag{4.33}$$

We therefore obtain the stability condition

$$\mathbf{f}^T(\mathbf{x}, \mathbf{r}(\mathbf{x}))\mathbf{P}\mathbf{x} + \mathbf{x}^T\mathbf{P}\mathbf{f}(\mathbf{x}, \mathbf{r}(\mathbf{x})) < 0 \tag{4.34}$$

This is a stability condition for the unknown function $\mathbf{r}(\mathbf{x})$, i.e. for the transfer characteristic of the fuzzy controller. All we still have to do is to check whether or not this transfer characteristic $\mathbf{r}(\mathbf{x})$ meets this condition for all vectors $\mathbf{x}$ of the state space region we are interested in. As this is not possible analytically, we have to check this condition of stability in a numerical way, i.e. we have to determine the output vector $\mathbf{r}(\mathbf{x})$ of the fuzzy controller—for a sufficiently large number of vectors $\mathbf{x}$—and check whether it meets the inequality (4.34). In very simple cases, it might even be possible to resolve this inequality for $\mathbf{r}(\mathbf{x})$. If we replace the $<$ by $=$ in the stability condition, we obtain a response characteristic at the border of stability, which we can directly compare to the transfer characteristic of the fuzzy controller ([29]).

If inequality (4.34) is not fulfilled for one or more state vectors, we would not be able to make any statement concerning the stability of the system. We would be confronted with the question whether the Lyapunov function, especially the matrix $\mathbf{P}$, was chosen disadvantageously or whether the system is indeed instable. In order to avoid such a situation, we are interested in methods, which help us to choose a suitable Lyapunov function or the matrix $\mathbf{P}$.

The most well-known method which does this is the one originated by Aisermann. As a prerequisite for this method it has to be possible to split up the state equation (4.32) of the entire system up into a linear, stable part, which should be as large as possible, and a nonlinear component:

$$\dot{\mathbf{x}} = \mathbf{A}\mathbf{x} + \mathbf{n}(\mathbf{x}) \tag{4.35}$$

We then obtain a positive definite matrix $\mathbf{P}$ from the Lyapunov equality for the linear part (cf. theorem D.1):

$$\mathbf{A}^T\mathbf{P} + \mathbf{P}\mathbf{A} = -\mathbf{I} \tag{4.36}$$

We use this matrix to define the Lyapunov function $V = \mathbf{x}^T\mathbf{P}\mathbf{x}$. The probability that we can prove the system's stability with this function is the higher (if the system is not unstable), the smaller the nonlinear part $\mathbf{n}(\mathbf{x})$ is. An example which illustrates this method can be found in [20].

4.2.2 Application to Takagi-Sugeno-Kang Controllers

Stability Criteria

Another procedure has to be used if the fuzzy controller is a TSK controller or at least understood as a TSK controller. As we have already explained in chapter 4.1, the other parts of the system, and therefore the entire system, can be approximated by a TSK model. So we can base the stability analysis on the TSK model of the closed loop (4.28) or its discrete version (4.29).

We start with the discrete version. The following theorem holds (cf. [189]):

Theorem 4.1 *Let a discrete system in the form*

$$\mathbf{x}(k+1) = \sum_i k_i(\mathbf{z}(k))\mathbf{A}_i\mathbf{x}(k). \tag{4.37}$$

be given. This system has a globally asymptotically stable position of rest $\mathbf{x} = \mathbf{0}$ *if a common positive definite matrix* $\mathbf{P}$ *exists for all subsystems* $\mathbf{A}_i$ *so that*

$$\mathbf{M}_i = \mathbf{A}_i^T\mathbf{P}\mathbf{A}_i - \mathbf{P} \tag{4.38}$$

is negative definite ($\mathbf{M}_i < 0$*) for all* i*.*

In order to prove this, we assume that a positive definite matrix $\mathbf{P}$ exists to define a Lyapunov function $V = \mathbf{x}^T(k)\mathbf{P}\mathbf{x}(k)$. Then we get:

$$\begin{aligned}
\Delta V(\mathbf{x}(k)) &= V(\mathbf{x}(k+1)) - V(\mathbf{x}(k)) \\
&= \mathbf{x}^T(k+1)\mathbf{P}\mathbf{x}(k+1) - \mathbf{x}^T(k)\mathbf{P}\mathbf{x}(k) \\
&= \left(\sum_i k_i\mathbf{A}_i\mathbf{x}(k)\right)^T \mathbf{P}\left(\sum_j k_j\mathbf{A}_j\mathbf{x}(k)\right) - \mathbf{x}^T(k)\mathbf{P}\mathbf{x}(k) \\
&= \mathbf{x}^T(k)\left[\left(\sum_i k_i\mathbf{A}_i^T\right)\mathbf{P}\left(\sum_j k_j\mathbf{A}_j\right) - \mathbf{P}\right]\mathbf{x}(k) \\
&= \sum_{i,j} k_ik_j\mathbf{x}^T(k)(\mathbf{A}_i^T\mathbf{P}\mathbf{A}_j - \mathbf{P})\mathbf{x}(k) \\
&= \sum_i k_i^2\mathbf{x}^T(k)(\mathbf{A}_i^T\mathbf{P}\mathbf{A}_i - \mathbf{P})\mathbf{x}(k) \\
&\quad + \sum_{i<j} k_ik_j\mathbf{x}^T(k)(\mathbf{A}_i^T\mathbf{P}\mathbf{A}_j + \mathbf{A}_j^T\mathbf{P}\mathbf{A}_i - 2\mathbf{P})\mathbf{x}(k) < 0
\end{aligned} \tag{4.39}$$

According to the prerequisites of the theorem, the matrices of the first sum are negative definite, so every single summand is certainly negative. The matrices of the second sum can be rewritten as follows:

$$\begin{aligned}\mathbf{A}_i^T\mathbf{P}\mathbf{A}_j + \mathbf{A}_j^T\mathbf{P}\mathbf{A}_i - 2\mathbf{P} &= -(\mathbf{A}_i - \mathbf{A}_j)^T\mathbf{P}(\mathbf{A}_i - \mathbf{A}_j)\\ &\quad + \mathbf{A}_i^T\mathbf{P}\mathbf{A}_i + \mathbf{A}_j^T\mathbf{P}\mathbf{A}_j - 2\mathbf{P}\\ &= -(\mathbf{A}_i - \mathbf{A}_j)^T\mathbf{P}(\mathbf{A}_i - \mathbf{A}_j)\\ &\quad + (\mathbf{A}_i^T\mathbf{P}\mathbf{A}_i - \mathbf{P}) + (\mathbf{A}_j^T\mathbf{P}\mathbf{A}_j - \mathbf{P}) \qquad (4.40)\end{aligned}$$

The first summand is negative definite because $\mathbf{P}$ is positive definite, while for the other two summands this holds because of the prerequisite of the theorem. Therefore, all of the matrices of the second sum of equation (4.39) are also negative definite and so all of the summands are less than zero. The derivative or difference of the Lyapunov function is for sure negative definite. From this follows the stability of the system.

For continuous TSK systems according to (4.28) the conditions are even more simple. Here, the theorem of stability is:

Theorem 4.2 *Let a continuous system of the form*

$$\dot{\mathbf{x}} = \sum_i k_i(\mathbf{z}(t))\mathbf{A}_i\mathbf{x}(t). \qquad (4.41)$$

be given. This system has a globally asymptotically stable position of rest $\mathbf{x} = \mathbf{0}$ *if for all subsystems* $\mathbf{A}_i$ *there exists one common positive definite matrix* $\mathbf{P}$*, so that*

$$\mathbf{M}_i = \mathbf{A}_i^T\mathbf{P} + \mathbf{P}\mathbf{A}_i \qquad (4.42)$$

is negative definite ($\mathbf{M}_i < 0$ *) for all i.*

For the proof, we choose again a Lyapunov function $V = \mathbf{x}^T\mathbf{P}\mathbf{x}$ with positive definite matrix $\mathbf{P}$. For the derivative of this function we get with respect to time:

$$\begin{aligned}\dot{V} &= \dot{\mathbf{x}}^T\mathbf{P}\mathbf{x} + \mathbf{x}^T\mathbf{P}\dot{\mathbf{x}}\\ &= \sum_i k_i\mathbf{x}^T\mathbf{A}_i^T\mathbf{P}\mathbf{x} + \sum_i k_i\mathbf{x}^T\mathbf{P}\mathbf{A}_i\mathbf{x}\\ &= \sum_i k_i\mathbf{x}^T(\mathbf{A}_i^T\mathbf{P} + \mathbf{P}\mathbf{A}_i)\mathbf{x} < 0 \qquad (4.43)\end{aligned}$$

According to the prerequisites of the theorem, all matrices of the sum are negative definite and, therefore, every single summand less than zero. So the system is stable.

Both of the theorems can be directly used for the stability analysis of a TSK system. The analysis gets exceptionally simple if we formulate the question of the existence of the matrix $\mathbf{P}$ as an LMI problem (linear matrix inequality). In the appendix G we describe in detail how to transform a system of inequations

$$\mathbf{M}_i = \mathbf{A}_i^T\mathbf{P} + \mathbf{P}\mathbf{A}_i < \mathbf{0} \qquad (4.44)$$

into the form (G.1). Then we can apply an LMI algorithm which estimates the existence of a solution $\mathbf{P}$ and therefore also answers the question of the stability of the system. The LMI algorithm can even calculate a solution for $\mathbf{P}$ which is necessary for the LMI controller design. The corresponding steps will be explained later.

Reduction of the Stability Conditions

We have to pay attention to the fact, that the negative definiteness of all $\mathbf{M}_i$ is a sufficient but not a necessary criterion for the stability of the system. If all $\mathbf{M}_i$ are negative definite, this causes that every summand of equation (4.39) or (4.43) is for any $\mathbf{x}$ negative although we need only the total sum to be negative in order to guarantee stability. Single summands could actually be positive without causing an instable system. So a significant simplification or reduction of the stability conditions can be achieved if the coefficients k_i and their dependence of the input vector $\mathbf{z}$ of the equations (4.39) or (4.43) are taken into consideration for a stability criterion. But such approaches until now only exist for TSK control loops (cf. (4.26)) and will be presented in chapter 4.2.2.

Another possibility is to check the inequation (4.39) or (4.43) directly in a numerical way for a given matrix $\mathbf{P}$ and an adequately great number of vectors $\mathbf{x}$ and $\mathbf{z}$. This approach has, in contrast to the solution with LMI algorithms, the significant disadvantage that it checks only the stability for a single given matrix $\mathbf{P}$. If the result is negative, a difficult, unstructured and maybe unsuccessful search for a suitable matrix $\mathbf{P}$ is necessary in order to prove the stability of the system. And worst of all, we do not know whether such a matrix does exist.

In contrast to this, an LMI algorithm does answer to just this fundamental question of the existence of a matrix $\mathbf{P}$ that satisfies the inequation system (4.44). Therefore, the method with the LMI algorithm should be preferred in any case, although there will still be systems where the stability cannot be proved with this method. In this chapter, from now on we always assume the use of an LMI algorithm. Just at the end of the chapter, we will present other approaches.

In the discussed matter, we can also use Aisermann's method for reducing the stability conditions (4.38) and (4.42) (cf. chapter 4.2.1 and [218]). Since the fundamental approach is only slightly variated by this method, it leads also to linear matrix inequalities, and the solution can be achieved with an LMI algorithm. Below we demonstrate the method for continuous models. For that, in

$$\dot{\mathbf{x}}(t) = \sum_i k_i(\mathbf{z}(t))\mathbf{A}_i\mathbf{x}(t) \tag{4.45}$$

we have to split every subsystem $\mathbf{A}_i$ into a common stable part $\mathbf{A}$ and a residue $\Delta\mathbf{A}_i$ so that we obtain the new system representation:

$$\dot{\mathbf{x}} = \left[\mathbf{A} + \sum_i k_i \Delta\mathbf{A}_i\right]\mathbf{x} \tag{4.46}$$

Then we use the positive definite solution $\mathbf{P}$ of the Lyapunov equation (cf. theorem D.1)

$$\mathbf{PA} + \mathbf{A}^T\mathbf{P} = -\mathbf{I} \tag{4.47}$$

in order to form the positive definite Lyapunov function $V = \mathbf{x}^T\mathbf{Px}$ with the derivative

$$\begin{aligned} \dot{V} &= \dot{\mathbf{x}}^T\mathbf{Px} + \mathbf{x}^T\mathbf{P}\dot{\mathbf{x}} \\ &= \mathbf{x}^T \left[\mathbf{A} + \sum_i k_i \Delta\mathbf{A}_i\right]^T \mathbf{Px} + \mathbf{x}^T\mathbf{P}\left[\mathbf{A} + \sum_i k_i \Delta\mathbf{A}_i\right]\mathbf{x} \\ &= \mathbf{x}^T \left[\mathbf{A}^T\mathbf{P} + (\sum_i k_i \Delta\mathbf{A}_i)^T\mathbf{P} + \mathbf{PA} + \mathbf{P}(\sum_i k_i \Delta\mathbf{A}_i)\right]\mathbf{x} \end{aligned} \tag{4.48}$$

Using (4.47) from this we obtain the stability condition

$$\dot{V} = \mathbf{x}^T \left[-\mathbf{I} + \sum_i k_i(\mathbf{z})(\Delta\mathbf{A}_i^T\mathbf{P} + \mathbf{P}\Delta\mathbf{A}_i)\right]\mathbf{x} < 0 \tag{4.49}$$

and taking $\sum\limits_i k_i(\mathbf{z}) = 1$ into consideration we get

$$\dot{V} = \mathbf{x}^T \left[\sum_i k_i(\mathbf{z})(\Delta\mathbf{A}_i^T\mathbf{P} + \mathbf{P}\Delta\mathbf{A}_i - \mathbf{I})\right]\mathbf{x} < 0 \tag{4.50}$$

Thus, we can replace the condition (4.42) by

$$\Delta\mathbf{A}_i^T\mathbf{P} + \mathbf{P}\Delta\mathbf{A}_i - \mathbf{I} < \mathbf{0} \tag{4.51}$$

which gives us less conservative results for certain system structures, for example, if the system contains only slight nonlinearities.

Robustness

With the help of the theorems above we can not only check the stability but also the robustness of a system (cf. [28]). Starting point of the considerations is the very general, time-discrete representation of a system with time-varying system parameters and external disturbance and input values

$$\begin{aligned} \mathbf{x}(k+1) &= \sum_i k_i \left[\tilde{\mathbf{A}}_{1i}\mathbf{x}(k) + \tilde{\mathbf{B}}_{1i}\mathbf{v}(k) + \tilde{\mathbf{B}}_{2i}\mathbf{u}(k)\right] \\ \mathbf{y}(k) &= \sum_i k_i \left[\tilde{\mathbf{C}}_{1i}\mathbf{x}(k) + \tilde{\mathbf{D}}_{1i}\mathbf{v}(k) + \tilde{\mathbf{D}}_{2i}\mathbf{u}(k)\right] \end{aligned} \tag{4.52}$$

with the time-discrete state vector $\mathbf{x}$, the vector of the input values $\mathbf{u}$, an external disturbance vector $\mathbf{v}$ and the output vector $\mathbf{y}$.

$$\tilde{\mathbf{A}}_{1i} = \mathbf{A}_{1i} + \Delta\mathbf{A}_{1i}(k) \qquad \tilde{\mathbf{B}}_{1i} = \mathbf{B}_{1i} + \Delta\mathbf{B}_{1i}(k) \qquad \text{etc.} \tag{4.53}$$

are the system matrices extended by a time-variable part. The time-variable parts are defined by

$$\begin{pmatrix} \Delta\mathbf{A}_{1i}(k) & \Delta\mathbf{B}_{1i}(k) & \Delta\mathbf{B}_{2i}(k) \\ \Delta\mathbf{C}_{1i}(k) & \Delta\mathbf{D}_{1i}(k) & \Delta\mathbf{D}_{2i}(k) \end{pmatrix} = \begin{pmatrix} \mathbf{E}_{1i} \\ \mathbf{E}_{2i} \end{pmatrix} \mathbf{F}_i(k) \begin{pmatrix} \mathbf{H}_{1i} & \mathbf{H}_{2i} & \mathbf{H}_{3i} \end{pmatrix} \tag{4.54}$$

$\mathbf{E}_{1i}, \mathbf{E}_{2i}$ as well as $\mathbf{H}_{1i}, \mathbf{H}_{2i}, \mathbf{H}_{3i}$ are constant matrices of appropriate dimensions, and the matrix $\mathbf{F}_i(k)$ contains the time-variableness. All of these matrices are arbitrary, merely for $\mathbf{F}_i(k)$ the condition

$$\mathbf{F}_i^T(k)\mathbf{F}_i(k) \leq \mathbf{I} \tag{4.55}$$

has to be satisfied. To be sensible we choose for $\mathbf{F}_i(k)$ a diagonal matrix whose single main diagonal elements lie between -1 and 1. At first, the equations (4.54) and (4.55) seem to be very restrictive regarding the permitted parameter changes. But if we reduce the system (4.52) to a system without external input and disturbance (all matrices are $\mathbf{0}$ except of $\tilde{\mathbf{A}}_{1i}$), so (4.54) reduces to

$$\Delta\mathbf{A}_{1i}(k) = \mathbf{E}_{1i}\mathbf{F}_i(k)\mathbf{H}_{1i} \tag{4.56}$$

Obviously, any parameter change $\Delta\mathbf{A}_{1i}(k)$ can be represented in detail by the coefficients of $\mathbf{H}_{1i}$ if we choose $\mathbf{E}_{1i} = \mathbf{I}$ and

$$\mathbf{F}_i(k) = \begin{pmatrix} f_1 & 0 \\ 0 & f_2 \end{pmatrix} \qquad \text{with} \qquad -1 < f_1(t), f_2(t) < 1. \tag{4.57}$$

For a system of the type (4.52) in [28] the conditions are given, under which the stability of the system is guaranteed despite varying parameters, and under which the H_∞ norm of its disturbance transfer function is less than a choosen bound γ. The conditions are summarized in matrix inequalities so that we can use again an LMI algorithm in order to prove stability. Since these inequalities are very large, we omit them here.

Instead of this, we just outline how to prove it. Starting point of the proof is the H_∞ norm of the disturbance transfer function. According to equation (C.15) for this norm holds:

$$||\mathbf{G}||_{\infty,dis} = \sup_{\mathbf{v}\neq 0} \frac{||\mathbf{y}||_2}{||\mathbf{v}||_2} \tag{4.58}$$

So that this norm is less than γ, for any disturbance signal $\mathbf{v}$ and the resulting output signal $\mathbf{y}$ the inequation

$$||\mathbf{y}||_2 < \gamma||\mathbf{v}||_2 \tag{4.59}$$

has to be satisfied. So we can say that the disturbance multiplicated with γ is the upper boundary for the resulting output signal. In the discussed time-discrete case we obtain (cf. equation (C.7) with $p = 2$)

$$\sqrt{\sum_{k=0}^{N-1} \mathbf{y}^T(k)\mathbf{y}(k)} < \gamma \sqrt{\sum_{k=0}^{N-1} \mathbf{v}^T(k)\mathbf{v}(k)}$$

$$\sum_{k=0}^{N-1} \mathbf{y}^T(k)\mathbf{y}(k) < \gamma^2 \sum_{k=0}^{N-1} \mathbf{v}^T(k)\mathbf{v}(k)$$

$$\sum_{k=0}^{N-1} \left[\mathbf{y}^T(k)\mathbf{y}(k) - \gamma^2 \mathbf{v}^T(k)\mathbf{v}(k)\right] < 0. \tag{4.60}$$

If inequation (4.60) is satisfied, then the H_∞ norm of the disturbance transfer function is less than γ. A simple additive extension of the inequation leads to

$$\sum_{k=0}^{N-1} [\mathbf{y}^T(k)\mathbf{y}(k) + \mathbf{x}^T(k+1)\mathbf{P}\mathbf{x}(k+1) - \mathbf{x}^T(k)\mathbf{P}\mathbf{x}(k)$$
$$-\gamma^2 \mathbf{v}^T(k)\mathbf{v}(k)] - \mathbf{x}^T(N)\mathbf{P}\mathbf{x}(N) < 0 \quad \text{with} \quad \mathbf{x}(0) = \mathbf{0}. \tag{4.61}$$

Without any restriction of the universal validity we assume $\mathbf{x}(0) = \mathbf{0}$. Inserting (4.52) into (4.61) and appropriate combining of certain quantities leads to the condition

$$\sum_{k=0}^{N-1} \sum_i \sum_j k_i k_j \bar{\mathbf{x}}^T(k)[\bar{\mathbf{G}}_{ij}^T \mathbf{P} \bar{\mathbf{G}}_{ij} + \bar{\mathbf{C}}_{ij}^T \bar{\mathbf{C}}_{ij} - \bar{\mathbf{P}}]\bar{\mathbf{x}}^T(k) - \mathbf{x}^T(N)\mathbf{P}\mathbf{x}(N)$$
$$< 0 \tag{4.62}$$

with the state vector $\bar{\mathbf{x}}^T(k)$, extended by the disturbances, at time k. $\bar{\mathbf{G}}_{ij}$ is an extended system matrix which contains the original system matrix, the control feedback (cf. (4.26)) and all corresponding parameter uncertainties. Analogously, $\bar{\mathbf{C}}_{ij}$ is an extended output matrix. And finally, $\bar{\mathbf{P}}$ is defined by

$$\bar{\mathbf{P}} = \begin{pmatrix} \mathbf{P} & \mathbf{0} \\ \mathbf{0} & \gamma \mathbf{I} \end{pmatrix} \tag{4.63}$$

and, therefore, it is merely a matrix $\mathbf{P}$ extended on the main diagonal.

(4.62) is definitely satisfied and the H_∞ norm of the disturbance transfer function less than γ, if all

$$\bar{\mathbf{G}}_{ij}^T \mathbf{P} \bar{\mathbf{G}}_{ij} + \bar{\mathbf{C}}_{ij}^T \bar{\mathbf{C}}_{ij} - \bar{\mathbf{P}} < \mathbf{0} \tag{4.64}$$

are negative definite. A by-product of this proof is that all $\bar{\mathbf{G}}_{ij}^T \mathbf{P} \bar{\mathbf{G}}_{ij} - \bar{\mathbf{P}}$ are also negative definite and the system is stable with varying parameters according to theorem 4.1. Some further transformations and the consideration of (4.55) lead to the stability conditions in form of an LMI system that is not shown here because the inequalities are very extensive.

Systems with variable Delay Time

In [27] it is shown that the theorems 4.1 and 4.2 can even be extended for systems with variable delay time. For continuous systems, the following theorem holds:

Theorem 4.3 *Let a continuous system of the form*

$$\dot{\mathbf{x}} = \sum_i k_i(\mathbf{z}(t)) \left[\mathbf{A}_{1i}\mathbf{x}(t) + \mathbf{A}_{2i}\mathbf{x}(t - \tau(t))\right] \tag{4.65}$$

be given with variable delay time τ which is bounded by $|\dot{\tau}(t)| \leq \beta < 1$. This system has a globally asymptotically stable position of rest $\mathbf{x} = \mathbf{0}$, if common positive definite matrices $\mathbf{P}$ and $\mathbf{S}$ exist for all subsystems $(\mathbf{A}_{i1}, \mathbf{A}_{i2})$, so that the following matrix inequality is satisfied:

$$\mathbf{A}_{1i}^T\mathbf{P} + \mathbf{P}\mathbf{A}_{1i} + \mathbf{P}\mathbf{A}_{2i}\mathbf{S}^{-1}\mathbf{A}_{2i}^T\mathbf{P} + \frac{1}{1-\beta}\mathbf{S} < \mathbf{0} \tag{4.66}$$

that is, the left-hand side of the inequality has to be negative definite.

For the proof, the Lyapunov function

$$V(\mathbf{x}(t)) = \mathbf{x}^T(t)\mathbf{P}\mathbf{x}(t) + \frac{1}{1-\beta} \int\limits_{t-\tau(t)}^{t} \mathbf{x}^T(\sigma)\mathbf{P}\mathbf{x}(\sigma) d\sigma \tag{4.67}$$

is defined. Using (4.66) shows, that the derivative of the Lyapunov function is negative for any $\mathbf{x}(t)$. Since the fundamental idea of the proof is the same like in the proofs of the theorems 4.1 and 4.2, we do not present it here.

Instead, let us discuss the matrix inequality (4.66) which is not linear for the unknown matrices $\mathbf{P}$ and $\mathbf{S}$. Therefore, an LMI algorithm cannot be applied to inequalities of this type. But with the help of the Schur complement (G.2), we can transform (4.66) into

$$\begin{pmatrix} \mathbf{A}_{1i}^T\mathbf{P} + \mathbf{P}\mathbf{A}_{1i} + \frac{1}{1-\beta}\mathbf{S} & \mathbf{P}\mathbf{A}_{2i} \\ \mathbf{A}_{2i}^T\mathbf{P} & \mathbf{S} \end{pmatrix} < \mathbf{0}. \tag{4.68}$$

This is a matrix whose single components depend linearly on the required matrices $\mathbf{P}$ and $\mathbf{S}$. As explained in the appendix G, inequalities containing partial matrices of this type can be easily united to the basic form (G.1) of an LMI problem and solved with an LMI algorithm.

Finally, let us remark that in [27] the stability criterion for systems with variable delay time is not only derived in the above presented form (4.66), but as well for continuous as for time-discrete systems with controllers and even observers. But the principle is the same in all cases.

Systems with Controllers

In the most cases the system will not be represented in the form of (4.37) or (4.41) but in the form of (4.26) because the stability analysis is normally requested when or after designing the controller. For this case, $\mathbf{A}_i$ in equation (4.42) would have to be replaced by $\mathbf{G}_{ij}$, and the condition would have to be checked for all index pairs (i, j). In principle, this would be possible but the great number of inequalities would lead to a very extensive LMI problem. Much better is first to reformulate the representation (4.26) of the system and then to apply theorem 4.2 to the system (cf. [188]).

The starting point for the continuous case is equation (4.26) whose right side is splitted into two equal sums:

$$\begin{aligned}
\dot{\mathbf{x}}(t) &= \sum_i \sum_j k_i(\mathbf{z}(t)) k_j(\mathbf{z}(t)) \mathbf{G}_{ij} \mathbf{x}(t) \\
\dot{\mathbf{x}}(t) &= \sum_i k_i(\mathbf{z}(t)) k_i(\mathbf{z}(t)) \mathbf{G}_{ii} \mathbf{x}(t) \\
&\quad + 2 \sum_{i<j} k_i(\mathbf{z}(t)) k_j(\mathbf{z}(t)) \left[\frac{\mathbf{G}_{ij} + \mathbf{G}_{ji}}{2} \right] \mathbf{x}(t)
\end{aligned} \tag{4.69}$$

The same Lyapunov function and the same operation as in (4.43) result in

$$\begin{aligned}
\dot{V} &= \dot{\mathbf{x}}^T \mathbf{P} \mathbf{x} + \mathbf{x}^T \mathbf{P} \dot{\mathbf{x}} \\
&= \sum_i k_i^2 \mathbf{x}^T \mathbf{G}_{ii}^T \mathbf{P} \mathbf{x} + 2 \sum_{i<j} k_i k_j \mathbf{x}^T \left[\frac{\mathbf{G}_{ij}^T + \mathbf{G}_{ji}^T}{2} \right] \mathbf{P} \mathbf{x} \\
&\quad + \sum_i k_i^2 \mathbf{x}^T \mathbf{P} \mathbf{G}_{ii} \mathbf{x} + 2 \sum_{i<j} k_i k_j \mathbf{x}^T \mathbf{P} \left[\frac{\mathbf{G}_{ij} + \mathbf{G}_{ji}}{2} \right] \mathbf{x} && (4.70) \\
&= \sum_i k_i^2 \mathbf{x}^T \left(\mathbf{G}_{ii}^T \mathbf{P} + \mathbf{P} \mathbf{G}_{ii} \right) \mathbf{x} \\
&\quad + 2 \sum_{i<j} k_i k_j \mathbf{x}^T \left(\left[\frac{\mathbf{G}_{ij} + \mathbf{G}_{ji}}{2} \right]^T \mathbf{P} + \mathbf{P} \left[\frac{\mathbf{G}_{ij} + \mathbf{G}_{ji}}{2} \right] \right) \mathbf{x} && (4.71) \\
&< 0
\end{aligned}$$

and, thus, the following two stability conditions which have to be satisfied for all i and $j > i$:

$$\mathbf{G}_{ii}^T \mathbf{P} + \mathbf{P} \mathbf{G}_{ii} < \mathbf{0} \tag{4.72}$$

$$\left[\frac{\mathbf{G}_{ij} + \mathbf{G}_{ji}}{2} \right]^T \mathbf{P} + \mathbf{P} \left[\frac{\mathbf{G}_{ij} + \mathbf{G}_{ji}}{2} \right] < \mathbf{0} \quad \text{for} \quad i < j \tag{4.73}$$

Obviously, the number of inequalities to be checked have been reduced from $i \times j$ to approximately 50 percent. The steps for the discrete case are the same.

The LMI problem can be even more reduced if we take into account that as well the partial plant model $(\mathbf{A}_i, \mathbf{B}_i)$ as every partial controller F_j are only active in the environment of the corresponding supporting point i or j. Since the indices i and j describe the same supporting points it follows that the product $k_i(\mathbf{z}(t))k_j(\mathbf{z}(t))$ for indices i and j which are far enough from each other is always zero. Therefore, in condition (4.73) we have to check only the $\mathbf{G}_{ij}$ whose indices lie next to each other.

Another way to reduce the stability conditions (4.72) and (4.73) is to take the coefficients k_i into account and regard the sum in (4.71) as a whole to derive the stability condition, like already mentioned in this chapter. Condition (4.73) results from the demand for every summand of (4.71) being negative. In contrast to this, the following approach takes into account that the single summands are weighed by the k_i and that positive summands can be compensated by negative ones. For the stability of the system it is only important that the total sum is negative and not every single summand.

Let us start with equation (4.71) which has to be represented in a matrix form (cf. [88]):

$$\dot{V} = \begin{pmatrix} k_1\mathbf{x} \\ k_2\mathbf{x} \\ \vdots \\ k_r\mathbf{x} \end{pmatrix}^T \mathbf{X} \begin{pmatrix} k_1\mathbf{x} \\ k_2\mathbf{x} \\ \vdots \\ k_r\mathbf{x} \end{pmatrix} < 0 \tag{4.74}$$

with

$$\mathbf{X} = \begin{pmatrix} \mathbf{L}_{11}^T\mathbf{P} + \mathbf{P}\mathbf{L}_{11} & 2(\mathbf{L}_{12}^T\mathbf{P} + \mathbf{P}\mathbf{L}_{12}) & \cdots & 2(\mathbf{L}_{1r}^T\mathbf{P} + \mathbf{P}\mathbf{L}_{1r}) \\ \mathbf{0} & \mathbf{L}_{22}^T\mathbf{P} + \mathbf{P}\mathbf{L}_{22} & \cdots & 2(\mathbf{L}_{2r}^T\mathbf{P} + \mathbf{P}\mathbf{L}_{2r}) \\ \cdots & \cdots & \cdots & \cdots \\ \mathbf{0} & \mathbf{0} & \cdots & \mathbf{L}_{rr}^T\mathbf{P} + \mathbf{P}\mathbf{L}_{rr} \end{pmatrix} \tag{4.75}$$

and $\mathbf{L}_{ij} = \frac{\mathbf{G}_{ij}+\mathbf{G}_{ji}}{2}$. Obviously, this equation is always satisfied if the matrix $\mathbf{X}$ is negative definite. Since the matrix depends linearly on $\mathbf{P}$ here we can also use an LMI algorithm to check whether a $\mathbf{P}$ exists for which the matrix is negative definite and, thus, the system is stable.

Now we have to compare this stability condition and the other two conditions (4.72) and (4.73). At first, the negative definiteness of the matrix developed above requires that all matrices on its main diagonal are negative definite, that is, the condition $\mathbf{L}_{ii}^T\mathbf{P} + \mathbf{P}\mathbf{L}_{ii} < \mathbf{0}$ has to be satisfied for all i. This corresponds to condition (4.72). Therefore, only condition (4.73) is dropped by using $\mathbf{X}$ while condition (4.72) is still implicitly contained in the demand for the negative definiteness of $\mathbf{X}$.

But since not all entries outside the main diagonal have to be negative definite to make $\mathbf{X}$ negative definite, which would be equivalent to condition (4.73), the negative definiteness of this matrix seems to be the less strict condition and is, therefore, more favourable for the proof of stability.

As another option [188] suggests the introduction of a further, positive semidefinite matrix $\mathbf{Q}$ in order to obtain additional degrees of freedom for the search for a common, positive definite matrix $\mathbf{P}$. With this additional matrix, (4.72) and (4.73) turn into

$$\mathbf{G}_{ii}^T\mathbf{P} + \mathbf{P}\mathbf{G}_{ii} + (s-1)\mathbf{Q} < \mathbf{0} \tag{4.76}$$

$$\left[\frac{\mathbf{G}_{ij} + \mathbf{G}_{ji}}{2}\right]^T \mathbf{P} + \mathbf{P}\left[\frac{\mathbf{G}_{ij} + \mathbf{G}_{ji}}{2}\right] - \mathbf{Q} < \mathbf{0} \quad \text{for} \quad i < j. \tag{4.77}$$

s is the maximum number of fuzzy rules which are active at the same time or, for a representation of a characteristic surface, it is the maximum number of adjacent supporting points which have a contribution to the output value of the TSK system. $\mathbf{Q}$ as well as $\mathbf{P}$ are are the unknown quantities of the LMI algorithm. Then, the LMI algorithm answers the question whether matrices $\mathbf{P}$ and $\mathbf{Q}$ exist for which the system of inequalities (4.76) and (4.77) is satisfied.

The resulting set of solutions contains the solution set of the inequality system (4.72) and (4.73). Since $\mathbf{Q}$ has to be only positive semidefinite it can also be the null matrix. So the case of $\mathbf{Q} = \mathbf{0}$ and an arbitrary $\mathbf{P}$ is implicitly contained in the computation, and this corresponds directly to the equations (4.72) and (4.73).

The idea of extending the inequality system in order to obtain additional degrees of freedom is also taken up in [79]. The result is similar to (4.76) and (4.77) so that we do not need to show it here.

Instead of the described subsequently performed stability analysis of an already designed TSK controller, the design can also be integrated to the formulation of the LMI problem [188]. Since that is no stability analysis but a controller design method it is described in chapter 5.1.

All of the presented methods can also be applied to systems with observers. The state vector of such an entire system contains not only the state variables of the plant but also the ones of the observer. By appropriate combination of the state space equations the system can be brought back to the form (4.26) and directly determine the stability conditions (cf. [188, 90, 27]). The resulting equations are very extensive, so that we do not discuss them here.

Other Approaches

A totally different approach, which also uses the advantages of LMI algorithms, can be found in [3]. There the original system is not interpreted as the overlapping of partial system matrices $\mathbf{A}_i$ like in (4.41) but as a system whose system matrix depends continuously on the vector of the input values:

$$\dot{\mathbf{x}}(t) = \mathbf{A}(\mathbf{z}(t))\mathbf{x}(t). \tag{4.78}$$

Unfortunately, for this approach the class of systems has to be restricted to systems with a single input value Θ:

$$\dot{\mathbf{x}}(t) = \mathbf{A}(\Theta(t))\mathbf{x}(t). \tag{4.79}$$

An additional restriction is, that the derivative of Θ has to be less than a given bound v ($\dot{\Theta} \leq v$) and Θ must lie within the interval $[0, 1]$, but this is no restriction of the universal validity with an appropriate scaling.

The stability conditions for this system can be combined to a linear matrix inequality $\mathbf{F} < \mathbf{0}$ and are, in principle, comparable with the condition (4.44). Again, $\mathbf{F}$ contains the system matrix $\mathbf{A}$ as well as the positive definite matrix $\mathbf{P}$, and since $\mathbf{A}$ depends on Θ the same holds for $\mathbf{P}$. Therefore, the LMI system is not constant but depends on Θ:

$$\mathbf{F}(\mathbf{A}(\Theta), \mathbf{P}(\Theta)) < \mathbf{0} \tag{4.80}$$

Now we have to answer the question whether a matrix $\mathbf{P}(\Theta)$ exists for which the inequality (4.80) is satisfied and, thus, the system is stable. Unfortunately, an LMI algorithm can answer this question only for constant systems. For this reason the system shall be approximated by a sum of constant systems to which an LMI algorithm can be applied.

At first, $\mathbf{A}(\Theta)$ and $\mathbf{P}(\Theta)$, which are contained in $\mathbf{F}$, are approximated :

$$\mathbf{A}(\Theta) \approx \sum_{i=0}^{L_a} \Theta^i \mathbf{A}_i \quad \text{and} \quad \mathbf{P}(\Theta) \approx \sum_{i=0}^{L_p} \Theta^i \mathbf{P}_i \tag{4.81}$$

where the $\mathbf{P}_i$ have to be symmetrical. They are combined to a common matrix

$$\mathbf{P}_{ges} = (\mathbf{P}_0, ..., \mathbf{P}_{L_p}). \tag{4.82}$$

We have to point out that this approximation touches only the dependence of $\mathbf{P}$ or $\mathbf{A}$ on Θ but not the dependence on time of Θ itself.

With (4.81) and (4.82) we can then approximate $\mathbf{F}$, too:

$$\mathbf{F}(\mathbf{A}(\Theta), \mathbf{P}(\Theta)) \approx \sum_{i=0}^{L_f} \Theta^i \mathbf{F}_i(\mathbf{P}_{ges}) < \mathbf{0} \qquad \text{with} \quad L_f = L_p + L_a \tag{4.83}$$

The dependence of the coefficients of the $\mathbf{F}_i$ on the matrices $\mathbf{A}_i$ is not represented explicitly any more. Since for all further considerations only the dependence on $\mathbf{P}_{ges}$ is relevant.

Because of $\Theta \in [0, 1]$, the inequality is satisfied if

$$\mathbf{F}_0(\mathbf{P}_{ges}) + \sum_{i=1}^{L_f} p_i \mathbf{F}_i(\mathbf{P}_{ges}) < \mathbf{0} \tag{4.84}$$

holds for any $p_i \in \{0, 1\}$. This means that 2^{L_f} inequalities have to be checked. They can, at least theoretically, be combined to a common linear matrix inequality, so that we obtain a constant LMI system which depends affinely on the symmetric matrix $\mathbf{P}_{ges}$. Thus, with an LMI algorithm we can check whether a matrix $\mathbf{P}_{ges}$ exists which satisfies the inequality system (4.84). That would prove the stability of the system (4.79).

Other approaches for proving the stability based on the theorem 4.1 or 4.2 fade beside the smart and exact method using an LMI algorithm. Nevertheless, here we present some of these because some contain interesting ideas.

In [191] a stability condition is developed from equation (4.71) which is similar to (4.74). At first, the summands

$$\mathbf{x}^T\mathbf{Q}_{ij}\mathbf{x} = \mathbf{x}^T\left(\left[\frac{\mathbf{G}_{ij}+\mathbf{G}_{ji}}{2}\right]^T\mathbf{P}+\mathbf{P}\left[\frac{\mathbf{G}_{ij}+\mathbf{G}_{ji}}{2}\right]\right)\mathbf{x} \tag{4.85}$$

are estimated by the maximum eigenvalues of the matrices $\mathbf{Q}_{ij}$

$$\lambda_{max}(\mathbf{Q}_{ij})||\mathbf{x}||^2 \geq \mathbf{x}^T\mathbf{Q}_{ij}\mathbf{x} \qquad \text{for all } \mathbf{x} \tag{4.86}$$

and then instead of the matrix $\mathbf{X}$ in (4.74) an analogously structured matrix is formed from eigenvalues which has to be checked again for negative definiteness. The advantage of this method is that the dimension of the matrix which has to be checked is, of course, essentially smaller because the elements of the matrix are real numbers and no matrices. The significant disadvantage is that with this method the negative definiteness or the stability is checked only for one single given matrix $\mathbf{P}$. In the method corresponding to (4.74), using an LMI algorithm, we can answer the fundamental question whether a matrix $\mathbf{P}$ exists and the system is stable.

A totally different approach is presented in [90]. Starting with a TSK model of the plant (4.10) the means of the system matrices are determined:

$$\mathbf{A}_0 = \frac{1}{L}\sum_{i=1}^{L}\mathbf{A}_i \qquad \text{and} \quad \mathbf{B}_0 = \frac{1}{L}\sum_{i=1}^{L}\mathbf{B}_i \tag{4.87}$$

Using these means one single, linear controller is designed. Then for this controller robustness bounds are determined within which it can stabilize the nonlinear system. Even uncertainties of the model can be integrated.

In [77] it is shown how to calculate a positive definite matrix $\mathbf{P}_i$ for a single subsystem, that is, for a certain value i of equation (4.38), and then by inserting this matrix back into the other systems to obtain a common matrix $\mathbf{P}$ which satisfies the condition of theorem 4.1. But for this method we need any two system matrices $(\mathbf{A}_i, \mathbf{A}_j)$ to be pairwise commutative which should not happen often in practice.

4.2.3 Application to Facet Functions

Our last variant of the direct method requires another representation of the system, an approximation of the system's behavior by so-called *facet functions* ([86], [87]).

A facet function is given if the space of the function's input quantities is split up into convex polyhedrons, and if inside every polyhedron the function is defined by an affine function. A facet function for a simple nonlinear transfer element $\mathbf{u}(\mathbf{x})$, for example, a conventional fuzzy controller, is defined by

$$\mathbf{u} = \mathbf{d}_i + \mathbf{K}_i^T \mathbf{x} \qquad \text{for } \mathbf{x} \in P_i \tag{4.88}$$

where P_i is a convex, not necessarily bounded polyhedron in the space of the input variables $\mathbf{x}$. $\mathbf{K}_i$ and $\mathbf{d}_i$ are constant matrices and vectors, that can only be computed numerically, if for example a given transfer characteristic has to be approximated.

Obviously, we can use facet functions also to approximate any dynamic transfer element. A nonlinear state equation $\dot{\mathbf{x}} = \mathbf{f}(\mathbf{x})$, for example, can be approximated by affine functions

$$\dot{\mathbf{x}} = \mathbf{d}_i + \mathbf{K}_i \mathbf{x} \qquad \text{for } \mathbf{x} \in P_i \tag{4.89}$$

Let us compare this representation to a characteristic surface and a TSK controller. With facet functions, we divide the input space up into several regions (polyhedrons), and for each of these regions, the transfer characteristic is defined by one single affine function. In contrast to this, for characteristic surfaces and TSK controllers we define the transfer characteristic inside one region by interpolating the given characteristics of the adjacent supporting points.

For stability analysis, we now have to represent the transfer characteristic of the closed loop by facet functions. First of all, we have to transform the entire system according to figure 2.79, to make the reference vector $\mathbf{w}$ to be zero. In doing so we achieve that the system is free from any outer stimulation, and that all transfer elements therefore depend only on the system's state variables. Each transfer element can then be approximated by a facet function depending on the system's state variables, where the segmentation of the state space into polyhedrons can be different for any transfer element. But we can split up the state space into smaller polyhedrons by forming intersections, inside of which every transfer element is given by one affine function.

We then compose those affine functions of all the transfer elements which are valid inside a polyhedron, and obtain another affine function, which represents the transfer characteristic of the closed loop. This way, we approximate the transfer characteristic of the entire system by facet functions. We do not even need continuous facet functions to perform the following stability analysis, which is why we can use this method even for switching elements.

The starting point for all of our considerations is the representation of the closed loop by means of facet functions

$$\dot{\mathbf{x}} = (\dot{\mathbf{x}})_j = \mathbf{K}_j \mathbf{x} + \mathbf{d}_j \qquad \text{for } \mathbf{x} \in P_j \tag{4.90}$$

where $\mathbf{K}_j$ and $\mathbf{d}_j$ are defined inside a convex, not necessarily bounded polyhedron P_j, and where $\mathbf{x}$ is the state vector of the system. All points for which $\dot{\mathbf{x}} = \mathbf{0}$ holds are positions of rest. Obviously, even a complete polyhedron may consist of positions of rest only, if we have $\mathbf{K}_j = \mathbf{0}$ and $\mathbf{d}_j = \mathbf{0}$ inside this polyhedron.

Now, the union of all zero positions of rest may form a compact convex polyhedron E. Figure 4.4 shows several different constellations. The zero position of the first example lies at the corner of all four adjacent polyhedrons. In the second example, the borderline between P_2 and P_4 consists of positions of rest, and in the last example the black polyhedron which is situated in the middle of the five other polyhedrons represents the set of all positions of rest.

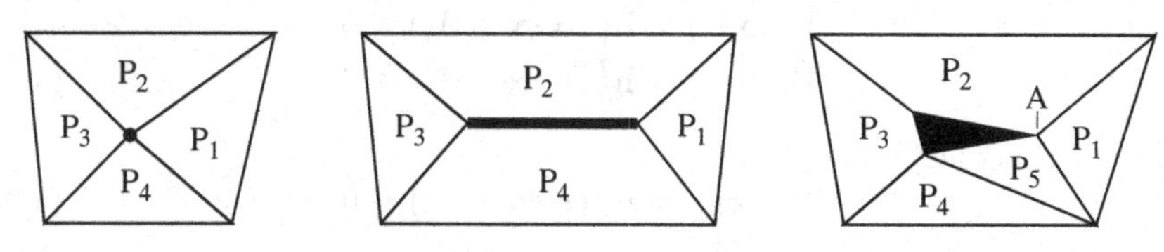

Fig. 4.4. Examples for positions of rest resulting from a facet-function-based system representation

As the facet function does not have to be a continuous function, it is possible at the border of two or more polyhedrons that the system shows a sliding-mode transfer characteristic, i.e. the system moves along this border by fast changes of the state vector from one polyhedron to the other. It is possible to show that for this case the system's state space equation is given by

$$(\dot{\mathbf{x}})_{sm} = \sum_j \mu_j(\mathbf{K}_j\mathbf{x} + \mathbf{d}_j) \qquad \text{where} \sum_j \mu_j = 1 \qquad \text{and } \mu_j \geq 0 \tag{4.91}$$

which is a weighed average of the state space equations of the adjacent polyhedrons.

We now define a Lyapunov function, given by a—this time continuous—facet function:

$$V(\mathbf{x}) = V_j(\mathbf{x}) = \mathbf{h}_j^T\mathbf{x} + c_j \qquad \text{for } \mathbf{x} \in P_j \qquad \text{and } V(\mathbf{x}) \begin{cases} = 0 : \mathbf{x} \in E \\ > 0 : \mathbf{x} \notin E \end{cases} \tag{4.92}$$

The polyhedrons P_j should be identical to the ones given in (4.90). Furthermore, the half-space $H := \{\mathbf{x} \mid V(\mathbf{x}) \leq c\}$ that is defined by V (where c a positive constant value) has to be compact. It contains the set of positions of rest E.

Next, we have to show that for all corners $\mathbf{x}_i$ of all polyhedrons $P_j \subset (H \cap \bar{E})$ (i.e. all polyhedrons of the half-space except the positions of rest) the condition

$$\dot{V}_\mu((\dot{\mathbf{x}}_i)_\eta) = \mathbf{h}_\mu^T(\mathbf{K}_\eta\mathbf{x}_i + \mathbf{d}_\eta) \begin{cases} \leq 0 : \mathbf{x}_i \in E \\ < 0 : \mathbf{x}_i \notin E \end{cases} \tag{4.93}$$

holds. (μ, η) are the indices of those polyhedrons whose joint border consists of this corner and at least one more point. This includes the case of $\mu = \eta$. For the corner $\mathbf{x} = A$, shown in figure 4.4, we would have to check the following constraints:

$$\begin{aligned}
\dot{V}_1((\dot{\mathbf{x}})_1) &= \mathbf{h}_1^T(\mathbf{K}_1\mathbf{x} + \mathbf{d}_1) \leq 0 \\
\dot{V}_2((\dot{\mathbf{x}})_2) &= \mathbf{h}_2^T(\mathbf{K}_2\mathbf{x} + \mathbf{d}_2) \leq 0 \\
\dot{V}_5((\dot{\mathbf{x}})_5) &= \mathbf{h}_5^T(\mathbf{K}_5\mathbf{x} + \mathbf{d}_5) \leq 0 \\
\dot{V}_1((\dot{\mathbf{x}})_2) &= \mathbf{h}_1^T(\mathbf{K}_2\mathbf{x} + \mathbf{d}_2) \leq 0 \\
\dot{V}_2((\dot{\mathbf{x}})_1) &= \mathbf{h}_2^T(\mathbf{K}_1\mathbf{x} + \mathbf{d}_1) \leq 0 \\
\dot{V}_1((\dot{\mathbf{x}})_5) &= \mathbf{h}_1^T(\mathbf{K}_5\mathbf{x} + \mathbf{d}_5) \leq 0 \\
\dot{V}_5((\dot{\mathbf{x}})_1) &= \mathbf{h}_5^T(\mathbf{K}_1\mathbf{x} + \mathbf{d}_1) \leq 0
\end{aligned} \tag{4.94}$$

but not this one:

$$\dot{V}_2((\dot{\mathbf{x}})_5) = \mathbf{h}_2^T(\mathbf{K}_5\mathbf{x} + \mathbf{d}_5) \leq 0 \tag{4.95}$$

as P_2 and P_3 have only one common point (A).

If (4.93) is true, then E is asymptotically stable and H is its domain of attraction.

For the proof, let us recall that an affine function defined inside a polyhedron P_j, takes its maximum value at one of the corners of the polyhedron. Furthermore, not only the Lyapunov function, which is defined inside a polyhedron, is affine, but also its derivative with respect to time:

$$\dot{V}_j(\mathbf{x}) = \mathbf{h}_j^T\dot{\mathbf{x}} = \mathbf{h}_j^T(\mathbf{K}_j\mathbf{x} + \mathbf{d}_j) \tag{4.96}$$

If we can therefore show that $\dot{V} < 0$ holds for all corners of any polyhedron, then $\dot{V}$ is negative also for all points inside any polyhedron. If the polyhedron shares a border with E, however, we can only get $\dot{V} \leq 0$ for the points of this border. But it still follows that $\dot{V} < 0$ for any point lying inside the polyhedron, if we consider the fact that the derivative of V is negative at the other corners.

The edges of the polyhedron are a crucial point, since there $\dot{V}$ may be discontinuous, because of the discontinuous facet function describing the system's transfer characteristic. Let us analyze an example where a state space trajectory starts inside a polyhedron P_μ and ends inside P_η. Since $\dot{V}_\mu((\dot{\mathbf{x}}_i)_\eta) < 0$, $\dot{V}_\mu((\dot{\mathbf{x}}_i)_\mu) < 0$, $\dot{V}_\eta((\dot{\mathbf{x}}_i)_\mu) < 0$ and $\dot{V}_\eta((\dot{\mathbf{x}}_i)_\eta) < 0$ was shown for the corresponding corners $\mathbf{x}_i$ at the border, the same has to be true for any other point lying on the border—because of the affinity of the functions $\dot{V}_\mu$ and $\dot{V}_\eta$. This, however, guarantees that the value for V decreases when the trajectory passes the border.

All that remains is the case of a sliding mode characteristic, i.e. a state space trajectory runs along the border of two or more polyhedrons. From (4.91) we get for every adjacent polyhedron P_j:

$$\dot{V}_j((\dot{\mathbf{x}})_{sm}) = \mathbf{h}_j^T\left(\sum_k \mu_k(\mathbf{K}_k\mathbf{x} + \mathbf{d}_k)\right)$$

$$= \sum_k \mu_k \mathbf{h}_j^T (\mathbf{K}_k \mathbf{x} + \mathbf{d}_k) = \sum_k \mu_k \dot{V}_j((\dot{\mathbf{x}})_k) < 0 \qquad (4.97)$$

But since we have shown that $\dot{V}_j((\dot{\mathbf{x}})_k) < 0$ holds for all the corners involved, the same has to be true for any other point lying on the border, because of the affinity of $\dot{V}$. So even in this case the value of the Lyapunov function will be continuously decreasing.

All in all, we have shown that the derivative of the Lyapunov function is negative for the entire region $H \cap \bar{E}$—inside of the polyhedrons as well as on their borders—independent of the shape of the trajectories. But this does not imply asymptotic stability of the system according to theorem 2.23, since the Lyapunov function which we have been using so far is not continuously differentiable. This, however, is a prerequisite of the theorem. The claim of stability therefore has to be proven in a different way.

First of all, because of $V(\mathbf{x}) \leq c$ inside of H and $V(\mathbf{x}) = c$ at the edge of H, all of the polyhedrons P_j which lie at the edge of H need to have a Lyapunov function whose values are either constant or increase in the direction of the edge. As furthermore we have $\dot{V} < 0$ inside of H, no trajectory can leave the half-space H.

Let us now prove simple stability for the region E. A ε-area S_ε around E shall be given. Then there has to exist a δ-area, so that for all initial states lying inside it the corresponding trajectory does not leave the region S_ε. In order to determine this δ-area, we have to find the smallest value V_ε of the Lyapunov function inside the compact remaining area $H \cap \overline{(E \cup S_\varepsilon)}$ of the half-space H. Because of (4.92), this is certainly a positive value. And because of V's continuity and E's compactness, we can then describe a δ-area around E such that $V(\mathbf{x})$ will be less than V_ε for all points lying inside this area. This area will be completely contained in S_ε, as otherwise V_ε would not be the smallest value of the remaining area. Since the derivative of V is always negative, a trajectory which has its origin inside the δ-area can never attain a value $V(\mathbf{x}) \geq V_\varepsilon$. Accordingly, the trajectory can never leave the S_ε-region, and we have proven E's stability.

E's asymptotical stability follows from the fact that the derivative of the Lyapunov function is negative for all points of $H \cap \bar{E}$, even when crossing the polyhedrons' borders or moving along them. A trajectory which starts inside H has a certain initial value $V > 0$. Since no trajectory can leave the half-space, the value of V decreases until $\dot{V} = 0$. But at that point E is reached, and we have also shown asymptotical stability.

This criterion of stability can only be applied numerically. First of all, the transfer characteristic of the system has to be approximated by a facet function, which of course can be done only in a numerical way. Next, the parameters of the Lyapunov function have to be fixed, i.e. the parameters $\mathbf{h}_j$ and c_j of every single polyhedron, paying attention to the boundary conditions (4.92) and the continuity of V. Additionally, the half-space H should contain that region of the state space which is of relevance to technical applications. This also has to be done in a numerical way.

Choosing the Lyapunov function is, as for all other direct methods, the crucial point of this procedure, since there is no algorithm available which would always provide a Lyapunov function suitable for the proof of stability. Finally, we have to check the stability condition (4.93). If it turns out not to be true, then another Lyapunov function must be searched or otherwise the procedure terminates unsuccessfully.

4.3 Describing Function

The describing function method, which we already discussed in detail in chapter 2.8.6, is based on a completely different idea. This method can be applied to control loops with one actuating variable and one control variable, although the (fuzzy) controller may have several input quantities like, for example, the output error e, its derivative $\dot{e}$ or its integral $\int e$. As another prerequisite, it has to be possible to split up the complete system into a linear part with a sufficient low-pass characteristic, and a nonlinear part with symmetrical transfer characteristic.

Our first task is to split up the system into a linear and a nonlinear part, as illustrated in figure 4.1 and 4.2. From figure 4.2 we can see that only the fuzzy controller itself belongs to the nonlinear part of the system, while all the other elements form the linear part. Of course, this is the most simple case. Normally, the nonlinearities of the plant together with the fuzzy controller form the nonlinear part of the system.

Next, we have to define one of the input quantities of the nonlinear part to be its main input quantity or, alternatively, to be the main output quantity of the linear part. This should always be the output quantity of the last integrator of the linear part, in figure 4.2 for example $e' = \int e$. Accordingly, we obtain

$$G_l(s) = \frac{e'}{\dot{u}} = \frac{1}{s^3}G'(s) = \frac{1}{s^2}G(s) \tag{4.98}$$

as the transfer function of the linear part of the system.

We then have to check if the linear part has a suitable low-pass characteristic, which means that all input quantities of the nonlinear part should be (more or less) pure sinusoidal oscillations. In figure 4.2, these input quantities are $\int e$, e and $\dot{e}$, and the transfer functions that have to be checked for sufficient low-pass characteristics are therefore $\frac{1}{s^2}G(s)$, $\frac{1}{s}G(s)$ and $G(s)$. Since integrators amplify low-pass effects, the low-pass characteristics of all transfer functions should be sufficient if the low-pass characteristic of $G(s)$ is.

The next step is to determine the describing function of the nonlinear part of the system. The easiest way to compute is the numerical solution. For this, we first have to define a sinusoidal oscillation with amplitude A and frequency ω for the main output signal $e' = \int e$ of the linear part of the system. The resulting derivatives e and $\dot{e}$ of this signal are therefore also harmonic oscillations. The input signals for the nonlinear part of the system are now set. If we feed these signals into the nonlinear part, we will receive a periodic oscillation at its output, which can be approximated by a sinusoidal function. If we compare this function to the sinusoidal oscillation e', we can determine the amplification and the phase shift of the nonlinear part for the pair of values (A, ω), which gives directly the value of the describing function $N(A, \omega)$. In this way, we obtain a point-wise representation of the describing function.

Of course, we can also determine the describing function analytically, but this requires an analytical description of the nonlinear part of the system. Since this is usually not given for a fuzzy controller, we first have to approximate its transfer characteristic by a simple function, which we then use to determine the describing function. However, attention has to be paid to the following fact: If the nonlinear part of the system consists not only of the fuzzy controller but also of other nonlinear transfer elements, then the approximation must not be made for the fuzzy controller only. Instead, the entire nonlinear part of the system has to be the subject of one overall approximation.

In order to illustrate the analytical computation steps, we continue with our example of figure 4.2. Let us assume that the transfer characteristic of the fuzzy controller can be approximated by an analytical function $f(\int e, e, \dot{e})$. If we set $e' = \int e$, this turns into $f(e', \dot{e}', \ddot{e}')$. From this, we can determine the coefficients A_1 and B_1 which we need in our computation of the describing function according to equation (2.262):

$$
\begin{aligned}
A_1 &= \frac{2}{T} \int_0^T f(e', \dot{e}', \ddot{e}') \cos(\omega t) dt \\
&= \frac{2}{T} \int_0^T f(A \sin(\omega t), A\omega \cos(\omega t), -A\omega^2 \sin(\omega t)) \cos(\omega t) dt \\
B_1 &= \frac{2}{T} \int_0^T f(e', \dot{e}', \ddot{e}') \sin(\omega t) dt \\
&= \frac{2}{T} \int_0^T f(A \sin(\omega t), A\omega \cos(\omega t), -A\omega^2 \sin(\omega t)) \sin(\omega t) dt
\end{aligned} \tag{4.99}
$$

where $T = \frac{2\pi}{\omega}$. Setting $C_1 = \sqrt{A_1^2 + B_1^2}$ and $\varphi_1 = \arctan \frac{A_1}{B_1}$, we obtain the describing function

$$
N(A, \omega) = \frac{C_1(A, \omega)}{A} e^{j\varphi(A,\omega)} \tag{4.100}
$$

We can now be sure that the describing function is available, determined either point-wise in a numerical way, or in an analytical way. For an analysis of stability, we can then represent the function $-\frac{1}{N(A,\omega)}$ of the variables A and ω as a family of curves in the complex plane. For any fixed ω_1, we obtain a curve $-\frac{1}{N(A,\omega_1)}$, which only depends on the amplitude A. Then we sketch the Nyquist plot $G_l(j\omega)$ of the linear part in the same diagram. Using the Nyquist criterion, we can draw conclusions concerning the stability from the relation between the Nyquist plot and the family of curves, as discussed in detail in chapter 2.8.6. Some concrete examples are given in [22] and [61].

The parameters A and ω of a possible steady oscillation result from the complex equation

$$G_l(j\omega) = -\frac{1}{N(A,\omega)} \tag{4.101}$$

The easiest way to determine them should be a numerical approach.

4.4 Popov Criterion

The Popov criterion represents an alternative to the method of harmonic balance, although only very few systems which contain fuzzy controllers meet its prerequisites. First of all, again we have to split up the system into a linear and a nonlinear part. The nonlinear part has to be without internal dynamics. If both parts have only one input and one output quantity, we can apply theorem 2.25 directly. If the transfer function of the linear part is unknown, we can determine the Nyquist plot by measuring. The characteristic curve of the nonlinear part is also easy to obtain, as the nonlinear part directly assigns one output value to one input value. An analytical form is not even required, since only the boundaries k_1 and k_2 of a sector are of relevance to the Popov criterion (cf. fig. 2.87).

We can then analyze the system in exactly the same way as described in chapter 2.8.7. A sector-transformation may be necessary, and as a result of it we would have to redefine the linear transfer function according to (2.275), but then we can sketch the Popov locus of the linear part in the complex plane (fig. 2.88). Drawing the axis of convergence gives the maximum upper bound of the sector, and all we still have to do then is to compare this upper bound to the actual boundary of the sector of the nonlinear characteristic curve. If it is not bigger than the maximum upper bound, then the system is absolutely stable. [20] gives a concrete example of this.

In principle, we can also think of applying the Popov criterion for MIMO systems (theorem 2.26) to systems which comprise fuzzy controllers, although it would be difficult to fulfill the prerequisite that the fuzzy controller (or the nonlinear part of the system) has the same number of input and output quantities and that every output quantity u_i is a function of exactly one single input quantity: $u_i = f_i(e_i)$. If this condition is not met, we can try to eliminate the influence which input quantities other than e_i have on function f_i, by means of a suitable transformation. In the next chapter, which is concerned with the circle criterion, we will give an example of such a transformation. We will show how the dependence of a function $f(e, \dot{e})$ on the variable $\dot{e}$ at the origin $e = 0$ can be eliminated by means of a suitable transformation of the input quantities e and $\dot{e}$. But for the MIMO Popov criterion, the problem is much more complex, as we have to find a transformation which eliminates all the dependencies on input quantities other than e_i for all the output quantities u_i at the same time. This is almost impossible. It will therefore be more reasonable to use another criterion instead of it.

4.5 Circle Criterion

In order to be able to use the circle criterion, we have split up the complete system into a linear and a nonlinear part, just like we had to do for the method of harmonic balance or the Popov criterion. In contrast to these two methods however, we can now explicitly include nonlinearities with internal dynamics and MIMO systems.

4.5.1 Controllers with One Input Quantity

Let us first focus on SISO systems, whose nonlinearities have only one input quantity and no internal dynamics. For such systems, we can use the most simple form of the circle criterion. We first have to measure or to compute the Nyquist plot of the linear part and draw it in the complex plane. Then, we have to determine the characteristic curve of the nonlinear part, which is also very easy to do since every input value has one output value directly assigned to it. The characteristic curve gives us the boundaries k_1 and k_2 of the sector, which then determines the forbidden region of the complex plane.

As we already mentioned in chapter 2.8.8, for this very simple case it is possible to derive the circle criterion directly from the Popov criterion by setting the parameter q of inequality (2.265) to zero. In cases like this, the circle criterion is therefore merely a special version of the Popov criterion. This is the reason why there exist systems whose stability can be proven by the Popov criterion, but not by the circle criterion. On the other hand, the circle criterion is easier to apply, since, without q, we only have to examine the plain Nyquist plot of the linear part—instead of the Popov locus. And another advantage is that the sector-transformation, which is often required when using the Popov criterion, is already included in the circle criterion.

4.5.2 Controllers of Multiple Input Quantities

The case of a fuzzy controller with several input quantities and one output quantity is more interesting, as this constellation occurs most frequently in practical applications. First we have to declare one input quantity to be the main input quantity e. Accordingly, we have to represent the transfer characteristic of the nonlinear part of the system by a characteristic curve $u(t) = f(e(t), t)$, that depends only on e, although it initially depended on several input quantities. But therefore it is now time-variant.

Then, we have to fix the boundaries k_1 and k_2 of a sector in such a way that

$$k_1 e(t) \leq u(t) \leq k_2 e(t) \tag{4.102}$$

is true at any time t. These are the boundaries of the sector which we have to use when applying the circle criterion. With these boundaries, it is unimportant whether u depends only on e, on a differential equation in e, or on other input quantities besides e. It just has to be made sure, that all possible combinations of input values are taken into account, so that eq. (4.102) is always true.

Finally, we have to find the transfer function of the linear part of the system, as we need its Nyquist plot for the circle criterion. In order to do so, we have to determine the transfer characteristic from the output quantity of the nonlinear part of the system to its main input quantity. For example: If we define e to be the main input quantity of the nonlinear part of the system given in figure 4.2, and $\dot{u}$ to be its output quantity, then the resulting transfer function of the linear part is given by $\frac{1}{s}G(s)$.

One problem arises: From inequality (4.102) it follows that the output quantity of the nonlinear part has to be zero if e is zero, independently of all other input values. Accordingly, $f(0, t) = 0$ has to be true for any t, and this is usually not the case for fuzzy controllers. But we can solve this problem by a coordinate transformation as suggested in [24]. We can use the closed loop given in figure 4.2 as an example. In order to keep things simple, however, let us neglect the fuzzy controller's dependence on $\int e$ and also the integration of the controller output variable. The resulting transfer characteristic of the fuzzy controller is then given by a function $f(e, \dot{e})$, and the transfer function of the linear part of the system by $G(s)$. This method can be also adjusted to cover fuzzy controllers of several input quantities.

Let us assume that the rule base of the fuzzy controller is the one given in figure 4.5. If we plot the values of f against an $e - \dot{e}-$plane, we see that all of f's values lying on an axis that is rotated from the $\dot{e}-$axis by α are zero. We can define an $e_t - \dot{e}_t$-coordinate system whose $\dot{e}_t-$axis coincides with this zero axis, and which therefore corresponds to the old coordinate system rotated by α. With respect to these coordinates, we can define a new function $f'(e_t, \dot{e}_t) = f(e, \dot{e})$ which obviously meets the condition $f'(0, \dot{e}_t) = 0$.

The rotation of a vector by α corresponds to a multiplication by the matrix

$$\mathbf{T} = \begin{pmatrix} \cos\alpha & -\sin\alpha \\ \sin\alpha & \cos\alpha \end{pmatrix} \tag{4.103}$$

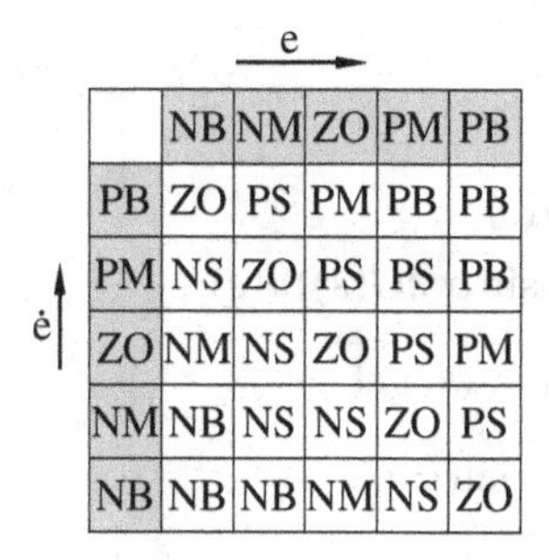

	NB	NM	ZO	PM	PB
PB	ZO	PS	PM	PB	PB
PM	NS	ZO	PS	PS	PB
ZO	NM	NS	ZO	PS	PM
NM	NB	NS	NS	ZO	PS
NB	NB	NB	NM	NS	ZO

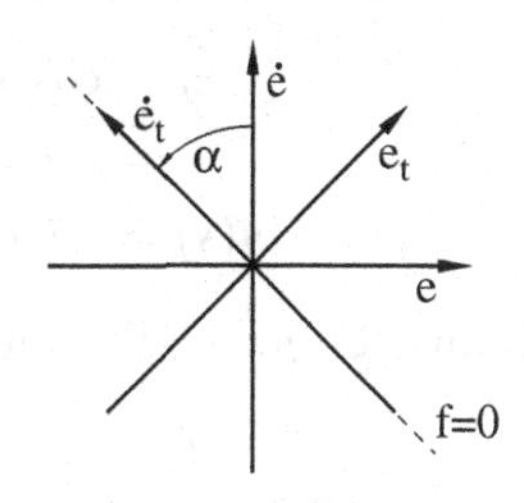

Fig. 4.5. Rule base

If we want to represent a vector whose coordinates are $[e, \dot{e}]^T$ (of the original coordinate system) by $[e_t, \dot{e}_t]^T$ (of the new system), we have to rotate it by $-\alpha$. This $-\alpha$-rotation, however, corresponds to a multiplication by $\mathbf{T}^{-1}$.

For stability analysis, we insert both matrices $\mathbf{T}^{-1}$ and $\mathbf{T}$ into the closed-loop system in such a way that they compensate each other and the system remains unchanged (cf. figure 4.6). A rotation of the vector $[e, \dot{e}]^T$ through $-\alpha$ produces $[e_t, \dot{e}_t]^T$, which is then re-transformed into $[e, \dot{e}]^T$ by a rotation through α. If we consider $\mathbf{T}$ to be a part of the nonlinear subsystem and $\mathbf{T}^{-1}$ to belong to the linear one, then the nonlinear part of the system is given by the function $f'(e_t, \dot{e}_t)$, which obviously complies the requirement $f'(0, \dot{e}_t) = 0$.

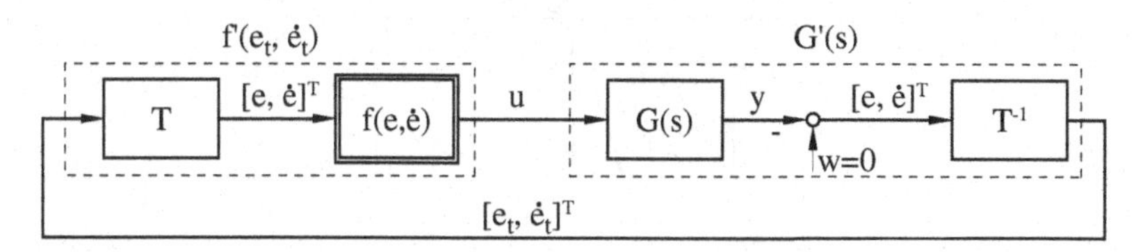

Fig. 4.6. Using a coordinate transformation in order to meet the requirements of the circle criterion

The multiplication by the matrix $\mathbf{S} = [1, s]^T$ is necessary to transform the scalar quantity e into the vector $[e, \dot{e}]^T$. It was inserted just for a precise drawing and does not cause any change of the system.

We can then apply the circle criterion to the system consisting of f' and G'. First, we have to compute the linear transfer function, i.e. the transfer behaviour from the output u of the nonlinear part to its main input quantity e_t:

$$G'(s) = -\frac{e_t}{u}(s) \tag{4.104}$$

The relation between $[e_t, \dot{e}_t]^T$ and $[e, \dot{e}]^T$ is given by

$$\begin{pmatrix} e_t \\ \dot{e}_t \end{pmatrix} = T^{-1} \begin{pmatrix} e \\ \dot{e} \end{pmatrix} = \begin{pmatrix} \cos\alpha & \sin\alpha \\ -\sin\alpha & \cos\alpha \end{pmatrix} \begin{pmatrix} e \\ \dot{e} \end{pmatrix} \tag{4.105}$$

which implies (with $e = w - y = -y$)

$$\begin{aligned} e_t &= \cos\alpha\ e + \sin\alpha\ \dot{e} \\ &= -\cos\alpha\ y - \sin\alpha\ \dot{y} \\ e_t(s) &= -(\cos\alpha\ G(s) + \sin\alpha\ sG(s))u(s) \end{aligned} \tag{4.106}$$

for e_t because of $y(s) = G(s)u(s)$. From this we get

$$G'(s) = \cos\alpha\ G(s) + \sin\alpha\ sG(s) \tag{4.107}$$

Next, we have to determine the sector for the characteristic curve of the nonlinear part. The easiest way to do this might just be to use several different values for e_t and $\dot{e}_t$. For every value of e_t we will get different values of u, depending on $\dot{e}_t$, so that we obtain a whole family of characteristic curves. (fig. 4.7). We have to choose the boundaries of the sector in a way that the entire family lies inside of it. With these boundaries, and the linear transfer function according to (4.107), we can directly perform a stability analysis according to the circle criterion.

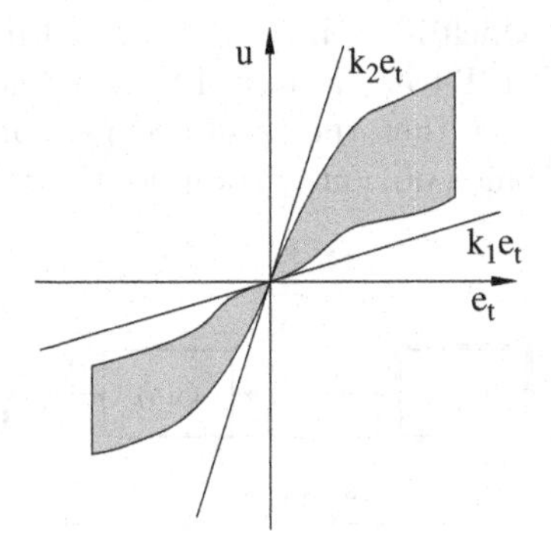

Fig. 4.7. Family of characteristic curves after transformation

4.5.3 MIMO Controllers

For true MIMO systems, i.e. where all parts of the system may have multiple input and output quantities, it is almost impossible to use the circle criterion on a base other than norms, as the circle criterion itself was also derived on this basis. A computation of a norm, however, requires an analytical description of the system's transfer characteristic. Accordingly, we have to define or approximate the fuzzy controller as in (4.5) or, even better, as a TSK controller (4.8). With the latter option, it is then also possible to represent the complete system by a TSK model (4.27) or (4.29).

However, with the direct analysis of the entire system we leave the framework of the circle criterion, that results from the split of the system into a linear and a nonlinear part. Therefore, we will represent these approaches in another (the following) chapter.

4.6 Norm-based Stability Analysis

Norms can be used for the stability analysis of continuous (eq. (4.28)) and discrete (eq. (4.29)) fuzzy systems as well. As the discrete case is more simple, we will present it first (c.f. [26]).

Starting point is the TSK model of a discrete system (4.29) without external activation:

$$\mathbf{x}(k+1) = \sum_i k_i(\mathbf{x}(k))\mathbf{A}_i\mathbf{x}(k) \tag{4.108}$$

The application of norms yields:

$$||\mathbf{x}(k+1)|| = ||\sum_i k_i(\mathbf{x}(k))\mathbf{A}_i\mathbf{x}(k)|| \leq \sum_i k_i(\mathbf{x}(k)) \; ||\mathbf{A}_i|| \; ||\mathbf{x}(k)|| \tag{4.109}$$

The system is stable in the sense of Lyapunov if the state vector converges towards zero. The condition for stability is therefore:

$$\sum_i k_i(\mathbf{x}(k))||\mathbf{A}_i|| < 1 \tag{4.110}$$

Because of

$$\sum_i k_i(\mathbf{x}(k)) = 1 \tag{4.111}$$

inequality (4.110) is met if

$$||\mathbf{A}_i|| < 1 \qquad \text{for all } i \tag{4.112}$$

If we use the ∞-norm, we get from (C.18)

$$||\mathbf{A}_i||_\infty = \sqrt{\lambda_{max}\left\{\bar{\mathbf{A}_i}^T\mathbf{A}_i\right\}} = \bar{\sigma}\left\{\mathbf{A}_i\right\} \tag{4.113}$$

where λ_{max} is the maximum eigenvalue or spectral radius of a matrix. We do not have to compute the least upper bound of all ω (as in (C.18)) since all the coefficients of the $\mathbf{A}_i$ are constant values. Therefore, in this case the ∞-norm is equal to the spectral norm $\bar{\sigma}\{\mathbf{A}_i\}$ according to (C.16). The resulting condition for stability is

$$\bar{\sigma}\left\{\mathbf{A}_i\right\} < 1 \qquad \text{for all } i \tag{4.114}$$

As this condition can be tested easily by a suitable software tool, the only problem which remains for this method is therefore to find the TSK model of the closed-loop system.

In addition to the condition for the spectral norm, in [192] another condition for the spectral radius $\lambda_{max}\{\mathbf{A}_i\}$ is derived. First, because of $\lambda_{max}\{\mathbf{A}_i\} \leq \bar{\sigma}\{\mathbf{A}_i\}$ the condition

$$\lambda_{max}\{\mathbf{A}_i\} < 1 \qquad \text{for all } i \tag{4.115}$$

is obviously a necessary condition for the stability of the system. It gets sufficient, if a common matrix $\mathbf{S}$ exists, so that $\mathbf{S}^{-1}\mathbf{A}_i\mathbf{S}$ is normal for all $\mathbf{A}_i$. A matrix $\mathbf{M}$ is normal, if $\mathbf{M}^T\mathbf{M} = \mathbf{M}\mathbf{M}^T$ holds. In [192] criteria for the existence of $\mathbf{S}$ are derived, but because of low importance for practical applications we will not present it here.

Another method (cf. [187]) is more interesting, which requires a continuous TSK model (4.27) of the closed-loop system and is based on the following theorem:

Theorem 4.4 *Let*

$$\dot{\mathbf{x}} = (\mathbf{A} + \mathbf{D}\mathbf{F}(t)\mathbf{E})\mathbf{x} \tag{4.116}$$

represent a system with given real-valued matrices $\mathbf{A}$*,* $\mathbf{D}$ *and* $\mathbf{E}$ *and a real-valued, time-variant uncertainty* $\mathbf{F}$*. The only thing which is known about* $\mathbf{F}$ *is that its norm is less than 1:* $||\mathbf{F}||_\infty \leq 1$*. This system is stable if all the eigenvalues of* $\mathbf{A}$ *have a negative real part and if furthermore*

$$||\mathbf{E}(s\mathbf{I} - \mathbf{A})^{-1}\mathbf{D}||_\infty < 1 \tag{4.117}$$

holds.

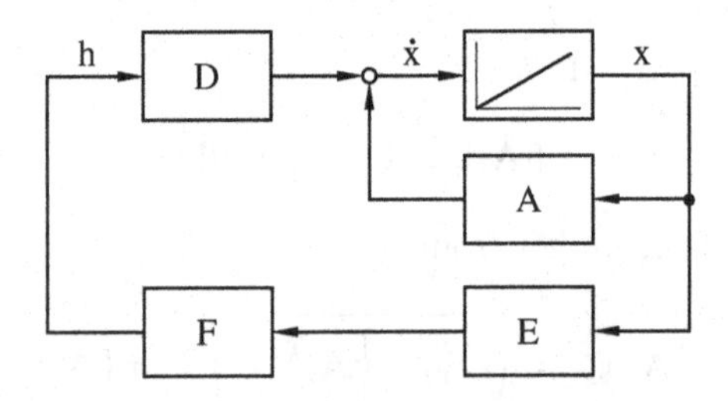

Fig. 4.8. Block diagram of the system

We can prove this theorem easily using the system's block diagram, which is shown in figure 4.8. We split the closed loop at the quantity $\mathbf{h}$. The resulting open-loop transfer function is $\mathbf{F}\mathbf{E}(s\mathbf{I} - \mathbf{A})^{-1}\mathbf{D}$. We can approximate its norm by

$$||\mathbf{F}\mathbf{E}(s\mathbf{I} - \mathbf{A})^{-1}\mathbf{D}||_\infty \leq ||\mathbf{F}||_\infty \, ||\mathbf{E}(s\mathbf{I} - \mathbf{A})^{-1}\mathbf{D}||_\infty \tag{4.118}$$

Because of $||\mathbf{F}||_\infty \leq 1$ and (4.117), the norm of the open-loop transfer function is certainly less than 1, which immediately implies the system's stability according to the Small Gain Theorem.

We can easily check inequality (4.117) in a numerical way. But the question remains how to convert the TSK model (4.27)

$$\dot{\mathbf{x}}(t) = \sum_i^r k_i(\mathbf{x}(t))\mathbf{A}_i\mathbf{x}(t) \tag{4.119}$$

into the form of (4.116). First, we have to split each matrix $\mathbf{A}_i$ up into a common part $\mathbf{A}_g$, that has only eigenvalues with a negative real part, and a residual part $\Delta\mathbf{A}_i$, which should be as small as possible. We obtain

$$\dot{\mathbf{x}}(t) = \left(\mathbf{A}_g + \sum_i^r k_i(\mathbf{x}(t))\Delta\mathbf{A}_i\right)\mathbf{x}(t) \tag{4.120}$$

We can then perform a singular value decomposition on the matrices $\Delta\mathbf{A}_i$, i.e.

$$\Delta\mathbf{A}_i = \mathbf{U}_i\mathbf{S}_i\mathbf{V}_i^T \tag{4.121}$$

where $\mathbf{U}_i$ and $\mathbf{V}_i$ are orthogonal matrices and $\mathbf{S}_i$ is a diagonal matrix, whose diagonal elements are the singular values of the matrix $\Delta\mathbf{A}_i$. Such a decomposition can be done numerically without any problems. As a result of this, (4.120) becomes

$$\dot{\mathbf{x}} = (\mathbf{A}_g + \mathbf{U}\mathbf{S}(t)\mathbf{V})\mathbf{x} \tag{4.122}$$

with

$$\begin{aligned}\mathbf{U} &= [\mathbf{U}_1, ..., \mathbf{U}_r]\\ \mathbf{V} &= [\mathbf{V}_1, ..., \mathbf{V}_r]^T\end{aligned} \tag{4.123}$$

and the diagonal matrix

$$\mathbf{S}(t) = \mathrm{diag}[k_1(t)\mathbf{S}_1 \; ... \; k_r(t)\mathbf{S}_r] \tag{4.124}$$

$\mathbf{S}$ contains the matrices $\mathbf{S}_i$ (multiplied with $k_i(t)$) on its main diagonal.

This form already corresponds to the one of the theorem (4.116). $||\mathbf{S}(t)||_\infty \leq 1$, however, is still not guaranteed, which is why we introduce a normalizing matrix

$$\mathbf{N} = \frac{1}{2}\mathrm{diag}[\mathbf{S}_1 \; ... \; \mathbf{S}_r] \tag{4.125}$$

We can then rewrite eq. (4.122) as

$$\dot{\mathbf{x}} = (\mathbf{A}_g + \mathbf{U}\mathbf{N}\mathbf{V} + \mathbf{U}\mathbf{N}\mathbf{N}^{-1}(\mathbf{S}(t) - \mathbf{N})\mathbf{V})\mathbf{x} \tag{4.126}$$

Setting

$$\begin{aligned}\mathbf{A} &= \mathbf{A}_g + \mathbf{U}\mathbf{N}\mathbf{V} = \mathbf{A}_g + \sum_i^r \frac{1}{2}\mathbf{U}_i\mathbf{S}_i\mathbf{V}_i^T = \mathbf{A}_g + \frac{1}{2}\sum_i^r \Delta\mathbf{A}_i\\ \mathbf{D} &= \mathbf{U}\mathbf{N}\\ \mathbf{F}(t) &= \mathbf{N}^{-1}(\mathbf{S}(t) - \mathbf{N})\\ \mathbf{E} &= \mathbf{V}\end{aligned} \tag{4.127}$$

leaves us with the desired representation (4.116). Even $||\mathbf{F}||_\infty \leq 1$ is now guaranteed, since any diagonal element of $\mathbf{S}$ consists of a product of a singular value σ and a factor $k_i(t)$, with $0 \leq k_i \leq 1$. As singular values cannot be negative, the diagonal element has to lie within the interval $[0, \sigma]$. The subtraction $\mathbf{S} - \mathbf{N}$ transforms this into $[-\frac{\sigma}{2}, \frac{\sigma}{2}]$, and a multiplication by $\mathbf{N}^{-1}$ into $[-1, 1]$. The absolute value of any element of the diagonal matrix $\mathbf{F}(t)$ is therefore at most 1, which is why the absolute value of $\mathbf{F}$'s output vector can never be greater than the absolute value of its input vector. Because of (C.17), the ∞-norm of $\mathbf{F}$ is therefore at most 1.

The matrices of (4.127) are the basis to apply theorem 4.4. The smaller we choose the matrices $\Delta\mathbf{A}_i$, the higher the probability is that $\mathbf{A}$ contains only eigenvalues with negative real part. It is obvious, that $\mathbf{F}$ fulfills the condition, so that it is eq. (4.117) that we have to check. But this is easy, because the necessary computation of the norm as well as the singular value decomposition in (4.121) can be performed with an adequate software tool. Therefore, the use of norms is a simple and elegant method for discrete and continuous systems as well to analyze the stability of the system. The precondition is a TSK model of the closed-loop system.

4.7 Hyperstability Criterion

Another method is based on the theory of hyperstability. Its application to fuzzy controllers requires almost no extension of the form given in chapter 2.8.9.

First of all, we have to separate the closed-loop system into a linear and a nonlinear part. We then have to structure the system in a way that both parts have the same number of input and output quantities. Let us use the system of figure 4.2 once more as an example, but this time we assume e and $\dot{e}$ to be the only input quantities of the fuzzy controller with the output quantity $\dot{u}$, i.e. it should not depend on $\int e$. We can regard this system as a SISO system or as one with two input and two output quantities.

If we treat it as a SISO system, we have to perform the same actions as with the circle criterion. First, we have to define one of the two input quantities to be the main input quantity. If we choose, for example, e, we have to interpret the transfer characteristic of the fuzzy controller $f(e, \dot{e})$ as a time-variant function $f(e, t)$. We obtain the transfer function of the linear part of the system from the transfer characteristic between the fuzzy controller's output quantity $\dot{u}$ and the quantity e, which is $\frac{1}{s}G(s)$ for our example. With $f(e, t)$ and $\frac{1}{s}G(s)$ we can perform the algorithm as described in chapter 2.8.9, where the first step, i.e. the insertion of the matrices $\mathbf{N}$ and $\mathbf{M}$, is not necessary any more, as both parts of the system have only one input and one output quantity.

If we want to treat this system as a two-input-two-output system, we get from

$$\mathbf{e} = \begin{pmatrix} e \\ \dot{e} \end{pmatrix} = \begin{pmatrix} \frac{1}{s}G(s) \\ G(s) \end{pmatrix} \dot{u} \tag{4.128}$$

the linear transfer matrix

$$\mathbf{G}(s) = \begin{pmatrix} \frac{1}{s}G(s) \\ G(s) \end{pmatrix} \tag{4.129}$$

We can see, that the linear system part has got two output quantities, while the nonlinear part possesses only one. Therefore, we have to define an additional, artificial output quantity with the constant value zero for the fuzzy controller. We get the output vector $\mathbf{u} = [u_1, u_2]^T = [\dot{u}, 0]^T$, and for the matrices $\mathbf{N}$ and $\mathbf{M}$ like in chapter 2.8.9

$$\mathbf{N} = \begin{pmatrix} 1 & 0 \end{pmatrix} \qquad \mathbf{M} = \begin{pmatrix} 1 \\ 0 \end{pmatrix} \tag{4.130}$$

For the transfer matrix of the linear system part it follows (see (2.313))

$$\mathbf{G}(s)\mathbf{N} = \begin{pmatrix} \frac{1}{s}G(s) & 0 \\ G(s) & 0 \end{pmatrix} = \mathbf{G}_{new} \tag{4.131}$$

With these definitions we make the linear and the nonlinear system part as well possessing each two input and two output quantities, so that the precondition for hyperstability analysis is fulfilled.

If necessary, we have to extend the system by the matrices $\mathbf{K}$ and $\mathbf{D}$ (cf. fig. 2.95) to make sure, that the linear part is stable or positive real respectively. The steps belonging to that were discussed in detail in chapter 2.8.9. Finally, we just have to check the condition (2.319)

$$\int_0^T \mathbf{u}'^T \mathbf{e}' dt \geq -\beta_0^2 \tag{4.132}$$

for the extended nonlinear part of the system. This can be done approximately by prooving the positiveness of the integrand in (2.320) or (2.326) for a suitable and sufficiently large set of vectors $\mathbf{e}$.

4.8 A Comparison with the Sliding Mode Controller

Significantly less problems arise if one analyzes the stability of a system by comparing a fuzzy controller to a sliding mode controller ([41], [155]). With this method, we do not have to separate the system into a linear and a nonlinear part. But unfortunately, it can be used in general for SISO systems only, although the controller may have several input quantities. We require that the entire system except the fuzzy controller can be described by the state equation (2.328)

$$x^{(n)}(t) = f(\mathbf{x}(t)) + u(t) + d(t) \tag{4.133}$$

Such a state equation can be obtained if the system is restructured according to figure 4.2, but this time, the plant may contain nonlinearities.

Starting point for our discussion is the control rule (2.342) of a sliding mode controller

$$u = -f_0(\mathbf{x}) + g_\lambda(\dot{\mathbf{e}}) + x_d^{(n)} + (F + D + \eta)\text{sgn}(q) \tag{4.134}$$

which we presented in chapter 2.8.10. To simplify the formula, let us assume that the n–th derivative of the reference value vanishes ($x_d^{(n)} = 0$), which imposes no severe restrictions on any real application. Let us furthermore assume that no nominal model of the plant exists: $f_0 = 0$. Of course, we then have to choose a sufficiently large value for F as an estimation of the uncertainty of the model. Finally, in order to achieve a continuous course of the reference value, we replace the sign function by the function $h(q)$, which we also already introduced in chapter 2.8.10. This leaves us with

$$u = g_\lambda(\dot{\mathbf{e}}) + (F + D + \eta)h(q) \tag{4.135}$$

as our control rule. In correspondence to the derivation given in chapter 2.8.10, we can say that any control rule

$$u = g_\lambda(\dot{\mathbf{e}}) + Uh(q) \tag{4.136}$$

with $U \geq F + D + \eta$ will force the system from any initial state into a zone $|q| < \Phi$ (with Φ from definition (2.345)) around the hyperplane defined by $q = 0$, and keep it there. Within this zone, the system will then converge towards the target point $\mathbf{e} = \mathbf{0}$, but it can not reach it exactly. The bigger a value we choose for Φ, the larger the region of tolerance will be, that we have to accept.

In order to present the relation between a sliding mode controller and a fuzzy controller more clearly, let us discuss the control rule (4.136) for a second-order system. The control rule changes to

$$u = \lambda \dot{e} + Uh(q) \tag{4.137}$$

We can represent this control rule as a characteristic surface in the $e-\dot{e}-$plane (fig. 4.9, right-hand side). The value of the actuating variable u is zero along a certain line, positive in the region above and negative in the region below this line. Without the first summand, this line would coincide with the line defined by $q = 0$, because of $h(0) = 0$.

If we compare this characteristic surface to the rule base of a typical fuzzy controller (which is also given in figure 4.9), the similarity between the output quantities of the two controllers is evident. This suggests—for stability analysis of a fuzzy controller—constructing a comparable sliding mode controller and to derive the stability conditions for the fuzzy controller from it.

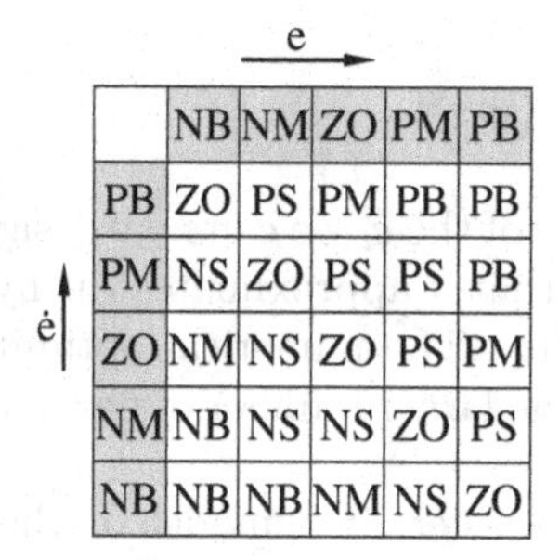

ė \ e	NB	NM	ZO	PM	PB
PB	ZO	PS	PM	PB	PB
PM	NS	ZO	PS	PS	PB
ZO	NM	NS	ZO	PS	PM
NM	NB	NS	NS	ZO	PS
NB	NB	NB	NM	NS	ZO

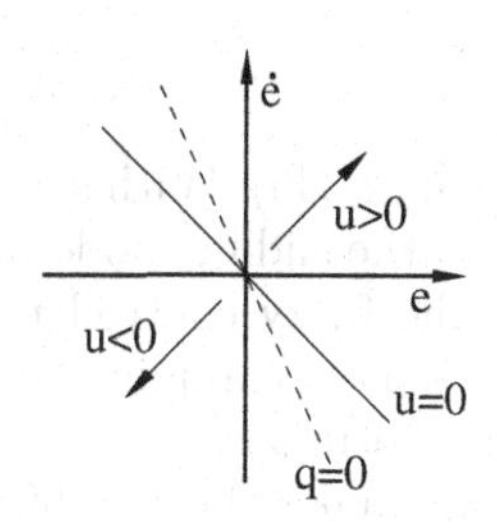

Fig. 4.9. Rule base of a fuzzy controller and zero line of a sliding mode controller

We obtain the following steps for the stability analysis of a given fuzzy controller: First we need to estimate an upper bound F for the model function f, and an upper bound D for the amplitude of the disturbance d. Since η only gives us a measure for the speed with which the system approaches the hyperplane $q = 0$, and the system is stable for any $\eta \geq 0$, we set $\eta = 0$ for reasons of simplicity. We then have to choose a value for the parameter Φ for the function $h(q)$, which at the same time represents a measure for the region of tolerance that has to be accepted. Hence, initially we pick a large value for Φ in order to obtain less restrictive conditions for the actuating variable of the fuzzy controller. The only remaining parameter of our sliding mode controller (4.136) that has to be fixed is λ. A numerical optimization will find a λ such that the sliding mode controller's hyperplane, defined by $u = 0$, matches the hyperplane of the fuzzy controller as much as possible.

Next, we need to select a sufficiently large number of vectors $\mathbf{e}$ and test the fuzzy controller's actuating variable numerically for two things: For all vectors $\mathbf{e}$, on one side of the hyperplane given by $u = 0$, its values have to be bigger than the values of the actuating variable of the sliding mode controller, and on the other side of this plane they have to be smaller. If this is true, then according to our explanations on (4.136) it can be guaranteed that the fuzzy controller will force the system from any initial state to a zone $|q| < \Phi$ around the hyperplane defined by $q = 0$, and keep it there. Within this zone, it converges towards the target point $\mathbf{e} = \mathbf{0}$ and finally reaches a region of tolerance, which will be the bigger, the larger a value was chosen for Φ. Since we picked a large value for Φ for a start, we can then repeat this calculation for smaller values of Φ, in order to establish not only stability, but also a small region of tolerance. However, smaller values for Φ produce more severe restrictions for the fuzzy controller.

It may often happen that the approximation of the fuzzy controller's hyperplane (given by $u = 0$) by the corresponding hyperplane of the sliding mode controller is insufficient. Hence, in order to allow a more flexible construction of the hyperplane, it may be necessary to introduce additional degrees of freedom. These can be received by replacing the definition of q corresponding to

(2.329)

$$q(\mathbf{e}) = (\frac{\partial}{\partial t} + \lambda)^{n-1} e \tag{4.138}$$

by equation (2.343). With a suitable choice of the c_i we can now design the hyperplane of the sliding mode controller so that it approximates the hyperplane given by the fuzzy controller nearly exactly. The numerical optimization, of course, gets more complicated caused by the larger number of parameters that have to be optimized.

We can make the stability condition easier, if a model of the plant is given. In this case, we have to add the summand $-f_0(x)$ of eq. (4.134) to the right-hand side of the equations (4.135) and (4.136), and then we can chose a smaller value for F. This will cause different values for λ or the c_i, but the computation steps remain the same. In addition to the weakening of the stability condition the addition of $-f_0(x)$ will make F to be a measure of the robustness of the fuzzy controller, as already discussed in chapter 2.8.10 .

Concrete examples for the comparison of fuzzy controllers with sliding mode controllers are given in [41] and [70], although no stability analysis of an already existing fuzzy controller is performed there, but the design procedure of a fuzzy controller following the requirements of a sliding mode controller.

In chapter 5.2.2 we will show how to adapt a fuzzy controller, that is based on a sliding mode controller, continuously to the plant with every single time-step. However, this fuzzy controller is given by a characteristic surface instead of fuzzy rules.

4.9 Direct Analysis in State Space

The methods which we have presented so far share the feature that they all avoid a direct computation of trajectories. Instead of it, they examine conditions which guarantee that the trajectories take a certain course, and thereby also ensure a certain stability behaviour of the system, without having to calculate specific trajectories. For example: The Lyapunov criterion examines the negative definiteness of the derivative of the Lyapunov function, the method of harmonic balance analyses the intersections of the describing function and the Nyquist plot, and the other methods focus on certain transfer characteristics of the individual parts of the system.

The methods we will describe in the following do not need such criteria. Here, we analyze possible trajectories directly in the state space.

4.9.1 Convex Factorization

A first step into this direction is the method of convex factorization, as presented in [85]. The underlying idea for this method is relatively simple. We require a closed-loop system whose time-discrete transfer characteristic is given by a facet function (cf. chapter 4.2.3), i.e. the state space is split up into convex polyhedrons P_j, and inside these polyhedrons the dynamic behavior of the system is approximated by an affine function:

$$\mathbf{x}(k+1) = \mathbf{f}_j(\mathbf{x}(k)) = \mathbf{K}_j\mathbf{x}(k) + \mathbf{d}_j \qquad \text{for } \mathbf{x}(k) \in P_j \tag{4.139}$$

The parameters $\mathbf{K}_j$ and $\mathbf{d}_j$ can be determined in a numerical way. Continuity of the facet function is not a necessary condition for the following stability analysis. However, we achieve an immense simplification if the polyhedrons are hyper-parallelepipeds parallel to the axis of the coordinate system. Furthermore, we need to now a certain area H around the zero position, which is guaranteed to belong to its domain of attraction. For reasons of simplicity, let us assume that H is the union of some of the polyhedrons, although this is not a necessary prerequisite.

We can then use the method of convex factorization to test whether a region G is also part of this area of attraction, where we assume that the polyhedron G is given in terms of the coordinates of its corners. First of all, we can exclude the part of G which is contained in H from any further investigation, as we already now that it is a part of the zero position's domain of attraction. We then split the remaining part of G up into sub-regions G_j, which are always entirely contained in a polyhedron P_j. This factorization requires some numerical effort, but can generally be done. Figure 4.10 shows an example for a second-order system. There, the polyhedrons P_j are rectangles, and the region H consists of the four inner rectangles. G extends into four rectangles and hence has to be split up into four sub-regions. G_4 is entirely contained in H, and can therefore be excluded from further considerations.

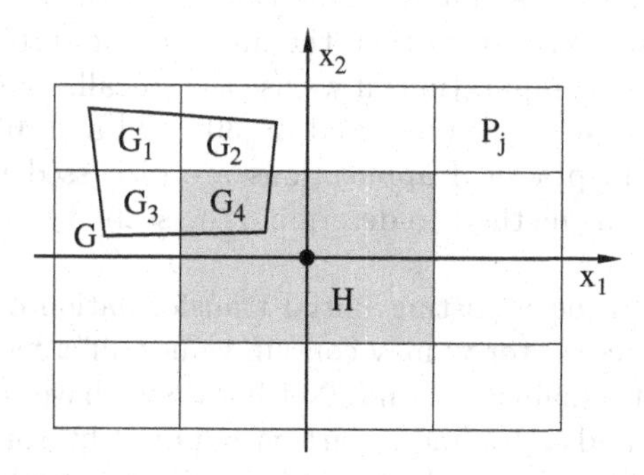

Fig. 4.10. Illustration to the method of convex factorization

Having split up the area into sub-regions, we apply the affine transformation $\mathbf{f}_j$ valid in P_j to the corresponding sub-region G_j, for any G_j that was not eliminated before. This means that we determine the images $\mathbf{f}_j(\mathbf{x}_{j,i})$ of the corners $\mathbf{x}_{j,i}$ of G_j. Due to the affinity of $\mathbf{f}_j$, these images are again corners of a convex bounded region $\mathbf{f}_j(G_j)$. However, this new region does not have to lie inside one polyhedron P_k only, but may extend into several polyhedrons. We then have to apply the algorithm once more to this new region, i.e. split it up, exclude sub-regions contained in H and apply the corresponding affine transformations to the remaining parts. In repeating these steps for several times, we obtain a tree structure as shown in figure 4.11. G is part of the zero position's domain of attraction if all the leaves of this tree are contained in H. If this is not the case, then no statement can be made. This latter case may occur in particular if the process is stopped after a certain maximum number of iterations, which is often the case for practical applications.

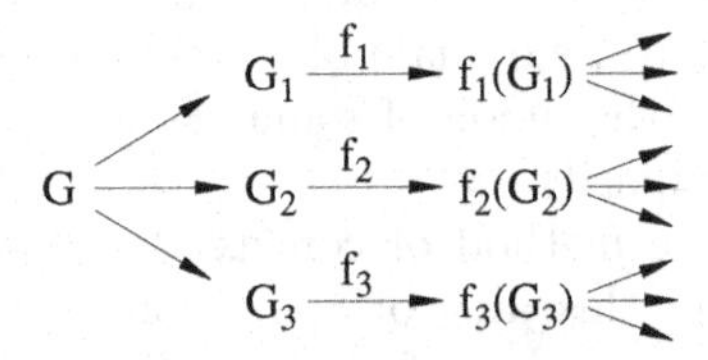

Fig. 4.11. Resulting tree for the method of convex factorization

This method obviously directly evaluates the courses of the trajectories in state space, even though it does not focus on single state points but always on entire regions in the state space.

4.9.2 Cell-to-Cell Mapping

The method of convex factorization still contains an analytical component: the characterization of the system's behavior inside the polyhedrons by affine transformations. In contrast to this, the analysis for stability is based completely on numerical computations if we use the so-called cell-to-cell mapping. Some theoretical aspects are presented in [30], [66] and [67], while problems which may occur in practical applications are discussed in [136], [127] and [126]. We can use this method to determine the stability behaviour of a given position of rest.

For this method, no adjusting initial transformation of the system is required. The reference vector $\mathbf{w}$ may contain values different from zero, as long as the vector itself remains constant. Neither do we have to separate the system into a linear and a nonlinear part as required by some other methods. Both the controller and the plant can be nonlinear, and can have internal dynamics.

The most important feature of this method, however, is that the information about the system does not have to be given in a special form, e.g. as a facet function, characteristic surface or TSK model. The only thing that has to exist for an application of this method are time-discrete functions which can be used to compute the current output values $\mathbf{u}(k)$ of the controller and $\mathbf{y}(k)$ of the plant from currently available quantities of the system $z_i(k)$ or even from past values $z_j(k-1)$. This allows us to use the controller for stability analysis in the same form as it was programmed for the real application. And the plant's transfer characteristic can be given by a TSK model or a facet function as well as by a classical, analytical model. Even a description which reflects only qualitative features, e.g. by a fuzzy relation or a neural network, is possible. This fact makes cell-to-cell mapping interesting especially for practical applications, as the main field of application for fuzzy controllers are plants whose behavior can be characterized only in a qualitative way.

If we have a description of the plant's and the controller's transfer characteristic, we first have to determine the state variables of the closed-loop system —which are the plant's states, but also possible states of the controller which are caused by integrations or differentiations of input and output quantities of the fuzzy controller. This process is usually the main problem of this method, especially if the user lacks sufficient control-engineering specific knowledge. This is, however, not a specific problem of cell-to-cell mapping, but one that occurs in any of the methods which are based on a state space description of the closed-loop system.

Knowing the states, we can sum up the functions for the controller and the plant into a time-discrete state space description of the system:

$$\mathbf{x}(k) = \mathbf{f}(\mathbf{x}(k-1)) \tag{4.140}$$

where the function $\mathbf{f}$ may actually be a combination of fuzzy relations, neural networks and approximating, analytical functions.

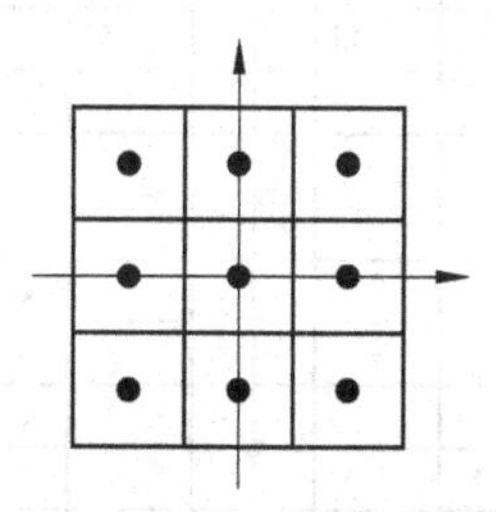

Fig. 4.12. Splitting up the state space into cells

For an analysis of stability, we then restrict the state space to a certain area of interest around the position of rest, and separate this area into cells by means of lines parallel to the axis of the coordinate system (fig. 4.12). Every cell is represented by its center. For every center, we compute—using the state equation (4.140)—its successor states until the first of these states lies in another cell. We record this cell as the successor cell of our starting cell. As an example, figure 4.13 shows a part of the state space. There, we first computed the successor states of the left upper center. The third of these states falls into the cell in the middle of the upper row, which is therefore the

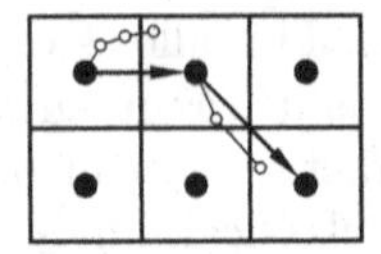

Fig. 4.13. Computing successor cells

successor cell of the upper left cell. In the same way, we obtain the lower right cell as the successor cell of this last cell.

As the result of this step, we finally obtain a cell-to-cell mapping as a description of the behavior of the system, i.e. a function which assigns every cell a successor cell (fig. 4.14). This function replaces the state equation, which defined a successor state for every state. It makes stability analysis a lot easier: Starting with the cell corresponding to the zero position of rest, we determine all those cells which are mapped onto this cell, then we focus on the predecessors of those cells etc. All the cells which are—in one or more steps—mapped onto the cell corresponding to the zero position form the domain of attraction of the position of rest. In figure 4.14, the zero position of rest lies in the center of the coordinate system. We can see that all of the inner cells form part of the zero position's domain of attraction.

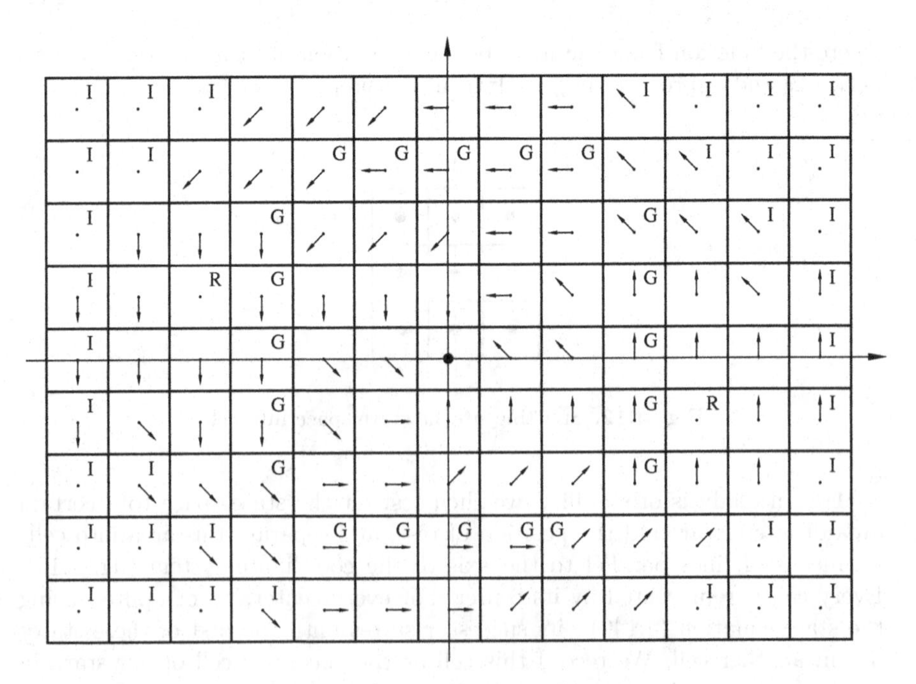

Fig. 4.14. Cell-to-cell mapping

Let us discuss some of the problems which may occur in practical applications. It is possible that for some regions of the state space no information about the behavior of the plant—and thereby also about the system—is available, especially if the model of the plant is a result of an automatic identification process. In this case, it is not possible to determine the successor cells for the cells corresponding to these regions, which makes any statements concerning the stability of these cells impossible. As, in general, a statement about stability better be restrictive, we will call these cells "instable", and will not consider them to be part of the zero position's domain of attraction. Another type of instable cells are those at the border of the analyzed area for which the trajectory of the successor states "leaves" this area of interest. And finally, we consider all the cells to be instable which are mapped (again in one or more steps) onto instable cells. In figure 4.14, all the instable cells are marked with I. We have to think of the corners as being instable since there is no information about the system's behavior available, while for the top and bottom row an instable cell exists whose successor cell would lie outside the analyzed area. Further instable cells are those which are mapped onto the ones just mentioned.

Additional zero positions also cause problems. We defined a position of rest as a point in state space whose corresponding state trajectory does not move away from it. If a cell contains a position of rest, then a trajectory computed for the center point of this cell may possibly run into this position of rest and remain there. Accordingly, all successor states for this center point would lie inside this cell. On the other hand, the algorithm attempts to compute successor states until a state outside this cell is reached, so that a successor cell for this cell can be named. Therefore, the position of rest in this cell would lead to an infinite process. In order to avoid such problems, we fix a lower bound for the minimum change of the state vector in one time step. If the change falls short of this border, we refrain from computing further successor states and define this cell to represent a position of rest, to be a "rest-cell". With this definition, however, it is now possible that we miss-classify cells as rest-cells just because their states change only very slowly. We cannot avoid this problem. If the algorithm defines other rest-cells besides the one in focus, we would have to check the corresponding region of the state space "by hand" in order to see whether the reason is an extremely slow change of the system's state or actually another position of rest.

The cells marked R in figure 4.14 represent rest-cells. Some other cells are mapped onto these cells. The additional rest-cells as well as the cells which are mapped onto them do, of course, not form part of the domain of attraction of the zero position of rest under consideration.

The last problem which we have to discuss here is caused by limit cycles. They can occur for cell-to-cell mapping just like for state equations, i.e. after several applications of the function, a cell is finally mapped onto itself. Limit cycles are easy to detect, since all the cells which are neither instable nor rest-cells nor mapped onto any of these have to be part of limit cycles or their domains of attraction. It follows that first of all we have to detect all the rest-cells and instable cells plus their domains of attraction. Next, we pick one of the remaining cells and keep on computing its successor cells until we finally reach a cell a second time. We can then immediately name the corresponding limit cycle. In figure 4.14, all the cells marked G form one limit cycle. In addition to this, cells exist which are not part of the limit cycle, but which are mapped onto it.

In order to estimate the validity of a statement concerning stability resulting from an application of this method, one has to be aware of the fact that it may be affected in two ways: On the one hand, the underlying state equation (4.140) may itself be a result of qualitative information only, and therefore reflect the actual relations in an incomplete way. Hence, if this model was used as a basis for a long-term simulation, then the computed behavior of the system would very soon differ immensely from the behavior of the real system, due to the accumulation of errors. However, we avoid such a long-term simulation. Starting from the center point of a cell, we compute only a few successor states until we reach an adjacent cell. Within these few steps of simulation, the deviation between the model and the real system cannot reach any significant size. Still, the inaccuracy of the state equation affects the quality of the results achieved using this method, of course. But on the other hand, such sources of error also exist for all other methods.

The other source of inaccuracy is specific to this method. It occurs at the transition from a state equation to a cell-to-cell mapping, i.e. it is caused by the discretization of the state space. In figure 4.13 for example, the cell in the lower right corner is considered to be the successor of the cell on the upper left. But if we continued to compute successor states starting from the center of the upper left cell, then the trajectory starting at this center point would probably pass through the cell in the middle, into the cell on the upper right. This would make the upper right cell to be the second successor of the upper left one. This illustrates how discretization may cause errors in the analysis. But these errors can be eliminated quite easy by repeating the complete stability analysis for several times with different discretizations. If the result remains the same for all cases, we can be sure that it is correct.

4.10 Conclusion

It is obvious, that all methods for stability analysis have different strong and weak points, that are mainly caused by their different preconditions. Therefore, it makes sense to draw a conclusion from exactly this aspect.

The most extreme example is the proof of stability by comparison with an existing sliding mode controller. A classical (sliding mode) controller has to be designed first to prove the stability of a fuzzy controller. But if such a classical controller design is possible the question arises why a fuzzy controller should be designed. A fuzzy controller whose stability can be proven in such a way does only make sense if it is understood and designed as a modification of the corresponding sliding mode controller, for example to adjust the course of the actuating variable to practical requirements.

The methods of the describing function, the Popov criterion, the circle criterion and the criterion of hyperstability have one essential defect, too. They need the separation of the closed loop into a linear and a nonlinear part. As the fuzzy controller is already nonlinear, the nonlinearities of the plant have to be located directly at its input or output, to enable their separation from the linear rest of the plant and their combination with the fuzzy controller to form the nonlinear part of the system. Otherwise the required separation of the system into a linear and a nonlinear part would not be possible. But if we have such a simply structured plant and even an analytical plant model available, it is normally no problem to design a controller in the classical (analytical) way. There are only a few cases thinkable, where under these conditions the design of a fuzzy controller could make more sense than the design of a classical one. And only in these cases the above named methods should be used to prove stability.

On the other hand, the direct method of Lyapunov does not require any special structure of the system. And if we have a TSK representation of the system available, we can even answer the question of the existence of the Lyapunov function—which is normally the critical point of this method—and therefore of the stability of the system easily with the help of LMI algorithms. Furthermore, the TSK representation is possible for any system without hysteresis, step functions and delay time. And even these three effects can often be approximated quite well by TSK systems. For these reasons we can state, that this method has got a huge potential for practical applications, which is verified by the increasing number of publications regarding this method throughout the last years.

The same range of application holds for the norm-based stability analysis, as this method requires TSK models, too. But it has the principle disadvantage, that it is based on the small gain theorem which causes very conservative results. For an existing TSK model the direct method of Lyapunov is therefore more promising.

Finally, we have to discuss the last two of the presented methods for the direct analysis in state space, which have to be seen separately from all other methods as they follow a completely different approach. They distinguish from the other methods, because here we reflect directly on the run of the trajectories in the state space in a numerical way, while for all other methods derived criteria are investigated to enable an analytical proof of stability. These criteria are for example the negative definiteness of the derivative of the Lyapunov function, the intersection of the curves of the describing function and the Nyquist plot, or certain transfer characteristics of the single parts of the system. In contrast to all the other methods, the two methods for the direct analysis in state space do not provide an (exact) analytical proof of stability at all.

On the other hand, we should see that the check of an analytical stability criterion can be done analytically only, if the system has a very simple structure, as this chapter has shown clearly. For the most real applications, we need numerical computations for the check of the analytical stability criterion, so that the analytical exactness of the proof gets lost. Besides that, a completely analytical proof often only simulates analytical exactness, as it is based on a model of the plant, that might have large differences from the real plant, especially if a linearization was made.

Therefore, there is no principal disadvantage of the two direct state space analysis methods compared to the other ones. And their decisive advantage lies in the fact, that there exist no restrictions or requirements regarding the structure of the system or the kind of information about the system. Analytical functions can be used as well as fuzzy models or neural networks. For the convex factorization these models are approximated by facet functions, and for cell-to-cell mapping they can even be integrated directly into the numerical computations.

The convex factorization is numerically much more demanding and therefore contains more possibilities for numerical problems, while the cell-to-cell mapping gives less safety regarding stability, as only the center of each cell is investigated, where for the convex factorization this is done for all states within every polyhedron. But this disadvantage of the cell-to-cell mapping can be compensated by performing the stability analysis for several times with different discretizations of the state space. If the results are similar for all cases, errors that might be caused by the discretization of the state space can be excluded.

Finally we can say, that the direct method of Lyapunov for TSK systems and the methods for the direct analysis in the state space are the methods, that promise the widest range of practical applications and should therefore receive more attention than the other ones. But this should not make someone to call the other methods useless, because for any method there exist systems for which exactly this method provides the optimal results.

5

Parametrization and Optimization of Fuzzy Controllers

One argument that is frequently used in favor of fuzzy controllers is that they can be constructed easily and fast. However, this is true only for relatively simple plants. With increasing complexity, the effort to develop a fuzzy controller for any system grows dramatically. In such cases, the heuristic approach to determine the membership functions and rules—which is an advantage for simple plants—turns into a time-consuming disadvantage. For this reason, several attempts emerged from the end of the 80s on in order to systematize the design and adaptation process of a fuzzy controller. We discuss the most important approaches in this chapter.

However, before we start with this discussion, let us once again present some of the methods of classical control engineering, which have been tried and tested for decades and which can simplify any controller design—including fuzzy controller design. Therefore, they should also be mentioned here.

The first of these methods is the principle of the multi-loop control (chapter 2.8.2). This type of control is useful if we can represent the plant as a series connection of separate components, and if the output quantities of these components can be measured. Every output quantity is fed back and controlled by a separate controller. This produces a system consisting of several loops, as shown in figure 2.64. We first have to design the controller for the innermost loop. We can then model this closed loop by a simple lag, and turn to the next—enclosing—controller. In this way, the controllers are designed from the inner- to the outermost loop. This method also offers the advantage of putting a system into operation in a step-by-step manner, from inner- to outermost loop, which reduces the risk of destruction by means of ill-designed controllers.

Decoupling is another (possibly helpful) classical method. It can be used to split MIMO systems up into single-input-single-output systems, which can then be controlled separately. In chapter 2.6.4, we gave a description of how linear systems can be decoupled if certain conditions are met. Unfortunately, for nonlinear systems neither an algorithm for computing the decoupling elements can be given nor is it possible to present the conditions which allow

K. Michels et al.: *Fuzzy-Control: Fundamentals, Stability and Design of Fuzzy Controllers*, StudFuzz **200**, 307–374 (2006)
www.springerlink.com

decoupling at all. It is therefore necessary first to detect the relationships between the individual quantities in a detailed analysis before a strategy for eliminating them can be developed. Some ideas concerning this topic with regard to fuzzy systems can be found in [49], [48] and [206]. However, these methods assume a genuine fuzzy system, i.e. that the transfer characteristics of both the plant and the controller are given by fuzzy relational equations. This emphasizes the set-theoretical aspect instead of the practical one, and we will therefore refrain from discussing them.

Gain scheduling is another principle of classical control engineering, which we already introduced in chapter 2.8.2. Its underlying idea is to design several linear controllers corresponding to different operating points of a nonlinear plant, and to activate these controllers according to the working point. [103] presents a method where a fuzzy controller is used to adapt the coefficients of PID controllers. There, the rules of the fuzzy system reflect the knowledge which is usually used to adjust PID controllers. Another version of gain scheduling are the TSK controllers, that were introduced in sections 3.2, 4.1.3 and 4.2.2. Here, in section 5.1 we will complete this by introducing a very smart design method for TSK controllers.

We can also use the methods for stability analysis, that were introduced in chapter 4, for the design of a fuzzy controller. If we use, for example, the direct method of Lyapunov for conventional fuzzy controllers, we first have to define a Lyapunov function which we can then use to compute a transfer characteristic for the controller at the border of stability (cf. chapter 4.2.1). Finally, we have to determine a fuzzy controller in a way that it obeys to the boundaries imposed by this transfer characteristic. [171] presents a similar process for the case that the system is characterized by a facet function. The Popov criterion, the circle criterion, the criterion of hyperstability and the concept of a sliding mode controller can be used in a similar way. For all methods, based on the plant we have to derive the conditions resulting from the stability criterion, that have to be fulfilled by the controller to get a stable closed-loop system. Then we can use these conditions to design a suitable controller.

A borderline case between classical control theory and fuzzy control is the adaptation of characteristic fields. There, the fuzzy controllers are given in the form of characteristic fields (cf. section 4.1.2), and these are adapted to the plant or to the current operating point with every time step. As fuzzy sets do not appear any more, it is a moot point if these methods belong to classical control or fuzzy control. But on the other hand, as there are many interesting contributions about these methods especially in fuzzy journals, we did not want to omit this subject (section 5.2).

However, the automatic modification of fuzzy rules in every time step, that is based on heuristic rules, is typical for fuzzy control. This method is widely spread and often treated in publications, so that we were forced to present it in this chapter. But the presentation could be kept very short, because this method has one essential defect, as we will explain in section 5.3. How this defect can be removed, will be shown in the following section 5.4.

Another typical application for fuzzy control is the simulation of a given transfer behaviour by a fuzzy controller. This approach often occurs in practical applications, for example if the human operator of a plant should be replaced by a fuzzy controller. In a case like this, first the operator's transfer characteristic has to be observed for a longer period of time, and the measured quantities have to be recorded. These values can then be used as a basis for computing a transfer element with approximately the same transfer characteristic as the human operator. This can be achieved by one of the traditional methods, using a characteristic surface whose parameters are determined by a regression algorithm. However, it is also possible to use the values to train a neural network until its reactions to the process are close enough to the ones of the operator. If linguistic interpretability of the transfer characteristic is desired, a fuzzy clustering algorithm seems to be a reasonable choice as discussed in section 5.5.

Another very interesting field of application for fuzzy control is to combine them with neural networks, to complete the advantages of a fuzzy controller (linguistic interpretability) by the ones of a neural network (ability to learn). Such combinations exist in many different varieties and are defined as Neuro Fuzzy Controllers (NFC). However, NFC's require suitable possibilities for the training of the neural network, that might not be given in any practical application. But if sufficient training can be ensured, NFC's represent a very interesting approach for the controller design. They are presented in section 5.6.

Finally, the possibilities of genetic algorithms in the design process of controllers and especially fuzzy controllers should not be neglected. We require an already fairly precise model of the plant. Next, we have to generate a certain number of possible fuzzy controllers (*population*), more or less at random. For each of these controllers, we run a simulation on the model and evaluate the results of the simulation according to a previously defined criterion. This value represents a measure (*fitness*) for the quality of the corresponding controller. Depending on the fitness of the controllers and the choice of values for the parameters of the genetic algorithm, the controllers will then be eliminated, mutated or combined. The resulting controllers form the population of the next generation, and the process starts all over again. After a certain number of repetitions, we hopefully end up with at least one reasonably "fit" controller.

However, two aspects of this method have to be taken in to account. One is the necessity of a fairly precise model of the plant. If such a model exists, it should be mentioned, if a classical controller design method is preferable. The other aspect is, that a controller, that is successful in a certain number of simulation runs, is not necessarily successful in any possible situation that might happen in a practical application. The stability of the control loop is not guaranteed by this method. But anyway, the use of genetic algorithms forms an interesting alternative for the controller design, as far as it is used well-thoughtout.

5.1 Design of TSK Controllers

For the design of a TSK controller also the plant has to be available as TSK system (cf. chapter 4.1.3). Provided that, analogously to the methods in chapter 4.2.2, the design of the TSK controller can be formulated as LMI problem and this problem can be solved with LMI algorithms (cf. appendix G) ([188]).

Starting point are the equations (4.72) and (4.73) for which we only have to define in an appropriate way some intermediate variables. With $\mathbf{G}_{ij} = \mathbf{A}_i + \mathbf{B}_i\mathbf{F}_j$ it results from (4.72) and (4.73) at first

$$\mathbf{A}_i^T\mathbf{P} + (\mathbf{B}_i\mathbf{F}_i)^T\mathbf{P} + \mathbf{P}\mathbf{A}_i + \mathbf{P}\mathbf{B}_i\mathbf{F}_i < \mathbf{0} \tag{5.1}$$

$$\mathbf{A}_i^T\mathbf{P} + (\mathbf{B}_i\mathbf{F}_j)^T\mathbf{P} + \mathbf{A}_j^T\mathbf{P} + (\mathbf{B}_j\mathbf{F}_i)^T\mathbf{P} + \mathbf{P}\mathbf{A}_i + \mathbf{P}\mathbf{B}_i\mathbf{F}_j + \mathbf{P}\mathbf{A}_j + \mathbf{P}\mathbf{B}_j\mathbf{F}_i < \mathbf{0} \quad \text{for} \quad i < j. \tag{5.2}$$

Then both inequations are multiplicated with $\mathbf{P}^{-1}$ as well from the left as from the right:

$$\mathbf{P}^{-1}\mathbf{A}_i^T + \mathbf{P}^{-1}\mathbf{F}_i^T\mathbf{B}_i^T + \mathbf{A}_i\mathbf{P}^{-1} + \mathbf{B}_i\mathbf{F}_i\mathbf{P}^{-1} < \mathbf{0} \tag{5.3}$$

$$\mathbf{P}^{-1}\mathbf{A}_i^T + \mathbf{P}^{-1}\mathbf{F}_j^T\mathbf{B}_i^T + \mathbf{P}^{-1}\mathbf{A}_j^T + \mathbf{P}^{-1}\mathbf{F}_i^T\mathbf{B}_j^T + \mathbf{A}_i\mathbf{P}^{-1} + \mathbf{B}_i\mathbf{F}_j\mathbf{P}^{-1} + \mathbf{A}_j\mathbf{P}^{-1} + \mathbf{B}_j\mathbf{F}_i\mathbf{P}^{-1} < \mathbf{0} \quad \text{for} \quad i < j \tag{5.4}$$

Finally, with the definitions $\mathbf{X} = \mathbf{P}^{-1}$ and $\mathbf{H}_i = \mathbf{F}_i\mathbf{P}^{-1}$ and in view of the symmetry of $\mathbf{P}$ and $\mathbf{X}$ we get

$$\mathbf{X}\mathbf{A}_i^T + \mathbf{H}_i^T\mathbf{B}_i^T + \mathbf{A}_i\mathbf{X} + \mathbf{B}_i\mathbf{H}_i < \mathbf{0} \tag{5.5}$$

$$\mathbf{X}\mathbf{A}_i^T + \mathbf{H}_j^T\mathbf{B}_i^T + \mathbf{X}\mathbf{A}_j^T + \mathbf{H}_i^T\mathbf{B}_j^T + \mathbf{A}_i\mathbf{X} + \mathbf{B}_i\mathbf{H}_j + \mathbf{A}_j\mathbf{X} + \mathbf{B}_j\mathbf{H}_i < \mathbf{0} \quad \text{for} \quad i < j. \tag{5.6}$$

This equation system is obviously linear for the known matrices $\mathbf{X}$ and $\mathbf{H}_i$, and therefore it can, according to appendix G, transferred to the basic form (G.1) of an LMI problem. The application of an LMI solution algorithm does not only give a statement about the existence of a solution but also a possible solution for $\mathbf{X}$ and $\mathbf{H}_i$. From this we get the controller matrices at the single supporting points from $\mathbf{F}_i = \mathbf{H}_i\mathbf{X}^{-1}$. Since these conroller matrices are computed directly from the stability conditions, we can be sure that the entire controller is stable.

As a final remark, we have to say that in [28] the presented method is even extended for systems with parameter uncertainties of the type (4.52).

Additionally to the stability, we can incorporate other criteria to the LMI controller design. For example, the demand of an adequately high control velocity can be expressed by the demand of a certain changing velocity of the

Ljapunov function which should be the higher the further a state is away from the origin of the state space (cf. [188]):

$$\dot{V}(\mathbf{x}(t)) \leq -\alpha V(\mathbf{x}(t)) \tag{5.7}$$

$\alpha > 0$ may be chosen arbitrarily. It should be the greater the higher the control velocity shall be. As a simplified version of (5.7) we can also use the condition of the *quadratical stability* :

$$\dot{V}(\mathbf{x}(t)) \leq -\alpha ||(\mathbf{x}(t))||^2 \tag{5.8}$$

With (5.7) the equation (4.43) becomes

$$\dot{V} = \sum_i k_i \mathbf{x}^T (\mathbf{A}_i^T \mathbf{P} + \mathbf{P}\mathbf{A}_i)\mathbf{x} < -\alpha V(\mathbf{x}) = -\alpha \mathbf{x}^T \mathbf{P} \mathbf{x}$$

$$\sum_i k_i \mathbf{x}^T (\mathbf{A}_i^T \mathbf{P} + \mathbf{P}\mathbf{A}_i)\mathbf{x} + \alpha \mathbf{x}^T \mathbf{P} \mathbf{x} < 0$$

$$\sum_i k_i \mathbf{x}^T (\mathbf{A}_i^T \mathbf{P} + \mathbf{P}\mathbf{A}_i + \alpha \mathbf{P})\mathbf{x} < 0 \qquad \text{with} \quad \sum_i k_i = 1 \tag{5.9}$$

and the stability condition (4.42) changes to

$$\mathbf{A}_i^T \mathbf{P} + \mathbf{P}\mathbf{A}_i + \alpha \mathbf{P} < \mathbf{0}. \tag{5.10}$$

So we just have to add the summand $\alpha\mathbf{P}$ to all of the inequations. The inequation system remains linear in $\mathbf{P}$ so that the LMI solution algorithms can still be used.

5.2 Adaptation of Characteristic Surfaces

The central point of these algorithms is a fuzzy controller given in the form of a characteristic surface (cf. chap. 4.1.2)

$$u = \sum_i k_i(\mathbf{x}) u_i = \mathbf{u}^T \mathbf{k}(\mathbf{x}) \tag{5.11}$$

where $\mathbf{k} = (k_1, k_2, ...)^T$ is the vector consisting of the weights and depending on the instantaneous state. $\mathbf{u} = (u_1, u_2, ...)^T$ represents the vector of the—normally constant—coefficients of the characteristic surface. These coefficients are continuously adapted in the following algorithms, to obtain a stable control behavior.

5.2.1 Adaptive Compensation Control

The first of the different classes of controllers which we present here is the one of adaptive compensation controllers. We follow the approach given in [201]. These algorithms can be applied to SISO plants of the form

$$x^{(n)} = f(\mathbf{x}) + bu \qquad\qquad y = x \tag{5.12}$$

where $b > 0$, $\mathbf{x} = [x, \dot{x}, ..., x^{(n-1)}]^T$ the state vector and y the output variable. After the definition of an error vector $\mathbf{e} = [e, \dot{e}, ..., e^{(n-1)}]^T$ (where $e = y_s - y$ and the reference variable y_s) and a suitable choice for a parameter vector $\mathbf{r} = [r_0, ..., r_{n-1}]^T$, we can define an ideal control rule

$$u^* = \frac{1}{b}\left[-f(\mathbf{x}) + y_s^{(n)} + \mathbf{r}^T\mathbf{e}\right] \tag{5.13}$$

If we substitute this in (5.12), we obtain

$$e^{(n)} + \mathbf{r}^T\mathbf{e} = e^{(n)} + r_{n-1}e^{(n-1)} + ... + r_1\dot{e} + r_0 e = 0 \tag{5.14}$$

for the error and its derivatives. Through an application of this control rule, we obviously achieve a compensation of the nonlinear function $f(\mathbf{x})$, which leaves us with a linear differential equation. If we choose $\mathbf{r}$ in a way that all zeros of the characteristic polynomial for this differential equation have a negative real part, then the error and its derivatives will converge towards zero even for a time-variable reference value.

Unfortunately, since the function f is not exactly known, it is not possible to realize such a control rule. For this reason, we define a control rule consisting of two parts:

$$u = u_c(t, \mathbf{x}) + u_s(\mathbf{x}) \tag{5.15}$$

Here, u_c is the output quantity of the fuzzy controller (5.11), which we would like to adapt through time in a way that it corresponds to the ideal actuating variable u^* as much as possible. Such an adaptation, however, requires that both the vector of errors and the state vector are bounded. To ensure this, we add a second controller with the output quantity u_s in parallel which will intervene only if the boundary is violated.

The differential equation for the output error changes due to this control rule. Starting from equation (5.12)

$$x^{(n)} = f(\mathbf{x}) + b(u_c + u_s) \tag{5.16}$$

the subtraction of bu^* together with (5.13) yields

$$x^{(n)} - \left[-f(\mathbf{x}) + y_s^{(n)} + \mathbf{r}^T\mathbf{e}\right] = f(\mathbf{x}) + b(u_c + u_s - u^*) \tag{5.17}$$

which implies

$$e^{(n)} = y_s^{(n)} - x^{(n)} = -\mathbf{r}^T\mathbf{e} + b(u^* - u_c - u_s) \tag{5.18}$$

for the error, or in vector notation:

$$\dot{\mathbf{e}} = \mathbf{R}\mathbf{e} + \mathbf{b}(u^* - u_c - u_s) \tag{5.19}$$

where $\mathbf{b} = [0, ..., 0, b]^T$ and the matrix

$$\mathbf{R} = \begin{pmatrix} 0 & 1 & 0 & \cdots & 0 \\ 0 & 0 & 1 & \cdots & 0 \\ \dots & \dots & \dots & \dots & \dots \\ 0 & 0 & 0 & \cdots & 1 \\ -r_0 & -r_1 & -r_2 & \cdots & -r_{n-1} \end{pmatrix} \tag{5.20}$$

describes a stable system. Now, let us develop the control rule for the actuating variable u_s. With a positive definite matrix $\mathbf{Q}$, which has to be provided, we obtain a symmetrical, positive definite matrix $\mathbf{P}$ from the Lyapunov equation (cf. theorem D.1)

$$\mathbf{R}^T\mathbf{P} + \mathbf{P}\mathbf{R} = -\mathbf{Q} \tag{5.21}$$

$\mathbf{P}$ can then be used to define another function

$$V_e = \frac{1}{2}\mathbf{e}^T\mathbf{P}\mathbf{e} \tag{5.22}$$

which we will later on use as a Lyapunov function. Furthermore, we have to determine an upper bound $F > f$ and a lower bound $0 < B \leq b$ for the plant parameters, while the threshold V_0 for the function V_e has to be defined. Thus, we obtain the following control rule:

$$u_s(\mathbf{x}) = \begin{cases} \text{sgn}(\mathbf{e}^T\mathbf{P}\mathbf{b}) \left[|u_c| + \frac{1}{B}(F + |y_s^{(n)}| + |\mathbf{r}^T\mathbf{e}|)\right] & \text{if } V_e \geq V_0 \\ 0 \quad \text{otherwise} \end{cases} \tag{5.23}$$

i.e. u_s is different from zero only if $\frac{1}{2}\mathbf{e}^T\mathbf{P}\mathbf{e}$ exceeds the threshold V_0.

We will show that in this case the derivate of the Lyapunov function V_e from (5.22) is always less than zero. The derivative of the Lyapunov function V_e is

$$\dot{V}_e = \frac{1}{2}\dot{\mathbf{e}}^T\mathbf{P}\mathbf{e} + \frac{1}{2}\mathbf{e}^T\mathbf{P}\dot{\mathbf{e}} \tag{5.24}$$

Inserting (5.19) yields

$$\begin{aligned} \dot{V}_e &= \frac{1}{2}(\mathbf{R}\mathbf{e} + \mathbf{b}(u^* - u_c - u_s))^T\mathbf{P}\mathbf{e} + \frac{1}{2}\mathbf{e}^T\mathbf{P}(\mathbf{R}\mathbf{e} + \mathbf{b}(u^* - u_c - u_s)) \\ &= \frac{1}{2}\mathbf{e}^T(\mathbf{R}^T\mathbf{P} + \mathbf{P}\mathbf{R})\mathbf{e} + \frac{1}{2}(u^* - u_c - u_s)(\mathbf{b}^T\mathbf{P}\mathbf{e} + \mathbf{e}^T\mathbf{P}\mathbf{b}) \end{aligned} \tag{5.25}$$

As $\mathbf{b}^T\mathbf{P}\mathbf{e}$ is a scalar and $\mathbf{P}$ is symmetrical, we get

$$\mathbf{b}^T\mathbf{P}\mathbf{e} = (\mathbf{b}^T\mathbf{P}\mathbf{e})^T = \mathbf{e}^T\mathbf{P}^T\mathbf{b} = \mathbf{e}^T\mathbf{P}\mathbf{b} \tag{5.26}$$

and therefore with (5.21)

$$\dot{V}_e = -\frac{1}{2}\mathbf{e}^T\mathbf{Q}\mathbf{e} + \mathbf{e}^T\mathbf{P}\mathbf{b}(u^* - u_c - u_s) \tag{5.27}$$

Using the control rule (5.23) for $V_e \geq V_0$ this becomes

$$\dot{V}_e = -\frac{1}{2}\mathbf{e}^T\mathbf{Q}\mathbf{e} + \mathbf{e}^T\mathbf{P}\mathbf{b}(u^* - u_c)$$
$$-|\mathbf{e}^T\mathbf{P}\mathbf{b}|\left[|u_c| + \frac{1}{B}(F + |y_s^{(n)}| + |\mathbf{r}^T\mathbf{e}|)\right] \tag{5.28}$$

Estimating the second summand with (5.13) yields

$$\dot{V}_e \leq -\frac{1}{2}\mathbf{e}^T\mathbf{Q}\mathbf{e} + |\mathbf{e}^T\mathbf{P}\mathbf{b}|\left[\frac{1}{b}(|f(\mathbf{x})| + |y_s^{(n)}| + |\mathbf{r}^T\mathbf{e}|) + |u_c|\right]$$
$$-|\mathbf{e}^T\mathbf{P}\mathbf{b}|\left[|u_c| + \frac{1}{B}(F + |y_s^{(n)}| + |\mathbf{r}^T\mathbf{e}|)\right]$$
$$\leq -\frac{1}{2}\mathbf{e}^T\mathbf{Q}\mathbf{e} < 0 \qquad \text{for } \mathbf{e} \neq \mathbf{0} \tag{5.29}$$

So we have proven that the Lyapunov function is negative definite. Using (5.22) it follows that the expression $\frac{1}{2}\mathbf{e}^T\mathbf{P}\mathbf{e}$ is reduced through u_s until it falls below the threshold V_0. This means that using u_s it is possible to transfer the error vector from any initial state into the region given by $V_e = \frac{1}{2}\mathbf{e}^T\mathbf{P}\mathbf{e} \leq V_0$ and to keep it there. Setting $V_0 = 0$ and dropping u_c in (5.15) would even guarantee that the error vector would converge towards zero, because of the continuous reduction of V_e. However, such a type of control would show the same disadvantage as a sliding mode controller: For any sign reversal of $\mathbf{e}^T\mathbf{P}\mathbf{b}$, the actuating variable would change its value in a comparatively big step, due to the sign function in (5.23). This would have negative effects on the actuator. Accordingly, as long as the system stays with the region given by $V_e \leq V_0$, an adaptive fuzzy controller should be preferred, which we are going to describe now.

We want to characterize this fuzzy controller by a characteristic surface according to equation (5.11). With a constant $\gamma > 0$, a suitable bound $U > 0$ and the n-th column $\mathbf{p}_n$ of matrix $\mathbf{P}$ from (5.21), we can define the following adaptation rule for the vector of coefficients:

$$\dot{\mathbf{u}} = \begin{cases} \gamma\left[\mathbf{e}^T\mathbf{p}_n\right]\mathbf{k}(\mathbf{x}) & \text{if } |\mathbf{u}| < U \text{ or } (|\mathbf{u}| \geq U \text{ and } \mathbf{e}^T\mathbf{p}_n\mathbf{u}^T\mathbf{k}(\mathbf{x}) \leq 0) \\ \mathbf{0} \text{ otherwise} \end{cases} \tag{5.30}$$

In order to prove that this adaptation rule actually delivers a stable controller, we first have to show the boundedness of the vector of coefficients $|\mathbf{u}| \leq U$. For this purpose, we define a Lyapunov function $V_u = \frac{1}{2}|\mathbf{u}|^2 = \frac{1}{2}\mathbf{u}^T\mathbf{u}$. Using (5.30), its derivative is

$$\dot{V}_u = \mathbf{u}^T\dot{\mathbf{u}} = \begin{cases} \gamma\mathbf{e}^T\mathbf{p}_n\mathbf{u}^T\mathbf{k}(\mathbf{x}) & \text{if } |\mathbf{u}| < U \text{ or} \\ & \quad (|\mathbf{u}| \geq U \text{ and } \mathbf{e}^T\mathbf{p}_n\mathbf{u}^T\mathbf{k}(\mathbf{x}) \leq 0) \\ 0 & \text{otherwise} \end{cases} \tag{5.31}$$

This implies: As long as $|\mathbf{u}|$ is less than U, the value of the Lyapunov function—and thus also the absolute value of the vector of coefficients—can change arbitrarily. If, however, $|\mathbf{u}| \geq U$ is true, then a change of the Lyapunov function (or of the absolute value) can happen only if $\mathbf{e}^T \mathbf{p}_n \mathbf{u}^T \mathbf{k}(\mathbf{x}) \leq 0$, and it changes exactly by this value $\mathbf{e}^T \mathbf{p}_n \mathbf{u}^T \mathbf{k}(\mathbf{x})$, multiplied by the positive constant γ. It is therefore guaranteed that for $|\mathbf{u}| \geq U$ the Lyapunov function is always negative semi-definite, and the absolute value of $|\mathbf{u}|$ is monotonically decreasing. The vector of coefficients $\mathbf{u}$ is driven into the region $|\mathbf{u}| < U$ and kept there by the adaptation rule (5.30). We can define a time t_1, from that on the absolute value of the coefficient vector $|\mathbf{u}|$ is always lower than the bound U.

The state vector is also bounded, for the following reason: u_s guarantees that the Lyapunov function $V_e = \frac{1}{2}\mathbf{e}^T \mathbf{P} \mathbf{e}$ is negative definite as long as its value exceeds the given boundary V_0. As mentioned before, u_s drives the error vector $\mathbf{e}$ into the region given by $V_e = \frac{1}{2}\mathbf{e}^T \mathbf{P} \mathbf{e} \leq V_0$ and keeps it there. It follows the boundedness of $|\mathbf{e}|$. And from this it follows, as long as the absolute value of the reference vector is bounded, the boundedness of the state vector $\mathbf{x}$ by a value $X \geq |\mathbf{x}|$, because of $e = y_s - y$ and $y = x$.

Now that we know that the existence of boundaries for the state vector and the vector of coefficients is guaranteed, we can define a vector of optimal values for the parameters within the given limits:

$$\mathbf{u}^* = \left\{ \mathbf{u} \mid \min_{|\mathbf{u}| \leq U} \sup_{|\mathbf{x}| \leq X} |u_c(\mathbf{x}, \mathbf{u}) - u^*(\mathbf{x})| \right\} \tag{5.32}$$

i.e. $\mathbf{u}^*$ denotes that parameter vector from the region $|\mathbf{u}| \leq U$, for which the output quantity u_c of the fuzzy controller attains the minimum deviation from the—in theory—optimal actuating variable u^* for all $|\mathbf{x}| \leq X$. We usually get a residual error different from zero,

$$w(\mathbf{x}) = u^*(\mathbf{x}) - u_c(\mathbf{x}, \mathbf{u}^*) \tag{5.33}$$

which will be the smaller, the more supporting points the characteristic surface contains. With this residual error, the differential equation for the control error (5.19) (with $u_c = \mathbf{u}^T \mathbf{k}$) changes to

$$\begin{aligned} \dot{\mathbf{e}} &= \mathbf{R}\mathbf{e} + \mathbf{b}(u_c(\mathbf{x}, \mathbf{u}^*) + w - u_c - u_s) \\ &= \mathbf{R}\mathbf{e} + \mathbf{b}\Delta\mathbf{u}^T \mathbf{k}(\mathbf{x}) + \mathbf{b}(w - u_s) \end{aligned} \tag{5.34}$$

where $\Delta\mathbf{u} = \mathbf{u}^* - \mathbf{u}$ represents the error of the model.

Let us now define a new Lyapunov function

$$V = \frac{1}{2}\mathbf{e}^T \mathbf{P} \mathbf{e} + \frac{b}{2\gamma} \Delta\mathbf{u}^T \Delta\mathbf{u} \tag{5.35}$$

If we can show that its derivative is negative definite, we could conclude that both the control error and the model error are permanently decreasing. In the end, this would imply asymptotic stability of the system as well as an optimal vector of coefficients.

First of all, we obtain

$$\dot{V} = \frac{1}{2}\dot{\mathbf{e}}^T\mathbf{P}\mathbf{e} + \frac{1}{2}\mathbf{e}^T\mathbf{P}\dot{\mathbf{e}} + \frac{b}{\gamma}\Delta\mathbf{u}^T\Delta\dot{\mathbf{u}} \tag{5.36}$$

as the derivative of the Lyapunov function. Inserting (5.34) yields

$$\begin{aligned}\dot{V} = &\frac{1}{2}(\mathbf{R}\mathbf{e} + \mathbf{b}\Delta\mathbf{u}^T\mathbf{k} + \mathbf{b}(w - u_s))^T\mathbf{P}\mathbf{e} + \\ &\frac{1}{2}\mathbf{e}^T\mathbf{P}(\mathbf{R}\mathbf{e} + \mathbf{b}\Delta\mathbf{u}^T\mathbf{k} + \mathbf{b}(w - u_s)) + \frac{b}{\gamma}\Delta\mathbf{u}^T\Delta\dot{\mathbf{u}}\end{aligned} \tag{5.37}$$

Using the fact, that the term $\Delta\mathbf{u}^T\mathbf{k}$ is a scalar, that may be moved arbitrarily within a vector product and remains unchanged by a transposition, leads to

$$\begin{aligned}\dot{V} = &\frac{1}{2}\mathbf{e}^T(\mathbf{R}^T\mathbf{P} + \mathbf{P}\mathbf{R})\mathbf{e} + \frac{1}{2}\mathbf{b}^T\mathbf{P}\mathbf{e}(\Delta\mathbf{u}^T\mathbf{k} + w - u_s) + \\ &\frac{1}{2}\mathbf{e}^T\mathbf{P}\mathbf{b}(\Delta\mathbf{u}^T\mathbf{k} + w - u_s) + \frac{b}{\gamma}\Delta\mathbf{u}^T\Delta\dot{\mathbf{u}}\end{aligned} \tag{5.38}$$

Using (5.21) and (5.26) yields

$$\dot{V} = -\frac{1}{2}\mathbf{e}^T\mathbf{Q}\mathbf{e} + \mathbf{e}^T\mathbf{P}\mathbf{b}(\Delta\mathbf{u}^T\mathbf{k} + w - u_s) + \frac{b}{\gamma}\Delta\mathbf{u}^T\Delta\dot{\mathbf{u}} \tag{5.39}$$

With $\Delta\dot{\mathbf{u}} = -\dot{\mathbf{u}}$ and the adaptation law (5.30) we get

$$\dot{V} = -\frac{1}{2}\mathbf{e}^T\mathbf{Q}\mathbf{e} + \mathbf{e}^T\mathbf{P}\mathbf{b}(\Delta\mathbf{u}^T\mathbf{k} + w - u_s) - b\Delta\mathbf{u}^T\left[\mathbf{e}^T\mathbf{p}_n\right]\mathbf{k} \tag{5.40}$$

As only the last component of $\mathbf{b}$ is different from zero, we have $\mathbf{e}^T\mathbf{P}\mathbf{b} = b\mathbf{e}^T\mathbf{p}_n$. This scalar may be moved to the beginning of the last vector product, so that the first summand in the brackets and the last summand of the equation compensate each other. Besides that, because of the sign function in (5.23) the term $-\mathbf{e}^T\mathbf{P}\mathbf{b}u_s$ is definitely negative, so that we obtain the estimation

$$\dot{V} \leq -\frac{1}{2}\mathbf{e}^T\mathbf{Q}\mathbf{e} + \mathbf{e}^T\mathbf{P}\mathbf{b}w \tag{5.41}$$

If the residual error w is so small, that the absolute value of the second summand is smaller than the absolute value of the first one, then the derivative of the Lyapunov function will be negative, and the error vector $\mathbf{e}$ as well as the deviation of the coefficient vector $\Delta\mathbf{u}$ converge towards zero. w can be made arbitrarily small by choosing a large number of supporting points for the fuzzy characteristic surface, but there will remain some very small values of $\mathbf{e}$, for which the absolute value of the first summand is smaller than the absolute value of the second one. It follows, that for very small error vectors $\mathbf{e}$ it cannot be guaranteed, that the derivative of the Lyapunov function is always negative. And therefore, it cannot be guaranteed, that $\mathbf{e}$ and $\Delta\mathbf{u}$ will converge towards zero. This controller would—even for a constant reference value—not possess steady-state accuracy.

On the other hand, we should take into account the following: The parameter vector $\mathbf{u}$ will be adapted by this algorithm in a way, that the controller output within a given region gets best possibly to the ideal control rule u^*. If, for a constant reference value, the system output has been driven close to this value, also the parameter vector will get close to an ideal value, that guarantees the controller output to be close to its ideal value. The parameter vector will be adapted nearly perfectly to this set point with time, and this will finally eliminate also the steady-state control error.

After this extensive computation we should shortly sum up: The control rule is defined by (5.15), where u_s is given by (5.23) and u_c by (5.11). The coefficient vector $\mathbf{u}$ is adapted as defined in (5.30) with every time step. $\mathbf{k}$ is the vector of weights, that depends on the actual state of the system.

We get the necessary parameters of the equations in the following way: At first, we have to define the coefficients r_i for the error differential equation (5.14) in a way, that the error converges towards zero, i.e. the characteristic polynomial possesses only zeros with negative real part. With this, the matrix $\mathbf{R}$ is defined by (5.20). After that, we have to choose a positive definite matrix $\mathbf{Q}$, and the easiest choice would be $\mathbf{Q} = \mathbf{I}$. With $\mathbf{R}$ and $\mathbf{Q}$ we can compute $\mathbf{P}$ as solution of the Lyapunov equation (5.21). The algorithm for that computation is part of any modern control software library.

Then, we have to choose a positive constant γ, which can be seen as an acceleration factor for the adaptation of the coefficient vector $\mathbf{u}$. For a large value of γ the adaptation is fast, but we have to deal with—stable—oscillations. For a small value of γ it can take a longer time until the right coefficient vector is found and the control works satisfactorily.

Furthermore, we have to define an upper bound U for the absolute value of the coefficient vector. It represents some kind of upper bound for the absolute value of the actuating variable.

As the vector of weights $\mathbf{k}$ results from the actual state of the system, the error vector $\mathbf{e}$ is measured and $\mathbf{p}_n$ is the last column of the matrix $\mathbf{P}$ computed before, the adaptation law (5.30) for $\dot{\mathbf{u}}$ and u_c is now defined completely.

For the computation of u_s there are still missing the estimations for the model parameters f and b in equation (5.12), i.e. an upper bound F for f and a lower bound B for b. In addition, we have to define an upper bound V_0 for V_e of equation (5.22), which represents a measure for the maximum permissible error. As long, as the error is smaller, the fuzzy controller is working, while for larger errors u_s will be active. But for that case, an evident stress of the actuator has to be accepted. The vector $\mathbf{b}$ does not have to be known because of $\text{sgn}(\mathbf{e}^T\mathbf{P}\mathbf{b}) = \text{sgn}(\mathbf{e}^T\mathbf{P}\mathbf{i})$ with the unit vector $\mathbf{i} = (0, .., 0, 1)^T$.

Now the control algorithm is defined completely. But if it is really able to keep the system output y always close enough to the reference value y_s, depends mainly on the fact, that the number of supporting points of the fuzzy characteristic surface (5.11) has to be large enough, so that the deviation w

from (5.33) between the ideal and the best possible actuating variable is sufficiently small. But we should not choose an extremely large number, as every additional supporting point causes an additional component of the coefficient vector $\mathbf{u}$ and therefore an additional parameter that has to be optimized by the algorithm, which has a negative impact on the convergence of the adaptation.

Let us make a note critically, as for the computation of u_s following (5.23) and for the adaptation of the fuzzy characteristic surface as well following (5.30) we need to know the complete error vector $\mathbf{e}$, i.e. especially its $(n-1)$ derivatives. In practical applications, it is not possible to compute these derivatives directly from e, because the unavoidable measurement noise of the measured variable e will already make the second derivative of e useless for any further computation. The only way to solve this problem is the use of an observer, which requires a relatively precise model of the plant, so that the question for the use of the complete algorithm arises, as the main reason for the adaptation was to approximate the plant model.

In [134] the discrete version of this algorithm is used for an example.

[42, 180] and [202] give an extension of this method for plants of the form

$$x^{(n)} = f(\mathbf{x}) + g(\mathbf{x})u \qquad\qquad y = x \tag{5.42}$$

where g is required to be either always positive or always negative. The structure of the control does not differ from the algorithm which we have presented here, although there both f and g are approximated by a characteristic surface in an adaptive process. The basic outline for the proof, however, remains the same, and—in the end—leads to the same statement regarding stability (5.41). Besides that, the presentation in [42] is attractive, because the principle of a compensation controller and the method itself are developed on the basis of Lie derivatives.

[107] also makes use of the method described above. They suggest to extend (5.15) by a part dependent on $\dot{e}$

$$u_d = \text{sgn}(\mathbf{e}^T \mathbf{p}_n b\dot{e}) k_d \dot{e} \tag{5.43}$$

(where $k_d > 0$). This should increase the reaction of the controller, in a way similar to the effect obtained by adding a differential part to a PI-controller. The statement about stability is not affected, and the associated proof in the end also leads to equation (5.41).

A different method for the adaptation of a compensation controller is presented in [179]. This method can also be used for plants of the form (5.42), but in contrast to the procedures mentioned above, it guarantees asymptotic stability, i.e. a Lyapunov function as in (5.35) can be given whose derivative will always be negative. As the prize for this, this method requires that boundary intervals are given for every single coefficient of the characteristic surface which approximates g. This is to ensure that the estimated value for g is always positive.

In [153] the method is extended even for systems of the form

$$\dot{\mathbf{x}} = \mathbf{A}(\mathbf{x}) + \mathbf{B}(\mathbf{x})\mathbf{u} \qquad\qquad \mathbf{y} = \mathbf{h}(\mathbf{x}) \tag{5.44}$$

But the requirement is, that the elements of the matrix $\mathbf{B}$ outside the main diagonal have only very small values in proportion to the main diagonal elements, what corresponds to a nearly completely decoupled system. Furthermore, the controller has a very large switching part to ensure stability, what should cause considerable problems in practical applications.

5.2.2 Adaptive Sliding Mode Control

[194] uses a different approach. They too adapt a characteristic surface, but they use the adaptation in order to improve a sliding mode controller (cf. chapter 2.8.10). Starting point of their considerations is the sliding mode control rule (2.338)

$$u = -f_0(\mathbf{x}) + g_\lambda(\dot{\mathbf{e}}) + x_d^{(n)} + U\mathrm{sgn}(q) \tag{5.45}$$

where

$$U \geq F + D + \eta \tag{5.46}$$

according to (2.341). The main disadvantage of this controller is that the course of the actuating variable along the hyperplane given by $q = 0$ is discontinuous. The associated step is obviously the bigger, the larger the value for U is.

F denotes the upper bound for the deviation between the real plant and the nominal model:

$$F \geq |\Delta f(\mathbf{x})| = |f(\mathbf{x}) - f_0(\mathbf{x})| \tag{5.47}$$

This quantity is usually set to a constant value, since in most cases the function $\Delta f(\mathbf{x})$ is not even approximately known. If we knew this function, we could replace U in (5.45) —which is a constant—by

$$U^*(\mathbf{x}) = |\Delta f(\mathbf{x})| + D + \eta \tag{5.48}$$

which is a state-dependent function that would also ensure asymptotical stability of the system. A discontinuous course of the actuating variable still can not be avoided, but the discontinuities will, in general, be much smaller than for a constant value of F or U.

The idea now is to approximate U^* by a characteristic surface (5.11), and to adapt this surface using the adaptation rule

$$\dot{\mathbf{u}} = \gamma|q|\mathbf{k}(\mathbf{x}) \tag{5.49}$$

in a way that $U(\mathbf{x})$ corresponds to $U^*(\mathbf{x})$ as much as possible. $\mathbf{u}$ is the parameter vector, $\mathbf{k}$ is the vector of weights, γ is a positive constant and q the variable defined in (2.329). Due to the discrete structure of the characteristic surface however, we can only obtain a rough approximation of U^*. Let

$$U_{opt}(\mathbf{x}) = U^*(\mathbf{x}) + \varepsilon \tag{5.50}$$

(where ε is a positive constant) be the best possible approximation. Correspondingly, there exists an optimal parameter vector $\mathbf{u}_{opt}$ with

$$U_{opt}(\mathbf{x}) = \mathbf{u}_{opt}^T \mathbf{k}(\mathbf{x}) \tag{5.51}$$

and a parameter error $\Delta\mathbf{u} = \mathbf{u}_{opt} - \mathbf{u}$. This establishes the following relation

$$U = U^* - \Delta\mathbf{u}^T \mathbf{k} + \varepsilon \tag{5.52}$$

which will be used later on.

In order to show that the entire system is still asymptotically stable if we approximate $U^*(\mathbf{x})$ by a characteristic surface and the adaptation rule (5.49), we define the Lyapunov function

$$V = \frac{1}{2}(q^2 + \frac{1}{\gamma}\Delta\mathbf{u}^T \Delta\mathbf{u}) \tag{5.53}$$

and show that it converges towards zero. Its derivative with respect to time is

$$\dot{V} = q\dot{q} - \frac{1}{\gamma}\Delta\mathbf{u}^T \dot{\mathbf{u}} \tag{5.54}$$

because of $\Delta\dot{\mathbf{u}} = -\dot{\mathbf{u}}$. Substituting (2.335) and (5.45) yields

$$\begin{aligned} \dot{V} &= q(-\Delta f(\mathbf{x}) - d - U \mathrm{sgn}(q)) - \frac{1}{\gamma}\Delta\mathbf{u}^T \dot{\mathbf{u}} \\ &= q(-\Delta f(\mathbf{x}) - d) - |q| U - \frac{1}{\gamma}\Delta\mathbf{u}^T \dot{\mathbf{u}} \end{aligned} \tag{5.55}$$

and using (5.48) and (5.52) we obtain

$$\begin{aligned} \dot{V} &= q(-\Delta f(\mathbf{x}) - d) - |q|(|\Delta f(\mathbf{x})| + D + \eta - \Delta\mathbf{u}^T \mathbf{k} + \varepsilon) \\ &\quad - \frac{1}{\gamma}\Delta\mathbf{u}^T \dot{\mathbf{u}} \end{aligned} \tag{5.56}$$

Because of $-q\Delta f(\mathbf{x}) - |q||\Delta f(\mathbf{x})| \leq 0$ and $-qd - |q|D \leq 0$ we can estimate

$$\dot{V} \leq -|q|(\eta + \varepsilon) - \frac{1}{\gamma}\Delta\mathbf{u}^T(\dot{\mathbf{u}} - \gamma|q|\mathbf{k}) \tag{5.57}$$

The first summand is certainly negative, and the second equal to zero because of the adaptation rule (5.49). Therefore, the derivative of the Lyapunov function is negative for $q \neq 0$, i.e. q and the error vector $\Delta\mathbf{u}$ converge towards zero. The convergence of q towards zero is equivalent to the asymptotic stability of the system, as already proven in chapter 2.8.10. And a decreasing error vector $\Delta\mathbf{u}$ means, that the output variable $\mathbf{u}$ of the characteristic surface approximates to the optimum value U^* best possibly. Taking into account the things we mentioned at the beginning, we have reached minimum discontinuities of the actuating variable now.

In [10, 11, 12] the complete method is analyzed in detail and a very clear example is presented.

In [207] an adaptive sliding-mode controller for systems of the form

$$x^{(n)} = f(\mathbf{x}) + b(\mathbf{x})u + d(t) \tag{5.58}$$

is developed. There, not the actuating variable is adapted, but fuzzy models for f and b. The corresponding adaptation rules and the proof of stability do not differ essentially from the method presented here, so that we can omit an explicit discussion.

As for the adaptive compensation controllers we have the essential drawback here, that for the computation of the actuating variable according to (5.45) we need $(n-1)$ derivatives of the error e. But these cannot be computed without an observer, that requires a relatively precise model of the plant, which causes the question, if the method is useful at all, because the use of the adaptation was to approximate this model.

As the final remark of this chapter, we wish to emphasize that in an adaptation of characteristic surfaces as presented here, fuzzy sets do not appear at all. It is therefore reasonable to ask whether this topic actually belongs to the field of fuzzy controllers, or whether it would better be considered to be part of the techniques of classical control engineering. On the other hand, however, we have to accept the fact that a fuzzy controller is nothing but a characteristic surface controller. In its effects, it does not differ from a classical characteristic surface controller at all, only the way it is developed is different. The methods discussed in this chapter therefore represent the borderline between classical control engineering and fuzzy controllers, and should therefore not be neglected. In the following sections, however, we will return to fuzzy controllers based on fuzzy sets.

5.3 Adaptation of Fuzzy Rules

This method is based on an idea, that seems to be very plausible at first sight. But on closer inspection, it can be seen that this approach contains the danger of instability and should therefore be avoided for plants of higher order.

We want to present the most simple approach, following for example [166, 178, 184, 190]. The approach is introduced here for SISO plants, but it can easily be extended also for MIMO plants.

The only prerequisite is an existing fuzzy controller, which is already in operation in a closed loop. We assign an adaptation unit to it (cf. fig. 5.1), which receives a measured quantity $e(k)$ at any moment $t = kT$, just like the fuzzy controller. Here, T is the sampling interval. The difference $\Delta e(k) = e(k) - e(k-1)$ between two successive measurements can be computed easily internally. Using the values $e(k)$ and $\Delta e(k)$, an adaptation can be carried out according to the following strategy: If both the error and the difference are zero, the fuzzy controller is obviously working fine. If, for example, the error is positive and the difference negative, then there is also no need to act, since the error is decreasing. However, if the error as well as the difference are both positive or both negative, i.e. if the absolute value of the error increases, then the adaptation intervenes.

The assumption is that the cause for an increasing error during the interval $(k-1)T < t \leq kT$ was a bad value of the actuating variable $u(k-1)$ at the time $t = (k-1)T$. The decisive fuzzy rule for this value of the actuating variable is reconstructed, and the output value of this rule is adjusted according to the rule base given in figure 5.1. This rule base is therefore not the fuzzy controller's rule base, but the one of the adaptation rule.

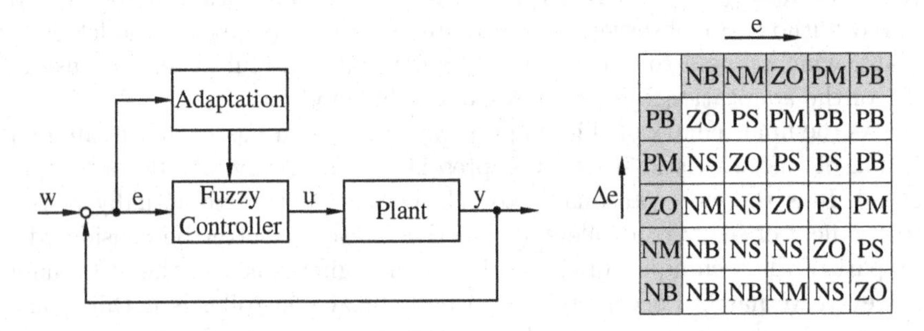

	NB	NM	ZO	PM	PB
PB	ZO	PS	PM	PB	PB
PM	NS	ZO	PS	PS	PB
ZO	NM	NS	ZO	PS	PM
NM	NB	NS	NS	ZO	PS
NB	NB	NB	NM	NS	ZO

Fig. 5.1. A simple adaptation of a fuzzy controller

If, for example, $e(k)$ as well as the difference $\Delta e(k)$ are positive, then we know that the output value of the system y is less than the desired value w, and that the difference between w and y is increasing. Obviously, the value of the fuzzy controller's actuating variable at $t = (k-1)T$ was too small, which is why the output value of the associated rule is being increased by the adaptation rule.

At first sight, this strategy seems reasonable, which is why it is frequently used—especially by non-control-engineers. However, this strategy can only be guaranteed to be successful for first order plants. In other cases, it may give rise to serious problems.

To illustrate this, think of a fuzzy controller with input quantities e and Δe, and the output quantity Δu, which is summed up as $u = \sum \Delta u$ to form the controller's actuating variable (see fig. 4.1). We obtain a controller comparable to a PI-controller. The rule base of the fuzzy controller shall be equal to the one of fig. 4.5. This controller controls for example a third-order linear plant, consisting of three first-order lags in series connection.

Now, let us assume that the actuating variable of the controller has too large absolute values and that the system is close to the borderline of stability. In the linear case, this would mean that the Nyquist plot of the open-loop transfer function runs close to the point -1, but does not go round it (cf. chap. 2.5.5). After an activation, this system performs large oscillations that hardly decrease.

$t = kT$ shall be that moment during an oscillation, when the control error is just becoming positive again, i.e. $e(k)$ and $\Delta e(k)$ are positive. The fuzzy controller will produce a positive actuating variable $\Delta u(k)$. In the next time step, as the control error has just become positive, $e(k+1)$ takes a little larger value than the time step before. This causes the adaptation law in the next time step according to fig. 5.1 to increase the output value of that rule, that was mainly responsible for the value of $\Delta u(k)$. Generally, in this case the adaptation law will increase the values of the actuating variable.

This means an increase of the control gain. If the fuzzy controller would be a linear controller, then the Nyquist plot of the open-loop transfer function would be stretched and probably go round the critical point -1. This would be a violation of the Nyquist criterion, and the system would be instable.

It should now be clear that this approach is not only unsuccessful, but for non-trivial plants even dangerous. A useful adaptation requires that the structure of the plant is taken into consideration, and not only the development of the control error during a certain time interval. For this reason, the method which we will present in the next chapter is based on a model of the plant.

5.4 Model-Based Control

5.4.1 Structure of the Model

From the measurements known at a certain time $t = kT$ (including past values), the model of the plant used in this method should facilitate predicting the vector of output quantities $\mathbf{y}(k+1)$ or the state vector $\mathbf{x}(k+1)$ at time $t = (k+1)T$. A state space model, for example, describes the relation between the current state vector $\mathbf{x}(k)$, the current actuating variables $\mathbf{u}(k)$ and the change of the state vector in the next time step $\Delta\mathbf{x}(k+1)$ resulting from them. The differential equation for such a model is:

$$\Delta\mathbf{x}(k+1) = \mathbf{x}(k+1) - \mathbf{x}(k) = \mathbf{f}(\mathbf{x}(k), \mathbf{u}(k)) \tag{5.59}$$

An alternative to the state space model is a model which describes the relation of the input to the output quantities of the plant, and which can therefore be compared to a transfer function:

$$\mathbf{y}(k+1) = \mathbf{f}[\mathbf{u}(k-n), ..., \mathbf{u}(k), \mathbf{y}(k-n), ..., \mathbf{y}(k)] \tag{5.60}$$

Both types can be used with the following algorithms; simpler solutions, however, should be obtained when using the state space model.

In general, the model of the plant will not be given by difference equations, particularly since in that case the design of a classical controller would be more appropriate. Instead, we have to assume that the information about the plant is given only in terms of a characteristic surface, a neural network or a fuzzy model. Here, we will not discuss characteristic surfaces or neural networks explicitly, as they are already described in chapter 4.1 and 5.6 respectively. We may not have to discuss a fuzzy model in detail either: On the one hand, it may be given by rules of the form

$$\begin{aligned} R: \quad & \text{If } x_1 \text{ is } \mu_R^{(1)} \text{ and } \dots \text{ and } x_n \text{ is } \mu_R^{(n)} \\ & \quad \text{and } u_1 \text{ is } \mu_R^{(n+1)} \text{ and } \dots \text{ and } u_m \text{ is } \mu_R^{(n+m)} \\ & \text{then } \Delta x_1 \text{ is } \nu_R^{(1)} \text{ and } \dots \text{ and } \Delta x_n \text{ is } \mu_R^{(n)}. \end{aligned} \tag{5.61}$$

where $\mathbf{x} = [x_1, ..., x_n]^T$ is the state vector, and $\mathbf{u} = [u_1, ..., u_m]^T$ is the vector of the actuating variables. On the other hand, it may also be given as a fuzzy relation.

Creating a Model

How can we actually obtain a model of the plant? A model in the form of difference equations can be constructed directly, in an analytical way, as long as sufficient knowledge about the laws of physics underlying the plant exists. If, however, only measured values about the plant are available, then statistical methods, which are described in detail in the corresponding literature, seem to be the obvious choice [71, 72, 110, 117, 199].

Characteristic surfaces and neural networks are developed on the basis of measured quantities. Characteristic surfaces or the coefficients of difference equations can be determined using classical methods from statistics, and hints on how to configurate and train a neural network can be found in chapter 5.6.

Fuzzy models allow different types of model construction. We can present the behavior of the plant linguistically in the form of fuzzy rules, based on the knowledge about the plant. We can also obtain a fuzzy model from measured quantities. Here, for example algorithms on fuzzy clustering may be a reasonable choice (cf. chap. 5.5). Linguistic model construction as well as model construction using clustering algorithms leads to a model which is linguistically interpretable, which can be stored in form of fuzzy rules or fuzzy relations.

Fuzzy Models

Finally, we also have the possibility to assign an individual fuzzy relation to every tuple of measured quantities we obtain during the identification process, and then form the disjunction of these relations ([125]). Of course, the resulting model will no longer be linguistically interpretable, but for a model-based control it has certain advantages, which is why we are going to discuss it in more detail.

For reasons of simplicity, we explain this algorithm for a static SISO plant without internal dynamics. For such a type of plant, we obtain one pair of measured quantities $(u(k), y(k))$ at every sample time $t = kT$ of the identification process. We therefore have a relation between the value of the actuating variable $u(k)$ and the one of the output quantity $y(k)$, which we can represent by a—still "crisp"—relation $R_k = \{(u(k), x(k))\}$, that means, we just have to store the pair of measured values $(u(k), y(k))$.

Now, we can expect, that, if the measured tuple $(u(k), y(k))$ can appear, similar pairs of measured values can appear as well. Using the similarity relation

$$\begin{aligned} E : &(\mathbb{R} \times \mathbb{R})^2 \to [0,1]\,, \\ &((u_1, x_1), (u_2, x_2)) \to \\ &\min\{1 - \min(|u_1 - u_2|, 1)\,, 1 - \min(|x_1 - x_2|, 1)\} \end{aligned} \tag{5.62}$$

we can replace R_k by a fuzzy relation, the extensional hull of R_k:

$$\begin{aligned} \mu_{R_k} : &\mathbb{R} \times \mathbb{R} \to [0,1]\,, \\ &(u, y) \to \\ &\min\{1 - \min(|u(k) - u|, 1)\,, 1 - \min(|y(k) - y|, 1)\} \end{aligned} \tag{5.63}$$

The point $(u(k), y(k))$ in the $u - y-$ plane becomes the fuzzy set μ_{R_k} (cf. fig. 5.2). From a set-theoretical point of view, μ_{R_k} is the set of all points which are similar to $(u(k), y(k))$, where "similarity" is defined by (5.63).

We can also think of this fuzzy relation as the set of all pairs of values (u, y) which are possible for this plant. The pair $(u(k), y(k))$, as a pair of values which were actually measured, is certainly possible, and its degree of membership to this set is therefore 1. For other pairs of values, the degree of membership to this set decreases with an increasing distance from $(u(k), y(k))$. The assumption is that pairs which are close to the measured pair of values are also possible; in fact, they are more possible, the smaller the distance to this pair is. It should be mentioned that we would obtain a different fuzzy relation μ_{R_k} if we chose another similarity relation. With regard to the computational effort however, it seems reasonable to choose a relation which is as simple as possible for the construction of the model.

The disjunctive combination of all the fuzzy relations μ_{R_k} which arose during the identification delivers the fuzzy model of the plant:

$$\mu_R = \bigcup_k \mu_{R_k} \tag{5.64}$$

Figure 5.2 shows such a model, as a result of two pairs of measured quantities $(u(1), y(1))$ and $(u(2), y(2))$.

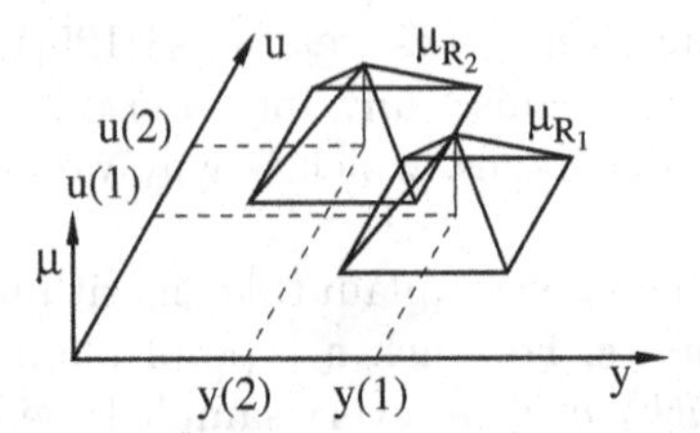

Fig. 5.2. Fuzzy model of a plant as a result of an identification process

Instead of a disjunctive composition of similarity relations, we can also think of a conjunctive composition of implications to construct a model. From a theoretical point of view, this would be even better, as every new pair of measured quantities would reduce the relation that describes the transfer behavior of the plant, i.e. it would make it more crisp and hence more precise. And this is what we would actually like to happen when adding new information to the model. In contrast to this, for a disjunctive combination of the single relations the resulting relation becomes wider and more and more blurred with every new pair of measured values and every new partial relation μ_{R_k}.

For a conjunctive composition, however, one practical problem exists which can easily be underestimated: With measuring noise, different values for the output quantity may result for the same value of the input quantity during the identification process. This would completely eliminate any information within the corresponding region of the model, and make it absolutely useless. For this reason, the disjunctive combination of similarity relations should be preferred.

As a firm footing for our further explanations, let us now describe how to compute the expected output value $y_m(k)$ for a given input value $u(k)$ using this model. First, we have to define a fuzzy set representing the input ("singleton"):

$$\mu_u : \mathbb{R} \to [0,1]\,,\ u \to \begin{cases} 1 & : \quad u = u(k) \\ 0 & : \quad \text{otherwise} \end{cases} \tag{5.65}$$

Then, using this singleton and the given relation describing the model μ_R, we compute the relational equation $\mu_y = \mu_u \circ \mu_R$. We obtain a fuzzy set as the output quantity:

$$\begin{aligned} \mu_y : \mathbb{R} \to [0,1], & \\ y \to & \sup \{\min [\mu_u(u), \mu_R(u,y)] \mid u \in U\} \\ = & \ \mu_R(u(k), y) = \mu_y(y) \end{aligned} \tag{5.66}$$

This procedure corresponds to making a cut parallel to the y-axis through the relation μ_R at $u = u(k)$, and projecting it onto the output variable y (fig. 5.3). A defuzzification of this fuzzy set yields the value which should be expected as the output value $y_m(k)$. These are obviously the same steps which have to

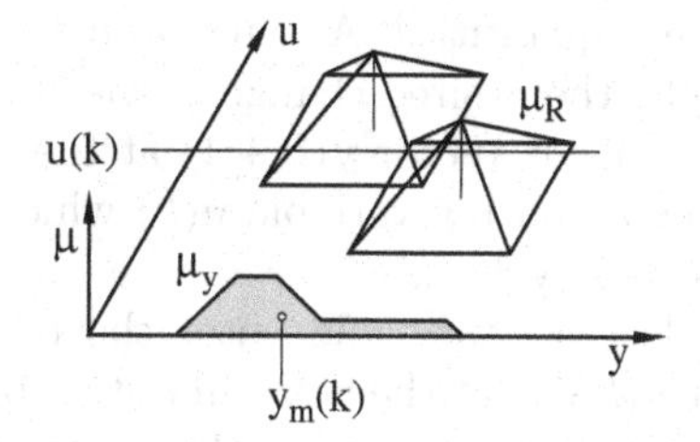

Fig. 5.3. Computing the output value of a fuzzy model

be carried out to compute the output value of a conventional fuzzy controller, if the rules are stored in form of a fuzzy relation.

The entire method can be adjusted to MIMO plants of higher degrees without any problems. A first order plant, for example, can be characterized by a triple of measured values $(u(k), x(k), \Delta x(k+1))$ (actuating variable, state, resulting change of state), where the actuating variable and the state form the input quantities of the model, and the resulting change of state is the output quantity. Therefore, the fuzzy model needs to have an additional dimension. In general, for arbitrary plants we may obtain multi-dimensional fuzzy models. But as any of the equations involved can easily be extended to any dimension, the method itself is not affected by adding further dimensions.

We should finally discuss possibilities of how such a model can be stored. It seems reasonable to discretize the entire space spanned by the involved quantities. At every supporting point which results from this process, the degree of membership valid at that point is recorded (fig. 5.4). The relation describing the model is then defined in terms of an interpolation between these recorded degrees. Of course, this causes differences between the stored relation and the original one, but we can adjust this difference according to our needs by choosing a grid of sufficiently high resolution.

5.4.2 Single Step Control

The heart of the single step control as shown in figure 5.5 is an inverted model of the plant, instead of a conventional one. Such a model describes the relation between input and output quantity just like a conventional model,

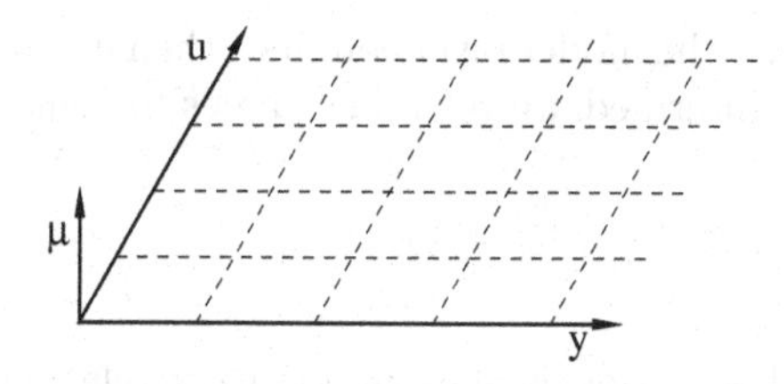

Fig. 5.4. Illustration of how to store a fuzzy model

but it interchanges these quantities. At time step $t = kT$, we can feed this model with, for example, the desired change of the state vector $\Delta\mathbf{x}(k+1)$ or a desired value for the output vector $\mathbf{y}(k+1)$ at the next time step, and we obtain that value of the actuating variable $\mathbf{u}(k)$ which is required at $t = kT$ to force the system to these values.

For the most simple case, we could enter the reference value $\mathbf{w}$ as the required output vector $\mathbf{y}(k+1)$ at the next time step, but one problem occurs: If the desired value cannot be reached within a single time step (which is usually the case), then the inverse model can not provide any value for the actuating variable either. We therefore need another unit, which computes an intermediate reference value $\mathbf{z}$ from $\mathbf{w}$ and $\mathbf{y}(k)$ for the next time step, such that this reference value can actually be reached within it. This intermediate reference value is then used as the input quantity of the inverse model.

The computation of the actuating variable on the basis of the plant's model still causes another problem. As long as the model of the plant is an exact image of the plant itself, the actuating variable computed from this model will also lead exactly to the desired (intermediate) reference value. However, if differences between the model and the plant occur—maybe because of inaccuracies of the model or some changes within the plant—then this is no longer true, and the controller is no longer steady-state accurate. The only solution to this problem is to continuously adjust the model to a possibly changing plant. In order to be able to do so, the measured quantities of the plant have to be fed back to the model, as shown in figure 5.5.

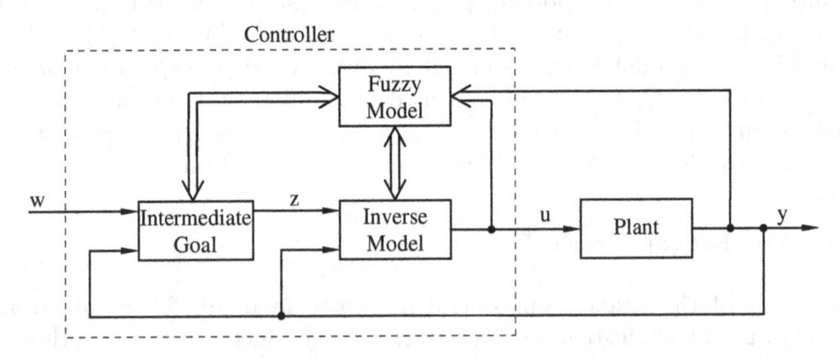

Fig. 5.5. Single step control with an inverse model

Let us now discuss the model-inversion first, then its adaptation and finally the computation of intermediate reference values in some more detail for the different model types.

Model Inversion

The easiest type of inversion is the one of fuzzy relations (fig. 5.2). As the fuzzy model is symmetrical regarding its input and output quantities, we can invert the model in such a case simply by interchanging the two quantities. For a static SISO system, the desired quantity $z = y(k)$ becomes the input quantity, and the associated actuating variable $u_m(k)$ will be the output of the model. The cut through the relation is then taken in parallel to the u-axis at $y = y(k)$ (cf. fig. 5.3). First, we obtain a fuzzy set μ_u, and after defuzzification the actuating variable u (cf. [115], [125]). As it is easy to construct fuzzy relations also for multi-dimensional systems without any difficulties, this type of model-inversion should cause no problems for higher order plants or MIMO systems. The models are constructed in the same way as usual, and then just used invertedly.

It is also possible to use another method for the model-inversion of fuzzy relations ([156],[196]), which is based on results due to *Sanchez* [174]. With these results, we can compute the maximum fuzzy set μ_u for two given sets μ_y and μ_R, such that the relational equation $\mu_y = \mu_u \circ \mu_R$ is still true. But this is just the given task: We know $\mathbf{y}$ and $\mathbf{R}$, and we have to find $\mathbf{u}$. However, one problem exists: The solution μ_u will be a non-empty set only for a sufficiently large μ_y. So if we choose mu_y to be a singleton (which is the common attempt), it is highly unlikely that we will ever find a solution mu_u for the relational equation. As a result of this, it will not be possible to determine an actuating variable for the controller in most of the time steps. As another disadvantage, the algorithm requires more computation time. So all in all, the initially presented solution should be preferred.

We have to use an altogether different way if the model is given in form of fuzzy rules, as a characteristic surface or as a neural network. While an inverted use of a characteristic surface is possible—at least in principle—by interpolation between the values at the supporting points, this is completely impossible for neural networks or fuzzy rules. As a remedy, [157] suggests to construct the model from the very beginning as an inverse model with interchanged input and output quantities. That means, the model is constructed with $\mathbf{y}$ and $\mathbf{x}$ as input quantities and the actuating variable $\mathbf{u}$ as output quantity.

If the model of the plant is given in terms of difference equations, there are only two alternatives for an inversion: It is either possible to resolve the function $\mathbf{f}$ of the difference equations (5.59) or (5.60) for the desired quantity $\mathbf{u}(k)$, or it is not. For the first case, we can determine an inverted model from these equations in an analytical way, which has $\mathbf{u}(k)$ as the output quantity. For the second case, however, we can only numerically compute the actuating variable corresponding to a given output quantity from the difference equations. This is the reason why [78], for example, presupposes that an analytical solution is possible.

In general, however, one possibility exists for all the different types of models in order to avoid an inversion. This possibility is very simple, but it requires a lot of computation time: A sufficiently large number of different values for $\mathbf{u}$ is used as the input for the original, uninverted model. The value of $\mathbf{u}$, for which the output quantity $\mathbf{y}(k+1)$ reaches the minimum distance to the intermediate reference value $\mathbf{z}$ is used as the actuating variable ([113], [165]).

Model Adaptation

Let us now close our discussion of model inversion and turn to the next point, the adaptation of a model. For a fuzzy relation, the model adaptation is relatively simple. Let us use the static SISO plant once more for an illustration: At every time step, we use $\mathbf{u}(k)$, the actuating variable, as the input to the plant as well as to the uninverted model of the plant. We obtain the expected value $\mathbf{y}_m(k)$ for the output quantity, and compare it to the real, measured value of the output quantity $\mathbf{y}(k)$. If a difference between these two values occurs, the corresponding part of the model relation μ_R has to be shifted parallel to the y-axis, until the modified relation reflects the current conditions of the plant (fig. 5.6). This means, at every time step the model is used a.) invertedly, for the computation of the actuating variable and b.) uninvertedly, for adaptation purposes.

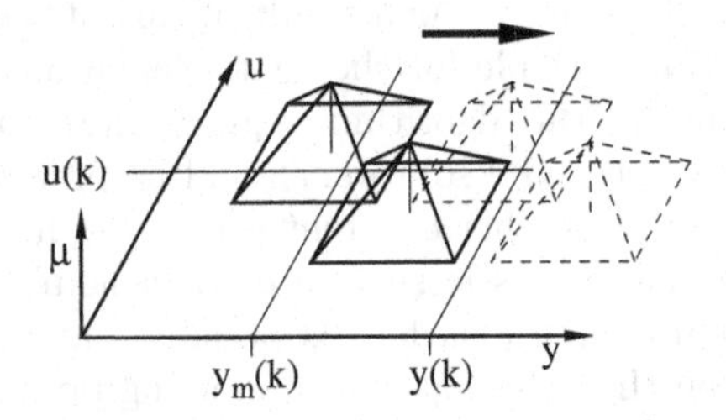

Fig. 5.6. Adaptation of a fuzzy relation

The same principle has to be used if the fuzzy model is given in the form of fuzzy rules or a characteristic surface. At every time step, the output quantity of the model has to be compared to the measured output quantity of the plant, and if any deviation occurs, the model has to be adjusted. For a fuzzy rule, we have to modify the output fuzzy set of the rule, and for a characteristic surface the values recorded at the supporting points. The adaptation is even easier for a neural network: as long as the net is in the training mode, it will automatically adjust to changes within the plant.

If the model is given by difference equations, we have to use one of the classical methods of identification. During a longer period of time, a sufficiently large number of measurements of the plant is collected and then used to compute a new model in a numerical way using a regression algorithm. The model can be updated only at a certain rate, not at every time step. Furthermore, such a recomputation may be numerically problematic if the measured values do not contain sufficient information. With regard to the adaptation, models which are not based on difference equations should be preferred.

Computation of Intermediate Reference Values

We should finally discuss how intermediate reference values can be determined, which will reveal the crucial point of a single step control. [54] suggests a very elegant solution to this problem for models given by fuzzy relations. The fact that the given reference value w may not be reached within a single time step is taken into account by representing the desired output value $\mathbf{y}(k+1)$ by a fuzzy set μ_y, instead of a singleton. The membership function of this fuzzy set is zero at $\mathbf{y} = \mathbf{y}(k)$, and increases with decreasing distance to $\mathbf{y} = \mathbf{w}$. Figure 5.7 shows such a function for a system of first order. This fuzzy set is then used as the input quantity for the relational equation $\mu_u = \mu_y \circ \mu_R$. The membership function of the output quantity μ_u will therefore be different from zero for those values of $\mathbf{u}$ which could produce an output value $\mathbf{y}(k+1)$ lying between $\mathbf{y}(k)$ and $\mathbf{w}$. The degree of membership depends on the distance between the expected output value and $\mathbf{w}$, as well as on the corresponding values of the model-relation. The degree of membership of a certain value of the actuating variable to the set μ_u will be the higher, the closer it could drive the system to the reference value $\mathbf{w}$. The defuzzification of μ_u will then certainly produce a suitable value for the actuating variable.

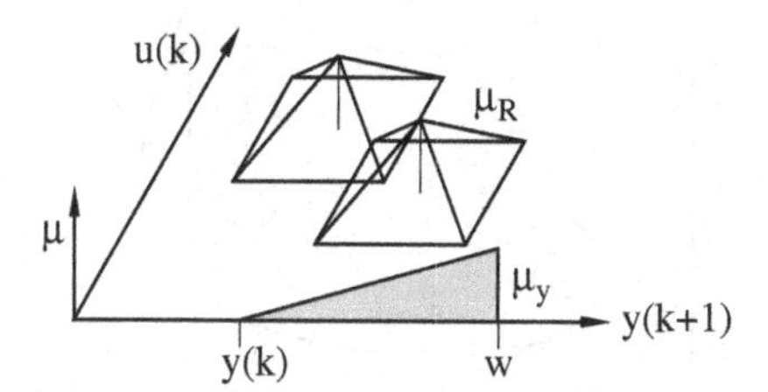

Fig. 5.7. Extended input quantity for the inverse model

For models given as fuzzy rules, characteristic surfaces, neural networks or difference equations, this algorithm can of course not be used. In cases like that, [78] suggests using a simple I-controller with the input quantity $\mathbf{e} = \mathbf{w} - \mathbf{y}$ in order to compute the intermediate reference values. The output quantity of this controller, i.e. the intermediate reference value $\mathbf{z}$, is—at the stationary state—equal to the output variable $\mathbf{y}$, as only the inverse model and the plant lie in between $\mathbf{z}$ and $\mathbf{y}$, and they should compensate each other (provided the model is sufficiently exact). Furthermore, in the stationary state $\mathbf{y}$ has to be equal to $\mathbf{w}$, as otherwise the input $\mathbf{w} - \mathbf{y}$ of the integrator would be different from zero, and its output would therefore still be changing, so that no stationary state would be reached.

Accordingly, we get $\mathbf{w} = \mathbf{z} = \mathbf{y}$ at the stationary state. After a change of the reference value $\mathbf{w}$, $\mathbf{z}$ as the output quantity of the I-controller will change continuously until $\mathbf{y}$ is equal to $\mathbf{w}$ and the control deviation $\mathbf{e}$ has disappeared. At this new, stationary state, $\mathbf{w} = \mathbf{z} = \mathbf{y}$ will again be true. $\mathbf{z}$, that means the sequence of intermediate reference values, changes continuously—and with an appropriate design of the I-controller also slowly enough—from the old reference value to the new one. This is to ensure that for any given intermediate reference value there exists one actuating variable, that can drive the plant to this value within one time step.

Both methods assume that an actuating variable exists which can be used to drive the output vector $\mathbf{y}$ at least into the direction of the reference value $\mathbf{w}$, if $\mathbf{w}$ cannot be reached within the following sample. With $\mathbf{y}(k)$ as the vector at a certain time step,

$$\begin{aligned} y_i(k) \leq y_i(k+1) \leq w_i & \qquad \text{if } w_i \geq y_i(k) \\ y_i(k) \geq y_i(k+1) \geq w_i & \qquad \text{if } w_i \leq y_i(k) \end{aligned} \tag{5.67}$$

should be true for all the single components of the output vector at the following time step. For two dimensions (fig. 5.8), this means that from a given point $\mathbf{y}(k)$ the system can be driven into the reference point $\mathbf{w}$ in such a way, that all successing vectors $\mathbf{y}(k+j)$ of $\mathbf{y}(k)$ lie inside the rectangle defined by the corners $\mathbf{y}(k)$ and $\mathbf{w}$.

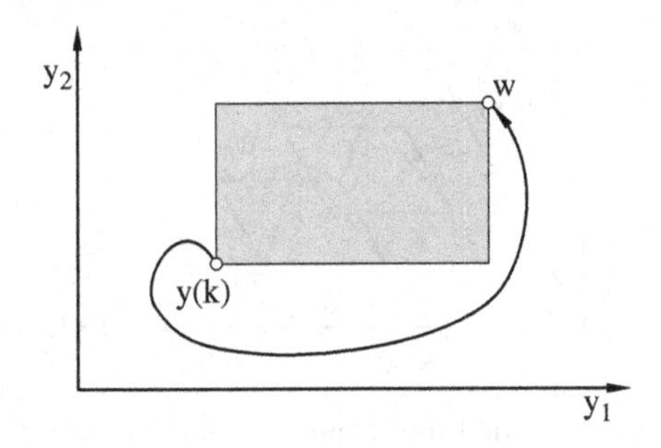

Fig. 5.8. Permissible region for successing output vectors and possible trajectory of a non-minimum phase system

This precondition, however, is true only for minimum phase systems (cf. appendix B). For systems with a non-minimum phase (which occur, in particular, among MIMO systems) it may happen that we can drive the system from $\mathbf{y}(k)$ to $\mathbf{w}$ only by following a trajectory which leaves this region, or which even runs entirely outside this area (fig. 5.8). A single step control can therefore not be used for such systems.

The permissible field of application is even smaller if we use a state space model according to (5.59), independent of whether it is given by differential equations or as a fuzzy model. Here, the single step control would fail in any case where the degree of the system is higher than 1. As an example, let us assume we are given an acceleration plant, whose differential equation is $F = m\ddot{l}$, with states l and v (cf. chapter 2.7.1). We would like to achieve a transformation of the system from the initial state $(l_0, v_0) = (0, 0)$ into the final state $(l_1, v_1) = (w, 0)$. Figure 2.48 shows the required trajectory in state space. We obviously need an intermediate velocity v different from zero in order to achieve that the body can reach its final state. The velocity at the initial state as well as at the final state, however, is zero. The region of possible intermediate states according to fig. 5.8 has degenerated from a rectangle into a line lying on the l-axis which connects the initial and the final state. The velocity is zero for all the states on this line. Here, both methods would create only intermediate reference values that lie on the l-axis with a velocity value of zero. This would be equal to the demand, that the body moves its position from $l = 0$ to $l = w$ with a velocity of zero, which is obviously impossible.

Because of this serious deficiency of the single step algorithm, [4] suggests using the model of the plant for a prediction over several time steps. First of all, upper and lower bounds and maximum change rates have to be fixed for the actuating variables with respect to the constraints of the plant and the actuator. The space for the actuating variables has to be discretized, i.e. only a finite number of values for the actuating variables is taken in to account. And finally a number, r, has to be chosen which represents how many time steps should be predicted. On these conditions, there are only a finite number of possible sequences for the actuating variables within the prediction interval.

For every time step $t = kT$, all the possible actuating variable sequences for the interval $kT \leq t \leq (k+r)T$ are simulated (based on the plant model), and the resulting course for the output variable is evaluated. The first value of the sequence which is considered to be the best is then used as the actuating variable at $t = kT$. The entire procedure is repeated for the next time step at $t = (k+1)T$, with the interval of prediction being shifted by one step. [40], [35] and [175] make similar suggestions, with a somewhat poorer quality in detail.

The fundamental problem with this approach is fixing r, the number of time steps which should be predicted, and the fitness function for the output variable. If the value for r is too small and the output variable does not reach the reference value within the interval of prediction, then the success or failure of an actuating variable sequence can not be estimated for a system of a non-minimal phase. This again prevents the selection of a suitable fitness function. On the other hand, due to the computational effort, one will be interested in an r which is as small as possible. Finally, r and the fitness function have to be adjusted to suit the plant. For an initially unknown plant, this can be achieved by means of an iterative process only.

5.4.3 Optimal Control

In contrast to this, an optimal control according to [125] follows a certain strategy in the computation of intermediate reference values, that guarantees an optimal course of the output variable and reaching the reference value. The basic structure of this control corresponds to the single step control as given in figure 5.5. A state space model of the plant in form of a fuzzy relation is the central part. It is used inversely in the computation of the actuating variable. The adaptation of the model—to ensure steady-state accuracy—is also carried out in the same way as for the single step control. Therefore, the difference between the optimal control and the single step control is only in the computation of the intermediate reference values.

Here, the strategy for the computation of intermediate reference values is based on the following idea: As a reaction to a change of the reference value, first the possible state space trajectories to the final state have to be detected, and then one of these trajectories is used to drive the system into this point during the following time steps. Due to the computational effort in the calculation of trajectories, this method is suitable especially for systems with a constant reference value. For a continuously varying reference value, however, the computational effort might exceed reasonable limits. Let us explain this algorithm for a second order system with one actuating variable and a fixed reference value or final state. The model-relation μ_R has the two states $x_1(k)$ and $x_2(k)$ and the regulated quantity $u(k)$ as its input quantities. The output quantities are given by the changes $\Delta x_1(k+1)$ and $\Delta x_2(k+1)$ of the states during the following time step, resulting from the input quantities. μ_R is therefore a five-dimensional relation.

First, we have to choose an operating region around the final state, where we are sure of that the system will never leave it. We discretize this limited state space, which leaves us with a finite number of discrete states inside this region. Figure 5.11 shows such a discretization for a second order system. There, the origin of the coordinate system is the final state.

As the second step, for every single state we have to detect the possibility with which it can be transfered into one of its adjacent states by means of a suitable actuating variable, within one time step. Figure 5.9 gives an illustration of this for one state of our example. The central point is the state under examination. It is surrounded by eight adjacent states. We are now looking, for example, for the possibility with which this state may be transfered into the state on the upper right. We know the coordinates (x_{1m}, x_{2m}) of the central state, and those of the state on the upper right, (x_{1r}, x_{2r}). We then define the state in the center to be the current state $(x_1(k), x_2(k)) = (x_{1m}, x_{2m})$, and the difference between this state and the one on the upper right $(\Delta x_1(k+1), \Delta x_2(k+1)) = (x_{1r} - x_{1m}, x_{2r} - x_{2m})$ to be the desired state difference for the next sample.

We then have to define membership functions (singletons) for x_1 and x_2 according to (5.65). For Δx_1 and Δx_2 as the desired output quantities, however, we need membership functions according to figure 5.7, which will be zero for $\Delta x_i = 0$ and 1 for $\Delta x_i = x_{ir} - x_{im}$. Using the fuzzy sets defined by these functions and the model-relation mu_R, we can then compute the relational equation

$$\mu_u = \mu_{x_1} \circ \mu_{x_2} \circ \mu_{\Delta x_1} \circ \mu_{\Delta x_2} \circ \mu_R \tag{5.68}$$

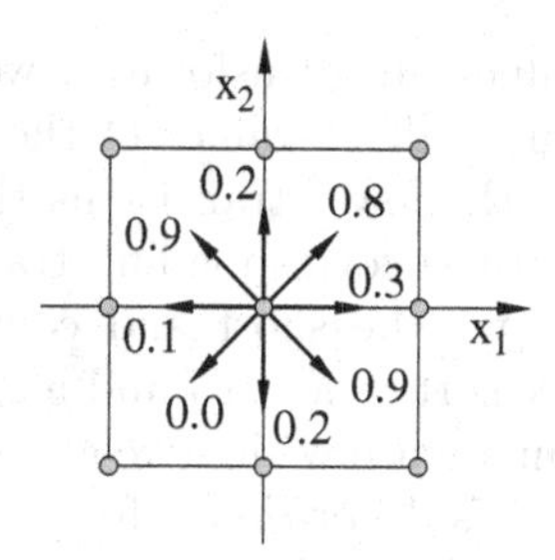

Fig. 5.9. Possibilities for a transition from one state to one of its adjacent states

which leaves us with a fuzzy set mu_u for the actuating variable. This fuzzy set contains all the actuating variables that can drive the system from the center state closer to the upper right state. A defuzzification is not necessary, since we are only interested in a measure of the possibility of the transition from one state to another one—and not in the actuating variable which would be required to obtain this transition. This measure is the highest degree of membership

$$P = \sup_u \{\mu_u(u)\} \tag{5.69}$$

which occurs within the fuzzy set mu_u (0.8 in our example). It would make little sense to use a different measure, for example $\int \mu_u(u)du$, since we are not interested in the power of the set, but only in whether there actually exists an actuating variable which can be used to enforce the transition.

As the result of this step, we now know the possibilities for all the transitions from any state of the limited, discretized state space to any of its adjacent states that can occur within one time step. We can think of this as having two directed lines connecting any two adjacent states, which are labeled with a measure describing the possibility of the corresponding transition. In fig. 5.9 only the lines beginning at the center state are sketched.

Now, we would like to find one trajectory from every point of the discretized state space into the final state, passing through several other discrete states if necessary. On the one hand, we would like this trajectory to be as short as possible. On the other hand, it should contain only those transitions which have a high value of possibility, since this increases the possibility that the real system can actually follow this trajectory. In order to be able to present this task as an optimization problem of a closed form, we transform the possibility values of all the transitions according to

$$P' = 1.0 - P^a \qquad \text{where } a > 0 \tag{5.70}$$

The value for P' will be the smaller, the higher the possibility for the corresponding transition is. a can be used to manipulate the relation of the P'-values of large to small possibility values. This again affects the probability of having transitions with small or large possibility values occur within the computed trajectories.

If all the possibility values are transformed, we can define the weight of a trajectory to be the sum of all P'-values of the transitions involved. This weight will be the smaller, the fewer transitions the trajectory contains and the smaller the values P' of the corresponding transitions are. Now this just means that the trajectory will be short and contain transitions with high possibility values. Our task is therefore to find a trajectory from every point of the state space to the final state with a weight as small as possible.

This is a task for *Dijkstra's Algorithm*, which we are now going to discuss. It is defined recursively, which is why we have to presuppose a connected domain of the discretized state space, where all the optimal trajectories from any state to the final state, which lie completely inside this region, are already known. We now want to add another (adjacent) state to this region, i.e. the optimal trajectory inside this region has to be computed for the new state. As the state is adjacent to the region, some transitions between this state and states inside the region will exist, with corresponding weights P'.

Since we already know the optimal trajectory of each of these adjacent old states, all we have to do is find the state whose trajectory is also optimal from the new state to the final state. And this merely requires adding the weight for the transition from the new to the old state to the weight of the optimal trajectory of the old state. The optimal trajectory for the new state will be the one passing through that old state, for which the addition produced the smallest value. We then have to store that old state as successor of the new state, and also store the weight of the new state's optimal trajectory.

Having done that, we also have to check if adding this new state changes the optimal trajectory of an old state, i.e. if it is "cheaper" for any of the old states to reach the final state via this new state. For all of the old states, for which this holds, we have to store the new state as successor and change the weight of their optimal trajectory. Then, we have finished one recursion, add the new state to the (extended) region, and repeat the entire algorithm for the next new state all over again.

Let us illustrate this algorithm with the simple example given in figure 5.10. The final state is S. The region for which we already know all the optimal trajectories consists of A, B and S. We would like to add N as the new state. We know the weights for the transitions between N and its neighbors A and B in both directions. The weight for the trajectory $N-A-S$ is $0.3+0.4=0.7$, the one for $N-B-S$ is $0.1+0.9=1.0$. The optimal trajectory from N to S therefore runs via A. We see that the weight for the trajectory $B-N-A-S$ $(=0.8)$ is less than the one of the originally optimal trajectory from B to S $(=0.9)$. We therefore define a new optimal trajectory for B, too, which now runs from B to S via N and A. At the end of this recursion step, we define A as the successor of N, and N as the successor of B.

With Dijkstra's algorithm, we can now compute the optimal trajectories from all the discrete states of the limited state space to the final state, starting from this final state (fig. 5.11). With the help of these trajectories, we can always find a suitable intermediate reference value for the control.

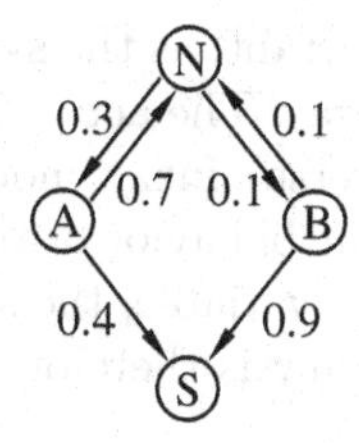

Fig. 5.10. An illustration to Dijkstra's algorithm

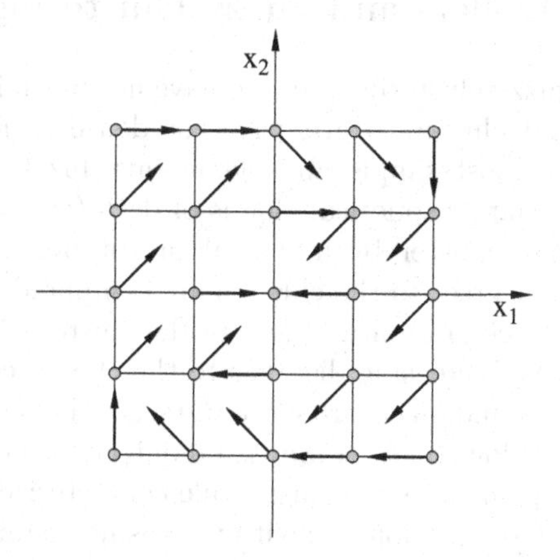

Fig. 5.11. Trajectories in a discretized, limited state space

The algorithm for any time step is the following: we have to measure the current state of the plant, and determine the discrete state which is closest to it. The successor state of this state with respect to the computed trajectory is then used as the intermediate reference value for the inverse model. From the inverse model we get the necessary actuating variable to drive the system to the intermediate reference value within the next time step at least approximately. The state that is reached at the end of this time step, is the starting point for the algorithm at the next time step.

The question arises, how far the only approximative reaching of the intermediate reference value within one time step affects the entire control algorithm, as this will be the normal case. Through accumulation of these errors, it is even possible that the system continues to move away from the computed trajectory, and that it finally reaches a state which is closer to another computed trajectory. However, this does not effect the overall result, as all that happens is that the system then reaches the final state via this other trajectory.

Because of this redundancy, it suffices if the trajectories are only approximately correct. And it is therefore even sufficient for slowly time-varying plants to recompute the trajectories with large distances in time. With such a basis, the system still gets close enough to the final state. And once there, the trajectories become unimportant, as the system could then reach the final state within one time step anyway. The actuating variable required to achieve this is obtained from the (inverse) fuzzy model, which—in contrast to the trajectories—reflects the current behavior of the plant, due to the continuous adaptation. With this actuating variable, the system will reach the final state exactly, and steady-state accuracy is therefore guaranteed.

5.5 Fuzzy Controllers and Fuzzy Clustering

Fuzzy control and fuzzy clustering seem to have not much in common. Cluster analysis tries to find clusters – groups or or dense regions of data – in a dataset. One of the most simple and elementary fuzzy clustering methods searches for approximately spherical clouds of data (clusters) in datasets and represents each single cluster by a typical prototype, the weighted cluster centre. Membership degrees to the clusters are computed for the data points. The membership degree of a data object in the cluster centre is 1 and, more or less, decreases with increasing distance to the cluster centre.

Considering the Mamdani controller in the sense that each rule represents a fuzzy point in the product space of input and output variables, the connection to the fuzzy cluster analysis is obvious: a rule characterizes a typical point on the graph of the control function and with increasing distance to this point the membership degree to the rule decreases. In this way, a rule can be interpreted as a cluster. The cluster centre is the point on the graph of the control function characterized by the rule. I the case of triangular membership functions it is normally the point at which all triangular functions assume the value 1.

When a process is controlled by a human operator, we can record the measured variables and the control actions chosen by the operator. Fuzzy cluster analysis is offers one way to derive fuzzy rules from these data. The result of a cluster analysis can be transformed into a rulebase based on the above mentioned analogy between fuzzy clusters and fuzzy rules. Each cluster induces a corresponding rule. Before we discuss this idea in detail, we first introduce some basic concepts of fuzzy cluster analysis. It is not our aim to provide a complete introduction to fuzzy clustering. We restrict our considerations to what is needed for understanding how fuzzy cluster analysis can applied to construct fuzzy rules from data. A more detailed discussion on various fuzzy clustering methods and further techniques for rule induction based on fuzzy cluster analysis can be found in [63, 64].

5.5.1 Fuzzy Cluster Analysis

For our purposes it is sufficient to assume fuzzy cluster analysis should be applied to a dataset $X = \{\mathbf{x}_1, \ldots, \mathbf{x}_k\} \subseteq \mathbb{R}^p$ consisting of k p-dimensional data tuples/vectors. For the moment, we assume that we specify the number of clusters c into which the dataset should partitioned in advance. How to determine the number of clusters automatically will be explained later on. Each of the c clusters is characterised by a set of parameters, the cluster prototype, $\mathbf{w}_i$. In the simplest case $\mathbf{w}_i$ could be the corresponding cluster, that is $\mathbf{w}_i \in \mathbb{R}^p$. Besides the cluster parameters $\mathbf{w}_i$, for every data object $\mathbf{x}_j$ a membership degree $u_{ij} \in [0,1]$ to each cluster $\mathbf{w}_i$ must be determined. The cluster parameters $\mathbf{w}_i$ and the membership degrees u_{ij} should be chosen in such a way that the data are as close as possible to the cluster to which they are assigned. Technically speaking, the following objective function has to be minimized.

$$F(\mathbf{U}, \mathbf{W}; X) = \sum_{i=1}^{c} \sum_{j=1}^{k} u_{ij}^m d(\mathbf{w}_i, \mathbf{x}_j) \tag{5.71}$$

$\mathbf{U} = (u_{ij})$ is the matrix of the membership degrees and $\mathbf{W} = (\mathbf{w}_1, \ldots, \mathbf{w}_c)$ is the matrix of the cluster prototypes. $d(\mathbf{w}_i, \mathbf{x}_j)$ is the distance of the data object $\mathbf{x}_j$ to the cluster $\mathbf{w}_i$. A precise definition of this distance will follow later on. The distances, weighted by the corresponding membership degrees, of the data objects to the clusters are summed up in the objective function F. Later in this section we discuss the function of the parameter m which has to be chosen in advance. If we want to minimize the function F without further constraints, the solution is obvious: we just choose $u_{ij} = 0$ for all $i = 1, \ldots, c$, $j = 1, \ldots, k$, that is, no data object is assigned to any cluster. In order to exclude this trivial, but undesirable solution, we introduce the constraints

$$\sum_{i=1}^{c} u_{ij} = 1 \qquad (j = 1, \ldots, k). \tag{5.72}$$

These constraints require that every data object $\mathbf{x}_j$ has a membership degree of 1 to whole cluster partition. This total membership degree can be distributed to single clusters. The constraint (5.72) enables us to interpret the membership degrees also as probabilities. Therefore, this clustering approach is also called probabilistic clustering.

The simplest fuzzy clustering algorithm based on the above described concept is the fuzzy c-means algorithm (FCM). Here the prototypes represent the cluster centres, i.e. $\mathbf{w}_i \in \mathbb{R}^p$. For measuring the distance between a data object $\mathbf{x}_j$ and a cluster prototype $\mathbf{w}_i$ we use the squared Euclidean distance

$$d(\mathbf{w}_i, \mathbf{x}_j) = \| \mathbf{w}_i - \mathbf{x}_j \|^2 .$$

In order to have a minimum the objective function (5.71) under the constraints (5.72) we have to fulfil the following necessary conditions.

$$\mathbf{w}_i = \frac{\sum_{j=1}^{k} u_{ik}^m \cdot \mathbf{x}_j}{\sum_{j=1}^{k} u_{ik}^m} \tag{5.73}$$

$$u_{ij} = \frac{1}{\sum_{\ell=1}^{c} \left(\frac{d(\mathbf{w}_i, \mathbf{x}_j)}{d(\mathbf{w}_\ell, \mathbf{x}_j)} \right)^{\frac{1}{m-1}}} \tag{5.74}$$

If the membership degrees u_{ij} were crisp, i.e. $u_{ij} \in \{0, 1\}$, then in the formula (5.73) $\mathbf{w}_i$ would be the centre of gravity of the vectors which are assigned to the ith cluster. In this case, according to the u_{ij}-values the numerator will only add up those data vectors that belong to the cluster. The denominator simply counts the number of data objects assigned to the cluster. For membership degrees $u_{ij} \in [0, 1]$ we obtain the centre of gravity weighted membership degrees.

Equation (5.74) states that the membership degrees depend on the relative distances of the data objects to the clusters. Therefore, for a data object we obtain the greatest membership degree for the cluster to which is has the smallest distance.

Taking this equation into account, we can explain the influence of the parameter m, called fuzzifier. $m \to \infty$ implies $u_{ij} \to \frac{1}{c}$, that is, each data object is assigned to each cluster with (almost) the same membership degree. For $m \to 1$ the membership degrees tend the (crisp) values 0 and 1. The smaller m (with $m > 1$) is chosen the "'less fuzzy"' is the cluster partition. A very common choice for the fuzzifier is $m = 2$.

Unfortunately, the minimization of the objective function (5.71) leads to a non-linear system of equation. In the two equations (5.73) and (5.74) on the right-hand side appear parameters which have to be optimized: for the prototypes $\mathbf{w}_i$ the membership degrees u_{ij} and vice versa. For this reason the strategy of alternating optimization is usually applied. At first, the cluster centres $\mathbf{w}_i$ are chosen randomly and with these fixed cluster centres the membership degrees u_{ik} are computed according to equation (5.74). With these membership degrees the cluster centres are updated using equation (5.73). This alternating scheme is repeated until convergence is reached, that is, until the change of the prototypes or the membership degrees is neglectable.

For this method we have to consider a special case for applying equation (5.74). If the distance of a data object to a cluster is 0, the denominator in equation (5.74) is also 0. In this case the membership degree of the data object to the cluster with distance 0 should be chosen as 1, the membership degree to all other clusters as 0. In case the distance of a data object to several clusters is 0, which for FCM would mean the pathological case that two or more cluster centres are identical, then the membership degrees of the data object, the total membership degree of 1 is divided equally among the clusters with distance 0.

In addition to FCM there are several other fuzzy clustering methods that are also applied for generating a fuzzy controllers from data. Here we only want to introduce one more technique and a combination with FCM. Other fuzzy clustering methods and their relation to fuzzy controllers are summarized in [63, 64].

Let us consider a variant of the fuzzy c-varieties algorithm (FCV) [15, 17] which we simplify slightly. This algorithm does not search clusters in the shape of clouds of data like FCM, but clusters in the form of hyperplanes, that is, for two-dimensional data clusters in the form of straight lines, for three-dimensional data, clusters in the form of planes. In this way a dataset can be locally approximated by linear descriptions which are suitable for the application of TSK models.

FCV is based on the same objective function (5.71) as FCM, but the distance function is modified. FCV describes a cluster by a hyperplane, that is, with a position vector $\mathbf{v}_i$ and a normalized normal vector $\mathbf{e}_i$ which is orthogonal to the (hyper-)plane which the cluster represents. The distance used in equation (5.71) is the squared distance of a data object to the corresponding hyperplane.

$$d((\mathbf{v}_i, \mathbf{e}_i), \mathbf{x}_j) = \left((\mathbf{x}_j - \mathbf{v}_i)^\top \mathbf{e}_i\right)^2 \tag{5.75}$$

$(\mathbf{x}_j - \mathbf{v}_i)^\top \mathbf{e}_i$ is the dot product of the vectors $(\mathbf{x}_j - \mathbf{v}_i)$ and $\mathbf{e}_i$.

Using FCV, the membership degrees are computed by equation (5.74), the same as for FCM. However, in this equation we have to use the distance function (5.75). The position vectors in FCV are determined in the same way as the cluster centres (5.73) of FCM. The normal vector $\mathbf{e}_i$ of a cluster is the normalized eigenvector of the matrix

$$\mathbf{C}_i = \frac{\sum_{j=1}^{k} u_{ij}^m (\mathbf{x}_j - \mathbf{v}_i)(\mathbf{x}_j - \mathbf{v}_i)^\top}{\sum_{j=1}^{k} u_{ij}^m}$$

corresponding to the smallest eigenvalue.

A disadvantage of FCV is that the linear clusters have infinite extension. This is the reason why a single cluster can cover two line segments (data clusters) with a gap in between, provided that the two line segments lie more or less on one straight line. Since this effect is not desirable in most cases and it is preferred that such line segments will be separated into different clusters, we do not use the original form of FCV but a combination with FCM. In this combination, the distance function is a convex combination of the FCV and FCM distances.

$$d((\mathbf{v}_i, \mathbf{e}_i), \mathbf{x}_j) = \alpha \cdot \left((\mathbf{x}_j - \mathbf{v}_i)^\top \mathbf{e}_i\right)^2 + (1 - \alpha) \cdot \parallel \mathbf{x}_j - \mathbf{v}_i \parallel$$

For $\alpha = 0$ we obtain FCM and for $\alpha = 1$ FCV. The modified distance function has only a direct effect on the calculation of the membership degrees u_{ij}. Using this distance function, the clustering scheme is called elliptotype clustering, because the clusters – depending on the choice of the parameter α – have the form of long stretched ellipses [15].

Probabilistic clustering under the constraints (5.72) has the disadvantage that the membership degrees depend only on the relative distances of the data objects to the clusters. This leads to sometimes undesirable properties of the membership functions of the data to the clusters. In the cluster centre the membership degree is 1 and with increasing distance the membership degree decreases, at least in the neighbourhood of the cluster centre. Nevertheless, later on the membership degree can increase again. For example, for data lying very far from all clusters we obtain a membership degree of about $1/c$ to all clusters.

Noise clustering [34] avoids this effect by maintaining the probabilistic constraint (5.72), but an additional noise cluster is added to the clusters. Noise data or outliers are supposed to have a high membership degree to the cluster. The noise cluster does not have any prototype that must be updated during the alternating optimisation scheme. All data have a fixed large distance to the noise cluster which is not change during the process of cluster analysis. In this way, data which are far from all clusters obtain a high membership degree to the noise cluster. A modification of noise clustering where the percentage of noise data can be specified is described in [93].

Possibilistic clustering [101] drops the probabilistic constraint completely and introduces an additional term in the objective function which puts a penalty on membership degrees near 0. The disadvantage of possibilistic clustering is that each cluster is optimised independently from all others, so that a good initialization (i.e. with the result of a probabilistic cluster analysis) is required.

Up to now we have assumed that the number c the clusters is chosen in advance for the cluster analysis. When the number of clusters should be determined automatically, there are two principle:

1. Using a global validity measure that evaluates the total result of a cluster analysis. One of many of such validity measures is the separation [205]

$$S = \frac{\sum_{i=1}^{c}\sum_{j=1}^{k} u_{ij}^{m} d(\mathbf{w}_i - \mathbf{x}_j)}{c \cdot \min_{i \neq t}\{\| \mathbf{v}_i - \mathbf{v}_t\} \|^2\}}$$

which relates the mean distance of the date to the assigned clusters to the smallest distance between two cluster centres. The less the value of S the better is the clustering result. Clustering is started with only two clusters and then the number of clusters is increased step by step up to a desired maximum number of clusters. We choose the clustering (including the number of clusters) for which we obtained the best (lowest) value of S.
2. Using local validity measures we rather start with an overestimated number of clusters and apply compatible cluster merging (CCM) [100]. Very

small clusters are deleted and similar clusters are joint together. In this way the number of clusters is reduced. Based on this result as initialisation, the cluster analysis is carried out again with the reduced number of clusters. The procedure is repeated until no more clusters can be deleted or joint together.

For a detailed overview on validity measures and their application to determine the number of clusters we refer to [63, 64].

5.5.2 Generating Mamdani Controllers

A very simple way to generate a Mamdani controller from measured data by fuzzy cluster analysis is to cluster the entire dataset with FCM and to interpret each cluster as a rule. The clusters are projected onto the single input variables and the output variable and then approximated by suitable fuzzy sets (i.e. triangular or trapezoidal membership functions). Figure 5.12 shows a typical projection of a fuzzy cluster onto the variable x. The projection is obtained by projecting the x coordinate of each datum onto the x axis. The height of the lines at these points indicate the respective membership degree of the corresponding data object to the considered cluster. The approximation of such a discrete fuzzy set by a triangular or trapezoidal membership functions is usually achieved by a heuristic optimization strategy. Data with small membership degrees are not taken into at all for the approximation and for the remaining data data with higher membership degrees, the sum of squared errors is minimized by the triangular or trapezoidal functions. A possible iterative method is described in [182]. In general, fuzzy sets obtained by approximation of cluster projections do not satisfy the conditions which are usually required for fuzzy partitions. For example, fuzzy sets can have very different degrees of overlap. Therefore, projected fuzzy sets which are very similar are normally replaced by a single fuzzy set. Furthermore, projection and approximation cause a loss of information. Fuzzy rules resulting from fuzzy cluster analysis are frequently used in order to get an idea about possible rules or to initialise other learning system like neuro fuzzy system [138] (see also section 5.6).

An alternative approach is to cluster the input variables and the output variable separately in order to obtain better fitting fuzzy sets [182].

5.5.3 Generating TSK Models

Elliptotype clustering is suitable for determining a TSK model from data. Just as with the Mamdani controller every cluster induces a rule and using projections suitable fuzzy sets are derived for the input variables. For the functions in the conclusion parts of the rules, functions of the form

$$f(x_1, \dots, x_n) = a_0 + a_1 x_1 + \dots + a_n x_n$$

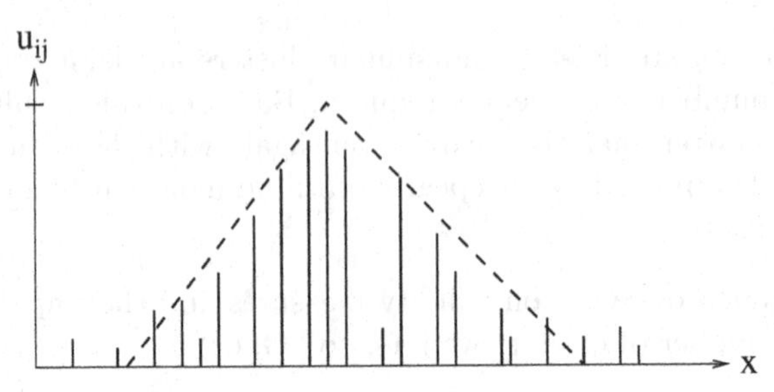

Fig. 5.12. Projection of a fuzzy cluster and approximation by a triangular membership function

are used which correspond to the lines, planes or hyperplanes that the clusters represent. The functions f are directly derived from the representation of the lines or (hyper-)planes by position and normal vectors.

5.6 Neuro Fuzzy Control

Promising approaches for optimization of existing fuzzy controllers and to learn a fuzzy system from scratch are approaches that combine fuzzy systems and learning methods of artificial neural nets. These methods are usually described with the terms *neural fuzzy* or *neuro fuzzy* systems. Nowadays, there are many specialized models. Beside the models for system control, especially systems for classification and more general models for function approximation have been developed. For a detailed introduction to this broad field see, e.g., [114, 75, 140]. A summary of applications can be found in, e.g., [21, 219, 154].

In this section we only introduce systematics and discuss some approaches. Furthermore, we give a short introduction to the fundamental principles of neural nets to the extent which is necessary to understand the following discussion about control engineering models. A detailed introduction to the fundamentals of artificial neural nets can be found in, e.g., [39, 58].

5.6.1 Neural Networks

The development of artificial neural networks was motivated by research in the area of biological nerve cells. The response time of such nerve cells is quite long compared to the clock speed of modern microprocessors. Nevertheless, they are able to solve efficiently complex tasks because nerve cells are massively interconnected. A biological neural net consists of several neurons (nerve cells or processing units) which are linked together to a complex network. The single neuron has the function of simple automaton or processor which calculates from its current input and state (activation) its new state and its output. With artificial neural nets we try to copy this behavior. The circuits of a biological neural net can be very complex, but in artificial neural nets we usually confine ourselves to networks with a hierarchical structure. For that, the neurons are grouped to layers. The neurons of a layer can only be influenced by neurons of layers which are next to it. Neurons of the same layer are not linked to each other. The two outer layers serve for the communication of the net with its environment and are called input and output layer, like we can see in figure 5.13. A net with such a structure is also called feed-forward net. If we also allow feedback between the layers or the single neurons, we call it a feedback or recurrent net.

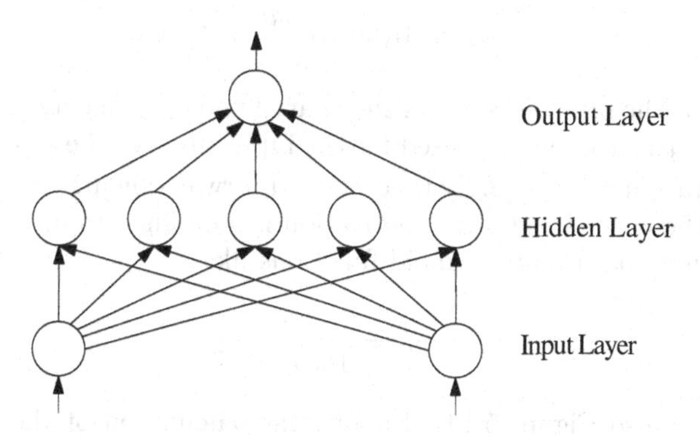

Fig. 5.13. A neural net of two layers consisting of two input neurons, five inner neurons and one output neuron

Most models of artificial neural nets can be described by the following general definition (according to [140]).

Definition 5.1 *A neural net is a tuple* $(U, W, A, O, \mathrm{NET})$ *wheras holds:*

1. *U is a finite set of processing units (neurons),*
2. *W, the network structure, is a matrix defining the mapping of the cartesian product $U \times U$ in $\mathbb{R}$,*
3. *A is a mapping which assigns to each $u \in U$ a* activation function $A_u : \mathbb{R}^3 \to \mathbb{R}$,
4. *O is a mapping which assigns to each $u \in U$ a* output function $O_u : \mathbb{R} \to \mathbb{R}$ *and*
5. NET *is a mapping which assigns to each $u \in U$ a* net input function $\mathrm{NET}_u : (\mathbb{R} \times \mathbb{R})^U \to \mathbb{R}$.

Computation of the output of a feed-forward net is done by successive computation of the outputs from the bottom to the top layer. This procedure is called progagation .

Propagation means that each neuron u_i calculates its own activation and output based on the outputs of the previous layer. At first, it calculates with the net input function NET_i the net input

$$\text{net}_i = \text{NET}_i(o_1, \ldots, o_n) \tag{5.76}$$

from the given inputs o_j (these are the outputs of the previous layer or the input to the net). In most neural network models the net input functions computes a weighed sum over the input values of the neurons u_i with the input weights w_{ji}, i.e.

$$\text{net}_i = \sum_j w_{ji} \cdot o_j. \tag{5.77}$$

Then the current activation a_i is calculated with the activation function

$$a_i = A_i(\text{net}_i, a_i^{(alt)}, e_i) \tag{5.78}$$

where $a_i^{(alt)}$ is the previous state of the neuron and e_i is an external input. The external input is usually only used for the input units of the input layer. Also the old activation $a_i^{(alt)}$ is rarely used in (feed-forward) neural network models. Therefore, the calculation of the activation is simplified to $a_i = A_i(\text{net}_i)$. As activation function usually sigmoid functions like

$$a_i = \frac{1}{1 + e^x} \tag{5.79}$$

are used (see also Figure 5.14). Finally, the calculation of the output o_i is done by the output function O_i. The output function can be used for scaling the output. However, most neural network models realize it by the identity, i.e.

$$o_i = O_i(a_i) = id(a_i) = a_i. \tag{5.80}$$

A schematic representation of how a neuron operates is given in Figure 5.15.

The most popular neural network model is the multilayer perceptron . A net of this type consists of an arbitrary amount of neurons which, analogously to Figure 5.13, can be arranged in the form of a directed graph which consists of an input layer, an output layer and an arbitrary amount of hidden layers. No feedback between the neurons is allowed, i.e., the output of the net can be calculated from the input to the output layer by a single propagation based on the given inputs. The input layer of this model forwards the inputs to the

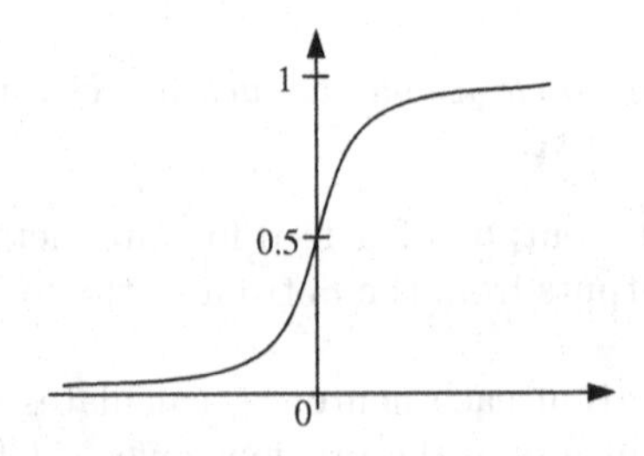

Fig. 5.14. A sigmoid activation function

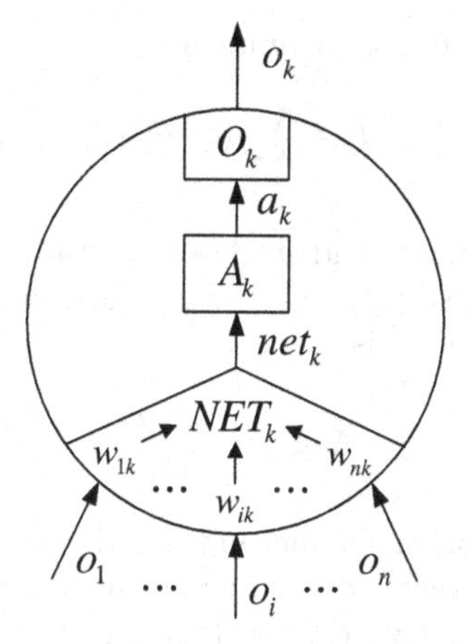

Fig. 5.15. Schematic representation of a neuron

next – the first hidden – layer without a transformation of the input values, i.e., the activation function is replaced by the identity. Therefore, it is possible to forward an input to several neurons. For the multilayer perceptron it was shown that it is an universal approximator, i.e., in principle, every continuous function can be approximated to any degree of accuracy (see e.g. [47, 65]).

Neural networks learn by modification of the network weights. The aim of the learning process is to train the net in a way that it will react to certain inputs with a certain output. In this way, it is also capable to respond with a suitable output to unknown inputs, that is, to inputs which were not part of the training process (capability of generalization).

The existing learning methods for neural networks can be divided into supervised and unsupervised learning methods. Unsupervised methods can be used if we know merely the inputs to the net. After the learning process the net shall map similar inputs to similar outputs. Therefore, mostly competitive learning methods are applied. A well-known network model which applies such a learning method is the self-organizing map, which is also called Kohonen network after its inventor [98].

In supervised learning problems the inputs to the network and the belonging outputs are known. The idea of this learning method is to first propagate the input through the network and then compare its output with the desired output. Finally, weights and threshold values are adjusted so that a re-propagation of this pattern through the network will lead to an output that is closer to the desired output. This process will be repeated with the available training data until the network has a certain accuracy. The probably best-known learning method applying this principle is the backpropagation method for the multilayer perceptron [172]. The fundamental idea of the method is the iterative reduction of the output error along the gradient of an error function. After each propagation the output error E is calculated based on the errors of the single output units A_i:

$$E = \sum_{A_i} (p_i - o_i)^2 \tag{5.81}$$

where p_i the desired output and o_i describes the actual output of the unit A_i. After every learning step the net weights are changed proportional to the negative error gradient, that is

$$\Delta w_{uv} \propto -\frac{\partial E}{\partial w_{uv}}. \tag{5.82}$$

A detailed description of the method and diverse variants which speed up the convergence of the method (like e.g. Quickprop [38] and resilient backpropagation (RPROP) [168]) can be found in the books mentioned at the beginning of this chapter. The fundamental principle of this learning method is also applied in the neuro fuzzy models described below.

Applications in System Control

Neural networks can be applied as controller and for modelling or identification of the plant. A fundamental requirement for their application is that enough data is available that can be used to train the net. Furthermore, these data have to cover the complete state space of the plant.

If pairs $(\mathbf{x}(k), \mathbf{u}(k))$ of measured data (in general the output and state variables of the plant or the input variables used by an existing controller) $\mathbf{x}(k) = [x_1(k), ..., x_n(k)]^T$ and the corresponding control values $\mathbf{u}(k) = [u_1(k), ..., u_m(k)]^T$ are known, a neural net, e.g. a multilayer perceptron, can be trained directly with this values. In this case, the used network has to consist of n input neurons and m output neurons. The training data can be obtained by, e.g., logging the input an output values of an existing controller – although in this case the net would just replace the controller – or by logging the control values of a human that is controlling the plant manually. Because of its generalization capability, after the training the neural net should be capable to provide suitable control values even for (intermediate) values that were not part of the training process.

A neural net can also be used to create a plant model (also an inverse model). For that, we have to know sufficient state vectors $\mathbf{x}(k) = [x_1(k), ..., x_n(k)]^T$ and output values $\mathbf{y}(k) = [y_1(k), ..., y_{n'}(k)]^T$ of the plant as well as the corresponding control values $\mathbf{u}(k) = [u_1(k), ..., u_m(k)]^T$. Based on this values we can design a neural net for plant modelling. For that, we use the vectors $\mathbf{x}(k)$ and $\mathbf{u}(k)$ as input values and $\mathbf{y}(k+1)$ as output values for training the net. An inverse plant model is obtained by simply changing the input and output values. We have to pay attention that feed-forward neural nets have a static input/output behavior because they merely carry out a mapping of the input values to the output values. Therefore, the input values of the net chosen for train the net have to describe unambiguously the mapping onto the output values. A feed-forward net is not able to learn a dynamic behavior.

Detailed discussion about control-engineering applications of neural nets can be found in e.g. [114, 128, 147, 68].

5.6.2 Combination of Neural Nets and Fuzzy Controllers

The combination of neural nets and fuzzy systems in so-called neuro fuzzy systems is supposed to combine the advantages of the two structures. Fuzzy systems provides the possibility of interpreting the controller and to introduce prior knowledge, and neural nets contribute its capability of learning and therefore the possibility of automatic optimization or automatic generation of the whole controller.

By the possibility to introduce a-priori knowledge in form of fuzzy if-then rules into the system we expect to strongly reduce the time and amount of training data required to train the system in comparison with pure neural network based controllers. If we have only some training data available, the prior knowledge might be even necessary in order to be able to create a controller. Furthermore, with neuro fuzzy systems we have, in principle, the possibility to learn a controller and then to interpret its control strategy by analyzing the learned fuzzy rules and fuzzy sets and, if necessary, revise them regarding the stability criteria discussed in the previous chapter.

Essentially, neural fuzzy controllers can be divided in cooperative and hybrid models. In cooperative models the neural net and the fuzzy controller operate separately. The neural net generates (offline) or optimizes (online, i.e., during control) some parameters (see, e.g., [99, 146, 160]). Hybrid models try to unite the structures of neural nets and fuzzy controllers. So, a hybrid fuzzy controller can be interpreted as neural net and can even be implemented with the help of a neural net. Hybrid models have the advantage of an integrated structure which does not require communication between the two different models. Therefore, the system is, in principle, able to learn online as well as offline and thus these approaches have become much more accepted than the cooperative models (see, e.g., [9, 57, 74, 141, 197]).

The fundamental idea of hybrid methods is to map fuzzy sets and fuzzy rules to a neural network structure. This principle is explained in the following. For that, we consider the fuzzy rules R_i of a Mamdani controller

$$\begin{aligned} R_i: \quad & \text{If } x_1 \text{ is } \mu_i^{(1)} \text{ and } \dots \text{ and } x_n \text{ is } \mu_i^{(n)} \\ & \text{then } y \text{ is } \mu_i, \end{aligned} \tag{5.83}$$

or R_i' of a TSK controller

$$\begin{aligned} R_i': \quad & \text{If } x_1 \text{ is } \mu_i^{(1)} \text{ and } \dots \text{ and } x_n \text{ is } \mu_i^{(n)} \\ & \text{then } y = f_i(x_1, \dots, x_n). \end{aligned} \tag{5.84}$$

Like described in section 3.1 or 3.2, the activation $\tilde{a}_i$ of these rules can be calculated by a t-norm. With the given input values $x_1, \ldots, x_n$ we obtain for $\tilde{a}_i$ with the t-norm min:

$$\tilde{a}_i(x_1, \ldots, x_n) = \min\{\mu_i^{(1)}(x_1), \ldots, \mu_i^{(n)}(x_n)\}. \tag{5.85}$$

One possibility to represent such a rule with a neural net is to replace each real-valued connection weight w_{ji} from an input neuron u_j to an inner neuron u_i by a fuzzy set $\mu_i^{(j)}$. Therefore, the inner neuron represents a rule and the connection from the input units represent the fuzzy sets in the premises of the rules. In order to calculate the rule activation with the inner neuron, we just have to modify its net input function. If we, for example, choose the minimum as t-norm, we define as net input (cf. 5.76) and (5.77)):

$$\text{net}_i = \text{NET}_j(x_1, \ldots, x_n) = \min\{\mu_i^{(1)}(x_1), \ldots, \mu_i^{(n)}(x_n)\}. \tag{5.86}$$

If we finally replace the activation function of the neuron by the identity, the activation of the neuron corresponds to the rule activation in (5.85) and therefore the neuron can be used directly to compute the rule activity of any fuzzy rule. A graphical representation of these structure for a rule with two inputs is shown in Figure 5.16 (on the left).

We obtain an alternative representation if the fuzzy sets of the premise are modelled as separate neurons. For that, the net input function is replaced by the identity and the activation function of the neuron by the characteristic function of the fuzzy set. Thus, the neuron calculates for each input the degree of membership to the fuzzy set represented by the activation function. In this representation we need two neuron layers in order to model the premise of a fuzzy rule (see figure 5.16 (on the right)). The advantage of this representation is that the fuzzy sets can be used directly in several rules in order to ensure the interpretability of the entire rule base. In this representation the net weights w_{ij} in the connections from the fuzzy sets to the rule neuron are initialized with 1 and are regarded as constant. The weights of the input values to the fuzzy sets can be used for scaling the input values.

If also the evaluation of a entire rule base has to be modelled, we have to decide whether we use a Mamdani or a TSK controller. For the TSK controller various realizations are possible. But, in principle, for each rule we get one more unit for evaluating the output function f_i – which will then be implemented as net input function – and it will be connected to all of the input units $(x_1, \ldots, x_n)$. In a output neuron the outputs of these units will be combined with the rule activations $\tilde{a}_i$ which are calculated by the rule neurons. This output neuron will finally calculate the output of the TSK controller with the help of the net input function (also see 3.9):

$$\text{NET}_{out} := \frac{\sum_i (\tilde{a}_i \cdot f_i)}{\sum_i \tilde{a}_i} = \frac{\sum_i (\tilde{a}_i \cdot f_i(x_1, \ldots, x_n))}{\sum_i \tilde{a}_i}. \tag{5.87}$$

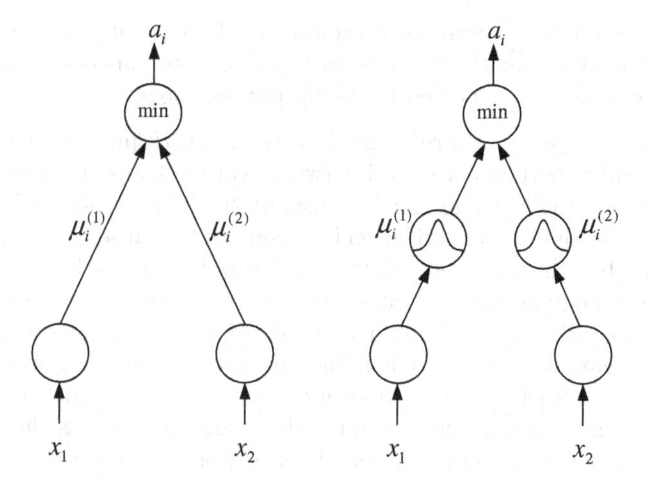

Fig. 5.16. Example of a neural net for calculating the activation of a fuzzy rule (If x_1 is $\mu_i^{(1)}$ and x_2 is $\mu_i^{(2)}$ then ...): Modelling of the fuzzy sets as weights (on the left) and as activation functions of a neuron (on the right)

The connection weights between the neurons are again constantly 1 and the identity is used as activation function.

For the Mamdani controller the concrete implementation depends on the chosen t-conorm and the defuzzification method (also see section 3.1). In every case, a common output neuron combines the activations of the rule neurons and calculates a crisp output value with the help of a modified net input function based on the corresponding fuzzy sets in the conclusion of the rules.

The transfer of a fuzzy rule base into a net structure can be summarized by the following steps:

1. For every input variable x_i we create a neuron of the same denotation in the input layer.
2. For every fuzzy set $\mu_i^{(j)}$ we create a neuron of the same denotation and connect it to the corresponding x_i.
3. For every output variable y_i we create a neuron of the same denotation.
4. For every fuzz rule R_r we create an inner (rule) neuron of the same denotation and we determine a t-norm for calculating the activation.
5. Every rule neuron R_r is connected according to its fuzzy rule to the corresponding neurons that represent the fuzzy sets of the premises.
6. *Mamdani controller:* Every rule neuron is connected to the output neuron according to the conclusion of its fuzzy rule. As connection weight we have to choose the conclusion of the corresponding fuzzy set. Furthermore, a t-conorm and the defuzzification method have to be integrated adequately into the output neurons.
   TSK controller: For every rule unit one more neuron is created for calculating the output function. These neurons are connected to the corre-

sponding output neuron. Furthermore, all of the input units are connected to the neurons for the calculation of the output function and all of the rule neurons are connected to the output neuron.

After the mapping of a rule base into the network presentation described above, learning methods of neural networks can be transferred to this structure. However, the learning methods usually have to be modified because of the changed net input and activation functions and because not the real-valued net weights but the parameter of the fuzzy sets have to be learned.

With the mapping onto a neural network structure we are able – provided that there are adequate training data – to optimize the parameters of a fuzzy set. What remains is the problem that we have to set an initial rule base which can be transferred to a network structure. This structure has to be created manually or we have to apply other methods, such as the fuzzy clustering method mentioned in the previous section, heuristics or evolutionary algorithms (see section 5.7) in order to learn an initial fuzzy rule base.

In the following sections we discuss two hybrid neuro fuzzy systems in more detail. Furthermore, we give further explanation to the principles and problems of neuro fuzzy architectures, especially with regard to applications in system control.

5.6.3 Models with Supervised Learning Methods

Neuro fuzzy models with supervised learning methods try to optimize the fuzzy sets and – for a TSK model – the parameters of the output function of a given rule base with the help of known input and output values.

Therefore, applying supervised models is convenient if we already have a description of the plant with fuzzy rules but need the control behavior – and thus the rule base – to be more exact. If measured data of the plant to be approximated exist (tuple of state, output and control variables), they can be used to retrain the system. This is possible for the normal as well as for the inverse plant model. The advantage in contrast to the application of neural nets is that fuzzy models already describe an approximation of the plant and, therefore, the training needs less time and less data than the training of neural nets. The neural net has to learn the transfer function of the plant completely from the data, which leads to two main problems: The learning procedure of a neural network can always end up in a local minimum, and if for some parts of the state space no or not sufficient training data was available the plant behavior may not be appropriately approximated.

Neuro fuzzy models with supervised learning methods are also convenient if an existing controller is to be replaced by a fuzzy controller, that is, that measured data of the control behavior of the real controller are available. Here an existing rule base is also presupposed. The learning methods can then be used in order to optimize the approximation of the original controller.

If no initial fuzzy rule base is available describing the system that should be approximated, i.e. the controller or the plant, and if an approximate rule base also cannot be created manually, we might use fuzzy clustering methods or evolutionary algorithms in order to obtain an initial rule base if measured data for training is available.

In the following, we discuss a typical example for a neuro fuzzy system with supervised learning, the ANFIS model. Beside this, there are several other approaches which are based on similar principles. For an overview of other models see, e.g. [21, 114, 140].

The ANFIS Model

In [74] the neuro fuzzy system ANFIS[1](Adaptiv-Network-based Fuzzy Inference System) was presented which by now has been integrated in many controller development and simulation tools.

The ANFIS model is based on a hybrid structure, that is, it can be interpreted as neural network and as fuzzy system. The model uses the fuzzy rules of a TSK controller (see also section 3.2). Figure 5.17 shows an example for a model with the three fuzzy rules

$$\begin{aligned} R_1: &\quad \text{If } x_1 \text{ is } A_1 \text{ and } x_2 \text{ is } B_1 \text{ then } y = f_1(x_1, x_2) \\ R_2: &\quad \text{If } x_1 \text{ is } A_1 \text{ and } x_2 \text{ is } B_2 \text{ then } y = f_2(x_1, x_2) \\ R_3: &\quad \text{If } x_1 \text{ is } A_2 \text{ and } x_2 \text{ is } B_2 \text{ then } y = f_3(x_1, x_2) \end{aligned}$$

where A_1, A_2, B_1 and B_2 are linguistic concepts which are assigned to the corresponding fuzzy sets $\mu_j^{(i)}$ in the premises. The functions f_i in the conclusion of the ANFIS model are defined by linear combination of the input variables, that is, in the example above by

$$f_i = p_i x_1 + q_i x_2 + r_i. \tag{5.88}$$

The structure of the model for the computation of the rule activation corresponds to the one discussed in the previous section (layer 1 and 2 in Figure 5.17). However, here we use the product as t-norm for the evaluation of the premise, that is, the neurons of layer 2 calculate the activation $\tilde{a}_i$ of a rule i by

$$\tilde{a}_i = \prod_j \mu_i^{(j)}(x_j). \tag{5.89}$$

In the ANFIS model the evaluation of the conclusion and the calculation of an output value is split up to the layers 3 till 5. Layer 3 calculates the relative

[1] In [74] the discussed TSK-controller-based model is called *type-3* ANFIS. The authors use the type in order to distinguish rule bases with different conclusions. Models with monotonic fuzzy sets in the conclusions are called *type-1* and Mamdani-controller-based models are called *type-2*. In general, ANFIS refers to the *type-3* ANFIS model because the authors suggest a learning method only for *type-3* models explicitly.

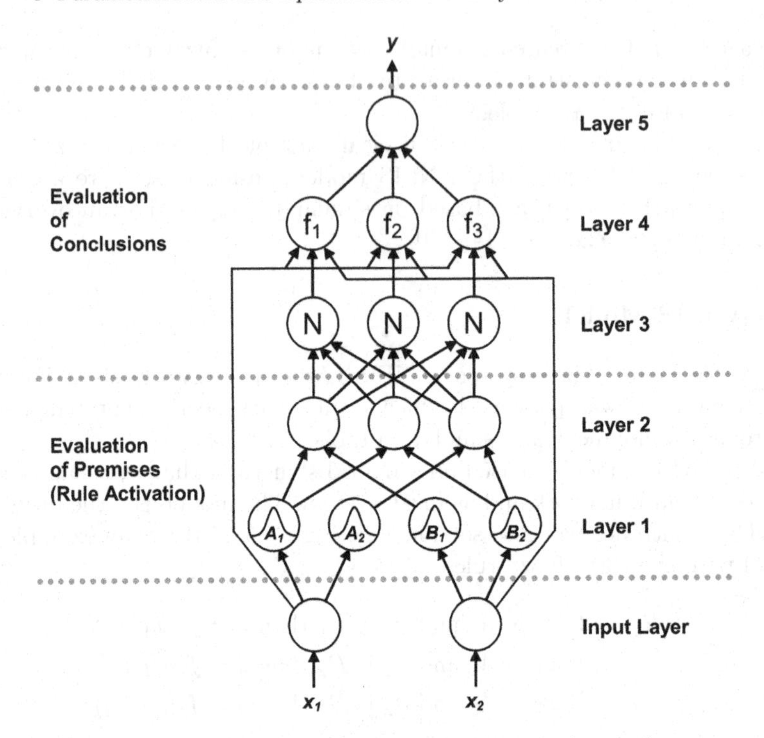

Fig. 5.17. Structure of an ANFIS net with three rules

contribution $\bar{a}_i$ of each rule to the total output based on the rule activations $\tilde{a}_i$. Therefore, the neurons of layer 3 compute

$$\bar{a}_i = a_i = net_i = \frac{\tilde{a}_i}{\sum_j \tilde{a}_j}. \tag{5.90}$$

Then the neurons of layer 4 compute the weighted control outputs based on the input variables x_k and the relative rule activations $\bar{a}_i$ of the previous layer:

$$\bar{y}_i = a_i = net_i = \bar{a}_i f_i(x_1, \ldots, x_n). \tag{5.91}$$

Finally, the output neuron u_{out} of layer 5 calculates the total output of the net or the fuzzy system:

$$y = a_{out} = net_{out} = \sum_i \bar{y}_i = \frac{\sum_i \tilde{a}_i f_i(x_1, \ldots, x_n)}{\sum_i \tilde{a}_i}. \tag{5.92}$$

For learning the ANFIS model needs a fixed learning problem. Therefore, for training it is necessary that a sufficient amount of pairs of input and output values exists. Based on these training data the model parameters – the parameters of the fuzzy sets and the parameters of the output functions f_i – are determined.

As learning method different approaches are suggested in [74]. Besides the pure gradient descent method which is analogous to the backpropagation method for neural nets (see also section 5.6.1) also combinations with methods for solving overdetermined linear equation systems (i.e. the least square (estimate) method (LSE) [23]) are suggested. Here, the parameters of the premises (fuzzy sets) are determined with a gradient descent method and the parameters of the conclusions (linear combination of the input variables) with a LSE method. Learning is performed in several separated steps, where in each case the parameters of the premises or of the conclusions are assumed to be constant.

In the first step, all input vectors are propagated through the net to layer 3 and for every input vector the rule activations are stored. Based on these values, for the parameters of the functions f_i in the conclusion a overdetermined equation system is created.

Let r_{ij} be the parameters of the output function f_i, $x_i(k)$ the input values, $y(k)$ the output value of the kth training pair and $\bar{a}_i(k)$ the relative rule activation, then we obtain

$$y(k) = \sum_i \bar{a}_i(k) y_i(k) = \sum_i \bar{a}_i(k) (\sum_{j=1}^{n} r_{ij} x_j(k) + r_{i0}), \forall i, k. \tag{5.93}$$

Therefore, with $\hat{x}_i(k) := [1, x_1(k), \ldots, x_n(k)]^T$ we obtain for a sufficient amount m of training data ($m > (n+1) \cdot r$, where r is the amount of rules and n the amount of input values) the overdetermined linear equation system

$$\mathbf{y} = \bar{\mathbf{a}} \mathbf{R} \mathbf{X}. \tag{5.94}$$

The parameters of this linear equation system – the parameters of the output function f_i in the matrix $\mathbf{R}$ – can be determined with an LSE method after the propagation of all training data. Finally, the error at the output units is determined based on the newly derived output functions, and by use of a gradient descent method the parameters of the fuzzy sets are optimized. The combination of the two methods leads to an improved convergence because the LSE already provides an optimal solution (with respect to least square error) for the parameters of the output function with regard to the initial fuzzy sets.

Unfortunately, the ANFIS model has no restrictions for the optimization of the fuzzy sets in the premises. Thus, it is not ensured that the input range will still be covered completely with fuzzy sets after the optimization and thus gaps in the definition of the system can appear. This has to be checked after optimization. Fuzzy sets can also change independently from each other and can also exchange their order and thus their importance and "linguistic" meaning. We have to pay attention to this, especially if an initial rule base was created manually and the controller has to be interpreted afterwards.

5.6.4 Models with Reinforcement Learning

The fundamental idea of the models that use reinforcement learning [7] is to determine a controller with as little as possible knowledge about the plant. The aim is to learn with a minimum of information about the control goal. In the extreme case, the learning method gets merely the information whether the plant is still stable or the controller had failed (controlling an inverse pendulum, this can happen when, for example, the pendulum fell or the slider bumped to the limit stop).

One main problem of reinforcement learning approaches is to find an evaluation measure of the control action that can be used in order to learn and optimize the controller. Like shown in section 5.3, a direct use of a error signal for optimization can even lead to a divergence of the learning process, in the extreme case. This is because of the fundamental problem that the current state of the plant is not only influenced by the last control action but by all of the previous states. In general, we cannot assume that the last control action has the greatest influence on the current system state. This problem is also called *credit assignment problem* [7], that is, the problem of assigning to a control action the (long-term) effect which it has on the plant.

By now many models have been suggested in the field of reinforcement learning. All of these are essentially based on the principle of dividing the learning problem into two systems: A criticizing or evaluating system (critic) and a system which stores a description of the control strategy and applies this to the plant (actor). The critic evaluates the current state considering the previous states and control action, evaluates as well the output of the actor based on these information and, if necessary, adapts its control strategy. An integration of such an approach to a control loop is shown in figure 5.18.

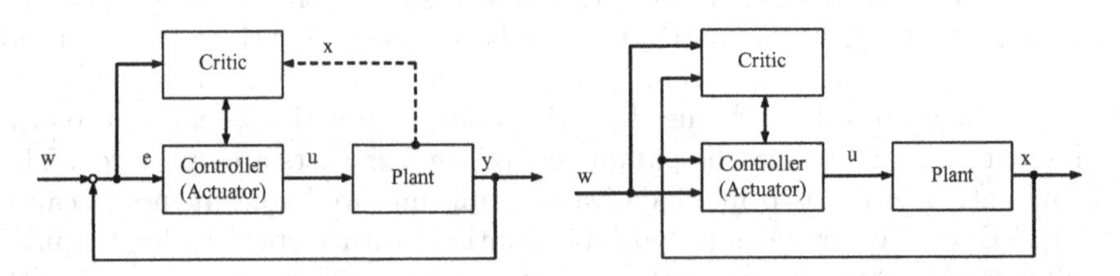

Fig. 5.18. Controller adaption with the help of a critic: based on the deviation (on the left) and as state controller (on the right)

The methods suggested by now for using the principles of reinforcement learning are mostly based on the combination with neural nets (see, e.g. [6, 80]). Very promising are the methods which use dynamic programming [8, 13, 14] in order to determine a optimal control strategy. For a more detailed discussion about this topic see e.g. [167, 185].

In the field of neuro fuzzy systems there are many approaches. But by now none of these have achieved the quality of the neural-net-based systems.

Examples of such approaches are GARIC [9], FYNESSE [169] and the NEFCON model [137, 150] which is briefly presented below.

The NEFCON Model

The principal aim of the NEFCON model (NEural Fuzzy CONtroller) is to detect online an appropriate and interpretable rule base with as few training cycles as possible. Furthermore, it should be possible to introduce prior knowledge into the training process in a simple way in order to speed up the learning process. This is different to most reinforcement learning approaches which try to generate an optimal controller (with as few prior knowledge as possible) and therefore loose much time in usually very long learning phases. Furthermore, the NEFCON model has also heuristic approaches for learning a rule base. In this point is is different to most other neuro fuzzy systems which, in general, can only be used for optimizing a rule base.

The NEFCON model is a hybrid model of a neural fuzzy controller. If we assume the definition of a Mamdani controller and we obtain the net structure if we – analogously to the description in section 5.6.2 – interpret the fuzzy sets as weights and the input and output variables as well as the rules as operating units (neurons). Then the net has the structure of a multilayer perceptron and can be interpreted as three-layer fuzzy perceptron [142]. The fuzzy perceptron is thus created based on a perceptron (see figure 5.13) by modelling the weights, the net inputs and the activation of the output unit as fuzzy sets. An example for a fuzzy controller with 5 rule units, two input (state) variables and one output (control) variable is shown in figure 5.19.

The inner units $\mathrm{R}_1, \ldots, \mathrm{R}_5$ represent the rules, the units x_1, x_2 and y the input and output variables and $\mu_r^{(i)}$ and ν_r the fuzzy sets for the premises or conclusions. The connection with joined weights denote equal fuzzy sets. A changing of these weights makes it necessary that all connections with this weight have to be adapted in order to ensure that the same fuzzy sets keep represented by the same weights. Thus the rule base defined by the net structure can also be formulated in form of the fuzzy rules listed in table 5.1.

$$
\begin{aligned}
R_1 &: \text{ if } x_1 \text{ is } A_1^{(1)} \text{ and } x_2 \text{ is } A_1^{(2)} \text{ then } y \text{ is } B_1\\
R_2 &: \text{ if } x_1 \text{ is } A_1^{(1)} \text{ and } x_2 \text{ is } A_2^{(2)} \text{ then } y \text{ is } B_1\\
R_3 &: \text{ if } x_1 \text{ is } A_2^{(1)} \text{ and } x_2 \text{ is } A_2^{(2)} \text{ then } y \text{ is } B_2\\
R_4 &: \text{ if } x_1 \text{ is } A_3^{(1)} \text{ and } x_2 \text{ is } A_2^{(2)} \text{ then } y \text{ is } B_3\\
R_5 &: \text{ if } x_1 \text{ is } A_3^{(1)} \text{ and } x_2 \text{ is } A_3^{(2)} \text{ then } y \text{ is } B_3
\end{aligned}
$$

Table 5.1. The rule base of the NEFCON system shown in figure 5.19

The learning process of the NEFCON model can be divided into two main phases. The first phase is designed to learn an initial rule base, if no prior

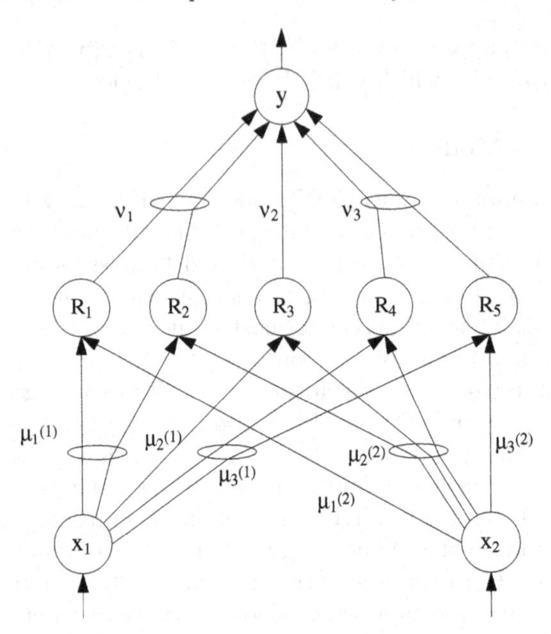

Fig. 5.19. A NEFCON system with two input variables and 5 rules

knowledge about the system is available. Furthermore it can be used to complete a manually defined rule base. The second phase optimizes the rules by shifting or modifying the fuzzy sets of the rules. Both phases use a fuzzy error e, which describes the quality of the current system state, to learn or to optimize the rule base. The fuzzy error plays the role of the critic element in reinforcement learning models. In addition the sign of the optimal output value must be known. Thus, the extended fuzzy error E is defined as

$$E(x_1, ..., x_n) = sgn(y_{opt}) \cdot e(x_1, ..., x_n),$$

with the crisp input $(x_1, ..., x_n)$. At the end of this section we briefly discuss some methods to describe the system error.

Learning a Rulebase

If for the system which has to be controlled an adequate rule base does not exist or cannot be determined manually, it has to be created by an appropriate rule learning method. Methods to learn an initial rulebase can be divided into three classes: Methods starting with an empty rule base, methods starting with a "full" rule base (combination of every fuzzy set in the antecedents with every consequent) and methods starting with a random rule base. In the following we briefly present algorithms of the first two classes. The algorithms do not require a given fixed learning problem but they try to determine a suitable rule base based on the extended system error E (see also [150]). However, both methods require an appropriate fuzzy partitioning of the input and output variables (see also the discussion in section 3.4.2).

An Elimination Method for Learning a Rule Base

The elimination method starts with a complete overdetermined rule base, that is, the rule base consists of all rules which can be defined by the combination of all fuzzy sets in the initial fuzzy partitionings of the input and output variables.

The algorithm can be divided into two phases which are executed during a fixed period of time or a fixed number of iteration steps. During the first phase, rules with an output sign different from that of the optimal output value are removed. During the second phase, a rule base is constructed for each control action by selecting randomly one rule from every group of rules with identical antecedents. The error of each rule (the output error of the whole network weighted by the activation of the individual rule) is accumulated. At the end of the second phase from each group of rule nodes with identical antecedents the rule with the least error value remains in the rule base. All other rule nodes are deleted. In addition, rules used very rarely are removed from the rule base.

A disadvantage of the elimination method is that it starts with a very large rule base and, thus, for systems with many state variables or fine grained partitions, i.e. many fuzzy sets are used to model the input and output values, it requires much memory and is computationally expensive. Therefore it is advisable to use the incremental rule learning procedure for larger rulebases.

Incremental Learning the Rulebase

This learning methods starts with an empty rule base. However, an initial fuzzy partitioning of the input and output intervals must be given. The algorithm can be divided into two phases. During the first phase, the rules' premises are determined by classifying the input values, i.e. finding that membership function for each variable that yields the highest membership value for the respective input value. Then the algorithm tries to "guess" the output value by deriving it from the current fuzzy error with a heuristics. It is assumed that input patterns with similar error values require similar output (control) values. The so found rule is inserted into the rule base. In the second phase we try to optimize the rule conclusion by applying the learned rule base to the plant and, based on the detected error values, exchanging the fuzzy sets in the conclusions, if necessary.

The used heuristics maps the extended error E merely linearly to the interval of the control variable. Thus, we assume a direct dependency between the error and the control variable. This is might be critical, especially for plants which would require an integrator for control, that is, which require a control value unequal 0 in order to achieve the control goal or to keep it. This heuristic also assumes that the error in not determined only from the control deviation but that the error also take the following plant states into account (see also the discussions in the section about the determination of a system error as well as the discussions in section 5.3).

Through the incremental learning method it is easily possible to introduce previous knowledge into the rule base. Missing rules are added during learning. However, because of the already discussed problems, both of the presented heuristics cannot provide a rule base which is appropriate for all plants.

The rule bases learned with the methods presented above should be checked manually for its consistence – at least if the optimization method, which we discuss in the following section, is unable to provide a satisfying solution. In any case, the possibility to introduce prior knowledge should be considered, that is, known rules should be inserted into the rule base before learning and the rule learning methods should not change the manually defined rules afterwards.

Optimization of a Rule base

The NEFCON learning algorithm for optimizing a rule base is based on the idea of the backpropagation algorithm for the multilayer perceptron. The error is, beginning with the output unit, propagated backwards trough the network and used locally for adapting the fuzzy sets.

The optimization of the rule base is done by changing the fuzzy sets in the premises and conclusions. Depending on the contribution to the control action and the resulting error the fuzzy sets of a rule are 'rewarded' or 'punished'. Thus, the principle of the reinforcement learning is applied. A 'reward' or 'punishment' can be done by displacing or reducing/enlarging the *support* of the fuzzy set. These adaptations are made iteratively, that is, during the learning process the controller is used for controlling the plant, and after every control action an evaluation of the new state is made as well as a incremental adaption of the controller.

The principal problem of this approach is that the system error has to be determined very carefully in order to avoid the problems of evaluating a control action which we discussed at the beginning of this section. In many cases, this can be very difficult or even impossible. Nevertheless, the discussed methods can be applied to simple plants, and they can very helpful for designing fuzzy controllers for more complex plants. We have to pay attention to the fact that then such a controller should be checked very carefully for its stability.

Determining a System Error

The system error of the NEFCON model has the function of the critic, that is, the error is directly responsible for the success or failure of the learning process. A good definition of the system error is a necessary condition for the discussed learning methods.

Actually, there are many possibilities for defining the system error which also include the use of plant models. If such a (inverse) model exists, it can be used for evaluating the control action of the fuzzy controller (see discussion about model-based controlling in section 5.4). So to speak, this approach would provide the 'perfect' critic. But the aim of the reinforcement learning method is to control a priori unknown systems. Therefore, for the NEFCON model approaches were suggested which need as little information as possible about the plant.

A very simple approach for the evaluation of the quality of a system state is to distinguish two cases of 'good' states:

- The state values are the optimal or desired values.
- The state values deviate from the optimum but tend to the optimal state (compensatory situation).

These states obtain the goodness $G = 1$. 'Bad' states (e.g. great deviation from the optimal values and tendency to a greater deviation from the optimum) obtain $G = 0$. Based on the goodness $G \in [0, 1]$ of the current system state the system error E can be determined directly:

$$E(x_1, ..., x_n) = 1 - G(x_1, \ldots, x_n).$$

However, this very simple approach can – analogously to the example discussed in section 5.3 – lead to an increasing error for higher order systems. Thus, when defining a goodness function one has to carefully consider the behavior of the plant.

In [139] an alternative approach was suggested where the system error is described directly by fuzzy rules. This has the advantage that with the use of state variables we can implicitly suggest a control strategy for the plant with the error description. This strategy can be used by the controller for optimizing. For creating a rule base we need to have an understanding how the plant react to certain control actions, which is often almost impossible for complex systems. However, as shown in [139] with the example of the inverse pendulum, this approach can be nevertheless successfully applied for the learning of a fuzzy controller.

5.6.5 Recurrent Approaches

In contrast to pure feed-forward architectures that have a static input-output behavior, recurrent models are able to store information of the past, e.g. prior system states, and can be thus more appropriate for the analysis of dynamic systems (see, for example, discussions concerning the approximation and emulation capabilities of recurrent neural networks [161, 122, 203]). If pure feed-forward architectures are applied to these types of problems, e.g. prediction of time series data or physical systems, the obtained system data usually has to be preprocessed or restructured to map the dynamic information appropriately, e.g. by using a vector of prior system states as additional input. If we apply a fuzzy system, this may lead to an exponential increase of the parameters – if we want to cover the whole system state space – that soon becomes intractable.

Recurrent neuro-fuzzy systems (RNFSs) can be constructed in the same way as discussed above for feed-forward neuro-fuzzy systems. So, they are based either on a recurrent fuzzy system or a recurrent neural network structure. However, the design and the optimization of (hierarchical) recurrent systems is, due to the dynamics introduced by the feed back connections, more difficult than that of feed forward systems. In Fig. 5.20 an example of a hierarchical RNFS is shown.

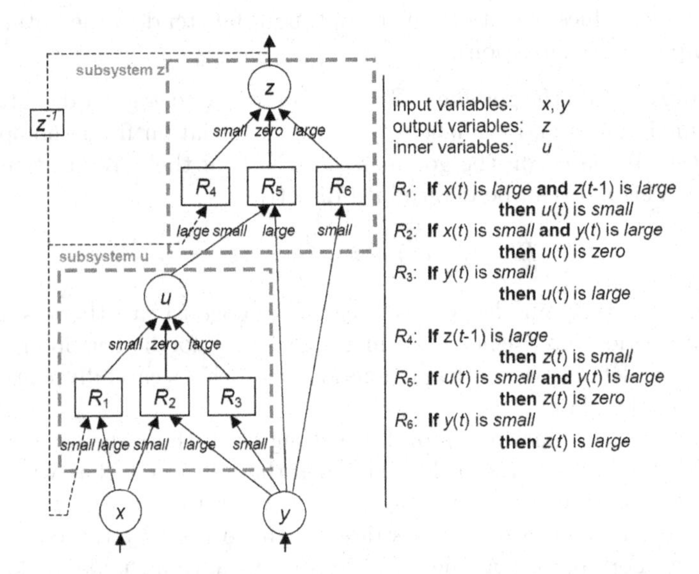

Fig. 5.20. Example of a simple hierarchical recurrent rule base consisting of two subsystems. The output of the system is reused by each subsystem as time-delayed input.

Probably the first recurrent fuzzy system that was combined with a (neural network motivated) learning method was proposed by Gorrini and Bersini in 1994 [53]. The proposed system is a Takagi-Sugeno-Kang fuzzy system and uses fuzzy rules with a constant conclusion. The internal variables of this system may be defined manually, if the designer has sufficient knowledge of the system that should be modelled. No learning method for the rule base itself was proposed except to initialize the rule base randomly. However, the authors propose a learning approach to optimize the parameters of a recurrent rule base, which was motivated by the real time recurrent learning algorithm [204].

According to Gorrini and Bersini the results for the approximation of a third order non-linear system for a given rule base was comparable to the approximation by a recurrent neural network. Unfortunately, a detailed discussion of the results was not given. Furthermore, the model had some insufficiencies. First of all, the structure has to be defined manually, since no learning methods for the rule base have been proposed. Furthermore, the learning is restricted to symmetric triangular fuzzy sets and the interpretability is not ensured, since the fuzzy sets are modified independently during learning. However, an extension to arbitrary (differentiable) fuzzy sets is easily possible.

Surprisingly, after this first model, for some time not many work had been published on recurrent systems that are also able to learn the rule base itself. Most likely the first models that were successfully applied to control – which, however, do not implement generic hierarchical recurrent models as described above – were proposed in [193, 217, 106]. For example, Lee and Teng proposed in [106] a fuzzy neural network, which implements a modified Takagi-Sugeno-Kang fuzzy system with Gaussian-like membership functions in the premises and constant conclusions. However, this model did not implement a fully recurrent system as shown in Fig. 5.20, but they restricted themselves to integrate feed back connections in the membership layer as depicted in Fig. 5.21.

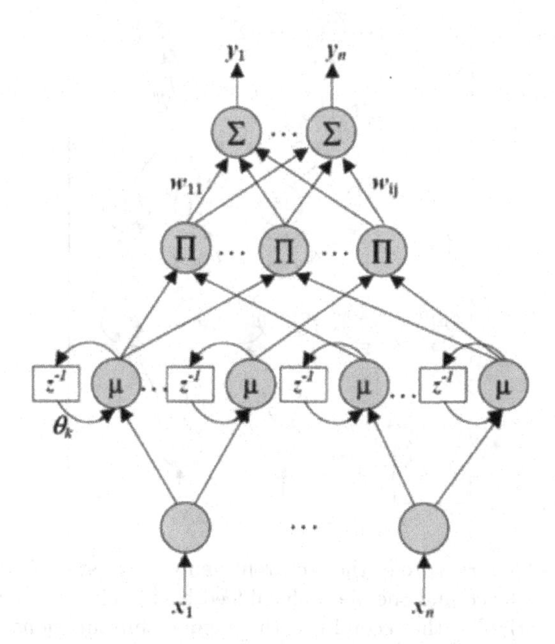

Fig. 5.21. Fuzzy neural network with simple feedback units as proposed by Lee and Teng.

Approaches to learn hierarchical recurrent fuzzy system were presented in [183], where a genetic algorithm was used, and in [148], where a template based approach to learn a structured rule base and a gradient descent based method motivated by the real time recurrent learning algorithm [204] to optimize the parameters of the learned rule base were proposed. The interpretability of the fuzzy sets of this model is ensured – similar to the NEFCON approach discussed above – by use of coupled weights in the conclusions (fuzzy sets, which are assigned to the same linguistic terms share their parameters) and in the premises. Furthermore, constraints can be defined, which have to be observed by the learning method, e.g. that the fuzzy sets have to cover the considered state space. An example of the network structure is given in Fig. 5.22. However, the template based learning approach has insufficiencies due to the use of a heuristics that creates inner fuzzy sets. Therefore, in [149] a slightly modified approach was proposed, that learns the rule base using a genetic algorithm.

Further, more recent models have been proposed in, e.g., [112, 200] that tackle specific problems of the learning process or specific applications.

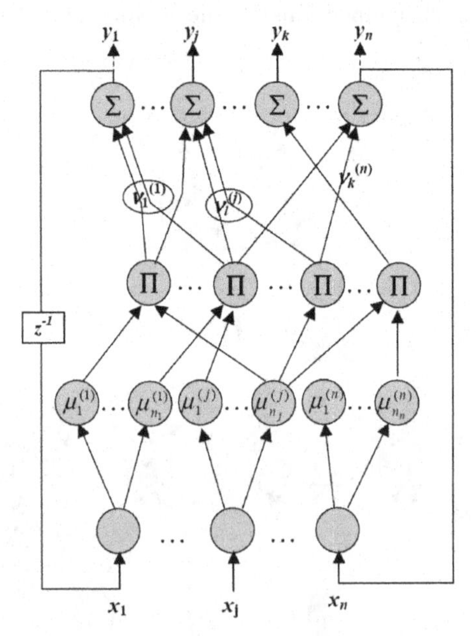

Fig. 5.22. Possible structure of the recurrent neuro-fuzzy system proposed in [148] (using one time-delayed and one hierarchical feed-back). The first row of neurons defines the input variables, the second row the membership functions of the premises, the third row the fuzzy rules, and the fourth row the output variables. The membership functions of the conclusions that are shared by rules are represented by coupled links from the rule to the output layer.

5.6.6 Discussion

Like shown in this section, neuro fuzzy systems can be used in order to optimize a fuzzy controller or the fuzzy model of a plant with the help of measuring data. Especially the methods based on supervised learning methods can also be applied to complex plants.

The existing neuro fuzzy systems which are based on reinforcement learning methods can (by now) only be applied successfully for the learning of controllers for very simple plants. But we can expect that in the near future there will be suggestions which – analogously to neural-net-based models – can be applied successfully for the learning of control strategies.

5.7 Fuzzy Controllers and Evolutionary Algorithms

The term evolutionary algorithms encompasses optimization strategies which are inspired by biological evolution. Evolutionary algorithms start with a random population of possible solutions and apply some basic evolutionary operators to the population and aim to obtain better solutions over the generations. A suitable definition of the objective function to be optimised as well as how possible solutions are coded is a crucial first step for applying an evolutionary algorithm. In the following, we describe the basic underlying ideas of using evolutionary algorithms in the context of fuzzy controllers.

In order to optimise (i.e. minimize or maximise) a given objective function, as a first step, we have to create a *population* of random solutions (*chromosomes*). This family of random initial solutions is, in general, far away from an optimum of the objective function. From this initial population we generate new solutions by applying genetic operators. The most important genetic operators are *mutation* and *crossover*. Mutation carries out small random changes of the parameters (*gens*) of an *individual* or chromosome (a solution within a population). Crossover just mixes the parameter sets of two individuals. Then selection is applied to the enlarged population in order to have a population of the same size as the initial population. The whole procedure is repeated with this new *generation* and the following generations. *Selection* is based on the principle of the survival of the fittest. The *fitness* of an individual is the greater the better it optimizes the objective function. The evolution process is stopped when a suitable solution is found or after several generations no improvement could achieved. There are various evolutionary algorithms based on these fundamental principles [5, 51, 124, 144].

In our context of fuzzy control, we will only introduce the basic ideas of two paradigms of evolutionary algorithms: the evolution strategies and genetic algorithms.

5.7.1 Evolution Strategies

Evolution strategies are suitable for the optimization of objective functions with real-valued parameters. A potential solution of the optimization problem, that is, an individual within a population, is defined as a real vector with a fixed dimension or fixed number of components. For instance, if we want to approximate measured data $\{(x_i, y_i, z_i) \mid i = 1, \ldots, n\}$ by a plane $f(z) = a + bx + cy$ which minimizes the sum of the absolute errors in z-direction, we would use

$$Z(a, b, c) = \sum_{i=1}^{n} |a + b \cdot x_i + c \cdot y_i - z_i|$$

as the objective function. An individual, that is, a possible solution, is a three-dimensional vector (a, b, c).

In evolution strategies the population size – the number of individuals within a generation – is usually denoted by μ (although a fuzzy set is not meant here!). From an initial population of μ random solutions (a_1, b_1, c_1), $\ldots, (a_\mu, b_\mu, c_\mu)$ mutation generates ν descendants. Mutation is carried out by adding three independent normally distributed random values $\varepsilon_1, \varepsilon_2, \varepsilon_3$, each one with mean value 0 and chosen small variance, to a solution (a_j, b_j, c_j), so that we obtain $(a_j + \varepsilon_1, b_j + \varepsilon_2, c_j + \varepsilon_3)$ as descendant. In this way, from a population we obtain ν descendants.

The selection within evolution strategies follows the *elite principle*. The μ best individuals are selected for the following generation. We can distinguish between two ways to do this.

- The μ best individuals are chosen among the $\mu + \nu$ parents and children together. This is called $(\mu + \nu)$ strategy or simply plus strategy.
- The μ descendants are selected from the ν children only. This approach is called (μ, ν) strategy or simply comma strategy.

The advantage of the plus strategy is that the best solution of the new generation will be at least as good as the best solution of the previous generation. In this, over the generations only improvements can take place. But the plus strategy tends to get stuck in local optima. Comma strategies can easier escape from a local optimum, but it is possible that good solutions are lost during the evolution process. Using the comma strategy we should always store the best solution that was found so far in order to be able recover it at the end of the evolution process.

Although this simple form of an evolution strategy will sometimes yield quite satisfactory solutions, it is recommended to use an adaptive strategy that adapts the step width of mutation. In order to achieve this, another component is included in the parameter vector of each individual. This additional component specifies the variance of the normal distribution of for mutation. If mutations lead to improvements quite frequently, the step width for mutation, i.e. the variance, can be increased. If almost all of the children are worse than the parents, we should choose a more careful step width, i.e. a smaller variance for mutation. A rule of thumb says that the step width is chosen well, if about one out of five children have a better fitness than their parents. If the amount of successful mutations is greater than 20%, the step width can be increased, if it is smaller, it should be decreased.

5.7.2 Genetic Algorithms

In contrast to evolution strategies genetic algorithms, in their original form, use a purely binary coding, in a more general form an arbitrary, discrete coding of the solution. Instead of the real parameter vectors of the evolution strategies genetic algorithms operate with binary vectors or vectors with components which can only take values from a finite domain. For example, a typical discrete problem is the assignment of jobs to machines. The jobs could be photocopies in different quantities. The machines could be to Xerox machines. The jobs should be assigned to the machines in such a way that the time needed to finish all jobs is as short as possible. If we have n jobs, we would use a binary vector with n components. A 1 at point i means that order i is assigned to machine 1, and if it is 0, it would be assigned to the other machine.

In genetic algorithms, mutation is carried out with a small probability for each variable. For a binary coding mutation would switch 1 to 0 and vice versa. If a variable can have more than just two values, the mutated value is often randomly chosen based on a uniform distribution over the possible values. In this case, it is recommended – provided that the structure of the problem or the domain enables this – to allow only mutations to similar or adjacent values.

Essentially for genetic algorithms is crossover. For crossover two individuals (chromosomes) are chosen randomly from a population and a random crossover point for the two parameter vectors is determined. The part before the crossover point of the first chromosome is concatenated with the part after the crossover point of the second chromosome and vice versa. In this way we obtain two new individuals. For example, if crossover is applied after the forth point to the two chromosomes 110011101 and 000010101, we obtain as a result the two chromosomes 110010101 and 000011101. Crossover is supposed to join partly optimal parameter configurations of two solutions to one improved one. If one chromosome has found a quite good configuration for the first parameters and the other one a good combination for the last parameters, crossover might lead to an overall better solution. This effect can only occur, if the single parameters do not depend too much on each other. Otherwise, crossover is just a massive mutation. The described crossover operator is also called one-point crossover, because the chromosomes are crossed at one point. In most cases, two-point crossover is better. Here two points are chosen in the chromosome and the part in between them is exchanged for two chromosomes.

For genetic algorithms, selection always involves a random mechanism, in contrast to elitist selection applied in the context of evolution strategies. In the original version of the genetic algorithm roulette wheel selection is used. Every chromosome obtains an individual probability to be selected which is proportional to its fitness. There are many variants of this selection method. Usually, at least the best chromosome is taken over to the next generation, even if it had, by chance, not been selected despite its higher probability.

The basic structure of genetic algorithms resulting from the operators described above is shown in figure 5.23.

```
begin
    t:=0;
    initialize(P(t));       // determine an initial population
    evaluate(P(t));         // compute the fitness values
    while (not StoppingCriteria(P(t),t)) do
        t:=t+1;
        select P(t) from P(t-1);
        crossover(P(t));
        mutate(P(t));
        evaluate(P(t));
    end;
end;
```

Fig. 5.23. Basic structure of a genetic algorithm

In principle, genetic algorithms can always be reduced to a binary coding. A finite amount of values can be represented by a suitable number of bits which merely leads to longer chromosomes. But then an additional, undesired mutation effect can occur during crossover. If one variable, which can have eight values, is coded by three bits, now the crossover can also take place within these three bits, so that after cross over, the coding of the corresponding variable with originally eight values is more or less random. Besides the desired combination of parameters of two solutions, crossover causes an additional change of one variable value.

Genetic algorithms can also be applied to problems with real-valued parameter vectors. In this case, we use for each real parameter a suitable number of bits, so that the real number can be presented with sufficient precision. This does not make a true difference between genetic algorithms and evolution strategies, because the latter ones also represent the real parameters only in a binary way in the computer. But mutation obtains a completely different meaning in genetic algorithm that actually encode real values. If a higher bit mutates, the coded real value will change extremely. Only mutation of a lower bit is more similar to mutation in small steps as is done with high probability in evolution strategies. Therefore, it is recommended to use different probabilities for mutation for the single bits. For higher bits we should, in contrast to lower bits, choose a very low mutation rate.

5.7.3 Evolutionary Algorithms for Optimising Fuzzy Controllers

If a fuzzy controller should be generated automatically by an evolutionary algorithms, at first, we have to define the objective function which has to be optimised by the evolutionary algorithm. If measure data of a controller are given – e.g., measuring data obtained from observing a human operator – the fuzzy controller should approximate the underlying control function as good as possible. As the error function, which has to be minimized, we could use the mean square error or the absolute error as well as the maximum deviation of the control function from the measured output. When the measured data come from different operators, an approximation can lead to a very bad behaviour of the controller. If the single operators use effective, but different control strategies and the controller is forced to approximate the data as good as possible, it will interpolate between the different strategies at each point. In the worst case, this mixed or interpolated control function might not work at all. If, for example, a car shall drive around an obstacle, and in the data for half of the cases an operator (driver) chose to avoid the obstacle to the right side and for the other half to the left side, interpolation will result in a strategy that keeps on driving straight ahead. If possible, the observed data should always be checked for consistency.

When we have a simulation model of the plant or process to be controlled, we can define various criteria that a good controller should satisfy, e.g. the time or the energy the controller needs to bring the process from different initial states to the desired state, some kind of evaluation of overshoots etc. If the evolutionary algorithm uses a simulation model with such an objective function, it is often better to slowly tighten the conditions of the objective function. In a random initial population it is very probable that no individual (controller) at all will be able to bring the process very close to a desired state. Therefore, at first the objective function might only measure the time how long a controller is able to keep the process in a larger vicinity of the desired state [62]. With an increasing number of generations the objective function is chosen more strict until it reflects the actual criterion.

The parameters of a fuzzy controller which can be learned with an evolutionary algorithm can be divided into three groups, described in more detail in the following.

The Rulebase

Let us first assume that the fuzzy sets are specified in advance or are optimised at simultaneously by another method. For instance, if the controller has two input variables for which n_1, respectively n_2 fuzzy sets are defined, for each of the possible $n_1 \cdot n_2$ combinations an output can be defined. For a

Mamdani controller with n_o fuzzy sets for the output variable, we could use a chromosome with $n_1 \cdot n_2$ parameters (gens), where each of these gens can have one of n_o values. But the coding of the rule table as a linear vector with $n_1 \cdot n_2$ components, which is required for the genetic algorithm, is not optimal in terms of crossover. Crossover should enable the genetic algorithm to combine to solutions (chromosomes) that have optimized different parameters (genes) to one good solution.

For the optimization of a rulebase of a fuzzy controller the conditions needed to benefit from crossover are satisfied, i.e. that the parameters are independent to a certain degree. Adjacent rules in a rulebase operate on overlapping regions, so that they contribute to the output of the controller at the same time, meaning that they interact and are dependent. Non-adjacent rules do not interact and never fire simultaneously. In this sense they are independent. If there are two fuzzy controllers in a population, which found, each one in a different part of the rule table, well-performing entries for the output of the rules, the combination of the two parts of the table will result in a better controller. However, a (fuzzy) region does not correspond to a row or a column in a rule table, but to a rectangular region in the table. A traditional genetic algorithm would only exchange linear parts in the crossover process. In the case of optimising the rule base of a controller with two input variables it makes sense, to deviate from the linear chromosome structure and to use a planar structure for the chromosome. Crossover should then exchange smaller rectangles within the rule table [89]. Here we discussed the case of only two input variables. This idea can be generalised to more input variables in a straight forward manner. For k input variables, we would use a k-dimensional hyperbox as the chromosome structure.

In order to guarantee small changes in the mutation process, an output fuzzy set should not be replaced by an arbitrary other fuzzy set, but by an adjacent one.

For a TSK model for the rulebase we have to determine output functions instead of output fuzzy sets. Usually, these functions are given in parameterized form, e.g.

$$f(x, y; a_R, b_R, c_R) = a_R + b_R x + c_R y$$

for the input variables x and y as well as three parameters a_R, b_R and c_R that have to be determined for each rule R. For a rule table with – like above – $n_1 \cdot n_2$ entries we had to determine, in total, $3 \cdot n_1 \cdot n_2$ real parameters for the rule table. In this case, we should apply an evolution strategy, since the parameters are not discrete, but continuous.

If we do not want to fill the rule table completely and want to limit the number of rules, we could assign to each rule an additional binary gene (parameter) which tells us whether the rule of the controller is used or not. For a TSK model, we have a proper evolutionary algorithm, because we have to deal with continuous and discrete parameters at the same time. The number of active rules can be fixed in advance, where we have to ensure that this number is not changed by mutation or crossover. Mutation could always activate one rule and deactivate another one at the same time, so that the number of active rules is not changed by mutation. For crossover a repairing algorithm is needed. If we have too many active rules after crossover, then, for example, rules could be deactivated randomly, until the desired number of active rule is reached again.

A better strategy is not to fix the number of active rules in advance. Since fuzzy controllers with a smaller number of rules are to be preferred due to better interpretability, we could introduce an additional term in the objective function penalising higher numbers of rules. This additional term should have a suitable weight. If the weight is too large, the emphasis is put more or less completely on keeping the number of rules small, ignoring the performance of the controller. A weight chosen too small will not contribute enough to keep the rule base small.

The Fuzzy Sets

Usually, the fuzzy sets are represented in a parameterized form like triangular, trapezoidal or Gaussian membership functions. The corresponding real parameters are suitable for an optimisation based on evolution strategies. However, giving complete freedom to the evolution to optimise these parameters, seldom to meaningful results The optimised fuzzy controller might perform perfectly, but the fuzzy sets overlap completely, so that it is hardly possible to assign meaningful linguistic terms to them and to formulate interpretable rules. In this case the fuzzy controller corresponds to a black box, as a neural network, without any chance to explain or interpret its control strategy.

It is recommended to choose the parameter set in such a way that the interpretability of the fuzzy controller is always guaranteed. One possible way would be the restriction to triangular functions which are chosen in such a way that the left and the right neighbour of a fuzzy set get the value 1 at the point where the membership degree of the fuzzy set in the middle drops to 0. In this case, the evolution strategy would have for each input or output variable as many real parameters as fuzzy sets are used for the corresponding domain. The respective real parameters indicate where the corresponding triangular function assume the value 1.

Even with this parameterisation undesired effects can occur, for example, if the fuzzy set *approximately zero*, because of mutation, overtakes the fuzzy set *positive small*. A simple change in the coding of the parameters can avoid this effect: the value of the parameter k determines no longer the point where the triangular membership function is 1, but how far it is shifted to the left relative to its left neighbour. The disadvantage of this coding is that a change (mutation) of the first value leads to a new position for fuzzy sets and therefore a quite great change of the total behaviour of the controller. If the fuzzy sets are parameterized independently, a mutation only has local effect. Therefore, we should stick to the independent parameterisation, but prohibit mutations which lead to overtaking fuzzy sets. In this way, mutations causes small changes and the interpretability of the fuzzy controller is preserved.

Additional Parameters

With evolutionary algorithms we can – if we want to – adjust also other parameters of a fuzzy controller. For example, we can use a parameterized t-norm for the aggregation of the rule premises and for each rule adjust the parameter of the t-norm individually. The same approach can also be used for a parameterized defuzzification strategy. Such parameters tend to cause problems in the interpretability of a fuzzy controller and will not be further pursued here.

So far, we have not answered the question, whether the rulebase and the fuzzy sets should be optimized at the same time or separately. As long as the rulebase can change drastically, it does not make sense to fine-tune the fuzzy set. The rulebase functions as the skeleton of a fuzzy controller. The concrete choice of the fuzzy sets is responsible for fine-tuning. In order to keep the number of parameters to be optimised by the evolutionary algorithm small, it is often better to learn at first the rulebase on the basis of standard fuzzy partitions and then optimise the fuzzy sets with a fixed rulebase.

5.7.4 A Genetic Algorithm for Learning a TSK Controller

In order to illustrate the principle of parameter coding, in the following section we briefly present a genetic algorithm for learning a TSK controller which was proposed in [108]. The algorithm tries to optimize all parameters of the controller, that is, the rulebase, the parameters of the fuzzy sets and the parameters of the conclusions at the same time.

In order to learn the rules

$$R_r : \text{If } x_1 \text{ is } \mu_R^{(1)} \text{ and } \dots \text{ and } x_n \text{ is } \mu_R^{(n)} \text{ then } y = f_r(x_1, \dots, x_n),$$

of a Takagi-Sugeno-Kang controller with

$$f_r(x_1, \dots, x_n) = p_0^r + x_1 \cdot p_1^r + \dots + x_n \cdot p_n^r,$$

we have to code the fuzzy sets of the input values and the parameters $p_0, \dots, p_n$ of each rule.

In this approach a triangular fuzzy set is described by three binary coded parameters (*membership function chromosome, MFC*):

leftbase	center	rightbase
10010011	10011000	11101001

The parameters *leftbase*, *rightbase* and *center* are no absolute values but denote the distances to the reference point. *Leftbase* and *rightbase* refer to the *center* of a fuzzy set and the *center* refers to the distance between the center and the left neighbour of the fuzzy set. If these parameters are positive, passing and abnormal fuzzy sets can be avoided.

The parameters $p_0, \ldots, p_n$ of a rule are coded directly by binary numbers and result in the *rule-consequent parameters chromosome* (RPC):

p_0	...	p_n
10010011	...	11101001

Based on these parameter codings, the complete rulebase of a TSK controller is coded in the form of a bit string:

variable 1	...	variable n	parameters of the conclusions
$\text{MFC}_{1 \ldots m_1}$	...	$\text{MFC}_{1 \ldots m_n}$	$\text{RPC}_{1 \ldots (m_1 \cdot \ldots \cdot m_n)}$

Besides the parameter optimization, the algorithm tries to minimize the amount of fuzzy sets assigned to a variable and, thus, also tries to minimize the amount of rules in a rulebase. For that, we assume a maximum amount of fuzzy sets. We eliminate fuzzy sets which are no longer in the permitted domain of a variable. Furthermore, among controllers of the same performance the selection process prefers the ones with less rules.

In [108] this approach was tested with an inverted pendulum (pole balancing problem). For a rulebase with five fuzzy sets for each input variable (2) and 8-bit binary numbers we obtain a chromosome length of $2 \cdot (5 \cdot 3 \cdot 8) + (5 \cdot 5) \cdot (3 \cdot 8) = 840$. For the evaluation of the learning process, the controllers were tested with eight different starting conditions and the time which the controller needed to bring the pendulum to the vertical position was measured. For this, we have to distinguish three cases:

1. If the controller brings the pendulum to the vertical position within a certain period of time, it obtains the more points the faster it was done.
2. If the controller is not able to bring the pendulum to the vertical position within this period of time, it obtains a predetermined amount of points which is in any case lower than the one in the first case.
3. If the pendulum falls over within the period of simulation, the controller obtains the more points the longer the pendulum did not fall but less than in the two previous cases.

The authors of [108] report that for learning a 'usable' controller more than 1000 generations were needed. This quite big amount of generations results from the great chromosome length. Furthermore, for coding neighbour relations are hardly used. Thus, it can happen that the premise of a rule is determined by the fuzzy set which is coded at the beginning of the chromosome but the corresponding conclusion is at the end of the chromosome. Therefore, the probability that the crossover destroys a good rule is quite big.

The interesting point of this approach is the ability of minimizing the required amount of rules. Here, the main idea is not only to optimize the controller's behaviour but to determine the important rules for the controller.

5.7.5 Discussion

We have discussed different approaches how to construct and optimize fuzzy controllers using evolutionary algorithms. A proper coding of the fuzzy controller is essential for the evolutionary algorithm. On the one hand, this coding should be chosen in such a way that the interpretability of the fuzzy controller can be ensured, no matter which parameter configuration will result from the evolutionary algorithm within the framework of possible configurations for the corresponding coding. And on the other hand, we should make sure that the evolutionary algorithm exploit its capabilities. For example, mutation should cause only small changes of the coded parameters and the complete control behaviour.

Altogether, we should try to use reasonable heuristics which facilitate the interpretability of the result and support the optimization strategy of the evolutionary algorithm. An overview about the large amount of publications in the field of fuzzy systems and evolutionary algorithms can b found in [32, 31, 59, 60, 159].

A

Correspondence Table for Laplace Transformations

$f(t)$ with $f(t<0)=0$	$f(s)=\mathcal{L}\{f(t)\}$
Impulse $\delta(t)$	1
Step function $s(t)$	$\frac{1}{s}$
$\frac{t^n}{n!}$	$\frac{1}{s^{n+1}}$
$\sin\omega_0 t$	$\frac{\omega_0}{s^2+\omega_0^2}$
$\cos\omega_0 t$	$\frac{s}{s^2+\omega_0^2}$
$h(t)e^{-at}$	$h(s+a)$

B

Non-Minimum Phase Systems

From a control engineer's point of view, non-minimum phase systems have certain disadvantageous features, which cause several methods of control engineering, e.g. different types of adaptive fuzzy controllers, to fail for such plants. It is therefore important to know about these systems.

To simplify the definition of such systems, let us introduce a special type of rational transfer function, the *all-pass element*. Its characteristic feature is that the gain of its frequency response is the same for all frequencies. A delay element would be a very simple example of such an element, but usually the term all-pass element refers only to purely rational stable transfer functions with a constant gain for all frequencies. Let us now examine some of the features of these transfer elements. If we write the frequency response of a transfer element with a general, rational transfer function as

$$G(j\omega) = \frac{b_m}{a_n} \frac{\prod_{\mu=1}^{m} (j\omega - n_\mu)}{\prod_{\nu=1}^{n} (j\omega - p_\nu)} \tag{B.1}$$

we can interpret the numerator and denominator polynomials as a product of complex-valued vectors. The gain, i.e. the absolute value of the frequency response, can be constant for all frequencies only if both the polynomial for the numerator and denominator are of the same degree, and if the denominator contains a vector of the same absolute value for every vector of the numerator. If we assume coprime polynomials, this can only be true if $n_\nu = -\bar{p_\nu}$, i.e. if all the poles and zeros are symmetrical with respect to the axis of imaginaries in the complex plane (fig. B.1). Since we required the transfer function to be stable, all the poles have to lie to the left of the axis of imaginaries.

As an example, let us have a closer look at a first order all-pass

$$G(s) = -\frac{s - p_1}{s + p_1} \tag{B.2}$$

or its frequency response

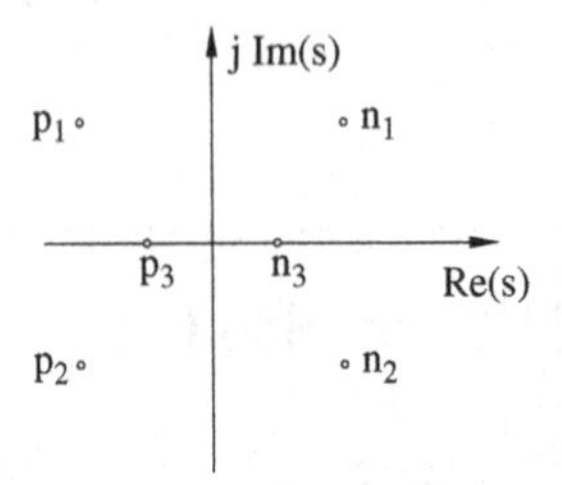

Fig. B.1. Poles and zeros of an all-pass element

$$G(j\omega) = -\frac{j\omega - p_1}{j\omega + p_1} = e^{-j2\arctan\frac{\omega}{p_1}} \tag{B.3}$$

Figure B.2 shows the corresponding Nyquist plot. It also gives the step response, where the final value is of opposite sign to the initial value. This is a characteristic feature of an all-pass element, and this is what makes automatic control of an all-pass element so difficult. We only have to think of a human operator who has to control an all-pass element: If he finds out that his actuating variable apparently seems to be the cause that the plant produces just the contrary of the desired reaction, he will probably adjust the actuating variable to the opposite direction—but, in fact, he only should have waited. It is plain to see that a controller of a primitive type will not be able to control an all-pass element.

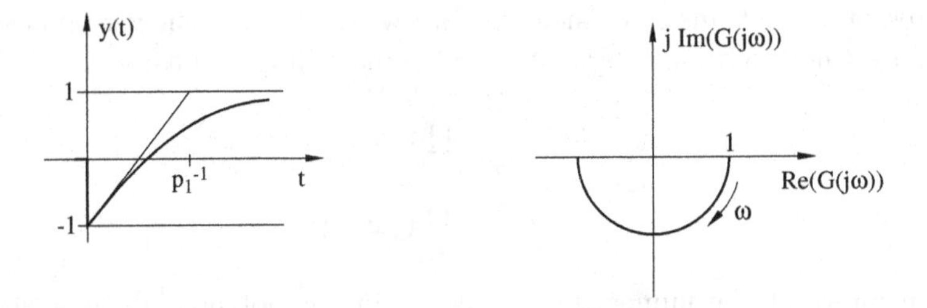

Fig. B.2. Step response and Nyquist plot of a first order all-pass

If we look at the frequency response of an all-pass element, it should be clear that such a component may cause serious problems to the stability of a system. For example, if we add an all-pass element to a given open-loop transfer function, then the phase lag of the frequency response will increase for larger values of ω, but the gain will remain unchanged. The Nyquist plot gets twisted around the origin depending on ω. This obviously increases the danger that it may revolve around the crucial point -1 of the Nyquist criterion, which would result in an instable closed-loop system.

Figure B.3 gives an example where — due to an additional all-pass element — the Nyquist plot of a first order lag is twisted in a way that it revolves around the point -1. Looking at the open-loop transfer function

$$G(s)K(s) = -\frac{V}{Ts+1}\frac{s-p_1}{s+p_1} \tag{B.4}$$

we can see that the number of poles which lie either on the axis of imaginaries or to the right of it is zero. Therefore, the permissible phase shift with respect to -1 according to the Nyquist criterion is zero, too. This is fulfilled for the plot on the left hand side and therefore, the closed-loop system with a first-order lag would be stable. But this is not true for the Nyquist plot on the right-hand side. There, the phase shift with respect to -1 is -2π, and the closed-loop system would therefore be unstable.

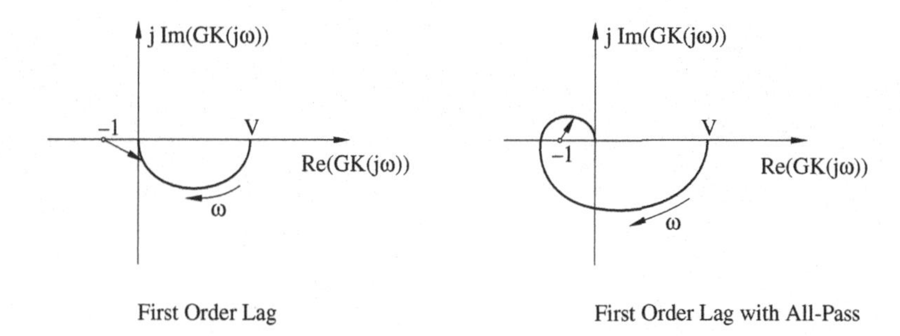

Fig. B.3. First order lag with all-pass

We conclude: Adding an all-pass element does not affect the gain of the frequency response, but it increases its phase lag, i.e. its absolute value of the phase. Thus, a first order lag with all-pass element compared to a pure first order lag will have phases with larger absolute values for the same gains. The same increase of the phase lag will happen for any stable, rational transfer element. On the other hand, it is possible to show that a rational transfer element which does not contain any all-pass will have the smallest absolute values for the phase possible for a given gain curve. Such a transfer element is called a *minimum phase element*.

Obviously, only transfer elements which have neither zeros nor poles with a positive real part can be minimum phase elements, since for any other rational transfer function it is possible to split of an all-pass element, as the following example shows:

$$G(s) = \frac{s-2}{s+1} = \frac{s+2}{s+1}\,\frac{s-2}{s+2} = G_{min}(s)G_{all}(s) \tag{B.5}$$

A rational transfer function, that contains at least one zero with positive real part, is containing an all-pass element and therefore, it is a non-minimum phase element.

Plants which contain an all-pass element occur more frequently in practical applications as one would be inclined to think. One example of such a plant is a motor-cyclist that is about to follow a left-hand bend. He first has to make a little turn to the right in order to be able to get into the inclined position required for the left turn. Another example is the tail of an airplane, which first drops a bit before starting to move upwards when the plane changes from horizontal to rising flight.

C

Norms

The following definitions are only intended to provide an introduction to norms of transfer elements. A more detailed presentation can be found, for example, in the appendix of [18].

Definition C.1 *A set X is a linear space, if an addition and a scalar multiplication is defined for it, so that the following conditions are satisfied (with $x, y, z \epsilon X$, and a, b complex numbers):*

$$
\begin{aligned}
x + y = y + x \qquad &\textit{(Commutativity)} \\
x + (y + z) = (x + y) + z \qquad &\textit{(Associativity)} \\
0 + x = x \qquad &\textit{(Existence of a null element)} \\
x + \hat{x} = x + (-x) = 0 \qquad &\textit{(Existence of an inverse element)} \\
a(x + y) = ax + ay \qquad &\textit{(Distributivity)} \\
(a + b)x = ax + bx \qquad &\textit{(Distributivity)} \\
(ab)x = a(bx) \qquad &\textit{(Associativity)} \\
1x = x \qquad &\textit{(Neutral element)}
\end{aligned}
\tag{C.1}
$$

Definition C.2 *X is a normed space if X is a linear space and if a real-valued norm $|| \bullet ||$ exists which satisfies the following conditions (where $x, y \epsilon X$, and a is a complex number):*

$$
\begin{aligned}
||x|| > 0 \qquad &\textit{if } x \neq 0 \\
||x|| = 0 \qquad &\textit{if } x = 0 \\
||x + y|| \leq ||x|| + ||y|| \qquad &\textit{(triangle axiom condition)} \\
||ax|| = |a| \; ||x||
\end{aligned}
\tag{C.2}
$$

The set of all time-variant functions $f(t)$, which are absolutely integrable for $-\infty < t < \infty$ at the p-th power ($1 \leq p < \infty$), form a linear space L_p. This space can be normed by the so-called *p-norm*:

$$||f||_p := \left(\int_{-\infty}^{\infty} |f(t)|^p dt \right)^{\frac{1}{p}} \tag{C.3}$$

For $p = \infty$, we obtain L_∞, the space of all bounded functions, with

$$||f||_\infty := \sup_t |f(t)| \tag{C.4}$$

as the associated norm.

For $p = 2$ we get the *2-norm*. If $f(t)$ is an input or output signal of a system, we can interpret the 2-norm as a measure of the energy content of the signal, that means, the energy that is put into or taken from the system. For example, let $f(t) = u(t)$ be the voltage at a resistor R. For the 2-norm we get

$$||f||_2 = ||u||_2 = \sqrt{\int_{-\infty}^{\infty} |u(t)|^2 dt} = \sqrt{R \int_{-\infty}^{\infty} \frac{1}{R} u^2(t) dt}$$

$$= \sqrt{R \int_{-\infty}^{\infty} P(t) dt} = \sqrt{RW} \tag{C.5}$$

with the electrical power P that is converted into heat in the resistor, and the electrical energy W. We can see, that the *2-norm* is proportional to the square root of the electrical energy.

The vector functions

$$\mathbf{f} = [f_1(t), ..., f_n(t)]^T \tag{C.6}$$

where $f_i \epsilon L_p$, form the L_p^n-space. For $1 \leq p < \infty$, a possible norm for this space is

$$||\mathbf{f}||_p := \left(\sum_{i=1}^{n} (||f_i||_p)^p \right)^{\frac{1}{p}} = \left(\sum_{i=1}^{n} \int_{-\infty}^{\infty} |f(t)|^p dt \right)^{\frac{1}{p}} \tag{C.7}$$

and for $p = \infty$, correspondingly:

$$||\mathbf{f}||_\infty := \max_{i=1,...,n} ||f_i||_\infty \tag{C.8}$$

Attention should be paid not to confuse function norms with norms for conventional complex-valued vectors, like

$$||\mathbf{x}||_k := \left(\sum_{i=1}^{n} |x_i|^k \right)^{\frac{1}{k}} \tag{C.9}$$

For $k = 2$, this is just the Euclidean norm. We will use the "single-vertical-bar" notation to facilitate a clear distinction from other norms:

$$|\mathbf{x}| := \sqrt{\sum_{i=1}^{n} |x_i|^2} \tag{C.10}$$

Definition C.3 *An operator T, defined for a linear space X, is a linear operator if*

$$T(ax + by) = a\,Tx \; + \; b\,Ty \tag{C.11}$$

is true for any $x, y \epsilon X$ and complex numbers a, b, where $Tx := T(x)$.

Definition C.4 *An operator T for a normed space X is said to be bounded if a real number c exists, such that*

$$||Tx|| \leq c\,||x|| \tag{C.12}$$

is true for all $x \epsilon X$. The lowest bound of this type is called operator norm, and is denoted by $||T||$:

$$||Tx|| \leq ||T||\;||x|| \tag{C.13}$$

For $T\,0 = 0$, it is obviously possible to compute the operator norm according to

$$||T|| = \sup_{x \neq 0} \frac{||Tx||}{||x||} \tag{C.14}$$

Using these definitions, let us now discuss a special case, which is of high importance for practical control applications. We are given the L_2^n-space of vector functions according to (C.6), together with the associated norm according to equation (C.7) for $p = 2$. If we interpret such a vector function as an input or output signal vector of a system, then the 2-norm gives us a measure for the energy content of this signal vector. For linear systems, the operator which maps one signal vector onto another one is given by the transfer matrix $\mathbf{G}(j\omega)$. The mapping of the input vector $\mathbf{x}$ onto the output vector $\mathbf{y}$ in the frequency domain is achieved through the multiplication of the input vector by the transfer matrix: $\mathbf{y} = \mathbf{G}\mathbf{x}$. It is possible to show that the operator norm associated with the operator $\mathbf{G}$ for the L_2^n-space normed by the 2-norm according to definition C.4 is just the ∞-norm of the transfer matrix $||\mathbf{G}(j\omega)||_\infty$. With the help of this norm, we can derive the following relations:

$$||\mathbf{G}(j\omega)||_\infty = \sup_{\mathbf{x}\neq\mathbf{0}} \frac{||\mathbf{y}||_2}{||\mathbf{x}||_2} \tag{C.15}$$

$$= \sup_{\omega} \bar{\sigma}\left\{\mathbf{G}(j\omega)\right\} \tag{C.16}$$

$$= \sup_{\omega} \sup_{\mathbf{x}\neq\mathbf{0}} \frac{|\mathbf{G}(j\omega)\mathbf{x}|}{|\mathbf{x}|} \tag{C.17}$$

$$= \sup_{\omega} \sqrt{\lambda_{max}\left\{\bar{\mathbf{G}}(j\omega)^T\mathbf{G}(j\omega)\right\}} \tag{C.18}$$

Equation (C.15) merely reflects what we already mentioned: $\mathbf{x}$ is the input vector function of the system, $\mathbf{y}$ the output vector function and $\mathbf{G}$ the associated operator. Equation (C.15) therefore corresponds to equation (C.14). Thus, the ∞-norm of a transfer matrix characterizes the maximum possible transfer ratio between the energy input and output of the system.

$\bar{\sigma}\left\{\mathbf{G}(j\omega)\right\}$ in equation (C.16) is the maximum singular value of the matrix $\mathbf{G}(j\omega)$, which is also called the *spectral norm*. The ∞-norm is therefore the maximum spectral norm possible for any frequency ω.

According to (C.17), the spectral norm of a matrix characterizes the maximum transfer ratio between the Euclidean norms of the input and the output vector of the matrix $\mathbf{G}(j\omega)$ for a fixed frequency ω. Now, because of (C.14), the spectral norm is exactly the associated operator norm of the operator $\mathbf{G}$ for the common vector space with the Euclidean norm as the associated vector norm.

The difference between equation (C.15) and (C.17) is that the 2-norm takes the entire integrated course of the signal into account, while in (C.17) only the absolute value of a vector (Euclidean norm) at a single moment is computed. As another consequence of equation (C.17), we find that the ∞-norm represents the largest possible transfer ratio of the Euclidean vector norm for a given system, for the following reason: On the one hand, the spectral norm is the maximum transfer ratio for one selected frequency ω, and on the other hand, the maximum of all frequencies has to be found for the ∞-norm.

According to (C.18), the spectral norm, and therefore also the ∞-norm can be computed from the maximum eigenvalue λ_{max} of the product of the matrix and its complex conjugated transpose.

From equation (C.17), we can immediately derive the following relation:

$$||G(j\omega)||_\infty = \sup_{\omega} |G(j\omega)| < \infty \tag{C.19}$$

Here, the ∞-norm is just the maximum distance between the Nyquist plot and the origin. According to definition C.4, this has to be a finite value, which implies that the transfer function must not have any poles on the axis of imaginaries.

While for linear systems, the output vector $\mathbf{y}$ is the result of a multiplication of the input vector and the transfer matrix, the output vector of a nonlinear system is a nonlinear function depending on the input vector, and possibly its derivatives: $\mathbf{y} = \mathbf{f}(\mathbf{x}, \dot{\mathbf{x}}, ...)$. As a result of it, equations (C.16) and (C.18) can not be used for nonlinear systems. Thus, we merely obtain the following relations for the ∞-norm:

$$||\mathbf{f}||_\infty = \sup_{\mathbf{x} \neq \mathbf{0}} \frac{|\mathbf{y}|}{|\mathbf{x}|} = \sup_{\mathbf{x} \neq \mathbf{0}} \frac{||\mathbf{y}||_2}{||\mathbf{x}||_2} \tag{C.20}$$

where the supremum is computed not only for all $\mathbf{x} \neq 0$, but also for its derivatives.

Interestingly enough, we are now able to present a more general version of the transfer stability (BIBO stability):

Definition C.5 *A system with the input quantity x and the output quantity y is called L_p-stable if the p-norm of the output quantity is bounded by the p-norm of the input quantity ($0 \leq c < \infty$):*

$$||y||_p \leq c\, ||x||_p \qquad \textit{for all}\, x \tag{C.21}$$

This definition is more restrictive than common BIBO stability. For BIBO stability, we only demanded that the output signal had to be bounded if the input signal was bounded. For an L_p-stable system, we also require that the output signal disappears if the input signal does. Definition C.5 obviously corresponds to the demand that the norm of the operator which characterizes the system is bounded:

$$||T|| \leq c < \infty \qquad \text{with } y = Tx \tag{C.22}$$

where $|| \bullet ||$ is the operator norm associated with the p-norm of the definition above.

For linear systems, a finite value of the ∞-norm therefore guarantees L_2-stability according to equation (C.15), and according to equation (C.17) also BIBO-stability. The same holds for nonlinear systems resulting from equation (C.20).

D

Lyapunov Equation

Theorem D.1 *Let* $\mathbf{Q}$ *be a symmetrical, positive definite matrix. The solution* $\mathbf{P}$ *of the Lyapunov equation*

$$\mathbf{A}^T\mathbf{P} + \mathbf{P}\mathbf{A} = -\mathbf{Q} \tag{D.1}$$

is positive definite if and only if the matrix $\mathbf{A}$ *has got only eigenvalues with a negative real part.*

Proof (cf. [84]): First, let us assume that all eigenvalues of $\mathbf{A}$ have negative real parts. Based on that, we want to show that this entails $\mathbf{P}$ to be positive definite.

Let $\mathbf{x}$ be the state vector of the system

$$\dot{\mathbf{x}} = \mathbf{A}\mathbf{x} \tag{D.2}$$

Using (D.1) gives

$$\begin{aligned}\frac{\partial}{\partial t}(\mathbf{x}^T\mathbf{P}\mathbf{x}) &= \dot{\mathbf{x}}^T\mathbf{P}\mathbf{x} + \mathbf{x}^T\mathbf{P}\dot{\mathbf{x}} \\ &= \mathbf{x}^T\mathbf{A}^T\mathbf{P}\mathbf{x} + \mathbf{x}^T\mathbf{P}\mathbf{A}\mathbf{x} = -\mathbf{x}^T\mathbf{Q}\mathbf{x}\end{aligned} \tag{D.3}$$

An integration of this equation yields

$$\mathbf{x}^T(\infty)\mathbf{P}\mathbf{x}(\infty) - \mathbf{x}^T(0)\mathbf{P}\mathbf{x}(0) = -\int_0^\infty \mathbf{x}^T(t)\mathbf{Q}\mathbf{x}(t)dt \tag{D.4}$$

Since all eigenvalues of $\mathbf{A}$ possess a negative real part, the system (D.2) is stable in the sense of Lyapunov, i.e. the state vector $\mathbf{x}$ will converge towards zero from any initial state: $\mathbf{x}(\infty) = \mathbf{0}$. Since $\mathbf{Q}$ is positive definite, we also have

$$\mathbf{x}^T\mathbf{Q}\mathbf{x} > 0 \tag{D.5}$$

for all $\mathbf{x}$. Altogether, this changes (D.4) into

$$\mathbf{x}^T(0)\mathbf{P}\mathbf{x}(0) = \int_0^\infty \mathbf{x}^T(t)\mathbf{Q}\mathbf{x}(t)dt > 0 \tag{D.6}$$

Since this inequality holds for any initial state, it follows that $\mathbf{P}$ has to be positive definite.

For the second part of the proof, we now assume that $\mathbf{P}$ is positive definite, and we have to prove that all eigenvalues of $\mathbf{x}$ have a negative real part.

For the system (D.2), we can find a Lyapunov function

$$V(\mathbf{x}) = \mathbf{x}^T\mathbf{P}\mathbf{x} \tag{D.7}$$

This function is obviously positive for all $\mathbf{x}$, because of the positive definiteness of $\mathbf{P}$. According to (D.3), the derivative of this Lyapunov function is:

$$\dot{V}(\mathbf{x}) = -\mathbf{x}^T\mathbf{Q}\mathbf{x} \tag{D.8}$$

This derivative is obviously negative for all $\mathbf{x}$, since $\mathbf{Q}$ is positive definite. From this it follows that the system is stable, which then implies that all eigenvalues of $\mathbf{A}$ have negative real parts.

E

Lie Derivative

Definition E.1 *Let* $\lambda(\mathbf{x})$ *be a scalar function of the vector* $\mathbf{x} = [x_1, ..., x_n]^T$*, and let* $\mathbf{f}(\mathbf{x}) = [f_1(\mathbf{x}), ..., f_n(\mathbf{x})]^T$ *be a vector function. The Lie derivative of* $\lambda(\mathbf{x})$ *with respect to* $\mathbf{f}(\mathbf{x})$ *is defined by the scalar function*

$$L_{\mathbf{f}}\lambda(\mathbf{x}) = \sum_{i=1}^{n} \frac{\partial \lambda(\mathbf{x})}{\partial x_i} f_i(\mathbf{x}) \tag{E.1}$$

Definition E.2 *The repeated Lie derivative—first with respect to* $\mathbf{f}(\mathbf{x})$ *and then with respect to* $\mathbf{g}(\mathbf{x})$*—is defined by*

$$\begin{aligned} L_{\mathbf{g}}L_{\mathbf{f}}\lambda(\mathbf{x}) &= \frac{\partial (L_{\mathbf{f}}\lambda(\mathbf{x}))}{\partial \mathbf{x}} \mathbf{g}(\mathbf{x}) \\ &= \left(\frac{\partial}{\partial \mathbf{x}} \sum_{i=1}^{n} \frac{\partial \lambda(\mathbf{x})}{\partial x_i} f_i(\mathbf{x}) \right) \mathbf{g}(\mathbf{x}) \\ &= \sum_{i=1}^{n} \left\{ \frac{\partial}{\partial x_i} \left[\sum_{i=1}^{n} \frac{\partial \lambda(\mathbf{x})}{\partial x_i} f_i(\mathbf{x}) \right] \right\} g_i(\mathbf{x}) \end{aligned} \tag{E.2}$$

Definition E.3 *The k-times Lie derivative of* $\lambda(\mathbf{x})$ *with respect to* $\mathbf{f}(\mathbf{x})$ *is the scalar function* $L_{\mathbf{f}}^k\lambda(\mathbf{x})$*, which is defined recursively by*

$$L_{\mathbf{f}}^k\lambda(\mathbf{x}) = L_{\mathbf{f}}L_{\mathbf{f}}^{k-1}\lambda(\mathbf{x}) \tag{E.3}$$

where $L_{\mathbf{f}}^0\lambda(\mathbf{x}) = \lambda(\mathbf{x})$.

F

Positive Real Systems

Depending on the dimension and representation of a system one of the following theorems can be used to evaluate, if the system is (strictly) positive real. The proofs of the theorems shall be omitted.

Theorem F.1 *A linear SISO system is strictly positve real, if and only if its transfer function has only poles with negative real parts and $Re(G(j\omega)) > 0$ holds for $\omega \geq 0$.*

A linear MIMO system with square transfer matrix $\mathbf{G}(s)$ is strictly positive real, if the elements $G_{ij}(s)$ of the matrix have only poles with negative real parts, and if the hermitian matrix

$$\mathbf{H}(j\omega) = \frac{1}{2}(\mathbf{G}(j\omega) + \bar{\mathbf{G}}^T(j\omega)) \tag{F.1}$$

is positive definite for all $\omega \geq 0$, that means, it has only positive eigenvalues.

Because all poles of all partial transfer functions have only negative real parts, every single partial transfer function is stable, and therefore also the entire system. It follows, that the stability of a system is a prerequisite for being positive real.

Theorem F.2 *A linear MIMO system*

$$\begin{aligned} \dot{\mathbf{x}} &= \mathbf{A}\mathbf{x} + \mathbf{B}\mathbf{u} \\ \mathbf{y} &= \mathbf{C}\mathbf{x} + \mathbf{D}\mathbf{u} \end{aligned} \tag{F.2}$$

is strictly positive real if and only if the following conditions hold:

- *The system must be completely controlable and observable.*
- *There must exist matrices $\mathbf{L}$, $\mathbf{P}$ and $\mathbf{V}$ of suitable dimension with*

$$\mathbf{A}^T\mathbf{P} + \mathbf{P}\mathbf{A} = -\mathbf{L}\mathbf{L}^T \tag{F.3}$$

$$\mathbf{L}\mathbf{V} = \mathbf{C}^T - \mathbf{P}\mathbf{B} \tag{F.4}$$

$$\mathbf{D} + \mathbf{D}^T = \mathbf{V}^T\mathbf{V} \tag{F.5}$$

- *$grad(\mathbf{L}) = grad(\mathbf{A}) = n$*
- **P** *is symmetrical and positive definite:* $\mathbf{P} = \mathbf{P}^T$ *and* $\mathbf{P} > 0$

Annotation: It follows from $grad(\mathbf{L}) = n$, that $\mathbf{L}\mathbf{L}^T$ is a symmetrical positive definite matrix. Furthermore it follows directly from the conditions, that **P** is also positive definite. Using that, it follows from (F.3) and theorem D.1, that **A** possesses only eigenvalues with negative real part. Therefore, we can derive from this theorem, too, that the stability of a system is a prerequisite for being positive real.

G

Linear Matrix Inequalities

The basic problem and starting point of all solution algorithms in the theory of linear matrix inequalities (LMI's) can be formulated in the following way (see [176]): A symmetrical matrix is given, whose coefficients depend linearly on free parameters. The question is, if these free parameters can be chosen in a way, so that the symmetrical matrix will be negative definite. With $\mathbf{p}$ is the vector of free parameters, and $\mathbf{F}(\mathbf{p})$ the symmetrical matrix, whose coefficients depend linearly on the free parameters, the formal definition of the basic problem is:

Does a vector $\mathbf{p}$ exist, that makes

$$\mathbf{F}(\mathbf{p}) < 0 \tag{G.1}$$

negative definite?

Because of the linearity of the matrix function $\mathbf{F}(\mathbf{p})$ the solution set for $\mathbf{p}$ is always convex, and therefore, the question of the existence of a solution can be answered with numerical algorithms clearly. One example for such algorithms is the Matlab LMI Toolbox, that can answer the question of the existence, but besides that, it can also compute a solution $\mathbf{p}$, if one exists.

With such a tool, a given problem just has to be transferred to the form (G.1). For this step, the following remarks give some support:

- An LMI of the form $\mathbf{G}(\mathbf{p}) > 0$ is equivalent to (G.1) with $\mathbf{G} = -\mathbf{F}$.
- In the matrix inequality $\mathbf{A}^T\mathbf{P} + \mathbf{P}\mathbf{A} + \mathbf{Q} < 0$ with the unknown matrix $\mathbf{P}$ we can set $\mathbf{F}(\mathbf{x}) = \mathbf{A}^T\mathbf{P} + \mathbf{P}\mathbf{A} + \mathbf{Q}$. The coefficients of $\mathbf{P}$ form the vector $\mathbf{x}$ of the unknown parameters. Obviously, the coefficients of $\mathbf{F}$ depend linearly on these parameters. To guarantee the symmetry of $\mathbf{F}$, $\mathbf{P}$ and $\mathbf{Q}$ have to be symmetrical.
- The negative definiteness of a block matrix can be defined by the negative definiteness of its blocks:

$$\begin{aligned}\begin{pmatrix} \mathbf{A} & \mathbf{C} \\ \mathbf{C}^T & \mathbf{B} \end{pmatrix} < 0 &\Leftrightarrow \mathbf{A} < 0 \quad \text{and} \quad \mathbf{B} - \mathbf{C}^T\mathbf{A}^{-1}\mathbf{B} < 0 \\ &\Leftrightarrow \mathbf{B} < 0 \quad \text{and} \quad \mathbf{A} - \mathbf{C}\mathbf{B}^{-1}\mathbf{C}^T < 0 \end{aligned} \tag{G.2}$$

The expressions $\mathbf{B} - \mathbf{C}^T\mathbf{A}^{-1}\mathbf{B}$ and $\mathbf{A} - \mathbf{C}\mathbf{B}^{-1}\mathbf{C}^T$ are the so-called *Schur complements* of the block matrix concerning to the blocks $\mathbf{A}$ and $\mathbf{B}$.

Using the Schur complements we can convert the *Riccati inequality*

$$\mathbf{A}^T\mathbf{P} + \mathbf{P}\mathbf{A} - \mathbf{P}\mathbf{B}\mathbf{B}^T\mathbf{P} + \mathbf{Q} < \mathbf{0}, \tag{G.3}$$

that depends quadratically on $\mathbf{P}$ and that can not be treated as LMI, into the form (G.1):

$$\begin{pmatrix} \mathbf{A}^T\mathbf{P} + \mathbf{P}\mathbf{A} + \mathbf{Q} & \mathbf{P}\mathbf{B} \\ \mathbf{B}^T\mathbf{P} & -\mathbf{I} \end{pmatrix} < \mathbf{0} \tag{G.4}$$

In this expression the single coefficients of the matrix depend only linearly on the coefficients of $\mathbf{P}$.

Setting $\mathbf{C} = \mathbf{0}$ in eq. (G.2), it follows, that a block diagonal matrix is negative definite if and only if every of its blocks is negative definite.

With this, the system of a finite number of linear matrix inequalities

$$\mathbf{F}_1(\mathbf{x}) < \mathbf{0} \quad ,..., \quad \mathbf{F}_n(\mathbf{x}) < \mathbf{0} \tag{G.5}$$

with $\mathbf{F}(\mathbf{x}) = \text{diag}(\mathbf{F}_1(\mathbf{x}), ..., \mathbf{F}_n(\mathbf{x}))$ can be converted to the form (G.1).

In particular, the inequality system

$$\mathbf{A}_i^T\mathbf{P} + \mathbf{P}\mathbf{A}_i < \mathbf{0} \qquad \text{mit} \quad i = 1, ..., n \tag{G.6}$$

for the unknown matrix $\mathbf{P}$ can be converted easily to the form (G.1):

$$\begin{pmatrix} \mathbf{A}_1^T\mathbf{P} + \mathbf{P}\mathbf{A}_1 & & \\ & \cdots & \\ & & \mathbf{A}_n^T\mathbf{P} + \mathbf{P}\mathbf{A}_n \end{pmatrix} < \mathbf{0} \tag{G.7}$$

The solution sets of (G.6) and (G.7) are obviously equivalent.

References

1. J. Ackermann. *Abtastregelung, Band I: Analyse und Synthese.* Springer-Verlag, Berlin, 1983.
2. M. A. Aisermann and F. R. Gantmacher. *Die absolute Stabilität von Regelsystemen.* Oldenbourg-Verlag, München, 1965.
3. T. Azuma, R. Watanabe, K. Uchida, and M. Fujita. A new LMI approach to analysis of linear systems depending on scheduling parameter in polynomial forms. *Automatisierungstechnik*, 48:199–204, 2000.
4. R. Babuska and H. B. Verbruggen. Fuzzy modeling and model-based control for nonlinear systems. In M. Jamshidi, A. Titli, S. Boverie, and L. A. Zadeh, editors, *Applications of Fuzzy Logic: Towards High Machine Intelligence Quotient Systems*, pages 49–74. Prentice Hall, New Jersey, 1997.
5. T. Bäck. *Evolutionary Algorithms in Theory and Practice.* Oxford University Press, New York, 1996.
6. A. G. Barto. Reinforcement learning and adaptive critic methods. In D. A. White and D. A. Sofge, editors, *Handbook of Intelligent Control. Neural, Fuzzy, and Adaptive Approaches*, pages 469–491. Van Nostrand Reinhold, New York, 1992.
7. A. G. Barto, R. S. Sutton, and C. W. Anderson. Neuronlike adaptive elements that can solve difficult learning control problems. *IEEE Transactions on Systems, Man, and Cybernetics*, 13:834–846, 1983.
8. R. E. Bellmann. *Dynamic Programming.* Princeton University Press, Princeton, NJ, 1957.
9. H. R. Berenji and P. Khedkar. Learning and tuning fuzzy logic controllers through reinforcements. *IEEE Transactions on Neural Networks*, 3:724–740, sep 1992.
10. R. Berstecher, R. Palm, and H. Unbehauen. Entwurf eines adaptiven robusten Fuzzy sliding-mode-Reglers, Teil 1. *Automatisierungstechnik*, 47:549–555, 1999.
11. R. Berstecher, R. Palm, and H. Unbehauen. Entwurf eines adaptiven robusten Fuzzy sliding-mode-Reglers, Teil 2. *Automatisierungstechnik*, 47:600–605, 1999.
12. R. Berstecher, R. Palm, and H. Unbehauen. Entwurf eines adaptiven robusten Fuzzy sliding-mode-Reglers, Teil 3. *Automatisierungstechnik*, 48:35–41, 2000.

13. D. P. Bertsekas. *Dynamic Programming.* Prentice-Hall, Englewood Cliffs, NJ, 1987.
14. D. P. Bertsekas. *Dynamic Programming and Optimal Control.* Athena Scientific, Belmont, MA, 1995.
15. J. C. Bezdek. *Pattern Recognition with Fuzzy Objective Function Algorithms.* Plenum Press, New York, 1981.
16. J. C. Bezdek. Fuzzy models - what are they, and why? *IEEE Transactions on Fuzzy Systems*, 1:1–5, 1993.
17. H. H. Bock. Clusteranalyse mit unscharfen Partitionen. In H. H. Bock, editor, *Klassifikation und Erkenntnis: Vol. III: Numerische Klassifikation*, pages 137–163. INDEKS, Frankfurt, 1979.
18. J. Böcker, I. Hartmann, and C. Zwanzig. *Nichtlineare und adaptive Regelungssyteme.* Springer-Verlag, Berlin, 1986.
19. H. W. Bode. *Network Analysis and Feedback Amplifier Design.* D. van Nostrand, Princeton/New Jersey, 1945.
20. R. Böhm and V. Krebs. Ein Ansatz zur Stabilitätsanalyse und Synthese von Fuzzy-Regelungen. *Automatisierungstechnik*, 41:288–293, 1993.
21. H.-H. Bothe. *Neuro-Fuzzy-Methoden.* Springer-Verlag, Berlin, 1997.
22. M. Braae and D. L. Rutherford. Theoretical and linguistic aspects of the fuzzy logic controller. *Automatica*, 15:553–577, 1979.
23. I. N. Bronstein and K. A. Semendjajew. *Taschenbuch der Mathematik.* Verlag Harri Deutsch, Frankfurt/Main, 1983.
24. H. Bühler. Stabilitätsuntersuchung von Fuzzy-Reglern. In *VDI-Berichte 1113*, pages 309–318. VDI-Verlag GmbH, Düsseldorf, 1994.
25. D. Butnariu and E.-P. Klement. *Triangular Norm-Based Measures and Games with Fuzzy Coalitions.* Kluwer Academic Publishers, Dordrecht, Netherlands, 1993.
26. S. G. Cao, N. W. Rees, and G. Feng. Stability analysis of fuzzy control systems. *IEEE Transactions on Systems, Man, and Cybernetics - Part B: Cybernetics*, 26:201–204, 1996.
27. Y.-Y. Cao and P. M. Frank. Analysis and synthesis of nonlinear time-delay systems via fuzzy control approach. *IEEE Transactions on Fuzzy Systems*, 8:200–211, 2000.
28. Y.-Y. Cao and P. M. Frank. Robust h_∞ disturbance attenuation for a class of uncertain discrete-time fuzzy systems. *IEEE Transactions on Fuzzy Systems*, 8:406–415, 2000.
29. Y. Chen. Stability analysis of fuzzy control - a lyapunov approach. *IEEE*, 1987.
30. Y. Y. Chen and T. C. Tsao. A description of the dynamical behavior of fuzzy systems. *IEEE Transactions on Systems, Man, and Cybernetics*, 19:745–755, 1989.
31. O. Cordón, F. Gomide, F. Herrera, F. Hoffmann, and L. Magdalena. Ten years of genetic fuzzy systems: Current framework and new trends. *Fuzzy Sets and Systems*, 141:5–31, 2004.
32. O. Cordon, F. Herrera, F. Hoffmann, and L. Magdalena. *Genetic Fuzzy Systems: Evolutionary Tuning and Learning of Fuzzy Knowledge Bases.* Advances in Fuzzy Systems. World Scientific Publishing, Singapore, 2001.
33. L. Cremer. Ein neues Verfahren zur Beurteilung der Stabilität linearer Regelsysteme. *Zeitschrift für angewandte Mathematik und Mechanik*, 25(27):161, 1947.

34. R. N. Davé. Characterization and detection of noise in clustering. *Pattern Recognition Letters*, 12:406–414, 1991.
35. J. V. de Oliveira and J. M. Lemos. Long-range predictive adaptive fuzzy relational control. *Fuzzy Sets and Systems*, 70:337–357, 1995.
36. J. C. Doyle, B. A. Francis, and A. R. Tannenbaum. *Feedback Control Theory.* Macmillan, New York, 1992.
37. J. C. Doyle, K. Glover, P. Khargonekar, and B. A. Francis. State-space solutions to standard h_2 and h_∞ control problems. *IEEE Transactions on Automatic Control*, 34:831–847, 1989.
38. S. E. Fahlmann. *An empirical study of learning speed in back-propagation networks, Technical Report CMU-CS-88-162.* Carnegie Mellon University, Pittsburgh, PA, 1988.
39. L. V. Fausett. *Fundamentals of Neural Networks.* Prentice Hall, Upper Saddle River, NJ, 1994.
40. D. P. Filev and P. Angelov. Fuzzy optimal control. *Fuzzy Sets and Systems*, 47:151–156, 1992.
41. D. P. Filev and R. R. Yager. On the analysis of fuzzy logic controllers. *Fuzzy Sets and Systems*, 68:39–66, 1994.
42. K. Fischle and D. Schröder. An improved stable adaptive fuzzy control method. *IEEE Transactions on Fuzzy Systems*, 7:27–40, 1999.
43. O. Föllinger. *Laplace- und Fourier-Transformation.* Elitera-Verlag, Berlin, 1977.
44. O. Föllinger. *Regelungstechnik.* Hüthig-Verlag, Heidelberg, 1992.
45. O. Föllinger. *Nichtlineare Regelungen, Band I.* Oldenbourg-Verlag, München, 1993.
46. O. Föllinger. *Nichtlineare Regelungen, Band II.* Oldenbourg-Verlag, München, 1993.
47. K. Funahashi. On the approximate realization of continuous mappings by neural networks. *Neural Networks*, 2:183–192, 1989.
48. A. E. Gegov and P. M. Frank. Decomposition of multivariable systems for distributed fuzzy control. *Fuzzy Sets and Systems*, 73:329–340, 1995.
49. A. E. Gegov and P. M. Frank. Hierarchical fuzzy control of multivariable systems. *Fuzzy Sets and Systems*, 72:299–310, 1995.
50. A. Gelb and W. E. V. Velde. *Multiple-Input Describing Functions and Nonlinear System Design.* McGraw-Hill, New York, 1968.
51. D. E. Goldberg. *Genetic Algorithms in Search, Optimization and Machine Learning.* Addison-Wesley, Reading, 1989.
52. K. Göldner and S. Kubik. *Mathematische Grundlagen der Systemanalyse.* Verlag Harri Deutsch, Frankfurt/Main, 1983.
53. V. Gorrini and H. Bersini. Recurrent fuzzy systems. In *Proc. of the 3rd Conference on Fuzzy Systems (FUZZ-IEEE 94)*, Orlando, 1994. IEEE.
54. M. B. Gorzalczany. Interval-valued fuzzy controller based on verbal model of object. *Fuzzy Sets and Systems*, 28:45–53, 1988.
55. S. Gottwald, editor. *A Treatise on Many-Valued Logic.* Research Studies Press, Baldock, 2003.
56. W. Hahn. *Stability of Motion.* Springer-Verlag, Berlin, 1967.
57. S. K. Halgamuge and M. Glesner. Neural networks in designing fuzzy systems for real world applications. *Fuzzy Sets and Systems*, 65:1–12, 1994.
58. S. Haykin. *Neural Networks: A Comprehensive Foundation.* Prentice Hall, Upper Saddle River, NJ, 1998.

59. F. Herrera and L. Magdalena. Genetic fuzzy systems. *Tatra Mountains Mathematical Publications*, 13:93–121, 1997.
60. F. Herrera and J. L. Verdegay. *Genetic Algorithms and Soft Computing*. Physica-Verlag, Heidelberg, 1996.
61. T. Hojo, T. Terano, and S. Masui. Stability analysis of fuzzy control systems based on phase space analysis. *Japanese Journal of Fuzzy Theory and Systems*, 4:639–654, 1992.
62. J. Hopf and F. Klawonn. Learning the rule base of a fuzzy controller by a genetic algorithm. In R. Kruse, J. Gebhardt, and R. Palm, editors, *Fuzzy Systems in Computer Science*, pages 63–74. Vieweg-Verlag, Braunschweig, 1994.
63. F. Höppner, F. Klawonn, and R. Kruse. *Fuzzy-Clusteranalyse: Verfahren für die Bilderkennung, Klassifikation und Datenanalyse*. Vieweg-Verlag, Braunschweig, 1997.
64. F. Höppner, F. Klawonn, R. Kruse, and T. Runkler. *Fuzzy Cluster Analysis*. Wiley, Chichester, 1999.
65. M. Hornik, M. Stinchcombe, and H. White. Multilayer feedforward networks are universal approximators. *Neural Networks*, 2:359–366, 1989.
66. C. S. Hsu. A theory of cell-to-cell mapping dynamical systems. *Journal of Applied Mechanics*, 47:931–939, 1980.
67. C. S. Hsu and R. S. Guttalu. An unravelling algorithm for global analysis of dynamical systems: An application of cell-to-cell mappings. *Journal of Applied Mechanics*, 47:940–948, 1980.
68. S. Huang, K. K. Tan, and K. Z. Tang. *Neural Network Control: Theory and Applications*. Research Studies Press, 2004.
69. A. Hurwitz. Über die Bedingungen, unter welchen eine Gleichung nur Wurzeln mit negativen reellen Teilen besitzt. *Math. Annalen*, 46:273, 1895.
70. G.-C. Hwang and S.-C. Lin. A stability approach to fuzzy control design for nonlinear systems. *Fuzzy Sets and Systems*, 48:279–287, 1992.
71. R. Isermann. *Identifikation dynamischer Systeme, Band I*. Springer-Verlag, Berlin, 1992.
72. R. Isermann. *Identifikation dynamischer Systeme, Band II*. Springer-Verlag, Berlin, 1992.
73. A. Isidori. *Nonlinear Control Systems*. Springer-Verlag, Berlin, 1995.
74. J.-S. R. Jang. Anfis: Adaptive-network-based fuzzy inference systems. *IEEE Transactions on Systems, Man, and Cybernetics*, 23:665–685, 1993.
75. J.-S. R. Jang, C.-T. Sun, and E. Mizutani. *Neuro Fuzzy and Soft Computing*. Prentice Hall, Upper Saddle River, NJ, 1997.
76. C. Jianqin and C. Laijiu. Study on stability of fuzzy closed-loop control systems. *Fuzzy Sets and Systems*, 57:159–168, 1993.
77. J. Joh, Y.-H. Chen, and R. Langari. On the stability issues of linear takagi-sugeno fuzzy-models. *IEEE Transactions on Fuzzy Systems*, 6:402–410, 1998.
78. T. A. Johansen. Fuzzy model based control: Stability, robustness, and performance issues. *IEEE Transactions on Fuzzy Systems*, 2:221–234, 1994.
79. M. Johansson, A. Rantzer, and K.-E. Arzen. Piecewise quadratic stability of fuzzy systems. *IEEE Transactions on Fuzzy Systems*, 7:713–723, 1999.
80. L. P. Kaelbling, M. H. Littman, and A. W. Moore. Reinforcement learning: A survey. *J. Artificial Intelligence Research*, 4:237–285, 1996.
81. J. Kahlert and H. Frank. *Fuzzy-Logik und Fuzzy-Control (2.Auflage)*. Friedr. Vieweg & Sohn Verlagsgesellschaft mbH, Braunschweig, Wiesbaden, 1994.

82. R. E. Kalman. On the general theory of control systems. In *Proc. 1st International Congress on Automatic Control 1960, Bd. 1*, pages 481–492, London, 1961. Butterworths.
83. H. Kang. Stability and control of fuzzy dynamic systems via cell-state transitions in fuzzy hypercubes. *IEEE Transactions on Fuzzy Systems*, 1:267–279, 1993.
84. H. Kiendl. Totale Stabilität von linearen Regelungssystemen bei ungenau bekannten Parametern der Regelstrecke. *Automatisierungstechnik*, 33:379–386, 1985.
85. H. Kiendl. Robustheitsanalyse von Regelungssystemen mit der Methode der konvexen Zerlegung. *Automatisierungstechnik*, 35:192–202, 1987.
86. H. Kiendl and J. Rüger. Verfahren zum Entwurf und Stabilitätsnachweis von Regelungssystemen mit Fuzzy-Reglern. *Automatisierungstechnik*, 41:138–144, 1993.
87. H. Kiendl and J. Rüger. Stability analysis of fuzzy control systems using facet functions. *Fuzzy Sets and Systems*, 70:275–285, 1995.
88. E. Kim and H. Lee. New approaches to relaxed quadratic stability condition of fuzzy control systems. *IEEE Transactions on Fuzzy Systems*, 8:523–534, 2000.
89. J. Kinzel, F. Klawonn, and R. Kruse. Modifications of genetic algorithms for designing and optimizing fuzzy controllers. In *Proc. IEEE Conference on Evolutionary Computation*, pages 28–33, Orlando, FL, 1994.
90. K. Kiriakidis. Fuzzy model-based control of complex plants. *IEEE Transactions on Fuzzy Systems*, 6:517–530, 1998.
91. F. Klawonn. On a lukasiewicz logic based controller. In *MEPP'92 International Seminar on Fuzzy Control through Neural Interpretations of Fuzzy Sets, Reports on Computer Science & Mathematics, Ser. B No 14*, pages 53–56, Turku, 1992. Åbo Akademi.
92. F. Klawonn. Fuzzy sets and vague environments. *Fuzzy Sets and Systems*, 66:207–221, 1994.
93. F. Klawonn. Noise clustering with a fixed fraction of noise. In A. Lotfi and J. Garibaldi, editors, *Applications and Science in Soft Computing*, pages 133–138. Springer, Berlin, 2004.
94. F. Klawonn and J. L. Castro. Similarity in fuzzy reasoning. *Mathware and Soft Computing*, 2:197–228, 1995.
95. F. Klawonn and R. Kruse. The inherent indistinguishability in fuzzy systems. In W. Lenski, editor, *Logic versus Approximation: Essays Dedicated to Michael M. Richter on the Occasion of his 65th Birthday*, pages 6–17. Springer, Berlin, 2004.
96. F. Klawonn and V. Novák. The relation between inference and interpolation in the framework of fuzzy systems. *Fuzzy Sets and Systems*, 81:331–354, 1996.
97. E.-P. Klement, R. Mesiar, and E. Pap, editors. *Triangular Norms.* Kluwer, Dordrecht, 2000.
98. T. Kohonen. Self-organized formation of topologically correct feature maps. *Biological Cybernetics*, 43:59–69, 1982.
99. B. Kosko, editor. *Neural Networks for Signal Processing.* Prentice Hall, Englewood Cliffs, NJ, 1992.
100. R. Krishnapuram. Fitting an unknown number of lines and planes to image data through compatible cluster merging. *Pattern Recognition*, 25:385–400, 1992.

101. R. Krishnapuram. A possibilistic approach to clustering. *IEEE Transactions on Fuzzy Systems*, 1:98–110, 1993.
102. R. Kruse, J. Gebhardt, and F. Klawonn. *Fuzzy-Systeme, 2. erweiterte Auflage*. Teubner-Verlag, Stuttgart, 1995.
103. T. Kuhn and J. Wernstedt. Ein Beitrag zur Lösung von Adaptionsproblemen klassischer Regler mittels optimaler Fuzzy-Logik. *Automatisierungstechnik*, 44:160–170, 1996.
104. J. LaSalle and S. Lefschetz. *Die Stabilitätstheorie von Ljapunow*. Bibliographisches Institut, Mannheim, 1967.
105. A. J. Laub. A schur method for solving algebraic riccati equation. *IEEE Transactions on Automatic Control*, 24:913–921, 1979.
106. C.-H. Lee and C.-C. Teng. Indentification and control of dynamic systems using recurrent fuzzy neural networks. *IEEE Transactions on Fuzzy Systems*, 8(4):349–366, august 2000.
107. C.-H. Lee and S.-D. Wang. A self-organizing adaptive fuzzy controller. *Fuzzy Sets and Systems*, 80:295–313, 1996.
108. M. Lee and H. Takagi. Integrating design stages of fuzzy systems using genetic algorithms. In *Proc. IEEE Int. Conf. on Fuzzy Systems 1993*, pages 612–617, San Francisco, mar 1993.
109. A. Leonhard. Ein neues verfahren zur stabilitätsuntersuchung. *Archiv für Elektrotechnik*, 38:17, 1944.
110. W. Leonhard. *Statistische Analyse linearer Regelsysteme*. Teubner-Verlag, Stuttgart, 1973.
111. W. Leonhard. *Control of Electrical Drives*. Springer-Verlag, Berlin, 1996.
112. C.-M. Lin and C.-F. Hsu. Identification of dynamic systems using recurrent fuzzy neural network. In *Proc. of Joint 9th IFSA World Congress and 20th NAFIPS International Conference*, pages 2671–2675, 2001.
113. C.-T. Lin. A neural fuzzy control system with structure and parameter learning. *Fuzzy Sets and Systems*, 70:183–212, 1995.
114. C.-T. Lin and C.-C. Lee. *Neural Fuzzy Systems. A Neuro-Fuzzy Synergism to Intelligent Systems*. Prentice Hall, New York, 1996.
115. S. Liu and S. Hu. A method of generating control rule model and its application. *Fuzzy Sets and Systems*, 52:33–37, 1992.
116. M. A. Ljapunov. Problème général de la stabilité du mouvement (translation from russian). *Ann. Fac. Sci.*, 9:203, 1907.
117. L. Ljung. *System Identification - Theory for the User*. Prentice Hall, Englewood Cliffs, New Jersey, 1987.
118. D. G. Luenberger. Observing the state of a linear system. *IEEE Transactions on Military Electronics*, 8:74–80, 1964.
119. D. G. Luenberger. An introduction to observers. *IEEE Transactions on Automatic Control*, 16:596–602, 1971.
120. E. H. Mamdani and S. Assilian. An experiment in linguistic synthesis with a fuzzy logic controller. *International Journal of Man Machine Studies*, 7:1–13, 1975.
121. A. Mayer, B. Mechler, A. Schlindwein, and R. Wolke. *Fuzzy Logic*. Addison-Wesley, Bonn, 1993.
122. L. R. Medsker and L. C. Jain. *Recurrent Neural Networks: Design and Applications*. CRC Press, 1999.
123. A. W. Michailow. Die Methode der harmonischen Analyse in der Regelungstheorie (russ.). *Automat. Telemek.*, 3:27, 1938.

124. Z. Michalewicz. *Genetic Algorithms + Data Structures = Evolution Programs.* Springer-Verlag, Berlin, 1996.
125. K. Michels. A model-based fuzzy controller. *Fuzzy Sets and Systems*, 85(2):223–232, 1997.
126. K. Michels. Numerical stability analysis for a fuzzy or neural network controller. *Fuzzy Sets and Systems*, 89(3):335–350, 1997.
127. K. Michels and R. Kruse. Numerical stability analysis for fuzzy control. *International Journal of Approximate Reasoning*, 16(1):3–24, 1997.
128. W. T. Miller, R. S. Sutton, and P. J. Werbos, editors. *Neural Networks for Control.* MIT Press, Cambridge, MA, 1990.
129. R. R. Mohler. *Nonlinear Systems, Vol. I, Dynamics and Control.* Prentice Hall, New Jersey, 1991.
130. R. E. Moore. *Interval Analysis.* Prentice Hall, Englewood Cliffs, 1966.
131. R. E. Moore. *Methods and Applications of Interval Analysis.* SIAM, Philadelphia, 1979.
132. K. Müller. *Ein Entwurfsverfahren für selbsteinstellende robuste Regelungen.* PhD thesis, Institut für Regelungstechnik, TU Braunschweig, 1991.
133. K. Müller. *Entwurf robuster Regelungen.* Teubner-Verlag, Stuttgart, 1996.
134. N. Muskinja, B. Tovornik, and D. Donlagic. How to design a discrete supervisory controller for real-time fuzzy control systems. *IEEE Transactions on Fuzzy Systems*, 5:161–166, 1997.
135. N. N. Ieee transactions on automatic control: Special issue on linear-quadratic-gaussian estimation and control problem. *IEEE Transactions on Automatic Control*, 16:527–869, 1971.
136. N. N. *FSM - Fuzzy Stability Manager, Handbook.* Transfertech GmbH, Braunschweig, 1996.
137. D. Nauck. NEFCON-I: Eine simulationsumgebung für neuronale fuzzy-regler. In *1. GI-Workshop Fuzzy-Systeme '93*, Braunschweig, oct 1993.
138. D. Nauck and F. Klawonn. Neuro-fuzzy classification initialized by fuzzy clustering. In *Proc. 4th European Congress on Intelligent Techniques and Soft Computing (EUFIT'96)*, pages 1551–1555, Aachen, 1996.
139. D. Nauck, F. Klawonn, and R. Kruse. *Neuronale Netze und Fuzzy-Systeme, 2. erweiterte Auflage.* Vieweg-Verlag, Wiesbaden, 1996.
140. D. Nauck, F. Klawonn, and R. Kruse. *Foundations of Neuro-Fuzzy Systems.* Wiley, Chichester, 1997.
141. D. Nauck and R. Kruse. A fuzzy neural network learning fuzzy control rules and membership functions by fuzzy error backpropagation. In *Proc. IEEE Int. Conf. on Neural Networks 1993*, pages 1022–1027, San Francisco, mar 1993.
142. D. Nauck and R. Kruse. NEFCON-I: An x-window based simulator for neural fuzzy controllers. In *Proc. IEEE Int. Conf. Neural Networks 1994 at IEEE WCCI'94*, pages 1638–1643, Orlando, FL, 1994.
143. O. Nelles and M. Fischer. Lokale Linearisierung von Fuzzy-Modellen. *Automatisierungstechnik*, 47:217–223, 1999.
144. V. Nissen. *Einführung in Evolutionäre Algorithmen.* Vieweg-Verlag, Braunschweig, 1997.
145. R. Noisser and E. Bodenstorfer. Zur Stabilitätsanalyse von Fuzzy-Regelungen mit Hilfe der Hyperstabilitätstheorie. *Automatisierungstechnik*, 45:76–83, 1997.
146. H. Nomura, I. Hayashi, and N. Wakami. A learning method of fuzzy inference rules by descent method. In *Proc. IEEE Int. Conf. on Fuzzy Systems 1992*, pages 203–210, San Diego, CA, 1992.

147. M. Norgaard, O. Ravn, N. Poulsen, and L. Hansen. *Neural Networks for Modelling and Control of Dynamic Systems: A Practitioner's Handbook*. Springer Verlag, London, 2003.
148. A. Nürnberger. A hierarchical recurrent neuro-fuzzy system. In *Proc. of Joint 9th IFSA World Congress and 20th NAFIPS International Conference*, pages 1407–1412. IEEE, 2001.
149. A. Nürnberger. Approximation of dynamic systems using recurrent neuro-fuzzy techniques. *Soft Computing*, 8(6):428–442, 2004.
150. A. Nürnberger, D. Nauck, and R. Kruse. Neuro-fuzzy control based on the NEFCON-model. *Soft Computing*, 2(4):182–186, feb 1999.
151. H. Nyquist. Regeneration theory. *Bell System Technical Journal*, 11:126, 1932.
152. H.-P. Opitz. *Entwurf robuster, strukturvariabler Regelungssysteme mit der Hyperstabilitätstheorie*. VDI-Verlag GmbH, Düsseldorf, 1984.
153. R. Ordonez and K. M. Passino. Stable multi-input multi-output adaptive fuzzy/neural control. *IEEE Transactions on Fuzzy Systems*, 7:345–353, 1999.
154. S. K. Pal and S. Mitra. *Neuro-Fuzzy Pattern Recognition: Methods in Soft Computing*. John Wiley and Sons, 1999.
155. R. Palm. Sliding mode fuzzy control. In *Proc. of IEEE International Conference on Fuzzy Systems*, San Diego, CA, 1992. IEEE.
156. C. P. Pappis and G. I. Adamopoulos. A computer algorithm for the solution of the inverse problem of fuzzy systems. *Fuzzy Sets and Systems*, 39:279–290, 1991.
157. Y.-M. Park, U.-C. Moon, and K. Y. Lee. A self-organizing fuzzy controller for dynamic systems using a fuzzy auto-regressive moving average (farma) model. *IEEE Transactions on Fuzzy Systems*, 3:75–82, 1995.
158. P. C. Parks and V. Hahn. *Stabilitätstheorie*. Springer-Verlag, Berlin, 1981.
159. W. Pedrycz, editor. *Fuzzy Evolutionary Computation*. Kluwer Academic Publishers, Boston, 1997.
160. W. Pedrycz and H. C. Card. Linguistic interpretation of self-organizing maps. In *Proc. IEEE Int. Conf. on Fuzzy Systems 1992*, pages 371–378, San Diego, CA, 1992.
161. F. J. Pineda. Recurrent backpropagation networks. In Y. Chauvin and D. E. Rumelhart, editors, *Backpropagation: Theory, Architectures and Applications*, pages 99–135. Lawrence Erlbaum Associates, Hillsdale, NJ, 1995.
162. V. M. Popov. The solution of a new stability problem for controlled systems. *Automatic and Remote Control*, 24:1–23, 1963.
163. V. M. Popov. *Hyperstability of Control Systems*. Springer-Verlag, Berlin, 1973.
164. E. P. Popow. *Dynamik automatischer Regelsysteme*. Akademie-Verlag, Berlin, 1958.
165. B. E. Postlethwaite. Building a model-based fuzzy controller. *Fuzzy Sets and Systems*, 79:3–13, 1996.
166. T. J. Procyk and E. H. Mamdani. A linguistic self-organizing process controller. *Automatica*, 15:15–30, 1979.
167. M. Riedmiller. *Selbständig lernende neuronale Steuerungen*. VDI-Verlag GmbH, Düsseldorf, 1997.
168. M. Riedmiller and H. Braun. A direct adaptive methode for faster backpropagation learning: The rprop algorithm. In *Proc. of IEEE Int. Conf. on Neural Networks (ICNN-93)*, pages 586–591, San Francisco CA, 1993.

169. M. Riedmiller, M. Spott, and J. Weisbrod. Fynesse: A hybrid architecture for selflearning control. In I. Cloete and J. Zurada, editors, *Knowledge-Based Neurocomputing*, pages 291–323. MIT Press, Cambridge, 1999.
170. E. J. Routh. *Stability of a Given State of Motion.* Adams Prize Essay, London, 1877.
171. J.-J. Rüger. Weiterentwicklung des Konzeptes der Facettenfunktionen zum Reglerentwurf und zur Stabilitätsanalyse. *Automatisierungstechnik*, 44:391–398, 1996.
172. D. E. Rumelhart, G. E. Hinton, and R. J. Williams. Learning internal representations by error propagation. In D. E. Rumelhart and J. L. McClelland, editors, *Parallel Distributed Processing: Explorations in the Microstructures of Cognition. Foundations*, volume 1, pages 318–362. MIT Press, Cambridge, MA, 1986.
173. T. Runkler and M. Glesner. A set of axioms for defuzzification strategies - towards a theory of rational defuzzification operators. In *IEEE International Conference on Fuzzy Systems*, pages 1161–1166, San Francisco, 1993. IEEE.
174. E. Sanchez. Resolution of composite relation equations. *Information and Control*, 30:38–48, 1976.
175. V. N. Sastry, R. N. Tiwari, and K. S. Sastri. Dynamic programming approach to multiple objective control problem having deterministic or fuzzy goals. *Fuzzy Sets and Systems*, 57:195–202, 1993.
176. C. W. Scherer. Lineare Matrixungleichungen in der Theorie der robusten Regelung. *Automatisierungstechnik*, 45:306–318, 1997.
177. H. Schwarz. *Nichtlineare Regelungssysteme.* Oldenbourg-Verlag, München, 1991.
178. S. Shao. Fuzzy self-organizing controller and its application for dynamic processes. *Fuzzy Sets and Systems*, 26:151–164, 1988.
179. J. T. Spooner and K. M. Passino. Stable adaptive control using fuzzy systems and neural networks. *IEEE Transactions on Fuzzy Systems*, 4:339–359, 1996.
180. C.-Y. Su and Y. Stepanenko. Adaptive control of a class of nonlinear systems with fuzzy logic. *IEEE Transactions on Fuzzy Systems*, 2:285–294, 1994.
181. M. Sugeno. An introductory survey of fuzzy control. *Information Sciences*, 36:59–83, 1985.
182. M. Sugeno and T. Yasukawa. A fuzzy logic-based approach to qualitative modelling. *IEEE Transactions on Fuzzy Systems*, 1:7–31, 1993.
183. H. Surmann and M. Maniadakis. Learning feed-forward and recurrent fuzzy systems: A genetic approach. *Journal of Systems Architecture*, 47(7):649–662, 2001.
184. R. Sutton and I. M. Jess. A design study of a self-organizing fuzzy autopilot for ship control. *Proceedings of the Institution of Mechanical Engineers*, 205:35–47, 1991.
185. R. S. Sutton and A. G. Barto. *Reinforcement Learning: An Introduction.* MIT Press, 1998.
186. T. Takagi and M. Sugeno. Fuzzy identification of systems and its applications to modeling and control. *IEEE Transactions on Systems, Man, and Cybernetics*, 15:116–132, 1985.
187. K. Tanaka, T. Ikeda, and H. O. Wang. Robust stabilization of a class of uncertain nonlinear systems via fuzzy control: Quadratic stabilizability, h_∞ control, and linear matrix inequalities. *IEEE Transactions on Fuzzy Systems*, 4:1–13, 1996.

188. K. Tanaka, T. Ikeda, and H. O. Wang. Fuzzy regulators and fuzzy observers: Relaxed stability conditions and LMI-based designs. *IEEE Transactions on Fuzzy Systems*, 6:250–265, 1998.
189. K. Tanaka and M. Sugeno. Stability analysis and design of fuzzy control systems. *Fuzzy Sets and Systems*, 45:135–156, 1992.
190. R. Tanscheit and E. M. Scharf. Experiments with the use of a rule-based self-organising controller for robotics applications. *Fuzzy Sets and Systems*, 26:195–214, 1988.
191. M. C. M. Teixeira and S. H. Zak. Stabilizing controller design for uncertain nonlinear systems using fuzzy models. *IEEE Transactions on Fuzzy Systems*, 7:133–142, 1999.
192. M. A. L. Thathachar and P. Viswanath. On the stability of fuzzy systems. *IEEE Transactions on Fuzzy Systems*, 5:145–151, 1997.
193. J. B. Theocharis and G. Vachtsevanos. Recursive learning algorithms for training fuzzy recurrent models. *International Journal of Intelligence Systems*, 11(12):10591098, 1996.
194. C.-S. Ting, T.-H. S. Li, and F.-C. Kung. An approach to systematic design of the fuzzy control system. *Fuzzy Sets and Systems*, 77:151–166, 1996.
195. H. Tolle. *Mehrgrößen-Regelkreissynthese, Band I: Grundlagen und Frequenzbereichsverfahren.* Oldenbourg-Verlag, München, 1983.
196. G. M. Trojan, J. B. Kiszka, M. M. Gupta, and P. N. Nikiforuk. Solution of multivariable fuzzy equations. *Fuzzy Sets and Systems*, 23:271–279, 1987.
197. N. Tschichold-Gürman. Generation and improvement of fuzzy classifiers with incremental learning using fuzzy rulenet. In K. M. George, J. H. Carrol, E. Deaton, D. Oppenheim, and J. Hightower, editors, *Applied Computing 1995. Proc. 1995 ACM Symposium on Applied Computing, Nashville, Feb. 26-28*, pages 466–470. ACM Press, New York, feb 1995.
198. H. Unbehauen. *Regelungstechnik II, Zustandsregelungen, digitale und nichtlineare Systeme.* Vieweg-Verlag, Braunschweig, 1993.
199. H. Unbehauen. *Regelungstechnik III, Identifikation, Adaption, Optimierung.* Vieweg-Verlag, Braunschweig, 1993.
200. J.-S. Wang and C. S. G. Lee. Self-adaptive recurrent neuro-fuzzy control of an autonomous underwater vehicle. *IEEE Transactions on Robotics and Automation*, 19(2):283–295, 2003.
201. L.-X. Wang. Stable adaptive fuzzy control of nonlinear systems. *IEEE Transactions on Fuzzy Systems*, 1:146–155, 1993.
202. L.-X. Wang. Stable adaptive fuzzy controllers with application to inverted pendulum tracking. *IEEE Transactions on Systems, Man, and Cybernetics - Part B: Cybernetics*, 26:677–691, 1996.
203. S. Wermter. Neural fuzzy preference integration using neural preference moore machines. *International Journal of Neural Systems*, 10(4):287–310, 2000.
204. R. J. Williams and D. Zipser. Experimental analysis of the real time recurrent learning algorithm. *Connection Science*, 1:87–111, 1989.
205. X. L. Xi and G. Beni. A validity measure for fuzzy clustering. *IEEE Transactions on Pattern Analysis and Machine Intelligence*, 13:69–78, 1991.
206. C.-W. Xu. Linguistic decoupling control of fuzzy multivariable processes. *Fuzzy Sets and Systems*, 44:209–217, 1991.
207. B. Yoo and W. Ham. Adaptive sliding mode control of nonlinear system. *IEEE Transactions on Fuzzy Systems*, 6:315–320, 1998.

208. D. C. Youla, H. A. Jabr, and J. J. J. Bongiorno. Modern wiener-hopf design of optimal controllers - part i: The single input-output case. *IEEE Transactions on Automatic Control*, 21:3–13, 1976.
209. D. C. Youla, H. A. Jabr, and J. J. J. Bongiorno. Modern wiener-hopf design of optimal controllers - part ii: The multivariable case. *IEEE Transactions on Automatic Control*, 21:319–338, 1976.
210. L. A. Zadeh. Fuzzy sets. *Information and Control*, 8:338–353, 1965.
211. L. A. Zadeh. Towards a theory of fuzzy systems. In R. E. Kalman and N. de Claris, editors, *Aspects of Networks and System Theory*, pages 469–490. Rinehart and Winston, New York, 1971.
212. L. A. Zadeh. A rationale for fuzzy control. *J. Dynamic Systems, Measurement and Control, Series 6*, 94:3–4, 1972.
213. L. A. Zadeh. Outline of a new approach to the analysis of complex systems and decision processes. *IEEE Transactions on Systems, Man, and Cybernetics*, 3(1):28–44, 1973.
214. L. A. Zadeh. The concept of a lingustic variable and its application to approximate reasoning, part i. *Information Sciences*, 8:199–249, 1975.
215. L. A. Zadeh. The concept of a lingustic variable and its application to approximate reasoning, part ii. *Information Sciences*, 8:301–357, 1975.
216. L. A. Zadeh. The concept of a lingustic variable and its application to approximate reasoning, part iii. *Information Sciences*, 9:43–80, 1975.
217. J. Zhang and A. J. Morris. Recurrent neuro-fuzzy networks for nonlinear process modelling. *IEEE Transactions on Neural Networks*, 10(2):313–326, 1999.
218. K. Zhou and P. P. Khargonekar. Stability robustness bounds for linear state-space models with structured uncertainty. *IEEE Transactions on Automatic Control*, 32:621–623, 1987.
219. H.-J. Zimmermann, editor. *Neuro + Fuzzy Technologien*. VDI-Verlag GmbH, Düsseldorf, 1995.
220. H.-J. Zimmermann and P. Zysno. Latent connectives in human decision making and expert systems. *Fuzzy Sets and Systems*, 4:37–51, 1980.

Index